Environmental and Microbial Biotechnology

Series Editor

Ram Prasad, Department of Botany, Mahatma Gandhi Central University, Motihari, Bihar, India

Innovative and novel advances in microbial biotechnology are providing great understandings in to the machineries of nature, presenting fascinating prospects to apply principles of biology to different arenas of science. Sustainable elucidations are emerging to address the concerns on improving crop productivity through microbes, depleting natural resources, environmental pollution, microbial degradation of pollutants, nanomaterials, nanotoxicity & safety issues, safety of food & agricultural products etc. Simultaneously, there is an increasing demand for natural bio-products of therapeutic and industrial significance (in the areas of healthcare, environmental remediation, microbial biotechnology). Growing awareness and an increased attention on environmental issues such as climate change, energy use, and loss of non-renewable resources have carried out a superior quality for research that provides potential solutions to these problems. Emerging microbiome approaches potentially can significantly increase agriculture productivity & human healthcare and henceforth can contribute to meet several sustainable development goals.

The main objectives have provided an impetus for research on plants and microorganisms that produce novel bio-products with variable properties and understanding their mechanisms of action at cellular and molecular level. Hence, research activities of the environmental and microbial Biotechnology are comprehensively focused up on major sectors viz., bioresources, biorefining, bioremediation of organic and inorganic pollutants, environmental risk analysis of microorganisms, environmental assessment using microbiological indicators, enzymes for environment, food & industrial applications, nanomaterials & nanotoxicity, sustainable ecobiotechnology, biofertilizer, biocontrol agents for agriculture improvement and natural products for healthcare applications.

This book series is a state-of-the-art for a wide range of scientists, researchers, students, policy makers and academician involve in understanding and implementing the knowledge on environmental and microbial biotechnology to develop biologics for proper health care to continue life in smooth and sustainable strategy without any adverse effect.

Naga Raju Maddela • Aransiola Sesan Abiodun •
Ram Prasad
Editors

Ecological Interplays in Microbial Enzymology

Springer

Editors
Naga Raju Maddela
Facultad de Ciencias de la Salud
Universidad Técnica de Manabí
Portoviejo, Ecuador

Aransiola Sesan Abiodun
Bioresources Development Centre
National Biotechnology Development Agency
Ogbomoso, Nigeria

Ram Prasad
Department of Botany
Mahatma Gandhi University
Bihar, India

ISSN 2662-1681 ISSN 2662-169X (electronic)
Environmental and Microbial Biotechnology
ISBN 978-981-19-0154-6 ISBN 978-981-19-0155-3 (eBook)
https://doi.org/10.1007/978-981-19-0155-3

This Springer imprint is published by the registered company Springer Nature Singapore Pte Ltd.
The registered company address is: 152 Beach Road, #21-01/04 Gateway East, Singapore 189721, Singapore

Foreword

Microbial technology is an important aspect to make the environment sustainable. The present edited volume entitled *Ecological Interplays in Microbial Enzymology* is the one that helps in the achievement of environmental sustainability as this book is highly instructive in the field of microbial enzymology. The most extensively studied and very interesting biocatalysts are "microbial enzymes" as they do offer undisputable advantages over chemical catalysts. Microbial enzymes have a greater selectivity in their action, can be used in mild reaction conditions and have specificity for wider substrates. For these reasons, microbial enzymes have a great importance in many fields, like chemical, fermentation, agricultural, pharmaceuticals and food production. However, little is known about the ecological roles of microbial enzymes in different environmental media such as soil, sediments, aquatic systems (freshwater and marine) and other extreme environments. Also, how microbial enzyme activities are regulated by environmental factors and what is the pollution burden on the microbial enzymes in different environments are poorly understood. Such knowledge gaps limit the usage of microbial enzymes for environmental sustainability. The book *Ecological Interplays in Microbial Enzymology* is a good collection of series of independent chapters which present in depth insights over several issues on microbial enzymes and their role in the environment. This book focuses on distribution of microbial enzymes in the environment, role of microbial enzymes in environmental sustainability and impact of environmental disturbances on microbial enzymes. I am of the opinion that the above focused areas do provide great insights over microbial enzymes and do fill the knowledge gaps that limit the implications of microbial enzymes for environmental sustainability.

Universidad Técnica de Manabí, Vicente Véliz Briones
Portoviejo, Ecuador
12 December 2021

Preface

This book titled *Ecological Interplays in Microbial Enzymology* was structured in providing updated scientific information in microbial enzymology. Products produced by microorganisms are leading producers of useful natural products. Natural products from microbes and plants make excellent products and solve vital problems. Significant portions of the microbial genomes are devoted to production of these useful metabolites including enzymes. Enzymes are natural catalysts, which are universally found in all living organisms. They may be used either for building more complex molecules from simple ones or for selective breakdown of a mixture of larger molecules. Hence, there are multiple ecological interplays that have to be reviewed. Therefore, this book in its present form has been designed so that researchers/scientists can be well aware about the interplays in microbial enzymology in the environments.

This book takes into consideration current interactions of microorganism with the environments. We are fully aware that the distinction of life on earth is vaguely connected to the general quality of the environment. The increasing awareness of the environment we live in is bringing about a more exhaustive search for alternative cleaner technologies. Currently, there are two fundamental pollution-related problems: the disposal of the large quantities of wastes that are continually being produced, and the removal of toxic compounds that have been accumulating at dump sites in the soils and in water systems over the last few decades. Just one microorganism can contain over 1000 different enzymes. Hence, the critical review of applications and ecological interplays of these enzymes are of necessity.

Considering the wide interplay of these microbial enzymes, the book echoes relevant new areas in the study and recent developments made by researchers across the globe on the interactions of microbial enzymes across all habitats. This book also has been premeditated to serve as a source of information hub about modern sciences of microbial enzymology and their general environmental relationship to students and researchers of this field. To unfold the importance of this book title, 3 parts have been designed with 18 chapters: *Part I: Microbial Enzymes—Distribution in the Environment; Part II: Microbial Enzymes—Role in the Environmental*

Sustainability; and Part III: Impact of Environmental Disturbances on Microbial Enzymes. Part I consists of 8 chapters. Chapter 1 provides a general overview on microbial enzymes. Chapter 2 reveals the diversity of microbial enzymes in the soil ecosystem. Chapter 3 of this part describes microbial enzymes of wastewater and sludge. Chapters 4, 5, 6, 7 and 8 inform about the occurrence and distribution of microbial enzymes of fresh water, an overview of marine microbial enzymes, hydrolytic enzymes producing bacteria from Algerian hot springs: attractive industrial molecules, enzymology of microbial biofilms and changes in the attributes of the oxisol "Arenito caiuá" after the use of the crop–livestock integration system, respectively. Part II consists of 7 chapters, microbial enzymes: role in soil fertility, microbial enzymes in the recycling of wastes, soil microbial enzymes and mitigation of heavy metals uptake by plants, communities of microbial enzymes and biodegradation of persistent environmental pollutants, implication of enzymes in the adaptation of extremophilic microbes, applications of microbial enzymes in industries and medicine and microbial enzymes in the biosynthesis of metal nanoparticles. Part III consists of 4 chapters, which focuses on the effect of agrochemicals on soil microbial enzymes, effects of aquatic (freshwater and marine) pollution on microbial enzyme activities, in silico analysis of biochemical pathways in bacterial enzyme synthesis and the last chapter of this book, i.e. Chap. 19, microbial enzymes for sustainable development—future guidelines. The chapters were contributed by 56 academicians/scientists from 10 different countries (Algeria, Brazil, Canada, Ecuador, Ghana, India, Malaysia, Nigeria, Pakistan and Turkey) across the world.

Portoviejo, Ecuador — Naga Raju Maddela
Ogbomoso, Nigeria — Aransiola Sesan Abiodun
Motihari, Bihar, India — Ram Prasad

Acknowledgements

It is our great honour to acknowledge the support of the authors for their valuable contribution and timely responses for the success of this project. All contributors are immensely appreciated for their eagerness and inordinate support for this volume to be ready in the scheduled time; we, therefore, appreciate their teamwork and partnership. We also thank the anonymous reviewers for their constructive criticism, which had helped us in improving the quality of this book by inviting experts to contribute the additional chapters. We greatly acknowledge the Springer Editorial and Production team for their respected support; without their guidelines, this project would not be finished in such a very short time. It is an honest honour and privilege to work with them all. Finally, yet importantly, we are very much thankful to the colleagues at Universidad Técnica de Manabí (Ecuador), National Biotechnology Development Agency (Nigeria) and Mahatma Gandhi Central University (India) for their unrestricted backing and for the establishment of treasured propositions at the time of book proposal and final book preparation.

Contents

Editors and Contributors

About the Editors

Naga Raju Maddela received his M.Sc. (1996–1998) and Ph.D. (2012) in Microbiology from Sri Krishnadevaraya University, Anantapuramu, India. During his doctoral program in the area of Environmental Microbiology, he investigated the effects of industrial effluents/insecticides on soil microorganisms and their biological activities and worked as a Faculty in Microbiology for 17 years, teaching undergraduate and postgraduate students. He received "Prometeo Investigator Fellowship" (2013–2015) from Secretaría de Educación Superior, Ciencia, Tecnología e Innovación (SENESCYT), Ecuador, and "Postdoctoral Fellowship" (2016–2018) from Sun Yat-sen University, China. He also received external funding from "China Postdoctoral Science Foundation" in 2017, internal funding from "Universidad Técnica de Manabí" in 2020, worked in the area of Environmental Biotechnology, participated in 20 national/international conferences, and presented research data in China, Cuba, Ecuador, India and Singapore. Currently, he is working as a full-time Professor at the Facultad de Ciencias de la Salud, Universidad Técnica de Manabí, Portoviejo, Ecuador. He has published 7 books, 40 chapters and 55 research papers.

Aransiola Sesan Abiodun obtained his first degree (B. Tech) and master's degree (M.Tech) from the Federal University of Technology, Minna, Niger State, Nigeria, in Microbiology in 2009 and Environmental Microbiology in 2014, respectively. Currently, Abiodun is concluding his Ph.D. studentship at the Federal University of Technology, Minna, working on "Microbial and Vermicompost assisted phytoremediation of heavy metals polluted soil". He has published over 40 referred articles in professional journals including five book chapters. Also, he is a senior scientific officer at Bioresources Development Centre, National Biotechnology Development Agency, Nigeria. His area of interest is environmental microbiology with research area in phytoremediation, biosorption and bioremediation of soil contaminated environment. Abiodun is a member of Nigerian Society for Microbiology and American Society for Microbiology.

Ram Prasad Ph.D., is associated with the Department of Botany, Mahatma Gandhi Central University, Motihari, Bihar, India. His research interest includes applied and environmental microbiology, plant–microbe interactions, sustainable agriculture and nanobiotechnology. Dr. Prasad has more than 200 publications to his credit, including research papers, review articles and book chapters, six patents issued or pending, and edited or authored several books. Dr. Prasad has 12 years of teaching experience and has been awarded the Young Scientist Award and Prof. J.S. Datta Munshi Gold Medal by the International Society for Ecological Communications; FSAB fellowship by the Society for Applied Biotechnology; the American Cancer Society UICC International Fellowship for Beginning Investigators, USA; Outstanding Scientist Award in the field of Microbiology; BRICPL Science Investigator Award and Research Excellence Award, etc. Previously, Dr. Prasad served as Assistant Professor at Amity University Uttar Pradesh, India; Visiting Assistant Professor at Whiting School of Engineering, Department of Mechanical Engineering at Johns Hopkins University, Baltimore, United States; and Research Associate Professor at School of Environmental Science and Engineering, Sun Yat-sen University, Guangzhou, China.

Contributors

O. P. Abioye Department of Microbiology, Federal University of Technology, Minna, Nigeria

D. O. Aboyeji Department of Microbiology, Federal University of Technology, Minna, Niger State, Nigeria

N. U. Adabara Department of Microbiology, Federal University of Technology, Minna, Nigeria

Binta Buba Adamu Department of Bioentrepreneurship and Extension Services, National Biotechnology Development Agency, Abuja, Nigeria

Femi Johnson Afolabi Bioresources Development Centre, National Biotechnology Development Agency, Nigeria

Aziz Ahmed Faculty of Science, Institute of Biological Sciences, University of Malaya, Kuala Lumpur, Malaysia
Faculty of Science, Center for Research in Waste Management, University of Malaya, Kuala Lumpur, Malaysia
Faculty of Marine Sciences, Lasbela University of Agriculture, Water and Marine Sciences, Uthal, Balochistan, Pakistan

Higo Forlan Amaral Departamento de Agronomia, Centro Universitário Filadélfia (UniFil), Londrina, Paraná, Brazil
Universidade Estadual de Maringá (UEM), Pós-Graduação em Agroecologia (PROFAGROEC), Maringá, Paraná, Brazil

Mounia Arab Laboratory of Applied Microbiology in Food, Biomedical and Environment, Department of Biology, Faculty of Nature Sciences and Life, University of Tlemcen, Tlemcen, Algeria
Faculty of Biological Sciences, University of Sciences and Technology Houari Boumediene, Algiers, Algeria

Aransiola Sesan Abiodun Bioresources Development Centre, National Biotechnology Development Agency (NABDA), Ogbomoso, Nigeria

H. S. Auta Department of Microbiology, Federal University of Technology, Minna, Niger State, Nigeria
Africa Centre of Excellence for Mycotoxin and Food Safety (ACEMFS), Federal University of Technology, Minna, Niger State, Nigeria

Y. T. Awe Environmental Management Programme, Pan African University Institute of Life and Earth Sciences (including Health and Agriculture), PAULESI, University of Ibadan, Ibadan, Nigeria

A. Aziz Faculty of Marine Sciences, Lasbela University of Agriculture, Water and Marine Sciences, Uthal, Balochistan, Pakistan

Babafemi Raphael Babaniyi Bioresources Development Centre, National Biotechnology Development Agency, Nigeria

Gabriel Gbenga Babaniyi Department of Agricultural Development and Management, Agricultural and Rural Management Training Institute (ARMTI), Ilorin, Nigeria

Oyemade Babatunde Bioresources Development Centre, National Biotechnology Development Agency, Ogbomoso, Nigeria

J. D. Bala Department of Microbiology, Federal University of Technology, Minna, Nigeria

Hafida Baoune Faculty of Natural and Life Sciences, Université Kasdi Merbah Ouragla, Ouragla, Algeria

Jayanthi Barasarathi Faculty of Health and Life Science (FHLS), INTI International University Pesiaran Perdana BBN, Nilai, Negeri Sembilan, Malaysia

Aisha Bisola Bello Department of Biological Sciences, The Federal Polytechnic Bida, Bida, Nigeria

Amira Bouaziz Laboratoire de Biotechnologie des Molécules Bioactives et de la Physiopathologie Cellulaire (LBMBPC), Faculté des sciences de la nature et de la vie, Université Batna 2, Batna, Algeria

Ahmet Çabuk Department of Biotechnology and Biosafety, Graduate School of Natural and Applied Sciences, Eskişehir Osmangazi University, Eskişehir, Turkey
Department of Biology, Faculty of Science and Letter, Eskişehir Osmangazi University, Eskişehir, Turkey

Pınar Aytar Çelik Department of Biotechnology and Biosafety, Graduate School of Natural and Applied Sciences, Eskişehir Osmangazi University, Eskişehir, Turkey
Environmental Protection and Control Program, Eskişehir Osmangazi University, Eskişehir, Turkey

Orija Damilola Federal University of Technology, Akure, Nigeria

Estyfany Kelle da Silva Kodaka Walichek Departamento de Agronomia, Centro Universitário Filadélfia (UniFil), Londrina, Paraná, Brazil

José Ozinaldo Alves de Sena Universidade Estadual de Maringá (UEM), Pós-Graduação em Agroecologia (PROFAGROEC), Maringá, Paraná, Brazil

Hâmara Milaneze de Souza Zaniboni Universidade Estadual de Maringá (UEM), Pós-Graduação em Agroecologia (PROFAGROEC), Maringá, Paraná, Brazil

Mordecai Gana Department of Microbiology, Federal University of Technology, Minna, Nigeria

Fauziah Shahul Hamid Faculty of Science, Institute of Biological Sciences, University of Malaya, Kuala Lumpur, Malaysia
Faculty of Science, Center for Research in Waste Management, University of Malaya, Kuala Lumpur, Malaysia

Idris Hannous Service de diabétologie, Hôpital Central de l'Armée, Alger, Algeria

Auwalu Hassan Faculty of Science, Center for Research in Waste Management, Institute of Biological Sciences, University of Malaya, Kuala Lumpur, Malaysia
Department of Microbiology, Federal University of Technology, Minna, Nigeria
Department of Biological Sciences, Faculty of Science, Federal University of Kashere, Kashere, Gombe State, Nigeria

Aicha Asma Houfani Michael Smith Laboratories and Department of Biochemistry and Molecular Biology, University of British Columbia, Vancouver, BC, Canada

Abdullahi Dabban Idris Department of Biological Sciences, The Federal Polytechnic Bida, Bida, Nigeria

U. J. J. Ijah Department of Microbiology, Federal University of Technology, Minna, Nigeria

A. A. Ikhumetse Department of Microbiology, Federal University of Technology, Minna, Nigeria

Femi Joseph Department of Microbiology, Federal University of Technology, Minna, Nigeria

Palakeerti Srinivas Kumar Center for Biotechnology, Institute of Science and Technology, JNTUH, Hyderabad, Telangana, India

Naga Raju Maddela Departamento de Ciencias Biológicas, Facultad la Ciencias de la Salud, Universidad Técnica de Manabí, Portoviejo, Manabí, Ecuador
Instituto de Investigación, Universidad Técnica de Manabí, Portoviejo, Ecuador

Enuh Blaise Manga Department of Biotechnology and Biosafety, Graduate School of Natural and Applied Sciences, Eskişehir Osmangazi University, Eskişehir, Turkey

Monday Peter Obagunwa Bioresources Development Centre, National Biotechnology Development Agency, Nigeria

Olusola David Ogundele Chemical Science Department, Achievers University, Owo, Nigeria

L. Ogunkanmi Department of Soil Science, College of Basic and Applied Sciences, University of Ghana, Accra, Ghana

Olaniran Victor Olagoke Department of Agricultural Development and Management, Agricultural and Rural Management Training Institute (ARMTI), Ilorin, Nigeria

Obagboye Fredrick Oluwole Chemistry Department, Federal University of Technology Akure, Akure, Nigeria

Innocent C. Ossai Faculty of Science, Institute of Biological Sciences, University of Malaya, Kuala Lumpur, Malaysia
Faculty of Science, Center for Research in Waste Management, University of Malaya, Kuala Lumpur, Malaysia

Olusegun Julius Oyedele Bioresources Development Centre, National Biotechnology Development Agency (NABDA), Ogbomoso, Nigeria

Oluwafemi Adebayo Oyewole Department of Microbiology, Federal University of Technology Minna, Minna, Nigeria

Agamuthu Pariatamby Jeffrey Sachs Center on Sustainable Development, Sunway University, Sunway, Malaysia

Pabbati Ranjit Center for Biotechnology, Institute of Science and Technology, JNTUH, Hyderabad, Telangana, India

Kondakindi Venkateswar Reddy Center for Biotechnology, Institute of science and Technology, JNTUH, Hyderabad, Telangana, India

Muhammed Muhammed Saidu Department of Microbiology, Federal University of Technology Minna, Minna, Nigeria

Nalam Renuka Satya Sree Center for Biotechnology, Institute of science and Technology, JNTUH, Hyderabad, Telangana, India

Samuel O. Thompson Chemistry Department, Federal University of Technology, Akure, Nigeria

Munachimso Odenakachi Victor-Ekwebelem Department of Microbiology, Alex Ekwueme Federal University Ndufu-Alike (AE-FUNAI), Abakaliki, Ebonyi State, Nigeria

Japhet Gaius Yakubu Department of Microbiology, Federal University of Technology Minna, Minna, Nigeria

Belma Nural Yaman Department of Biotechnology and Biosafety, Graduate School of Natural and Applied Sciences, Eskişehir Osmangazi University, Eskişehir, Turkey
Department of Biomedical Engineering, Faculty of Engineering and Architecture, Eskişehir Osmangazi University, Eskişehir, Turkey

Alex Eduardo Zaniboni Universidade Estadual de Maringá (UEM), Pós-Graduação em Agroecologia (PROFAGROEC), Maringá, Paraná, Brazil

S. S. Leh-Togi Zobeashia National Biotechnology Development Agency, Lugbe, Abuja, Nigeria

Part I
Microbial Enzymes: Distribution in the Environment

Ecological Interplays in Microbial Enzymology: An Introduction

Sesan Abiodun Aransiola, Femi Joseph, Olusegun Julius Oyedele, and Naga Raju Maddela

Abstract Microorganisms are ubiquitous and produce certain products, which could be beneficial or detrimental to either their survival or the survival of surrounding lives. Among these products are enzymes. These are specialized proteins that are responsible for respiration, digestion, and other metabolic activities in living bodies. Enzymes, especially microbial enzymes, have several uses in industries, such as the agricultural sector, in environmental fields, and many more. They have the capacity to degrade toxic chemical substances found in domestic and industrial wastes. The process of detoxifying toxic substances is either via conversion or via degradation. This chapter, however, deals with the editorial overview and the purpose of this book.

1 Environmental Occurrence of Microbial Enzymes

1.1 Microbes

Microorganisms are always present in the environment, and their presence affects their environment. Their interference with their environment could be of positive or negative influence. Their most important positive influence is the recycling of nutrients that are vital for living organisms. These nutrients include oxygen,

S. A. Aransiola · O. J. Oyedele
Bioresources Development Centre, National Biotechnology Development Agency (NABDA), Ogbomoso, Nigeria

F. Joseph
Department of Microbiology, Federal University of Technology, Minna, Nigeria

N. R. Maddela (✉)
Departamento de Ciencias Biológicas, Facultad de Ciencias de la Salud, Universidad Técnica de Manabí, Portoviejo, Ecuador

Instituto de Investigación, Universidad Técnica de Manabí, Portoviejo, Ecuador
e-mail: raju.maddela@utm.edu.ec

N. R. Maddela et al. (eds.), *Ecological Interplays in Microbial Enzymology*, Environmental and Microbial Biotechnology,
https://doi.org/10.1007/978-981-19-0155-3_1

nitrogen, and carbon (Ricardo et al., 2019). In the process of recycling nutrients, which involves the biodegradation of plant and animal biomass as well as breaking down or converting toxic chemicals into harmless substances, microbes clean up the environment, thereby making it less toxic, and lower the hazards posed by waste matter. These important roles of microbes are promoted by the secretion of various enzymes (Mariana et al., 2020). Several fungi and bacteria have the capability to efficiently transform high-weight molecules into lower substrates via the extracellular, synthesized enzymes. Strains of different organisms are used in combination or singly, to biodegrade unfriendly environmental wastes (Mariana et al., 2020).

1.2 Enzymes

Enzymes are generally known to be specialized proteins that are responsible for respiration, digestion, and other metabolic activities in living bodies. However, little is known about their role in the natural environment (Brett et al., 2017). The transformation of pollutants by enzymes into less toxic and useful products is considered to be better than completely removing the pollutants (Humberto et al., 2004). Enzymes are biological catalysts. They lower the energy required for a reaction to occur, thus increasing the rate of that reaction. Consequently, enzymes enable products to be formed in the shortest time possible owing to the use of the smaller amount of energy (Rajendra et al., 2016). Enzymes are either pure proteins or glycoproteins. There is an exact portion of an enzyme that is involved in the catalytic function; this part is called the active site. An enzyme may contain or need more than one group to carry out the process of catalysis. The protein portion of such a group is called an apoenzyme, whereas the nonprotein portion is called a prosthetic group. The combination of an apoenzyme and a prosthetic group forms a holoenzyme (Chandrakant & Shwetha, 2011).

The biodegradability potentials of the catalysts of microorganisms have been in use for centuries for the production of wine, bread, vinegar, and other products, without really understanding the biochemical principles of the ingredient responsible (Sindhu et al., 2018). Enzymes have gained attention for their widespread usage in medicine and industries owing to their catalytic activities, stability, and ease of production. The adoption of enzymes in industries is rapidly increasing due to the features they possess, such as reduced processing time, cost-effectiveness, low energy requirement, eco-friendliness, and nontoxic nature. Enzymes, especially microbial enzymes, have the capacity to degrade toxic chemical substances found in domestic and industrial wastes (Rajendra et al., 2016). The process of detoxifying toxic substances is either via conversion or via degradation (Chandrakant & Shwetha, 2011).

2 Classification of Enzymes

All enzymes fall into one of the following six categories:

1. Lyases, which carry out nonhydrolytic breakage by addition or elimination reactions.
2. Hydrolases, which promote the binding of carbon to carbon (C–C), carbon to oxygen (C–O), carbon to nitrogen (C–N), and other bonds by water.
3. Isomerases, which promote the structural rearrangement of molecules.
4. Ligases, which catalyze the joining of two molecules covalently, coupled with the hydrolysis of bonds in adenosine triphosphate (ATP) or other similar triphosphates.
5. Transferases, which increase the rate of functional group transfer from a donor to an acceptor.
6. Oxidoreductases, which catalyze electron and proton transfers from a donor to an acceptor (Rajendra et al., 2016).

3 Occurrence and the Role of Oxidoreductases and Amylases in the Environment

3.1 Oxidoreductases

The process of detoxification of organic substances via oxidative coupling is aided by oxidoreductases (microbial enzymes). Microbes harvest energy by energy-producing biochemical reactions, which are facilitated by oxidoreductases to cleave or break bonds, and assist electron transfer from a donor substrate to an acceptor. The oxidation–reduction reaction, consequently, results in the formation of harmless substances, thereby detoxifying the environment.

Lignin in the soil environment is decomposed into phenolic substances by oxidoreductases. In the same manner, these enzymes are involved in the detoxification of harmful xenobiotics, such as phenolic or anilinic compounds, via polymerization. Furthermore, microbial enzymes have been used in the decolorizing and degrading of azo dyes (Humberto et al., 2004). Numerous bacteria reduce metallic radioactive compounds from oxidized soluble forms to an insoluble reduced form. During energy production in biochemical processes, bacteria receive electrons from organic compounds and use the metallic radioactive elements as acceptors. Phenolic compound-polluted water bodies can be decontaminated by plants through enzyme exudates from their roots. Gramineae, Fabaceae, and Solanaceae are families of plants known to produce oxidoreductases responsible for the oxidative degradation of some soil pollutants or contaminants. An example of these kinds of enzymes is chrome reductase, which reduces dangerously toxic Cr(VI) to less harmful, insoluble Cr(III) (Humberto et al., 2004).

3.2 Amylases

Amylases are enzymes that facilitate the breakdown of starchy molecules into simpler sugars. They are present in abundance in the human saliva for the initiation of the digestion process. Foods like rice and potato, which contain much of starch, taste a bit sweet in the mouth because amylase converts part of the starch into sugar in the mouth. Certain bacteria and plants secrete amylase. Amylase acts on α-1,4-glycosidic bonds and are all glycoside hydrolases. In this age, almost all industrial processing of starch by chemical hydrolysis is carried out using amylases (Sindhu et al., 2018). In the last three decades, amylases from microbial sources have made better and satisfying contributions to the biotechnology industry compared to those produced from plants or animals. Amylases that are secreted by microbes have a broad spectrum in industrial applications. They are more stable compared to those produced by animals or plants (Kritika et al., 2017). Using microbes for mass production of amylases in industries is much cheaper and easier to manipulate so as to obtain enzymes with better desired qualities. Amylolytic enzymes are extremely vital in biotechnological processes such as fermentation and food production and in the textile and paper industries (Chandrakant & Shwetha, 2011).

3.3 α-Amylases Occur Either as α-Amylase or β-Amylase

α-Amylase is applied in the production of ethanol by the fermentation of sugars derived from grain starch. High-fructose corn syrup is produced by treating cornstarch with α-amylase, which produces a short chain of monomers called oligosaccharides. Termamyl, a type of α-amylase, is obtained from *Bacillus licheniformis* and is used in certain detergents, particularly starch-removing and dishwashing detergents (Sindhu et al., 2018).

3.4 β-Amylases

β-Amylase is also a form of amylase, alternatively called 1,4-α-D-glucan maltohydrolase. It is synthesized by fungi, bacteria, and plants. As a reducing agent, it fast-tracks the hydrolysis of the glycosidic bond (a second α-1,4 glycosidic bond), which cleaves off two maltoses at the same time. As the fruit ripens, β-amylase cleaves starch into maltose, thus consequently causing the sweetness of the ripened fruit. Some microbes secrete β-amylase extracellularly to degrade starch in the environment. β-Amylase does not occur in animals, except in microbes found in the gut (Sindhu et al., 2018).

Enzymatic processes can be manipulated to target specific compounds that are dangerous to the environment. Such compounds that are treated using this approach are those that cannot be ordinarily treated using traditional methods. However, application of enzymes may sometimes be a pretreatment approach, which makes the pollutant processable by other methods. For instance, toxic or inhibitory compounds can be selectively eliminated, thereby enabling the bulk of the remaining materials to be treated via biological means, thus minimizing the cost of treatment (Kritika et al., 2017).

4 The Role of Microbial Enzymes in Environmental Sustainability

4.1 Application of Enzymes

An increase in the awareness of our environment (the earth) results in the intensive search for a better and alternative cleaning technology. There are currently two basic problems related to pollution: the disposal of large amounts of wastes that are in continuous production and the elimination or escape of toxic substances found in accumulated wastes from dump sites into water bodies and into the soil over a long period of time. Biotechnology is an important instrument for curbing this problem since it can provide new fronts to understand, preserve, manage, and restore the environment, by transforming toxic pollutants into nonharmful substances (Maddela & García, 2021; Maddela et al., 2021), thereby producing biodegradable substances. Biotechnology offers a safe manufacturing environment. However, increased industrialization consequentially leads to increased environmental hazards (Humberto et al., 2004).

4.2 Bioremediation

Various kinds of enzymes from plants, animals, fungi, and bacteria are known to be responsible for the biodegradation of poisonous biological pollutants (Indu, 2019). Bioremediation, which is basically carried out by microbial enzymes, is cheap and environment-friendly. The sum of the quality of life on Earth is dependent on the sum of the environmental quality (Chandrakant & Shwetha, 2011). However, advancements in human activities via industry, science, and technology have resulted in the dumping of wastes into the environment, which, in turn, constitutes environmental problems to the survival of life on Earth. In the past, waste was disposed by digging holes and dumping the waste into them. This method of disposal was limited due to limitation in the availability of space. By advancements in technology, chemical decomposition and high-temperature incineration was adopted. This, however, creates and poses other environmental nuisances and is uneconomical. These problems, therefore, inform the use of bioremediation, which is the use of microbial enzymes to remove waste materials from the environment. Bioremediation is the degradation or transformation of toxic pollutants into nonhazardous or less toxic materials (Indu, 2019). The process of bioremediation is dependent on the enzymatic activities of microorganisms on pollutants. Thus, the process of bioremediation is affected by the growth of microorganisms in an environment. Environments that permit microbial activities and growth will be easily remediated. Thus, the occurrence of microbial enzymes in the environment is influenced by modifying the environment into a favorable condition, thereby increasing the rate of enzymatic activities in the environment (Abatenh et al., 2017). Enzymes carry out a highly important role in nature. Enzymes are efficient and specific. They are responsible for leading the biochemistry of living things with fidelity and great precision. This function is vital for living cells. Enzymes play a vital role in tapping energy from the sun via the process of photosynthesis. Fungi and bacteria synthesize enzymes that are important for environmental survival (Ricardo et al., 2019). These organisms survive in varied environments where they carry out their biological activities, most of which influence the environment and are performed by intracellular or extracellular enzymes. Microbes that survive under extreme environmental conditions secrete enzymes that are capable of functioning in such environments (Ricardo et al., 2019).

4.3 *Application of Enzymes in Medicine*

The application of enzymes in medicine is highly extensive just like in industry and is still in rapid growth. Presently, the majority of medical applications of microbial enzymes are for dead skin removal, burn removal by proteolytic enzymes, and busting clot by fibrinolytic enzymes. Dextranase acid, protease, and rhodanese can be used to treat tooth decay, alimentary dyspepsia, and cyanide poisoning, respectively (Rajendra et al., 2016).

4.4 *Application of Enzymes in Environmental Monitoring*

Monitoring the environment involves the chemical analysis of identified pollutants. The methods employed via advanced technologies to manage and monitor the environment have over the years proved to have inherent setbacks. For instance, just a few amounts of potential pollutants are monitored. At the same time, toxic pollutants can change their chemical properties. Therefore, taking measures to monitor indirect indicators like enzymes can be of great value in monitoring the environment (Brett et al., 2017).

4.5 *Application of Enzymes in Wastewater Treatment*

Numerous enzymes play a crucial role in the treatment of waste matter. Enzymes have the ability to work on and remove harmful pollutants via precipitation and transformation into other products. They can also change the features of certain wastes, enabling pollutants to be prone to treatment, or help convert them into valuable products. The thermal and mechanical stability of enzymes is enhanced by immobilization but decreases the chance of them being leached into solution (Kritika et al., 2017).

4.6 *Removal of Solid Wastes Using Enzymes*

Soils that are polluted with crude oil can be corrected using plant and/or microbial enzymes (Nedaa et al., 2020). Fermentation by living organisms that have the ability to produce useful enzymes serves as a more effective low-cost substitute compared to direct application of the enzymes. This is because direct application of free enzymes to recalcitrant materials may increase the problem of decreased enzyme activity and stability. Applying enzymes directly to the soil will result in direct interaction between the particles that are present and the enzymes. This reaction may consequently cause a change in the properties of the enzymes. This is the problem presented by contaminated soil and solid organic wastes when enzymes are directly applied (Mariana et al., 2020).

5 Impact of Environmental Pollution on Microbial Enzymes

The most common chemicals involved in soil pollution are petroleum hydrocarbons, polycyclic aromatic hydrocarbons (PAHs), solvents, pesticides, lead, and other heavy metals (HMs) (https://en.wikipedia.org/wiki/Soil_contamination). In this section, we discuss how these pollutants affect the activities of enzymes in the soil.

5.1 *Effects of Petroleum Hydrocarbons*

Petroleum hydrocarbons (PHs) are one of the most common pollutants in the soil, and the principal reasons for soil pollution by PHs include increased production and increased number of accidental oil spills on land (Banks et al., 2003). PHs are highly toxic to soil microorganisms, which is attributed to the presence of the high concentration of toxic substances in PHs and their long persisting nature. Compared to aquatic environments, PHs are more likely to accumulate in the soil and sediment matrices due to their (PHs') high hydrophobicity (Karthikeyan & Bhandari, 2001). Furthermore, high hydrophobicity enables the PHs to bind soil and sediment particles; likewise, PHs have limited bioavailability in the terrestrial environment (Luepromchai et al., 2007). Nevertheless, the effects of PHs on soil enzymes seem to be associated with the concentrations of PHs (Wu et al., 2014). Alrumman et al. (2015) reported that the activities of dehydrogenase and phosphatase are inhibited by PHs in the soil and this inhibition increases with increasing concentrations of the PHs in the soil. Similar results have also been found in some other studies. Urease activity is decreased in diesel oil-amended soil (10 mg g^{-1} soil) incubated for 10 weeks under laboratory conditions (Margesin et al., 2000). Achuba and Okoh (2014) found that the order of toxicity of different PHs to soil dehydrogenase and catalase is as follows: kerosene > diesel > petrol > engine oil. Overall, they concluded that the soil biochemistry is substantially influenced by PHs.

5.2 *Effects of PAHs*

Incomplete combustion of organic materials such as coal, oil, petrol, and wood generates large amounts of environmental pollutants in the form of polycyclic aromatic hydrocarbons (PAHs), which have a complex structure and are highly toxic to living systems (both flora and fauna). Chemically, PAHs are highly heterogeneous by having multiple benzene rings in different arrangements; the simplest PAH is naphthalene, which contains two benzene rings. In order to reveal the impact of PAHs on microbial enzyme activities, investigations have been carried out using either artificially contaminated soils or soils spiked with known concentrations of PAHs. For example, the activity of dehydrogenase was poor in soils spiked with PAHs (e.g., naphthalene, phenanthrene, anthracene, and pyrene) at a concentration of 1000–4000 mg kg^{-1} soil (Lipińska et al., 2014). In general, the effects of PAHs on enzyme activities depend on two factors, i.e., the content and the composition of PAHs. Extremely low enzyme activities have been reported in soils (at the military airfield in Deblin, SE Poland) with high PAH content (1986 μg kg^{-1}) (Baran et al., 2004). In several other investigations, it has been widely reported that PAHs are found to be highly toxic to the growth and metabolic activities of microorganisms in a dose-dependent manner (Boopathy, 2000; Wyszkowska & Kucharski, 2000; Loehr et al., 2001).

5.3 *Effects of Solvents*

Like other pollutants, organic solvents show significant negative effects on microbial activities. Evidently, it has been found that the nitrification process is negatively affected in soils treated with the organic solvents acetone and dichloromethane at a rate of 10–20 cm^3 kg^{-1} soil (Klimkowicz-Pawlas & Maliszewska-Kordybach, 2008). Recently, a new class of solvents has emerged, called deep eutectic solvents (DESs), which exist in liquid state at around room temperature. DESs are prepared by mixing two or more solid components, resulting in a decreased melting point over the starting material (Mateusz et al., 2021). Compared to traditional organic solvents, DESs are widely used due to the following benefits: low volatility and inflammability, easy preparation, and cost-effectiveness (Mateusz et al., 2021). However, certain types of DESs have been found to be toxic to microbial activities. For example, a DES made up of methyltriphenylphosphonium bromide (MTPB) and glycerol (in a 1:3 ratio) has shown significant toxic effects on both Gram-positive (*Bacillus subtilis*, *Staphylococcus*) and Gram-negative (*Escherichia coli*) bacteria (Hayyan et al., 2013). A DES of bacterial cellulose (BC) and acrylic acid (1:2) has shown toxic effects on both bacteria (*Staphylococcus aureus* NRS234, *Escherichia coli* ATCC 25922) and fungi (*Candida albicans* ATCC 18804) (Wang et al., 2020). Similarly, other DESs that have shown significant toxic effects on different microbial species (bacteria and fungi) are as follows: BC: methacrylic acid (1:2.5) (Wang et al., 2020), menthol: lactic acid (1:2) (Alsaud et al., 2021), CHCl: ethylene glycol (1:2) (Mao et al., 2016), betaine: malic acid (1:2) (Liang et al., 2020), CHCl: oxalic acid (1:1) (Radošević et al., 2018), CHCl: 1,2-propanediol (1:2) (Wojeicchowski et al., 2021), CHCl: $ZnCl_2$ (1:2) (Juneidi et al., 2018), and CHCl: urea (1:2) (Juneidi et al., 2018); however, additional investigations are needed to identify the exact biochemical pathway(s) and/or enzyme(s) that is/are inhibited by each of these DES chemicals.

5.4 *Effects of Pesticides*

The interaction effects of microbial enzymes with pesticides are highly varied, depending on the type and quantity of pesticides. Several pesticides do exhibit both stimulatory and inhibitory effects on microbial enzymes. Brominal (an herbicide) and Selecron (an insecticide) have been shown to inhibit soil cellulase activity after the majority of incubation periods (Omar & Abdel-Sater, 2001). Similar negative effects of the two pesticides have also been observed in acid phosphatase at higher application doses (Omar & Abdel-Sater, 2001). In contrast, Brominal and Selecron have increased the activities of alkaline phosphatase in soil at both field-level application doses and higher application rates (Omar & Abdel-Sater, 2001). Very recently, Riah et al. (2014) have reviewed the impact of pesticides (insecticides, fungicides, herbicides) on nine different enzymes in soil (e.g., acid phosphatase, alkaline phosphatase, arylsulfatase, cellulase, dehydrogenase, fluorescein diacetate hydrolase, phosphatase, urease, and β-glucosidase) and have reported that: (1) the activities of dehydrogenase have been found to be inhibited by pesticides in 61% of studies, (2) stimulation in the activities of cellulase has been found in 56% of studies, (3) fungicides have exhibited mainly negative effects on enzyme activities, and (4) insecticides and herbicides have exhibited both negative and positive effects on enzyme activities. There are also reports on the impact of pesticide combinations on soil enzyme activities in the presence of nutrient amendments (Maddela & Venkateswarlu, 2018a, b, c, d, e, f). Higher concentrations (7.5 and 10.0 μg g^{-1}) of two pesticides (acephate and buprofezin) in combination have shown significant antagonistic effects on amylase activity in soils received with or without an NPK fertilizer (Maddela & Venkatewarlu, 2013). Similarly, 5 or 7.5 μg g^{-1} concentration of the acephate + buprofezin combination has adversely affected the activities of proteases, ureases, and acid phosphatases in NPK-amended and NPK-unamended soils (Raju & Venkateswarlu, 2013). Furthermore, repeated applications (once, twice, or thrice) of acephate and buprofezin affected the activities of cellulases, amylases, and invertases in NPK-amended and NPK-unamended soils in a dose-dependent manner; the activities were decreased with increasing concentrations of two insecticides (Raju & Venkateswarlu, 2014). Mohiddin et al. (2015) found that acetamiprid and carbofuran are inhibitory to acrylamidase and myrosinase activities in soil. All the abovementioned insights clearly imply that the indiscriminate use of pesticides should be minimized as they do exhibit several adverse nontarget effects in the soil environment.

5.5 *Effects of Heavy Metals*

Intensive agricultural and industrial activities result in the extensive release of heavy metals (HMs) into the soil system. HMs have significant negative effects on soil fertility by inhibiting the activities of soil enzymes (Gao et al., 2010; Karaca et al., 2010; Xian et al., 2015). A 10-week greenhouse study has revealed that enzyme activities (e.g., urease, acid phosphatase, and dehydrogenase) were significantly ($p < 0.05$) decreased in the presence of Cd and Pb and that the activities were decreased with increasing concentrations of HMs (Pan & Yu, 2011). Similar negative effects of HMs on soil enzyme activities have also been found on other occasions, such as inhibition of microbial activity in arid soil upon application of Cd-contaminated sewage sludge (Moreno et al., 1999), decreased soil enzyme activities upon application of different rates of Cd and Pb (Sardar et al., 2007), etc. A recent meta-analysis of HM effects on soil enzyme activities has revealed that soil contamination with five different HMs (i.e., Pb, Zn, Cd, Cu, and As) has linearly reduced the activities of different enzymes in the following order: arylsulfatase > dehydrogenase > β-glucosidase > urease > acid phosphatase > alkaline phosphatase > catalase (Aponte et al., 2020). Since HMs remain mainly in top soils, the adverse effects of HMs on soil enzymes are decreased with the soil depth (Aponte et al., 2020).

6 The Purpose of this Book

Microorganisms and their products are highly important. Microbes are the leading producers of useful natural products. Natural products from microbes and plants are excellent and solve many vital problems. Significant portions of microbial genomes are devoted to the production of these useful metabolites including enzymes. Enzymes are natural catalysts, which are universally found in all living organisms. They may be used either for building more complex molecules from simple ones or for selective breakdown of a mixture of larger molecules. The eminence of life on planet Earth is indistinguishably connected to the overall quality of the environment. The increasing awareness of the environment we live in is bringing about a more intensive search for alternative cleaner technologies. Currently, there are two fundamental pollution-related problems: the disposal of large quantities of wastes that are continually being produced and the removal of toxic compounds that have been accumulating at dump sites in soils and in water systems over the last few decades. Just one microorganism can contain more than 1000 different enzymes. As already indicated by many authors, microbial enzyme research and application will increase because of the successful rate of their usage for industrial, pharmaceutical, medical, and environmental purposes, especially for bioremediation processes. Moreover, microbial enzyme technology opens new possibilities for cleanup methods at sites where other methods fail due to the toxicity or recalcitrant nature of the pollutant. Microbial enzyme technology provides some "cleaner" alternatives and replaces many chemical industrial processes that are less environment-friendly; they are directly used in waste management (solid or liquid) for enhanced efficiency in waste treatment plants and as analytical tools to assist in environmental monitoring. Microbial enzymes are utilized for environmental purposes in a number of industries including agro-food, oil, animal feed, detergent, pulp and paper, textile, leather, petroleum, and the specialty chemical and biochemical industry. The aquatic environment (freshwater and marine) consists of various kinds of discharged wastes that are toxic, especially to the microbial community, thereby affecting the microbial enzyme distribution of this environment. However, more of its roles and distributions in the aquatic world will majorly be in focus in this volume. Microbial enzymes help address ecological problems because they are involved in recycling of wastes and dead plants and animals (which are highly important in keeping all biotic agents intact), thus acting as a cleaning agent for the environment. It must be noted that microbial enzymes return essential nutrients, nitrogen, and sulfur back to the soil by decomposition of dead animals and plants. From the facts outlined above, this book publication is considered to fully unfold the relevant areas of the application of microbial enzymes in industrial frame, medicine, and remediation, most especially its applications in the ecology biota.

In order to finalize and shed more light on the this book title, *Ecological Interplays in Microbial Enzymology*, Part I subtitled "Microbial Enzymes: Distribution in the Environment" covers topics like Diversity of Microbial Enzymes in a Soil Ecosystem, Microbial Enzymes of Wastewater and Sludge, Marine Microbial

Enzymes: An Overview, Microbial Enzymes of Extremophiles, and Enzymology of Microbial Biofilms. Part II subtitled "Microbial Enzymes: Role in Environmental Sustainability" reveals the important areas and applications of microbial enzymes in making our environment habitable. This section revolves around the recent advances in microbial enzymes. These selected topics include Microbial Enzymes: Role in Soil Fertility, Soil Microbial Enzymes and Mitigation of Heavy Metal Uptake by Plants, Communities of Microbial Enzymes and Biodegradation of Persistent Environmental Pollutants, Role of Microbial Enzymes in Wastewater Treatment, Microbial Enzymes in the Recycling of Wastes, Applications of Microbial Enzymes in Industries and Medicine, and Microbial Enzymes in the Biosynthesis of Metal Nanoparticles. A section of this volume presents the dangers that environmental factors pose to microbial enzymes with the subtitle "Impact of Environmental Disturbances on Microbial Enzymes" with subtopics such as Environmental Impact on the Synthesis and Activity of Microbial Ectoenzymes and Endoenzymes, Effects of Agrochemicals on Soil Microbial Enzymes, Functional Effects of Rhizospheres on Microbial Enzyme Communities, Microbial Enzyme Responses to Soil Climate Change, Effects of Aquatic (Freshwater and Marine) Pollution on Microbial Enzyme Activities, and Microbial Enzymes for Sustainable Development: Future Guidelines. This section is the key to this book because it exposes and widens the scope of the readers with future perspectives of the said microbial enzymes and their prospecting applications in making life better.

7 Conclusions

Enzymes are generally known to be specialized proteins that are responsible for respiration, digestion, and other metabolic activities in living bodies. Enzymes are biological catalysts. They lower the energy required for a reaction to occur, thus increasing the rate of that reaction. The adoption of enzymes in industries is rapidly increasing due to the features they possess, such as reduced processing time, cost-effectiveness, low energy requirement, eco-friendliness, and nontoxic nature. Enzymes, especially microbial enzymes, have the capacity to degrade toxic chemical substances found in domestic and industrial wastes. Using microbes for mass production of enzymes in industries is much cheaper and easier to manipulate so as to obtain enzymes with better desired qualities. Enzymatic processes can be manipulated to target specific compounds that are dangerous to the environment. Numerous enzymes play a crucial role in the treatment of waste matter. Soils that are polluted with crude oil and other pollutants can be corrected using plant and/or microbial enzymes. Numerous bacteria can be used to reduce metallic radioactive compounds from oxidized soluble forms to an insoluble reduced form. The process of bioremediation is dependent on the enzymatic activities of microorganisms on pollutants. Fungi and bacteria synthesize enzymes that are important for environmental survival and maintenance of natural events. Monitoring the environment, which involves the chemical analysis of identified pollutants, is vital for effective use of enzymes in bioremediation. Biotechnology, using enzymes, offers a safer manufacturing environment. Biotechnology is an important instrument for curbing the problem of continuous release of toxic wastes into the environment since it can provide new fronts to understand, preserve, manage, and restore the environment, by transforming toxic pollutants into nonharmful substances. By the reduction and removal of toxic wastes, which have over time affected the global climatic conditions, the rate of climate change could be reduced to a bearable range.

References

Abatenh, E., Gizaw, Z. T., & Wassie, M. (2017). Microorganisms in bioremediation—A review. *Open Journal of Environmental Biology, 2*, 38–46. https://doi.org/10.17352/ojeb.000007

Achuba, F. I., & Okoh, P. N. (2014). Effect of petroleum products on soil catalase and dehydrogenase activities. *Open Journal of Soil Science, 4*(12), 399.

Alrumman, S. A., Standing, D. B., & Paton, G. I. (2015). Effects of hydrocarbon contamination on soil microbial community and enzyme activity. *Journal of King Saud University-Science, 27*(1), 31–41.

Alsaud, N., Shahbaz, K., & Farid, M. (2021). Antioxidant and antibacterial evaluation of Manuka leaves (Leptospermum scoparium) extracted by hydrophobic deep eutectic solvent. *Chemical Engineering Research and Design, 174*, 96–106.

Aponte, H., Meli, P., Butler, B., Paolini, J., Matus, F., Merino, C., Kuzyakov, Y., et al. (2020). Meta-analysis of heavy metal effects on soil enzyme activities. *Science of the Total Environment, 737*, 139744.

Banks, M. K., Mallede, H., & Rathbone, K. (2003). Rhizosphere microbial characterization in petroleum-contaminated soil. *Soil and Sediment Contamination, 12*, 371–385.

Baran, S., Bielińska, J. E., & Oleszczuk, P. (2004). Enzymatic activity in an airfield soil polluted with polycyclic aromatic hydrocarbons. *Geoderma, 118*(3–4), 221–232.

Boopathy, R. (2000). Factors limiting bioremediation technologies. *Bioresource Technology, 74*(1), 63–67.

Brett, M. T., Bunn, S. E., Chandra, S., Galloway, A. W. E., Guo, F., Kainz, M. J., Kankaala, P., Lau, D. C. P., Moulton, T. P., Power, M. E., Rasmussen, J. B., Taipele, S. J., Thorp, J. H., & Wehr, J. D. (2017). How important are terrestrial organic carbon inputs for secondary production in freshwater ecosystem?. *62*(5), 833–853.

Chandrakant, S. K., & Shwetha, S. R. (2011). Role of microbial enzymes in the bioremediation of pollutants: A review. *Enzyme Research, 2011*, 805187.

Gao, Y., Zhou, P., Mao, L., Zhi, Y. E., & Shi, W. J. (2010). Assessment of effects of heavy metals combined pollution on soil enzyme activities and microbial community structure: Modified ecological dose–response model and PCR-RAPD. *Environmental Earth Sciences, 60*(3), 603–612.

Hayyan, M., Hashim, M. A., Al-Saadi, M. A., Hayyan, A., AlNashef, I. M., & Mirghani, M. E. (2013). Assessment of cytotoxicity and toxicity for phosphonium-based deep eutectic solvents. *Chemosphere, 93*(2), 455–459.

Humberto, G. A., Miguel, A., & Antonio, B. (2004). Use and improvement of microbial redox enzymes for environmental purposes. *Microbial Cell Factories, 3*, 10.

Indu, S. (2019). *Bioremediation techniques for polluted environment: Concept, advantages, limitations, and prospects*. IntechOpen Book Series.

Juneidi, I., Hayyan, M., & Hashim, M. A. (2018). Intensification of biotransformations using deep eutectic solvents: Overview and outlook. *Process Biochemistry, 66*, 33–60.

Karaca, A., Cetin, S. C., Turgay, O. C., & Kizilkaya, R. (2010). Effects of heavy metals on soil enzyme activities. In *Soil heavy metals* (pp. 237–262). Springer.

Karthikeyan, R., & Bhandari, A. (2001). Anaerobic biotransformation of aromatic and polycyclic aromatic hydrocarbons in soil microcosms: A review. *Journal of Hazardous Substance Research, 3*(1), 3.

Klimkowicz-Pawlas, A., & Maliszewska-Kordybach, B. (2008). Effect of the selected organic solvents on the activity of soil microorganisms. *Roczniki Panstwowego Zakladu Higieny, 59*(1), 83–96.

Kritika, P., Brajesh, S., Ashutosh, K. P., & Ishrat, J. B. (2017). Application of microbial enzymes in industrial waste water treatment. *International Journal of Current Microbiology and Applied Sciences, 6*(8), 1243–1254.

Liang, Y., Pan, Z., Chen, Z., Fei, Y., Zhang, J., Yuan, J., Zhang, J., et al. (2020). Ultrasound-assisted natural deep eutectic solvents as separation-free extraction media for hydroxytyrosol from olives. *ChemistrySelect, 5*(35), 10939–10944.

Lipińska, A., Kucharski, J., & Wyszkowska, J. (2014). The effect of polycyclic aromatic hydrocarbons on the structure of organotrophic bacteria and dehydrogenase activity in soil. *Polycyclic Aromatic Compounds, 34*(1), 35–53.

Loehr, R. C., McMillen, S. J., & Webster, M. T. (2001). Predictions of biotreatability and actual results: Soils with petroleum hydrocarbons. *Practice Periodical of Hazardous, Toxic, Waste Management, 5*(2), 78–87.

Luepromchai, E., Lertthamrongsak, W., Pinphanichakarn, P., Thaniyavarn, S., Pattaragulwanit, K., & Juntongjin, K. (2007). Biodegradation of PAHs in petroleum-contaminated soil using tamarind leaves as microbial inoculums. *Biodegradation, 29*(2), 516.

Maddela, N. R., & García, L. C. (2021). *Innovations in biotechnology for a sustainable future* (p. VIII, 502). Springer Nature. ISBN: 978–3–030-80107-6. Retrieved from https://www.springer.com/gp/book/9783030801076

Maddela, N. R., & Venkateswarlu, K. (2018a). Impact of Acephate and Buprofezin on soil cellulases. In *Insecticides—soil microbiota interactions*. Springer. https://doi.org/10.1007/978-3-319-66589-4_4

Maddela, N. R., & Venkateswarlu, K. (2018b). Impact of Acephate and Buprofezin on soil amylases. In *Insecticides—soil microbiota interactions*. Springer. https://doi.org/10.1007/978-3-319-66589-4_5

Maddela, N. R., & Venkateswarlu, K. (2018c). Impact of Acephate and Buprofezin on soil invertase. In *Insecticides—soil microbiota interactions*. Springer. https://doi.org/10.1007/978-3-319-66589-4_6

Maddela, N. R., & Venkateswarlu, K. (2018d). Impact of Acephate and Buprofezin on soil proteases. In *Insecticides—soil microbiota interactions*. Springer. https://doi.org/10.1007/978-3-319-66589-4_7

Maddela, N. R., & Venkateswarlu, K. (2018e). Impact of Acephate and Buprofezin on soil urease. In *Insecticides—soil microbiota interactions*. Springer. https://doi.org/10.1007/978-3-319-66589-4_8

Maddela, N. R., & Venkateswarlu, K. (2018f). Impact of Acephate and Buprofezin on soil phosphatases. In *Insecticides—soil microbiota interactions*. Springer. https://doi.org/10.1007/978-3-319-66589-4_9

Maddela, N. R., & Venkatewarlu, K. (2013). Impact of pesticides combination on soil microorganisms. *Research and Reviews: Journal of Microbiology and Biotechnology, 2*(4), 29–36.

Maddela, N. R., Garcia, L. C., & Chakraborty, S. (2021). *Advances in the domain of environmental biotechnology* (p. XVIII, 717). Springer Nature Singapore. ISBN: 978-981-15-8999-7. Retrieved from https://www.springer.com/gp/book/9789811589980. https://doi.org/10.1007/978-981-15-8999-7

Mao, S., Yu, L., Ji, S., Liu, X., & Lu, F. (2016). Evaluation of deep eutectic solvents as co-solvent for steroids 1-en-dehydrogenation biotransformation by Arthrobacter simplex. *Journal of Chemical Technology & Biotechnology, 91*(4), 1099–1104.

Margesin, R., Walder, G., & Schinner, F. (2000). The impact of hydrocarbon remediation (diesel oil and polycyclic aromatic hydrocarbons) on enzyme activities and microbial properties of soil. *Acta Biotechnologica, 20*(3–4), 313–333.

Mariana, F., Mirela, N. D., Georgiana, M., Bianca, S. Z., & Gigel, P. (2020). Microorganisms and enzymes used in the biological pretreatment of the substrate to enhance biogas production: A review. *Sustainability, 12*(7012), 7205.

Mateusz, M., Hubert, C., & Grzegorz, B. (2021). Deep eutectic solvents microbial toxicity: Current state of art and critical evaluation of testing methods. *Journal of Hazardous Materials, 425*, 127963. https://doi.org/10.1016/j.jhazmat.2021.127963

Mohiddin, G. J., Srinivasulu, M., Maddela, N. R., Manjunatha, B., Rangaswamy, V., Rosel, A., Rueda, O. D., et al. (2015). Influence of the insecticides acetamiprid and carbofuran on arylamidase and myrosinase activities in the tropical black and red clay soils. *Environmental Monitoring and Assessment, 187*(6), 1.

Moreno, J. L., Hernández, T., & Garcia, C. (1999). Effects of a cadmium-contaminated sewage sludge compost on dynamics of organic matter and microbial activity in an arid soil. *Biology and Fertility of Soils, 28*(3), 230–237.

Nedaa, A., Narjes, D., Majida, K., Husain, A., & Samir, R. (2020). Bioremediation of soils saturated with spilled crude oil. *Scientific Reports, 10*, 1116.

Omar, S. A., & Abdel-Sater, M. A. (2001). Microbial populations and enzyme activities in soil treated with pesticides. *Water, Air, and Soil Pollution, 127*(1), 49–63.

Pan, J., & Yu, L. (2011). Effects of cd or/and Pb on soil enzyme activities and microbial community structure. *Ecological Engineering, 37*(11), 1889–1894.

Radošević, K., Čanak, I., Panić, M., Markov, K., Bubalo, M. C., Frece, J., Redovniković, I. R., et al. (2018). Antimicrobial, cytotoxic and antioxidative evaluation of natural deep eutectic solvents. *Environmental Science and Pollution Research, 25*(14), 14188–14196.

Rajendra, S., Manoj, K., Anshumali, M., & Praveen, K. (2016). Microbial enzymes: Industrial progress in 21st century. *US National Library of Medicine., 6*(2), 174.

Raju, M. N., & Venkateswarlu, K. (2013). Adverse effect of buprofezin and acephate on enzymatic activities in NPK amended and unamended cotton soils. *University Journal of Microbiological Research, 1*(3), 36–42.

Raju, M. N., & Venkateswarlu, K. (2014). Effect of repeated applications of buprofezin and acephate on soil cellulases, amylase, and invertase. *Environmental Monitoring and Assessment, 186*(10), 6319–6325.

Riah, W., Laval, K., Laroche-Ajzenberg, E., Mougin, C., Latour, X., & Trinsoutrot-Gattin, I. (2014). Effects of pesticides on soil enzymes: A review. *Environmental Chemistry Letters, 12*(2), 257–273.

Ricardo, C., William, J. R., & Nicole, S. W. (2019). Scientists' warning to humanity: Microorganisms and climate change. *Nature Reviews Microbiology, 17*, 569–586.

Sardar, K. H. A. N., Qing, C. A. O., Hesham, A. E. L., Yue, X. I. A., & He, J. Z. (2007). Soil enzymatic activities and microbial community structure with different application rates of Cd and Pb. *Journal of Environmental Sciences, 19*(7), 834–840.

Sindhu, R., Binod, P., Sabeela, B. U., Amith, A., Anil, K. M., Aravind, M., Sharrel, R., & Ashok, P. (2018). Applications of microbial enzymes in food industry. *Food Technology and Biotechnology, 56*(1), 16–30.

Wang, J., Xue, J., Dong, X., Yu, Q., Baker, S. N., Wang, M., & Huang, H. (2020). Antimicrobial properties of benzalkonium chloride derived polymerizable deep eutectic solvent. *International Journal of Pharmaceutics, 575*, 119005.

Wojeicchowski, J. P., Marques, C., Igarashi-Mafra, L., Coutinho, J. A., & Mafra, M. R. (2021). Extraction of phenolic compounds from rosemary using choline chloride–based deep eutectic solvents. *Separation and Purification Technology, 258*, 117975.

Wu, B., Lan, T., Lu, D., & Liu, Z. (2014). Ecological and enzymatic responses to petroleum contamination. *Environmental Science: Processes & Impacts, 16*(6), 1501–1509.

Wyszkowska, J., & Kucharski, J. (2000). Biochemical properties of soil contaminated by petrol. *Polish Journal of Environmental Studies, 9*(6), 479–486.

Xian, Y., Wang, M., & Chen, W. (2015). Quantitative assessment on soil enzyme activities of heavy metal contaminated soils with various soil properties. *Chemosphere, 139*, 604–608.

Diversity of Microbial Enzymes in a Soil Ecosystem

O. P. Abioye, S. S. Leh-Togi Zobeashia, A. A. Ikhumetse, and U. J. J. Ijah

Abstract Microbial enzymes are of utmost importance, especially in a soil ecosystem because biochemical processes in the soil ecosystem are catalyzed by enzymes. This chapter reviews the diversity of microbial enzymes, their importance, and their applications in a soil ecosystem, factors that enhance the production and the inhibition of microbial enzymes in a soil ecosystem.

1 Introduction

Soil ecosystem is a significant constituent of all terrestrial habitats as well as sites for different interactions and biochemical reactions. Its function is vital for the preservation of biogeochemical cycles. The processes in soils affect many chemical and physical changes of ecosystems, both living and nonliving. Most of these biochemical processes in a soil ecosystem are catalyzed by enzymes.

Enzymes are specialized organic substances composed of polymers of amino acids that are connected by amide bonds that vary between kilodalton (insulin) and megadalton (ribosome) in molecular mass. Enzymes are biocatalysts that regulate the speed of a large number of reactions involved in the metabolism of living organisms. Enzymes are highly efficient catalysts for biochemical reactions, are selective in nature, and can speed up the rate of a reaction and a metabolic reaction by lowering the barriers that normally prevent chemical reactions from occurring or can slow them down by decreasing the required activation energy, which is the energy required to carry out a reaction, and, thus, in the presence of enzymes, reactions proceed at a faster rate. Like all catalysts, they take part in the reaction and that is how they provide an alternative reaction pathway, but they do not undergo permanent changes and so remain unchanged at the end of the reaction. They can

O. P. Abioye (✉) · A. A. Ikhumetse · U. J. J. Ijah
Department of Microbiology, Federal University of Technology, Minna, Nigeria

S. S. L.-T. Zobeashia
National Biotechnology Development Agency, Lugbe, Abuja, Nigeria

N. R. Maddela et al. (eds.), *Ecological Interplays in Microbial Enzymology*, Environmental and Microbial Biotechnology,
https://doi.org/10.1007/978-981-19-0155-3_2

only alter the rate of a reaction, not the position of the equilibrium. Many enzymes consist of proteins and nonproteins (cofactors). The proteins in enzymes are usually globular. The intra- and intermolecular bonds that hold proteins in their secondary and tertiary structures, respectively, are disrupted by changes in environmental conditions (pH and temperature, among others) (Illanes & Valencia, 2017). Enzymes are basically of two types (endoenzymes and exoenzymes) but are generally classified according to the Enzyme Commission on the basis of the type of reactions that they catalyze. The six classes of enzymes based on the type of reaction catalyzed include oxidoreductases (EC 1), transferases (EC 2), hydrolases (EC 3), lyases (EC 4), isomerases (EC 5), and ligases (EC 6) (Webb, 1984).

Enzymes can be obtained from plants, animals, and microorganisms. More than half of the enzymes are from fungi and yeast, over a third of them are from bacteria that are derived through fermentation (e.g., amylase) (Burhan et al., 2003), and the remaining enzymes are divided between animals (8%) and plants (4%).

Commercial and industrial utilizations of enzymes for production purposes have increased and so has the demand. The shortage in supplies of enzymes from plant and animal sources has prompted a much closer and a more pragmatic evaluation of microbial enzymes. Microbial enzymes are alternative sources of enzymes because they are of different diversities and can be cultured in large quantities within a short time by fermentation and have diverse biochemical functions and susceptibility to gene manipulation (Nyamful et al., 2014). Many microbial enzymes are applied in various commercial processes and industries such as starch, food, brewing, textile, paper, and pharmaceutical (Nimkar et al., 2010; Krishna et al., 2011). They are currently utilized in various fields, for example, brewing industries, medicinal, analytical chemistry, and food processing (Nimkar et al., 2010).

A wide range of enzymes including amylase, lactase, and invertase among others have being isolated from microorganisms, such as bacteria, fungi, and yeast, for their economically viable formulation utilized in industrial production (Krishna et al., 2011). The use of microorganisms for the production of enzymes is economical because microorganisms can be easily manipulated to produce metabolites (Aiyer, 2005). However, fungi are preferred over bacteria for enzyme production because of their filamentous nature, which helps in their penetration through solid substrates (Ramachandran et al., 2004).

Enzymes in a soil ecosystem are derived from microorganisms, plant root exudates, and putrefaction products of remains in the soil (Sinsabaugh et al., 2009; Joniec, 2017) and are responsible for the decomposition and formation of organic matter, mineralization and recycling of nutrients, degradation of organic compounds, and soil ecosystem maintenance (Baddam et al., 2016). Diversity in microbial enzymes in a soil ecosystem is linked to the functional activity of the enzymes and the specificity of the enzymes to the substrate and also to several factors that determine the interaction, productivity, and activity of microbial enzymes in the soil ecosystem.

2 Microbial Enzymes

Microbial enzymes are enzymes that are obtained from different microorganisms. They are active, easy to control, predictable, reliable, and are more superior than animal and plant enzymes (Oyeleke & Oduwale, 2009). Microbial enzymes are readily available, which is due to the viability of mass culture and because their microbial cells can be genetically manipulated. The majority of enzymes used are produced from microorganisms and have functional activity with quite diverse catalytic potential. In recent times, microbial enzymes have been used in many large-scale industrial bioprocesses. Commercially, microbial enzymes are applied in the food, textile, biorefinery, pharmaceutical, medical, paper, brewing, pulp, and environmental sectors for bioremediation and in the leather industry (Brisibe & Helen, 2014).

Microbial enzymes from bacteria, fungi, and yeast are classified into six major groups including microbial oxidoreductases, microbial transferases, microbial hydrolases, microbial lyases, microbial isomerases, and microbial ligases.

2.1 Microbial Oxidoreductases

The enzyme oxidoreductase catalyzes oxidation–reduction (redox) reactions by the cleavage of chemical bonds and transfer of electrons between the reduced donor and the acceptor. The oxidoreductase enzymes involved in biochemical reactions include oxidases, dehydrogenases, peroxidases, and oxygenases.

Oxidases

Microbial oxidase mediates the transfer of an electron to an oxygen molecule, for example, glucose oxidase catalyzes the redox reaction between glucose and glucono-δ-lactone while using the oxygen atom as the electron acceptor and hydrogen peroxide as the product created. Glucono-δ-lactone produced from the oxidation of glucose is further hydrolyzed to gluconic acid by the lactonase enzyme, and the hydrogen peroxide thus created is cleaved into water and oxygen by the catalase enzyme.

Commercially, oxidases (glucose oxidase) have been utilized for gluconic acid production, which is used in the food industry as an antioxidant in beverages and for the production of wine and beer and in pharmaceuticals as sodium gluconate for drug synthesis (Nimkar et al., 2010). Glucose oxidases are used as biosensors in the environmental sector for monitoring pollutants, and, in the energy industry, the enzyme oxidase is used to enhance the production of renewable energy (biofuel) (Dubey et al., 2017). It is applied in baking to make bread to increase the size and quantity (Steffolani et al., 2010).

Laccases

Laccases, also known as phenol oxidases, are enzymes that catalyze the oxidation of aromatic and phenolic compounds; they also oxidize some ester amines and ether groups through a single electron method. Multicopper oxidases are produced intracellularly and extracellularly by microbes and are catalyzed via transfer of electrons, thereby leading to the reduction of molecular oxygen to water (Gianfreda et al., 1999), which is attributed to their specificity to a particular substrate. Their catalytic activity does not require peroxides or cofactors but uses oxygen as the final electron acceptor. Laccases have applications in various bioprocesses including the textile, food, agricultural, paper, cosmetics, refinery, and environmental fields. In agriculture, laccases are applied for the degradation of herbicides and pesticides, whereas in the environmental sector, laccases are used for bioremediation of hydrocarbon-contaminated environments and toxic pollutants (D'Annibale et al., 2006; Guimaraes et al., 2017).

Peroxidases

Peroxidases are oxidoreductase enzymes that promote the oxidation of various compounds (phenol, lignin) through the mechanism of free radicals into polymerized or oxidized lignin and other phenolic compounds. Their catalytic activity includes the donation of electrons, for example, electron donation to ascorbic acid and ferricyanide for degradation into less harmful products. Peroxidases are used for bioremediation of wastewater contaminated with chemical compounds. Peroxidases are classified into manganese-dependent peroxidase (MnP), lignin peroxidase (LiP), and versatile peroxidase (VP).

Oxygenases

Oxygenases are oxidoreductase enzymes involved in the oxidation of a compound by the removal of oxygen from the oxygen molecule (O_2) using flavin adenine dinucleotide (FAD)/nicotinamide adenine dinucleotide (NADH)/nicotinamide adenine dinucleotide phosphate (NADPH) as a co-chemical compound. Based on the number of oxygen atoms used for oxygenation, oxygenases are classified into monooxygenases and dioxygenases. Oxygenases are essential for the breakdown of organic compounds by increasing water solubility or reactivity. They are mostly used for degradation of environmental pollutants and bioremediation.

Microbial oxidoreductase enzymes and their sources

Enzymes	Microorganisms	References
Glucose oxidases	*Aspergillus* sp., *Penicillium* sp.	Bhat et al. (2013)
Peroxidases	*Bacillus* sp., *Pseudomonas* sp., *Escherichia coli* *Thanatephorus* sp., *Auricularia* sp., *Pleurotus ostreatus* *Candida tropicalis*, *Debaryomyces polymorphus*	Bansal and Kanwar (2013), Di-Gennaro et al. (2014) and Telke et al. (2010) Sugano et al. (2000), Liers et al. (2010) and Faraco et al. (2007) Yang et al. (2003)
Laccases	*Pleurotus* sp., *Trametes* sp., *Coriolopsis* sp., *Grifola* sp., *Lentinula* sp. *Bacillus* sp., *Streptomyces*, Pseudomonas sp., *Rhodococcus*	Fernandez-Fernandez et al. (2013), Couto and Herrera (2006) and Nitheranont et al. (2011) Chandra and Chowdhary (2015)

2.2 *Microbial Hydrolases*

Hydrolases are hydrolytic enzymes that catalyze the cleavage of chemical bonds, for example, they catalyze the cleavage of C–C, C–O, C–N, C–P, and other single bonds. Hydrolases are classified based on the kind of bond they hydrolyze (http://www.wiley-vch.de/publish/dt/). Hydrolytic enzymes include amylases, cellulases, proteases, lipases, L-asparaginase DNases, pullulanases, chitinases, β-galactosidases, xylanases, pectinases, keratinases, phytases, tannases, and inulinases and have diverse commercial applications in fields including the leather processing, food, chemical, feed additive, biomedical sciences, biomass degradation (cellulases), wastewater treatment, and the pulp industries (Sánchez-Porro et al., 2003; Kumar et al., 2017). Hydrolases are produced by different microorganisms, from bacterial strains, such as *Bacillus* sp., *Chromohalobacter* sp., *Halobacillus* sp., *Chromohalobacter* sp., *Cellulomonas* sp., *Thermomonospora* sp., *Lactobacillus* sp., *Ruminococcus* sp., *Streptomyces* sp., *Actinomycetes* sp., and *Thermoanaerobacter* sp., among others, from fungal strains, including *Aspergilli* sp., *Penicillium* sp., *Trichoderma* sp., *Rhizopus oryzae*, *Myceliophthora thermophila*, *Neurospora*, and *Mucor*, among others, and from yeast strains, such as *Candida* sp., *Kluyveromyces* sp., *Pichia* sp., *Saccharomycopsis crataegenesis*, *Torulaspora globosa*, *Trichosporon asteroides*, *Hansenula* sp., and *Debaryomyces* sp., among others (Sánchez-Porro et al., 2003; Kupski et al., 2014; Agrawal & Matkar, 2016).

2.3 Microbial Isomerases

Microbial isomerases are enzymes that facilitate the isomerization of intramolecular rearrangement of bonds, that is, conversion of a chemical compound from one isomer to another, for example, the isomerization of D-glucose to D-fructose by glucose isomerase. Isomerases are used in the pharmaceutical and food industries. Isomerase enzymes can be produced from bacteria and yeast. The bacterial strains reported to produce isomerases include *Pseudomonas* sp., *Flavobacterium* sp., *Lactobacillus* sp., *Streptomyces* sp., *Micromonospora* sp., *Brevibacterium* sp., *Nocardia* sp., *Staphylococcus* sp., *Bacillus* sp., *Micromonospora* sp., *Enterobacter* sp. and the yeast strains *Candida* sp., *Saccharomyces* sp., and *Pichia* sp. (Harner et al., 2015; Mert et al., 2017; Shakoor et al., 2018) (Fig. 1).

2.4 Microbial Lyases

Lyases are enzymes that catalyze the cleavage of chemical bonds between two molecules using a biochemical method instead of the redox means. They partake in removal reactions, in which a group of atoms is eliminated from the substrate. Lyases consist of aldolases, dehydratases, and decarboxylases. Lyases are produced by bacterial strains such as *Escherichia coli*, *Pseudomonas* sp., *Propionibacterium* sp., *Lactobacillus helveticus*, *Diplococcus pneumoniae*, *Micrococcus*, *Sarcina* sp., and *Salmonella enteritidis*, among others (Quastel & Woolf, 1926; Blasco et al., 2011).

3 Microbes and Microbial Enzymes in a Soil Ecosystem

Microbes and their bioprocesses help in the maintenance of the soil ecosystem by contributing to biogeochemical cycling, decomposition and formation of organic matter, formation of soil structure, mineralization and recycling of nutrients, circulation of elements, synthesis of proteins and nucleic acids, and transformation of

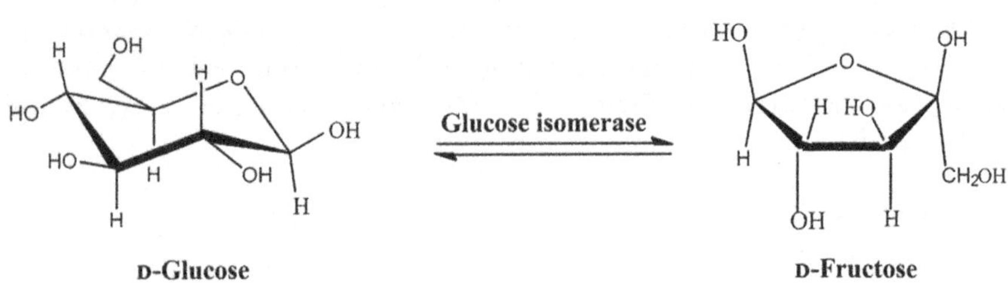

Fig. 1 Glucose isomerase conversion of D-glucose to D-fructose

compounds to other forms and also to increasing soil/plant health and resistance to pathogenic diseases (rhizosphere interaction) (Turco et al., 1994), whereas microbial enzymes are the biological catalysts of countless biochemical reactions necessary for the bioprocesses of microorganisms in the soil ecosystem.

3.1 Decomposition of Organic Matter

Decomposition of organic matter is the physical breakdown and biochemical transformation of complex organic molecules into simpler organic and inorganic molecules. Organic matter decomposition is an essential contributor to soil ecosystem respiration and controls the net carbon emission from soil (Juma, 1998). When residues consisting of deceased microorganisms, insects and earthworms, old plant roots, and crop residues are deposited in the soil, a variety of organic compounds undergo decomposition. Constant deposits of these residues onto the soil surface contribute to the process of carbon cycling and biological activity in the soil ecosystem. Decomposition of organic matter is a biological process that happens naturally, and its rate is determined by soil microbes and their enzymes and also the properties of organic matter (Brussaard, 1994). The decomposition of organic matter leads to the release of energy, carbon dioxide (CO_2), water, and nutrients among others. Organic matter decomposition is carried out by heterotrophic microflora and microfauna including bacteria, fungi, and protozoa. The heterotrophic organisms derive energy and carbon for their growth solely from organic compounds and also essential nutrients and elements needed for cell growth.

Microbial decomposition of organic matter is basically an enzymatic process. The enzymes are synthesized by microorganisms. The produced enzymes are formed in the presence of a specific substrate based on the type of enzyme (extracellular or intercellular) that is synthesized. The organic matter is broken down into its basic components by extracellular enzymes, and these basic components are subsequently utilized by intracellular enzymes.

Decomposition proceeds initially with amino acids, sugars, lipids, water-soluble nitrogenous compounds, and starches, whereas insoluble compounds, such as cellulose, hemicellulose, lignin, and proteins, which form the major fraction of organic matter, are slowly decomposed later.

Microbial Decomposition of Cellulose

Cellulose is an abundant carbohydrate that is present in organic matter in a soil ecosystem. Microbial enzymes (cellulase) hydrolyze cellulolytic substrates and their transformation into monomeric products in the soil ecosphere. They hydrolyze the β-1,4-glycosidic linkages of cellulose, which is the most abundant organic matter. Cellulases are synthesized by microbial strains during their growth on cellulosic materials. Celluloses are polysaccharides that contain glucose. For microbes to access glucose as an energy source, cellulose must be broken down by extracellular enzymes (cellulase) and transported into the cell for production of energy (catabolism) or generation of biomass (anabolism). Cellulases are important enzymes in the microbial degradation of polymers in soil. The bacterial producers of cellulases include *Acetivibrio cellulolyticus*, *Pseudomonas fluorescens*, *Ruminococcus albus*, *Bacillus subtilis*, *Bacillus amyloliquefaciens*, *Cellulomonas fimi*, *Thermotoga maritima*, and *Thermobifida fusca* (Robson & Chambliss, 1989). Fungi such as *Aspergillus* sp. and *Trichoderma* sp. are also effective cellulase producers. Additional fungal strains that have also shown possible cellulase activity include *Rhizopus oryzae* (Kupski et al., 2014), *Penicillium echinulatum* (Camassola & Dillon, 2014), and *Myceliophthora thermophila* (Pereira et al., 2015).

Microbial Decomposition of Hemicellulose

The decomposition of hemicellulose is carried out by hemicellulase enzymes (xylanases, mannanases, arabinofuranosides, and pectin lyases) synthesized by microbes. Hemicelluloses are water-soluble polysaccharides made up of uronic acids, hexoses, and pentoses, and they are sources of nutrients and energy to soil microbes. During the hydrolytic decomposition, hemicelluloses are converted to soluble monosaccharides (galactoses, mannose xyloses, arabinoses), which are further convened to organic acids, H_2O, and CO_2, and uronic acids are broken down to pentoses and CO_2. Several microbes consisting of bacteria and fungi are involved in the decomposition of hemicelluloses (Sylvia et al., 2005). Hemicellulases are excreted by bacteria, including *Bacillus* sp., other bacteria, such as *Cellulomonas*, *Arthrobacter*, *Micrococcus*, *Thermotoga*, *Paenibacillus*, *Staphylococcus*, *Rhodococcus*, *Microbacterium*, and *Pseudoxanthomonas*, fungi, including *Aspergillus*, *Trichoderma*, *Penicillium*, *Paecilomyces*, *Geotrichum*, *Paecilomyces*, *Cephalosporium*, and *Fusarium*, and actinomycetes, such as *Thermoactinomycets*, *Streptomyces*, and *Thermomonospora* (Kumar et al., 2017).

Microbial Decomposition of Lignin

The microbial enzymes lignin peroxidase and lignin oxidase are enzymes excreted by microbes, such as *Escherichia coli*, *Bacillus* sp., *Pseudomonas* sp., *Penicillium geastrovirus*, and *Phanerochaete chrysosporium*, among others, and yeasts such as *Candida tropicalis* and *Debaryomyces polymorphus* (Yang et al., 2003) for lignin decomposition. Lignin is a major constituent of soil organic matter and is among one of the most difficult organic compounds to decompose. The cleavages of lignin yield organic acids, methane, water, and carbon dioxide. Lignins have a unique chemical structure, which contains several aromatics compounds which can be released from the lignin structure by enzymes. The enzymes use H_2O_2 and OH radicals to cleave the bonds in lignin. When the aromatic compounds are released, they are integrated into the metabolic pathway as acetyl CoA, tricarboxylic acid (TCA) cycle, and pyruvate (Sylvia et al., 2005).

Microbial Decomposition of Lipids

Microbial decomposition of lipids is mediated by lipases. The lipid content of soil organic matter ranges from 2% to 20%. Lipases catalyze the hydrolysis of triglycerides to glycerol and free fatty acids over an oil–water interface. Under nonaqueous conditions, glycerides are produced from glycerol and fatty acids via reverse reactions, that is, interesterification, transesterification, and esterification. Lipases also catalyze acidolysis, alcoholysis, and aminolysis on triglycerides. Their enzymatic activity is substrate-specific, which makes them the most versatile biocatalyst.

Some lipid enzymes and their microbial strains include *Bacillus subtilis* and *E. coli* for carboxyl esterase (EC 3.1.1.1) (Sanishvili et al., 2003; Streit et al., 2008), *Acinetobacter calcoaceticus*, *B. subtilis*, *Chromobacterium viscosum*, *Micrococcus freudenreichii*, *Lactobacillus delbrueckii*, *Pseudomonas aeruginosa*, and *Streptococcus lactis* (Sharma et al., 2001) for triacylglycerol lipase (EC 3.1.1.3), and *Lactobacillus casei* and *Gluconobacter oxydans* for arylesterase (EC 3.1.1.2) (Navarro-Gonzalez et al., 2012).

Microbial Decomposition of Proteins

Proteins are complex organic substances consisting of nitrogen, carbon, sulfur, phosphorus, and oxygen. Proteins are decomposed by proteolytic enzymes known as proteases, also called proteinases, belonging to the hydrolases family. In the process of decomposition of organic matter, the protein peptide bond is hydrolyzed into polypeptides, which are further decomposed into amino acids and amines. The amines and amino acids, which are decomposed and converted into ammonia during ammonification to several organic acids, aldehydes, and alcohols among others, are finally decomposed to carbon dioxide and water that are required by the microbial cells in the soil.

Proteases modify proteins via cleaving them. Proteases are classified into two groups, namely, exopeptidases and endopeptidases. Exopeptidases split the peptide bond from the amino or carboxy terminal that is their active site at the N- or C-terminus, and they are subdivided into aminopeptidase and carboxypeptidase. While endopeptidases cleave the peptide bond internally based on the functional group present at the active site, endopeptidases are subdivided into cysteine, serine, metalloproteases, and aspartic acid.

Proteases are present in all living organisms and are essential for cell growth, cell signaling, metabolism, and differentiation. The bacteria and fungi synthesizing proteases include *Bacillus* sp., *Lactobacillus* sp., *Pseudomonas stutzeri*, *Microbacterium* sp., and *Engyodontium album*, whereas the fungal strains mainly include *Aspergillus* sp. and others such as *Cladosporium herbarum*, *Penicillium chrysogenum*, and *Entomophthora coronata* (Jisha et al., 2013; Sethi & Gupta, 2015).

3.2 Nutrient Cycling

In soil ecosystems, nutrients such as nitrogen (N), phosphorus (P), and sulfur (S) are organic molecules and are therefore slightly available for the soil biota. To gain access to these nutrients, other soil ecosystem biotas depend on bacteria and fungi with unique metabolism to mineralize and depolymerize the organic forms of N, P, and S into inorganic forms such as sulfate, nitrate, ammonium, and phosphate (Van der Heijden et al., 2008).

Microbial Cycling of Phosphorus

Phosphatases are responsible for hydrolytic catalysis of organic phosphorus into inorganic form (phosphate); they achieve this by the hydrolysis of ester–phosphate bonds to phosphates. Phosphatases are subdivided into phosphodiesterases, phosphomonoesterases, enzymes that hydrolyze phosphorus-containing anhydrides (EC 3.6.1.x), and P–N bonds (EC 3.9.1.x). The microbial producers of these enzymes are numerous, including *Escherichia coli*, *B. subtilis*, *Pseudomonas aeruginosa*, *Acidithiobacillus thiooxidans*, *Lactobacillus curvatus*, *Rhodobacter capsulatus*, *Geobacillus stearothermophilus*, and *Rhodopseudomonas palustris* (Klemme et al., 1971; Hachimori et al., 1975).

Microbial Cycling of Nitrogen

Ureases are extracellular enzymes that catalyze the hydrolysis of urea to ammonia (NH_3) and carbon dioxide (CO_2) in a soil ecosystem. Ureases account for up to 63% of the total enzyme activity in soil (Tabatabai, 1977).

$$(NH_2)_2CO + H_2O = CO_2 + 2NH_3$$

The urease enzyme is produced by microorganisms as well as plants. The bacterial producers include *Arthrobacter oxydans*, *Brevibacterium ammoniagenes*, *Aerobacter aerogenes*, *Brucella suis*, *Escherichia coli*, *Bacillus pasteurii*, *Helicobacter pylori*, *Proteus mirabilis*, *Selenomonas ruminantium*, *Sporosarcina pasteurii*, *Staphylococcus saprophyticus*, *Providencia stuartii*, and *Ureaplasma urealyticum* (Krajewska & Ureases, 2009).

Other enzymes that participate in nutrient cycling include sulfatases that convert organic sulfur to sulfate and amidases that convert carbon and nitrogen compounds to ammonium (NH_4).

4 Factors That Enhance the Production of Microbial Enzymes in a Soil Ecosystem

Enzymes catalyze most of the biochemical reactions in the soil. Their production and activity is influenced by some basic parameters, as given below.

4.1 pH of the Soil

In a soil ecosystem, the pH is necessary for microbial enzyme production and appropriate operations of enzyme catalytic actions (Sinsabaugh et al., 2008) and also for circuitously controlling the production and activity of enzymes via their effects on the microorganisms that synthesize them (Keeler et al., 2008). There are countless enzymes in a soil ecosystem, which contribute to biological conversion of various organic and inorganic compounds. In addition, enzymes are of diverse sources and have different requirements for microbial enzyme production, stabilization, and catalytic activity in a soil ecosystem. Enzymes are proteins made up of carboxyl groups of one or more amino acids called polypeptides. The chains of amino acids determine the structure, which is linked to their specificity, and a change in pH affects the protein structure, enzymatic activity, and synthesis. Every soil microbial enzyme has a pH for optimum activity. An alteration in H^+ ion concentration in the soil ecosystem influences the dynamics of the enzyme and modifies the cofactor by solubility and ionization and substrate digestion (Tabatabai, 1994). The pH required for optimal activity differs for different microorganisms that synthesize them. Microbes from uncultured soil have been observed to produce amylolytic enzymes with an optimal pH of 9.0 (Yun et al., 2004). However, at tremendously lower or higher pH, permanent denaturation and deterioration of microbial enzymes occur. The potent pH for production of bacterial lipases is within the range of 4–12 (Rathi et al., 2000; Hasan et al., 2007), whereas the optimum pH varies within the range of 6–11 (Ogino et al., 2004).

4.2 Temperature of the Soil

Microbial enzyme syntheses and activities are determined by the operating temperature in a soil ecosystem. An increase in temperature increases the rate of chemicals by about 10 °C. The rate of microbial enzyme formation and activities increases with temperature and then reaches its maximum level at an optimal temperature and drops suddenly with further decrease or increase in temperature. Microbial enzymes have a certain temperature at which they are more effective, for example, α-amylase isolated from the soil and produced by the bacterial strain *Bacillus* sp. had an optimum synthetic activity rate at 75–80 °C (Sajedi et al., 2005). Adejuwon (2010) reported that α-amylase produced by *Aspergillus niger* had an optimum temperature of 35 °C. Rodriguez et al. (2006) reported that α-amylase from *Bacillus licheniformis* is able to hydrolyze soluble starch within a temperature range of 60 °C–75 °C. Thermophilic *Thermus* sp. was reported to produce an extracellular α-amylase, which degrades starch at 70 °C (Shaw et al., 1995). *Bacillus stearothermophilus* was able to produce a thermophilic α-amylase with an optimum temperature range of 65 °C–73 °C (Ogasahara et al., 1970).

4.3 Organic Matter Concentration of a Soil Ecosystem

Organic matter concentration of a soil ecosystem is important for microbial growth, reproduction, and microbial enzyme production and enzymatic activity. Organic matter not only increases soil fertility but also serves as a source of nutrients, such as nitrogen, carbon, sulfur, phosphorus, and other micronutrients. Microorganisms need nutrients for their growth and activity, and, besides, organic matter is a source of substrates for diverse soil microbial enzymes synthesized by different soil microbial communities for their nutritional needs. An increase in the concentration of organic matter enhances the nutritional requirement for microbial enzyme catalytic action and synthesis (Torres & Castro, 2004; Iyer & Ananthanarayan, 2008). Further increase increases the organic matter to an extremely high concentration, which does not have any significant effect on the activity and synthesis of the enzyme.

4.4 Moisture Content of the Soil

Moisture content of the soil can determine the growth of the producing microorganism (Stevenson, 1985). An increase in moisture content of the soil improves the development of microorganisms and increases enzyme activity and the production of organic matter in the form of organic carbon, whereas a decrease in moisture content increases organic matter concentration, which supports microbial growth and enzymatic activity (Iyer & Ananthanarayan, 2008). However, the growth of some bacterial strains (*Escherichia coli*) and their enzymes is hindered when the moisture content is reduced.

4.5 The Presence of Inhibitors

Enzymes are proteinous in nature, and their synthesis requires linking together amino acids in a correct sequence. Enzymes can be inhibited by inhibitors such as ions of metals (Hg^{+}, Fe^{2+}, Cu^{+}, Pb^{2+}), large bioorganic molecules (peptides, proteins), and organic compounds (*N*-ethylmaleimide, diisopropyl phosphofluoridate, oligomycin) by binding the active site. Toxic compounds can also act as inhibitors, inhibiting enzyme synthesis and activity. A high concentration of a toxic compound inhibits microbial growth and thus the enzymes synthesis.

4.6 Soil Inorganic Matter

Inorganic matter in a soil ecosystem influences the diversity and the amount of microbial soil enzymes in the soil. A number of microbial enzymes proliferate with an increased concentration of inorganic soil nutrients. Addition of fertilizers to the soil increases the inorganic nutrient source for microbes, thus increasing their population, and enzyme production to maintain their growth, thereby leading to enhancement of enzymatic activity. In addition, enzymes are released extracellularly by microorganisms to obtain the elements that are lacking in the fertilizers added. An increase in nitrogen fertilizers accelerates the activity of some phosphorus, carbon, and nitrogen cycling enzymes such as phosphatases and cellulases (Sinsabaugh et al., 2005) among others. However, other studies reported that enzyme activities are not affected by the application of inorganic nitrogen fertilizers (Doran, 1987).

4.7 Cropping System

Rotation of crops increases organic materials to the soil, which usually increase the soil biomass and thus the microbial proliferation and their activities (Deng & Tabatabai, 2000). Organic acids enhance microbial diversity and population in the soil and therefore the enzymatic synthesis and activity. Soil enzymes are greatly affected by the type of vegetation on the soil. Knauff et al. (2003) reported that the maximum activities of arylsulfatase were observed in cruciferous crops.

References

Adejuwon, O. A. (2010). Synthetic production of amylase from *Aspergillus niger* isolated from citrus fruit. *African Journal of Basic and Applied Sciences, 2*(5–6), 158–160.

Agrawal, D., & Matkar, K. (2016). Enhancement of cellulase production by substrate manipulation in three aspergillus sp. *Indian Journal of Applied Research, 6*, 219–221.

Aiyer, P. V. (2005). Amylases and their applications. *African Journal of Biotechnology, 4*(13), 1525–1529.

Baddam, R., Reddy, G. B., Johnsely, C. R., & Cyrus, S. (2016). Activity of soil enzymes in constructed wetlands treated with swine wastewater. *Ecological Engineering, 91*, 24–30.

Bansal, N., & Kanwar, S. (2013). Purification and characterization of an extracellular peroxidase of a bacterial isolate *Bacillus* sp. F31. *Current Biotechnology, 2*, 155–161.

Bhat, S. V., Swathi, B. R., Rosy, M., & Govindappa, M. (2013). Isolation and characterization of glucose oxidase (GOD) from *Aspergillus flavus* and *Penicillium* sp. *International Journal of Current Microbiology and Applied Sciences, 2*, 153–161.

Blasco, L., Kahala, M., Tupasela, T., & Joutsjoki, V. (2011). Determination of aspartase activity in dairy propionic acid bacteria. *FEMS Microbiology Letters, 321*, 10–13.

Brisibe, E. A., & Helen, B. (2014). Biotechnological potential of A- amylase production by *Bacillus subtilis* using cassava peel powder as a substrate. *British Biotechnology Journal, 4*(11), 1201–1211.

Brussaard, L. (1994). Interrelationships between biological activities, soil properties and soil management. In D. J. Greenland & I. Szabolcs (Eds.), *Soil resilience and sustainable land use* (pp. 309–329). CABI.

Burhan, A., Nisa, U., Gokhan, C., Omer, C., Ashabil, A., & Osman, G. (2003). Enzymatic properties of a novel thermophilic, alkaline and chelator resistant amylase from an alkalophilic *Bacillus* sp. isolate ANT-6. *Process Biochemistry, 38*, 1397–1403.

Camassola, M., & Dillon, A. J. P. (2014). Effect of different pretreatment of sugar cane bagasse on cellulase and xylanases production by the mutant Penicillium echinulatum 9A02S1 grown in submerged culture. *BioMed Research International, 2014*, 720740.

Chandra, R., & Chowdhary, P. (2015). Properties of bacterial laccases and their application in bioremediation of industrial wastes. *Environmental Science. Processes & Impacts, 17*, 326–342. https://doi.org/10.1039/C4EM00627E

Couto, S. R., & Herrera, J. L. T. (2006). Industrial and biotechnological applications of laccases: A review. *Biotechnology Advances, 24*, 500–513.

D'Annibale, A., Quaratino, D., Federici, F., & Fenice, M. (2006). Effect of agitation and aeration on the reduction of pollutant load of olive mill wastewater by the white-rot fungus Panus tigrinus. *Biochemical Engineering Journal, 29*, 243–249.

Deng, S. P., & Tabatabai, M. A. (2000). Effects of cropping systems on nitrogen mineralization in soils. *Biology and Fertility of Soils, 31*, 211–218.

Di-Gennaro, P., Bargna, A., Bruno, F., & Sello, G. (2014). Purification of recombinant catalase-peroxidase HPI from *E. coli* and its application in enzymatic polymerization reactions. *Applied Microbiology and Biotechnology, 98*, 1119–1126.

Doran, J. W. (1987). Microbial biomass and mineralizable nitrogen distribution in no-tillage and plowed soils. *Biology and Fertility of Soils, 5*, 68–75.

Dubey, M. K., Zehra, A., Aamir, M., Meena, M., Ahirwal, L., Singh, S., Shukla, S., Upadhyay, R. S., Bueno-Mari, R., & Bajpai, V. K. (2017). Improvement strategies, cost effective production, and potential applications of fungal glucoseoxidase (GOD): Current updates. *Frontiers in Microbiology, 8*, 1032. https://doi.org/10.3389/fmicb.2017.01032

Faraco, V., Piscitelli, A., Sannia, G., & Giardina, P. (2007). Identification of a new member of the dye-decolorizing peroxidase family from Pleurotus ostreatus. *World Journal of Microbiology and Biotechnology, 23*, 889–893.

Fernandez-Fernandez, M., Sanroman, M. A., & Moldes, D. (2013). Recent developments and applications of immobilized laccase. *Biotechnology Advances, 31*, 1808–1825.

Gianfreda, L., Xu, F., & Bollag, J. M. (1999). Laccases: A useful group of oxidoreductive enzymes. *Bioremediation Journal, 3*(1), 1–25.

Guimaraes, L. R. C., Woicieehowski, A. L., Karp, S. G., Coral, J. D., Zandona Filho, A., & Soccol, C. R. (2017). Laccase. In A. Pandey, S. Negi, & C. R. Soccol (Eds.), *Current developments in biotechnology and bioengineering: Production, isolation and purification of industrial products* (Vol. 7, pp. 199–216). Elsevier.

Hachimori, A., Takeda, A., Kaibuchi, M., Ohkawara, N., & Samejima, T. (1975). Purification and characterization of inorganic pyrophosphatase from Bacillus stearothermophilus. *Journal of Biochemistry, 77*, 1177–1183.

Harner, N. K., Wen, X., Bajwa, P. K., Austin, G. D., Ho, C. Y., Habash, M. B., Trevors, J. T., & Lee, H. (2015). Genetic improvement of native xylose-fermenting yeasts for ethanol production. *Journal of Industrial Microbiology & Biotechnology, 42*, 1–20.

Hasan, F., Shah, A. A., & Hameed, A. (2007). Purification and characterization of a mesophilic lipase from Bacillus subtilis FH5 stable at high temperature and pH. *Acta Biologica Hungarica, 58*, 115–132.

Illanes, A., & Valencia, P. (2017). Penicillin acylase. In A. Pandey, S. Negi, & C. R. Soccol (Eds.), *Current developments in biotechnology and bioengineering: Production, isolation and purification of industrial products* (Vol. 7, pp. 267–305). Elsevier.

Iyer, P. V., & Ananthanarayan, L. (2008). Enzyme stability and stabilization—Aqueous and nonaqueous environment. *Process Biochemistry, 43*, 1019–1032.

Jisha, V. N., Smitha, R. B., Pradeep, S., Sreedevi, S., Unni, K. N., Sajith, S., Priji, P., Josh, M. S., & Benjamin, S. (2013). Versatility of microbial proteases. *Advances in Enzyme Research, 1*, 39–51.

Joniec, J. (2017). Enzymatic activity as an indicator of regeneration processes in degraded soil reclaimed with various types of waste. *International Journal of Environmental Science and Technology*. https://doi.org/10.1007/s13762-017-1602-x

Juma, N. G. (1998). *The pedosphere and its dynamics: A systems approach to soil science* (Vol. 1, p. 315). Quality Color Press Inc.

Keeler, S. L., Sinsabaugh, R. L., Crenshaw, C., et al. (2008). Pulse dynamics and microbial processes in aridland ecosystems. *Journal of Ecology, 96*(3), 413–420.

Klemme, J. H., Klemme, B., & Gest, H. (1971). Catalytic properties and regulatory diversity of inorganic pyrophosphatases from photosynthetic bacteria. *Journal of Bacteriology, 108*, 1122–1128.

Knauff, U., Schulz, M., & Scherer, H. W. (2003). Arylsufatase activity in the rhizosphere and roots of different crop species. *European Journal of Agronomy, 19*(2003), 215–223.

Krajewska, B., & Ureases, I. (2009). Functional, catalytic and kinetic properties: A review. *Journal of Molecular Catalysis B: Enzymatic, 59*, 9–21.

Krishna, P. R., Srivastava, A. K., Ramaswamy, N. K., Suprasanna, P., & Souza, S. F. D. (2011). Banana peel as substrate for alpa-amylase production using *aspergillus niger*NCIM 616 and process optimization. *Indian Journal of Biotechnology, 11*, 314–319.

Kumar, D., Kumar, S. S., Kumar, J., Kumar, O., Mishra, S. V., Kumar, R., & Malyan, S. K. (2017). Xylanases and their industrial applications: A review. *Biochemical and Cellular Archives, 17*, 353–360.

Kupski, L., Pagnussatt, F. A., Buffon, J. G., & Furlong, E. B. (2014). Endoglucanase and total cellulase from newly isolated Rhizopus oryzae and Trichoderma reesei: Production, characterization, and thermal stability. *Applied Biochemistry and Biotechnology, 172*, 458–468.

Liers, C., Bobeth, C., Pecyna, M., Ullrich, R., & Hofrichter, M. (2010). DyP-like peroxidases of the jelly fungus Auricularia auricula-judae oxidize nonphenolic lignin model compounds and high-redox potential dyes. *Applied Microbiology and Biotechnology, 85*, 1869–1879.

Mert, M. J., Rose, S. H., La Grange, D. C., Bamba, T., Hasunuma, T., Kondo, A., & Van Zyl, W. H. (2017). Quantitative metabolomics of a xylose-utilizing Saccharomyces cerevisiae strain expressing the Bacteroides thetaiotaomicron xylose isomerase on glucose and xylose. *Journal of Industrial Microbiology & Biotechnology, 44*, 1459–1470.

Navarro-Gonzalez, I., Sanchez-Ferrer, A., & Garcia-Carmona, F. (2012). Molecular characterization of a novel arylesterase from the wine-associated acetic acid bacterium Gluconobacter oxidans 621H. *Journal of Agricultural and Food Chemistry, 60*, 10789–10795.

Nimkar, M. D., Deogade, N. G., & Kawale, M. (2010). Production of alpha-amylase from *Bacillus subtilis* and *aspergillus Niger* using different agro-wastes by solid state fermentation. *Asiatic Journal of Biotechnology Resources, 1*, 23–28.

Nitheranont, T., Watanabe, A., & Asada, Y. (2011). Extracellular laccase produced by an edible basidiomycetous mushroom, Grifola frondosa: Purification and characterization. *Bioscience, Biotechnology, and Biochemistry, 75*, 538–543.

Nyamful, A., Moses, E., Ankudey, E., & Woode, M. (2014). Solid state fermentation of *aspergillus Niger* MENA1E and Rhizopus MENACO11A for Glucoamylase production on agricultural residues. *International Journal of Scientific and Research Publications, 4*(6), 5–8.

Ogasahara, K., Imanishi, A., & Isemura, T. (1970). Studies on thermophilic α-amylase from Bacillus stearothermophilus: I. Some general and physico-chemical properties of thermophilic α-amylase. *The Journal of Biochemistry, 67*(1), 65–75.

Ogino, H., Hiroshima, S., Hirose, S., Yasuda, M., Ishimi, K., & Ishikawa, H. (2004). Cloning, expression and characterization of a lipase gene (lip3) from Pseudomonas aeruginosa LST-03. *Molecular Genetics and Genomics, 271*, 189–196.

Oyeleke, S. B., & Oduwale, A. A. (2009). Production of amylase by *bacteria* isolated from cassava waste dumpsite in Minna, Niger state, Nigeria. *African Journal of Microbiology Research, 3*, 143–146.

Pereira, J. D. C., Marques, N. P., Rodrigues, A., Oliveira, T. B. D., Boscolo, M., Silva, R. D., Gomes, E., & Martins, D. B. A. (2015). Thermophilic fungi as new sources for production of cellulases and xylanases with potential use in sugarcane bagasse saccharification. *Journal of Applied Microbiology, 118*, 928–939.

Quastel, J. H., & Woolf, B. (1926). The equilibrium between L-aspartic acid, fumaric acid and ammonia in presence of resting bacteria. *Journal of Biochemistry, 20*, 545–555.

Ramachandran, S., Patel, A. K., Nampoothiri, K. M., Francis, F., & Nagy, V. (2004). A potential raw material for the production of alpa-amylase. *Bioresource Technology, 93*, 169–174.

Rathi, P., Bradoo, S., Saxena, R. K., & Gupta, R. (2000). A hyper-thermostable, alkaline lipase from *pseudomonas* sp. with the property of thermal activation. *Biotechnology Letters, 22*, 495–498.

Robson, L. M., & Chambliss, G. H. (1989). Cellulases of bacterial origin. *Enzyme and Microbial Technology, 11*, 626–644.

Rodriguez, V. B., Alamenda, E. J., Gellegos, J. F. M., & Lopez, A. I. G. (2006). Thermal deactivation of a commercial alpha-amylase from *bacillus licheniformis* used in detergents. *Biochemical Engineering Journal, 27*(3), 299–304.

Sajedi, R. H., Naderi-Manesh, H., Khajeh, K., Ahmadvand, R., Ranjbar, B., Asoodeh, A., & Moradian, F. (2005). A Ca-independent α-amylase that is active and stable at low pH from the Bacillus sp. KR-8104. *Enzyme and Microbial Technology, 36*(5–6), 666–671.

Sánchez-Porro, C., Martín, S., Mellado, E., & Ventosa, A. (2003). Diversity of moderately halophilic bacteria producing extracellular hydrolytic enzymes. *Journal of Applied Microbiology, 94*(2), 295–300.

Sanishvili, R., Yakunin, A. F., Laskowski, R. A., Skarina, T., Evdokimova, E., Doherty-Kirby, A., Lajoie, G. A., Thornton, J. M., Arrowsmith, C. H., Savchenko, A., Joachimiak, A., & Edwards, A. M. (2003). Integrating structure, bioinformatics, and enzymology to discover function. BioH, a new carboxylesterase from Escherichia coli. *The Journal of Biological Chemistry, 278*, 26039–26045.

Sethi, S., & Gupta, S. (2015). Optimization of protease production from fungi isolated from soil. *International Journal of Applied Biology and Pharmaceutical Technology, 3*, 49–154.

Shakoor, S., Singh, G., & Singh, M. (2018). Glucose isomerase production and its applications in various fields. *International Journal of Scientific Research in Science and Technology, 4*, 690–699.

Sharma, R., Chisti, Y., & Banerjee, U. C. (2001). Production, purification, characterization and applications of lipases. *Biotechnology Advances, 19*, 627–662.

Shaw, J. F., Lin, F. P., Chen, S. C., & Chen, H. C. (1995). Purification and properties of an extracellular alpha-amylase from *Thermus sp. Botanical Bulletin of Academia Sinica, 36*, 195–200.

Sinsabaugh, R. L., Gallo, M. E., Lauber, C., Waldrop, M. P., & Zak, D. R. (2005). Extracellular enzyme activities and soil organic matter dynamics for northern hardwood forests receiving simulated nitrogen deposition. *Biogeochemistry, 75*, 201–215.

Sinsabaugh, R. L., Lauber, C. L., & Weintraub, M. N. (2008). Stoichiometry of soil enzyme activity at global scale. *Ecology Letters, 11*(11), 1252–1264.

Sinsabaugh, R. L., Hill, B. H., & Follstad Shah, J. J. (2009). Ecoenzymatic stoichiometry of microbial organic nutrient acquisition in soil and sediment. *Nature, 462*, 795–798. https://doi.org/10.1038/nature08632

Steffolani, M. E., Ribotta, P. D., Perez, G. T., & Leon, A. E. (2010). Effect of glucose oxidase, transglutaminase and pentosanase on wheat proteins: Relationship with dough properties and bread-making quality. *Journal of Cereal Science, 51*, 366–373.

Stevenson, F. J. (1985). *Cycles of soils*. Wiley.

Streit, T. M., Borazjani, A., Lentz, S. E., Wierdl, M., Potter, P. M., Gwaltney, S. R., & Ross, M. K. (2008). Evaluation of the side door in carboxylesterase-mediated catalysis and inhibition. *Biological Chemistry, 389*, 149–162.

Sugano, Y., Matsushima, Y., & Shoda, M. (2000). Complete decolorization of the anthraquinone dye reactive blue 5 by theconcerted action of two peroxidases from *Thanatephorus cucumeri. Applied Microbiology and Biotechnology, 73*, 862–871.

Sylvia, D. M., Fuharmann, J. J., Hartel, P. G., & Zuberer, D. A. (2005). *Principles and applications of soil microbiology* (2nd ed.). Prentice Hall.

Tabatabai, M. A. (1994). Soil enzymes. In R. W. Weaver, S. Angle, P. Bottomley, D. Bezdicek, S. Smith, A. Tabatabai, & A. Wollum (Eds.), *Methods of soil analysis: Part 2 microbiological and biochemical properties, 5.2*. Soil Science Society of America Book Series.

Tabatabai, M. A. (1977). Effects of trace elements on urease activity in soils. *Soil Biology and Biochemistry, 9*, 9–13.

Telke, A. A., Joshi, S. M., Jadhav, S. U., Tamboli, D. P., & Govindwar, S. P. (2010). Decolorization and detoxification of Congo red and textile industry effluent by an isolated bacterium pseudomonas sp. SU-EBT. *Biodegradation, 21*, 283–296.

Torres, S., & Castro, G. R. (2004). Non-aqueous biocatalysis in homogeneous solvent systems. *Food Technology and Biotechnology, 42*(4), 271–277.

Turco, R. F. A., Kennedy, C., & Jawson, M. D. (1994). Microbial indicators of soil quality. In J. W. Doran, D. C. Coleman, D. F. Bedicek, & B. D. Stewart (Eds.), *Defining soil quality for a sustainable environment*. SSSA Spec. Publ. 35 SSSA and ASA.

Van der Heijden, M. G. A., Bardgett, R. D., & Van Straalen, N. M. (2008). The unseen majority: Soil microbes as drivers of plant diversity and productivity in terrestrial ecosystem. *Ecology Letters, 11*, 296–310. https://doi.org/10.1111/j.1461-0248.2007.01139.x

Webb, E. C. (1984). *Enzyme nomenclature*. Academic Press.

Yang, Q., Yang, M., Pritsch, K., Yediler, A., Hagn, A., Achloter, M., & Kettrup, A. (2003). Decolorization of synthetic dyes and production of manganese-dependent peroxidase by new fungal isolates. *Biotechnology Letters, 25*, 709–713.

Yun, J., Kang, S., Park, S., Yoon, H., Kim, M. J., Heu, S., & Ryu, S. (2004). Characterization of a novel amylolytic enzyme encoded by a gene from a soil-derived metagenomic library. *Applied and Environmental Microbiology, 70*(12), 7229–7235.

Microbial Enzymes of Wastewater and Sludge

Y. T. Awe and L. Ogunkanmi

Abstract The treatment of wastewater is a panacea for the protection of environmental and public health. Biological methods of wastewater treatment that are environmentally friendly are gradually replacing the other conventional methods of treatment. Among the biological methods that have been studied over the years, the use of microbial enzymes to develop environmentally friendly bioremediation processes remains a potential to be tapped into. Microorganisms are ubiquitous organisms and can be cultured from wastewater and are good sources of enzymes with great metabolic catalytic activity capable of reducing complex toxic compounds into less harmful ones. This chapter, therefore, discusses the different classes of microbial enzymes based on their modes of action and their distribution in activated sludge (AS), their characteristics (such as effect of temperature and pH), methods of extraction, and their applications in wastewater on municipal and industrial scales.

1 Introduction

Sewage sludge is the end product of treated wastewater. It is a concentrated suspension of solids, which is chiefly composed of organic matter and nutrient-rich organic solids. These organic compounds as well as heavy metals are the most common source of pollutants in sludge (Stevens et al., 2003). Other contents of sludge include a wide range of emerging pollutants, which largely still remain soluble in effluents (wastewaters) (Servos et al., 2005). Sewage sludge can significantly contribute to pollutants present in the environment, particularly in water

Y. T. Awe (✉)
Environmental Management Programme, Pan African University Institute of Life and Earth Sciences (including Health and Agriculture), PAULESI, University of Ibadan, Ibadan, Nigeria
e-mail: yawe0442@stu.ui.edu.ng

L. Ogunkanmi
Department of Soil Science, College of Basic and Applied Sciences, University of Ghana, Accra, Ghana

N. R. Maddela et al. (eds.), *Ecological Interplays in Microbial Enzymology*, Environmental and Microbial Biotechnology,
https://doi.org/10.1007/978-981-19-0155-3_3

bodies and soils, which are often bioaccumulated by plants and biomagnified up the food chain. Therefore, the treatment of wastewater is of utmost importance in the prevention of environmental pollution and protection of public health.

Wastewater, depending on its source, can differ considerably. Domestic wastewater contains approximately 80% of complex organic matter such as polysaccharides, proteins, and lipids (Raunkjær et al., 1994). The biological treatment process has been noted to be beneficial for the treatment of wastewater. This involves the primary sedimentation process, which precedes the biological treatment by the activated sludge (AS) process (Liu & Smith, 2021). It is interesting to also know that the residual sludge after treatment is a host to a large number of microorganisms, which develop in the suspended floc structures. The large sizes of the residual non-settable organic substrates are not easily degraded by these microorganisms; hence, these microbes are compelled to produce hydrolytic enzymes such as lipases, proteases, aminopeptidases, cellulases, galactosidases, glucosidases, phosphatases, and dehydrogenases to enhance degradation (Nabarlatz et al., 2010). Specifically, bacteria in the activated sludge are responsible for the degradation of many complex organic matters or polymeric substrates like lipids, carbohydrates, and proteins into intermediate compounds of low molecular weight by the action of extracellular hydrolases (Nybroe et al., 1992). The toxicity of many toxic compounds is reduced by these hydrolytic enzymes through the breaking down of the major chemical bonds within the molecules (Karigar & Rao, 2011). Most of these hydrolytic enzymes are found within and around the extracellular polymeric substances (EPSs) surrounding bacterial cells. EPSs consist mainly of proteins, humic compounds, uronic acids, carbohydrates, lipids, etc., which are mainly derived from cell lyses (Nabarlatz et al., 2010; Liu & Smith, 2021).

As more pollutants are emerging in wastewaters, the adoption of enzyme-based biocatalytic processes offers many advantages including low energy input, nontoxicity, ability to operate under mild reaction conditions, reduced amount of sludge generation, and can be applied over a wide range of pollutants (Unuofin et al., 2019). Enzymes can work independently of the substrate functional groups, have a long half-life, and work on unnatural substrates as well (Johnson, 2013). In addition, enzymes can be chemically modified to develop new hybrid biocatalysts that show close thermal stability as the free enzymes (Morsi et al., 2020), operation over a wide range of pH, temperature, and salinity, reduction in sludge volume, and the ease of controlling the process (Al-Kassim et al., 1993). In wastewater treatment, microbial enzymes can be utilized to develop bioremediation processes that are environmentally friendly than are the conventional methods (Pandey et al., 2017). Microbial enzymes play a vital role as metabolic catalysts; hence, they have a wide range of usage in industries. In the past, and in most recent times, microbes have been widely used and will continue to serve as one of the largest and useful sources of many enzymes (Demain & Adrio, 2008).

2 Microbial Enzymes

Enzymes are biological catalysts produced by living cells, which initiate specific biochemical reactions forming the various metabolic processes of the cells, and are essential to the maintenance and activity of life (Pandey et al., 2017). Enzymes are proteins consisting of at least one polypeptide moiety (Karigar & Rao, 2011), which may combine with a nonprotein coenzyme or a metal–ion cofactor (Liu & Smith, 2021). The enzyme active sites are directly involved in catalytic processes, which involve the substrate binding to the active site of the enzyme, thus leading to the formation of an enzyme–substrate complex (ES) (Price & Stevens, 1999), and this favors a level of mobility (Rupley et al., 1983). Enzymes are highly specific in their action on substrates and facilitate the conversion of substrates into products by providing favorable conditions that lower the activation energy of the reaction (Karigar & Rao, 2011). Thus, the use of microbial enzymes may be a beneficial method for reducing the volume of sewage sludge solids. A preliminary experiment conducted by Parmar et al. (2001) showed that sludge of about 3.1% solids, which was treated with a combination of commercial protease, lipase, cellulase, and hemicellulase, showed a 29% reduction of solids in enzyme-treated samples compared to a reduction of 6.1% in the controls. This implies that enzymes have the potential to reduce about 80% of the organic fraction of biosolids of sewage sludge. Previous research studies by Carlsson (1979) and Robertson et al. (1994) have proven that bacterial enzyme mixtures are effective in reduction of sludge.

There are six main classes of enzymes: hydrolases, oxidoreductases, transferases, lyases, isomerases, and ligases (synthetases) (Karigar & Rao, 2011). Of the six classes, hydrolases have proven to be the most relevant in wastewater treatment because of their high stability, broad substrate specificity, availability, and catalytic efficiency.

Hydrolases are a class of enzymes that commonly perform as biochemical catalysts that use water to break down a chemical bond. Their function is naturally digestive, to break down complex nutrient molecules (such polysaccharides, proteins, and lipids) into smaller units before being absorbed by cells for metabolism. Examples of hydrolases are described below.

Proteases They are the hydrolase class of enzymes that catalyze the breakdown of protein peptide bonds and the degradation of proteins in wastewater treatment (Karigar & Rao, 2011; Razzaq et al., 2019). They are capable of converting the remnants from meat and fish into a more soluble form, hence improving sludge dewatering (Sun et al., 1992). Protease enzymes have the potential to degrade α-ester bonds, poly(hydroxybutyrate) (PHB) depolymerase ß-ester bonds, and lipase c–ω bonds. An example of the protease enzyme is keratinase, which is capable of degrading keratin proteins present in poultry wastes, animal carcasses, and horns and nails of animals, which constitutes a major cause of environmental pollution (Bhandari et al., 2021). *Stenotrophomonas maltophilia* KB13 produces the enzyme keratinase, which is significantly involved in the biodegradation of poultry feathers (Bhange et al., 2016). Some other microorganisms that produce protease enzymes include *Bacillus* sp. and *Aspergillus* sp. Furthermore, they are widely used in wastewater treatment and also in the food industry amongst others due to their high efficient activity and low cost of production (Kumar & Sharma, 2019).

Lipases They are enzymes that catalyze the hydrolysis of triacylglycerols to glycerol and free fatty acids. Triglycerides being the major component of natural oil or fat can be hydrolyzed consecutively to monoacylglycerol, diacylglycerol, glycerol, and fatty acids. These hydrolyzed components, particularly glycerol and fatty acids, are widely used as raw materials (Karigar & Rao, 2011). Lipases are capable of degrading lipids derived from microbes, plants, and animals, thereby reducing the amount of hydrocarbons present in contaminated soils (Bhandari et al., 2021; Karigar & Rao, 2011). Microbial lipases have the potential to catalyze several reactions, which include hydrolysis, esterification, aminolysis, and alcoholysis (Prasad & Manjunath, 2011). Some examples of microorganisms that produce lipases are *Acinetobacter* sp., *Mycobacterium* sp., *Rhodococcus* sp., *Pseudomonas* sp., and *P. aeruginosa*, which are used at various stages of wastewater treatment.

Glycosidases They form a large group of enzymes that hydrolyze the glycosidic bond between two or more sugars or between a sugar and some other chemical residue. They are common enzymes with roles in nature including degradation of biomass such as cellulose (cellulase), hemicellulose, and starch (amylase).

Other classes of enzymes that can be found or utilized in wastewater treatment are as follows:

Oxidoreductases catalyze the oxidation–reduction reaction, exhibit broad substrate specificity, and are effective in the treatment of effluents that are usually resistant to bacterial degradation at elevated concentrations. They are responsible for the detoxification of organic compounds, toxic xenobiotics such as dyes and phenols (Park et al., 2006), and other related compounds through oxidative mechanisms. Oxidoreductases are essentially utilized in many industries due to their inherent ability to transfer electrons from one substrate molecule to another (Unuofin et al., 2019). During this reaction process, the contaminants are oxidized and rendered harmless. Oxidases, peroxidases, dehydrogenases, and oxygenases belonging to the oxidoreductases class are the major classes of enzymes that are extensively investigated for the bioremediation of wastewater (Bilal et al., 2019).

Peroxidases are ubiquitous enzymes presenting in almost all forms of life ranging from decomposers and producers to consumers (Unuofin et al., 2019). They catalyze the oxidation of lignin and other phenolic compounds (Karigar & Rao, 2011). This reaction is initiated by the reduction of hydrogen peroxide (H_2O_2) and is accompanied by the oxidation of chemically diverse compounds (Unuofin et al., 2019). Peroxidases can be heme or nonheme proteins; heme peroxidases are found in animals, fungi, and prokaryotes, whereas nonheme peroxidases are not readily found in nature. Some heme peroxidases include ascorbate peroxidase from plants, lignin peroxidase and manganese peroxidase from fungi, and horseradish or soybean peroxidase from secretory plants (Karigar & Rao, 2011). These classes of peroxidases are involved in plant cell wall formation and lignification processes (Hiner et al., 2002; Koua et al., 2009; Johnsy & Kaviyarasan, 2011). Lignin peroxidase (LiP), manganese-dependent peroxidase (MnP), and versatile peroxidase (VP) are widely studied due to their inherent ability to degrade toxic substances in nature. Versatile peroxidase (VP) is a hybrid of lignin peroxidase and manganese peroxidase (Wong, 2009), which has the ability to oxidize substrates with or without manganese as compared to other peroxidases (Karigar & Rao, 2011), and is highly effective in the bioremediation of recalcitrant pollutants (Wong, 2009). The increasing literature has shown that peroxidases are effective in the degradation and remediation of organic pollutants in wastewater (Rathner et al., 2017; Unuofin et al., 2019; Morsi et al., 2020).

Oxygenases are a class of enzymes that involve the oxidation of a reduced substrate via transfer of oxygen from molecular oxygen, which is in a triplet state ($^3O_2$), using flavin adenine dinucleotide (FAD)/nicotinamide adenine dinucleotide (NADH)/nicotinamide adenine dinucleotide phosphate (NADPH) as a cosubstrate (Karigar & Rao, 2011). They are capable of making the substrate readily available by substrate activation, which aids in acceleration of the reaction. Oxygenases are utilized for the biodegradation of hydrocarbons (Arora et al., 2009; Wang et al., 2019) and are also involved in biosynthetic and metabolic processes (Unuofin et al., 2019). They catalyze the ring cleavage of hydrocarbons, which is necessary for the total mineralization of the compounds (Arora et al., 2009). Oxygenases can be grouped as monooxygenases and dioxygenases based on their oxygen utilization. Monooxygenases are enzymes that require only one atom of the oxygen molecule for their activity, which is incorporated into the substrate and subsequently appears as the addition of a hydroxyl group; this reaction usually occurs in the presence of a cofactor (Arora et al., 2010; Karigar & Rao, 2011; Unuofin et al., 2019). Monooxygenases are versatile and act as biological catalysts in bioremediation processes due to being regio- and stereoselective in a broad range of substrates, which mediate the conversion of most endobiotic and xenobiotic compounds into more water-soluble and responsive forms (van Beilen et al., 2003; Arora et al., 2010; Unuofin et al., 2019). Furthermore, they catalyze the oxidative reaction of substrates from alkanes to complex molecules (fatty acids and steroids) (Karigar & Rao, 2011; Pandey et al., 2017). One major example of monooxygenases is methane monooxygenase, which is involved in the breakdown of hydrocarbons, which include alkanes, alkenes, ethers, and aromatic and heterocyclic hydrocarbons (Grosse et al., 1999; Karigar & Rao, 2011). Other examples include quinol monooxygenase (YgiN) from *Escherichia coli* (Adams & Jia, 2005) and tetracenomycin F1 monooxygenase (TcmH) from *Streptomyces glauscens* (Shen & Hutchinson, 1993). Dioxygenases utilize both atoms of oxygen (Arora et al., 2009). They serve the major function of oxidizing aromatic compounds reflecting their application in environmental remediation (Karigar & Rao, 2011; Pandey et al., 2017). An example of dioxygenases is catechol dioxygenases, which are found in soil bacteria and are responsible for the degradation of aromatic compounds that are present in the environment (Karigar & Rao, 2011).

Dehydrogenases are enzymes present in various organisms such as bacteria, yeast, plants, animals, and even humans (Bhandari et al., 2021). Aldehyde dehydrogenase, a form of dehydrogenase found in *Azoarcus evansii*, has been observed to be relevant to the metabolism of aromatic compounds (Gescher et al., 2006). Some examples of microorganisms producing the enzyme dehydrogenase include *Rhizobium* sp. involved in the degradation of alkanoate (Cairns et al., 1996), *Bacillus* sp. involved in the degradation of tribromophenol (TBP) (Liang et al., 2017), and, finally, *Rhodococcus* sp. involved in the bioremediation of steroids (Ye et al., 2019).

3 Microbial Enzyme Activities and Distribution in Activated Sludge

3.1 Microbial Abundance and Enzyme Activities

A number of studies have demonstrated that organic components in different wastewaters directly affect the growth and succession of dominant microorganisms in activated sludge (Dircks et al., 2001; Whiteley et al., 2003; Li & Chrost, 2006). Since enzymes are essential in the degradation of these organic components, it is highly important to understand the enzymatic distribution in sludge flocs because enzyme activities reflect their microbial activities when degrading organic matter into wastewater (Yu et al., 2007). In other words, enzyme activities are highly correlated with microbial abundance. Although there are many microorganisms in wastewater (such as viruses, bacteria, and protozoa), bacteria are the most important in the biodegradation of organic substances by producing substantial quantities of hydrolytic enzymes.

Bacteria generally inhabit ecological niches characterized by specific attributes such as pH, temperature, specific carbon or substrate availability, and the presence of salt and other chemical factors (solvents, inhibitors, toxicants, oxygen, etc.). These attributes are the determinants of the enzymes' arrangement and metabolic pathways, which are important for survival and are an essential aspect of screening. Hence, the concentration of enzymes, their location, and their product transport mechanisms are factors that influence the reaction rate in biological processes (Morgenroth et al., 2002). The measurement of enzyme activities is the most direct method to study mechanisms and biological reactions in wastewater and activated sludge. Specifically, it is an effective method to assess microbial biomass and activity of sludge and acts as an indicator of specific processes such as chemical oxygen demand (COD) and phosphorus removal (Richards et al., 1984; Nybroe et al., 1992).

There are various research studies on the relationship between microbial abundance and enzyme activities. Nybroe et al. (1992) evaluated the potential of selected enzyme activity assays (α-glucosidase, alanine aminopeptidase, esterase, and dehydrogenase) to determine microbial abundance and heterotrophic activity in wastewater and activated sludge. In wastewater, microbial abundance, which is measured as colony-forming units of heterotrophic bacteria, has been found to correlate with activities of esterase and dehydrogenase. The authors also reported that the enzymatic activity profiles were distinctly different, suggesting that microbial populations were different or had different physiological properties, in the two types of sludges assessed. The activity profiles of the enzyme in activated sludge from the four full-scale plants appear to be largely influenced by the organic composition of the inlet. For instance, addition of hydrolyzed starch was reflected in a high α-glucosidase activity. They concluded that in waste water, enzyme activities are highly correlated with bacterial abundance. This is consistent with the results of Yin Li and Chrost (2005), whose experimental results showed that enzymatic activities associated with microbial cells in activated sludge flocs accounted for 54.5–97.4% of the total enzyme activities in model reactors from their experiment on microbial enzymatic activities in aerobic activated sludge model reactors. They further reported in their study that the different components of sludge and wastewater in the system exerted a great influence on the secretion and activity of enzymes and, in turn, brought about differences in the enzyme distribution and nutrient removal system performance. This was deduced from their results of lipases being the most active enzymes in the three aerobic activated sludge model reactors of communal, dairy, and petroleum wastewaters. It was further explained that the major reason might be due to changes in dominant bacterial populations in response to the components of the sludge, wastewater, and environmental conditions because the complex wastewater organic matter was biodegraded simultaneously by a set of various enzymes, which were produced by mixed microbial communities forming aggregates.

3.2 Distribution of Enzymes in Extracellular Polymeric Substances: Ectoenzymes and Exoenzymes

Microbial extracellular polymeric substances (EPSs) are a complex high-molecular-weight mixture of polymers excreted by microorganisms and are produced from cell lysis and adsorbed organic matter from wastewater (Sheng et al., 2010). Their characteristics such as adsorption abilities, biodegradability, and hydrophilicity/hydrophobicity, and the contents of the main components (such as polysaccharides, proteins, humic substances, and deoxyribonucleic acids), exhibit crucial effects on microbial adhesion and aggregation processes and promote the formation and stability of a microbial community structure (Flemming & Wingender, 2010; Sheng et al., 2010). In other words, they are responsible for increased bridging flocculation that helps create good settling (Sponza, 2003) in both aerobic and anaerobic sludge treatment systems (Tchobanoglous et al., 2003). However, when in excess, Tchobanoglous et al. (2003) stated that EPSs may hinder the dewatering of sludge, bioflocculation, and sludge settling.

Extracellular enzymes in sludge flocs are mainly bound to the extracellular polymeric substance (EPS)–cell matrix (Frølund et al., 1995; Gessesse et al., 2003). EPSs in sludge flocs comprise both soluble EPSs (slime) and bound EPSs. While soluble EPSs represent the part that binds loosely to sludge flocs, which are liable to removal when washed, bound EPSs are referred to as a discrete covering layer with a distinct margin outside of the cell wall exhibiting a dynamic double-layered structure composed of loosely bound EPS (LB-EPS) and tightly bound EPS (TB-EPS) (Ramesh et al., 2006; Li & Yang, 2007; Yu et al., 2007). The cells in the residue after EPS extraction form the pellet fraction, which is composed of a variety of microorganisms (Yu et al., 2007, 2008). The LB-EPS fraction is considered to easily exchange substances with the bulk solution, having greater impact on numerous sludge processes like coagulation and dewatering (Li & Yang, 2007; Ramesh et al., 2007).

Although Cadoret et al. (2002) stated that the localization of extracellular enzymes is not clearly established, and that the distribution of extracellular enzymes between the cell surface and the EPSs is still quite unknown, enzymatic distributions have been widely studied in EPS sludge flocs by activity measurements. Cadoret et al. (2002) found that 17% of L-Leu-aminopeptidase, 5% of α-glucosidase, 23% of protease, and 44% of α-amylase activities were associated with the easily extractable extracellular polymeric substances (EPSs) from the flocs. Yu et al. (2007) quantified the activities of enzymes in extracellular polymeric substances (EPSs) and in pellets and reported that enzyme assay tests showed that the protease activity was localized mainly on the pellets and that the activities of α-amylase and α-glucosidase were mostly bound to LB-EPS and a few proteases, whereas α-amylase and α-glucosidase activities were associated with the TB-EPS fraction. Yu et al. (2008) collected sludge floc samples from 14 full-scale wastewater treatment plants (WWTPs), including those treating sewage, leachate, and industrial wastewaters, to examine extracellular enzyme distribution and discovered that significant quantities of α-amylase were bound to the pellet fraction and the remainder was uniformly dispersed over the sludge matrix, whereas alkaline phosphatase, acid phosphatase, and protease bound mainly to the pellet and TB-EPS fractions. Similarly, Szabolcs et al. (2009) analyzed the enzymatic activity parameters of α-amylase, α-glucosidase, and alkaline phosphatase in the function of chemical oxygen demand values of samples to determine the activity and distribution in the course of the whole process. They found that significant α-amylase and α-glucosidase enzyme activities were measured in the LB and TB-EPS fractions of the activated sludge flocs in every sample, alkaline phosphatase distribution was significant in all the three fractions of the sludge flocs, and thus concluded that all the three studied enzymes can be characterized and localized in the sludge flocs because during the whole treatment period of the batch system, activity values fluctuated but the enzyme localization was the same. However,

Table 1 Distributions of extracellular enzymes in sludge flocs

References	Sample source	Supernatant	Slime	LB-EPS	TB-EPS	Pellet
Whiteley et al. (2002)	Sewage sludge from a treatment plant	Pro (3–5%) Pho (0.4–0.6%)				Pro (96–97%) Pho (99.4–99.6%)
Li and Chrost (2005)	Activated sludge from communal, dairy, and petroleum wastewaters	L-amp, β-glu, alk pho, lip (2.6–45.4%)			L-amp, β-glu, alk pho, lip (54.5–97.4%)	
Yu et al. (2007)	Sludge from an aerated basin of a municipal wastewater treatment plant (WWTP)			α-Amy and α-glu (++)	Pro (+), α-amy, and α-glu (+)	Pro (++) α-Amy and α-glu (+)
Yu et al. (2008)	Sludge from 14 full-scale WWTPs	α-Amy (8.7–32.0%) Alk pho, aci pho, and pro (+)	α-Amy (6.1–25.8%) Alk pho, aci pho, and pro (+)	α-Amy (5.0–28.8%) Alk pho, aci pho and pro (+)	α-Amy (7.1–34.6%) alk pho, aci pho, and pro (++)	α-Amy (16.8–57.7%) Alk pho, aci pho, and pro (++)
Szabolcs et al. (2009)	A lab-scale Sequencing batch reactor (SBR)			α-Amy (49.2%) α-Glu, alk pho (31.3%)	α-Amy (43.8%) α-glu, alk pho (32.6%)	α-Amy (7.0%) α-Glu, alk pho (36.1%)

Pro protease, *alk pho* alkaline phosphatase, *Aci pho* acid phosphatase, *Amy* amylase, *Glu* glucosidase, *Lip* lipase, *L-ami* leucine aminopeptidase, (++) largely present, (+) slightly present

alkaline phosphatase was termed as a "universal enzyme" because it can be found in whole sludge flocs. Table 1 shows the percentage distribution of various enzymes in sludge flocs investigated in five studies.

Understanding how extracellular enzymes are distributed in sludge flocs provides information on how the organic pollutants in wastewater are biologically degraded and possibly lead to even higher removal efficiencies or better control over the wastewater treatment processes (Yu et al., 2007). It has been recommended that the enzymatic approach to treating wastewater will significantly contribute to the body of knowledge on the biochemical factors controlling the treatability of wastewater and sludge, and this will likely help in the optimization of hydrolyzing and mineralizing processes of organic pollutants. In addition, it may yield technological information on the production of a variety of exogenous enzymes, which can be applied to improve wastewater purification systems (Czerska et al., 1997; Kibret et al., 2000; Guellil et al., 2001; Barjenbruch & Kopplow, 2003; Li & Chrost, 2006).

Table 1 suggests that significant enzyme activities are distributed in the pellet, TB-EPS, and LB-EPS regions. Protease activities, for instance, were mostly localized in the pellet region, with only a minor proportion distributed in the TB-EPS region, and scarcely detected in the LB-EPS region. It can be inferred that protease is bound to the cell surface, i.e., ectoenzyme. Enzymes such as α-amylase that were largely distributed in the LB-EPS region can be said to be immobilized in the EPS matrix in the free form, i.e., exoenzymes.

The inconsistence of the enzymes distributed may be due to differences in the concentration of organic matters of the various sample sources, e.g., distribution of proteins and polysaccharides in the sludge flocs.

3.3 Factors Affecting Enzyme Activities in Wastewater

There are many factors that can affect the ideal conditions for microbial growth and enzyme activities in a wastewater treatment system. These factors can range from compositions of organic loadings, temperature, pH, and kinetic parameters to substrate availability. All of these conditions, individually or in combination, can either increase or decrease the activities of enzymes in the treatment plant.

Temperature A biological reactor does not contain a single, identical, bacterial population. There are numerous groups of microorganisms with dominance among any group constantly shifting in response to the change in temperature. In other words, as the temperature changes, one group of microorganisms slows down or dies off, whereas another group becomes highly active. Generally, bacteria are categorized into four classes based on their adaptability to temperature ranges: psychrophilic, mesophilic, thermophilic, and extreme thermophilic or hyperthermophilic. However, microorganisms in aerobic biological wastewater treatment systems are classified as mesophilic with an optimum temperature range of 25 °C–35 °C. Aerobic digestion and nitrification stop when the temperature rises to 50 °C, methane-producing bacteria become inactive when the temperature drops to about 15 °C, at about 5 °C, autotrophic nitrifying bacteria stop functioning, and, at 2 °C, chemoheterotrophic bacteria acting on carbonaceous material become dormant (Metcalf and Eddy, 2003). This implies that the diversity and adaptability of the microbial population in wastewater starts from a temperature of about 25 °C and continues up to a temperature of 35 °C. At a higher or lower temperature, the adaptability of the various microorganisms diminishes, and optimal conditions of the system (e.g., dissolved oxygen) will be affected. This can lead to deflocculation and high effluent total suspended solids (TSSs).

Gaddad and Rodgi (1987) investigated the effect of temperature on the growth and biochemical activities of *Escherichia coli* in sewage. The bacteria *Escherichia coli* was isolated from a stabilization pond and grown in sterile sewage at various temperatures, ranging from 10° to 50 °C, and its growth and associated biochemical activities were studied. A temperature of 30 °C was found to be optimum for the growth, biological oxygen removal (BOD), NH_3-N release, and the activities of protease and catalase. However, the optimal temperature range is a value that can be tolerated by all enzymes and microorganisms involved, and there is a proportion of them showing activity above the optimal level. For example, temperature optimization studies by Whiteley et al. (2002) demonstrated neutral proteases surviving temperatures of up to 70 °C, those at pH 5 and 10 with temperature optima at 50 and 60 °C, respectively, and phosphatases at 60 °C. It was observed that each of the enzymes demonstrated extensive heat stability for many hours at their individual optimum temperatures.

The temperature effect can be modeled using the modified Arrhenius equation (Burgess & Pletschke, 2008)

$$r(T) = r(20^{\circ}C) \cdot \theta^{(T-20)} \quad (1)$$

where $r(T)$ is the reaction rate at temperature T °C,

$r(20$ °C) is the reaction rate coefficient at temperature 20 °C,

θ is the temperature correction coefficient, which can be found from the literature or

determined experimentally for each biological process under consideration, and

$T =$ temperature (°C).

pH The pH of the wastewater treatment environment has a profound effect on the microbial growth and affects the function of the metabolic enzymes. The operating pH range of most units is controlled in the range 6.5–8.5, depending on the wastewater and the target pollutant (Burgess & Pletschke, 2008). However, since there are different enzymes involved in wastewater treatment, and all have varying optimal pH values, the operating pH range is just what can be tolerated by all microorganisms and their corresponding enzymes involved. A lot of microbial enzymes in wastewater can tolerate a more acidic or alkaline condition. For example, pH optimization studies by Whiteley et al. (2002) showed a broad range of proteolytic activities with prominent enzyme activity at pH 10, whereas phosphatases had the greatest activity at pH 4.5. A whole pH range of proteases were present in the sewage sludge with the most prominent being the alkaline proteases at pH 10, whereas phosphatases showed optimum activity in the acidic region between pH 4.5 and 5.5.

Kinetic Parameters Enzyme kinetics is the study of factors that influence the rates of enzyme-catalyzed reactions. Leonor Michaelis and Maud Menten demonstrated that these reactions occur in two stages (Engelking, 2015). First, the substrate (S) binds to its enzyme (E), forming an enzyme–substrate complex (ES) in a fast and reversible reaction. The enzyme–substrate complex is then transformed into a product (P) that is separated from the enzyme. The catalytic power of an enzyme on a given substrate involves two basic parameters: K_m, which is a measure of the affinity of the enzyme to its substrate, and V_{max}, which is a measure of the maximal velocity of enzymatic catalysis (Engelking, 2015). The Michaelis–Menten equations of enzyme kinetic are

$$\mathrm{E} + \mathrm{S} \leftrightarrow \mathrm{ES} \leftrightarrow \mathrm{E} + \mathrm{P} \tag{2}$$

$$V = \frac{S_c(V_{max})}{S_c + K_m} \tag{3}$$

where

V = reaction velocity,
S_c = substrate concentration,
V_{max} = maximum reaction velocity expressed in terms of a change in the concentration of the substrate or product per unit time (μmol/min or mol/s), and
$K_m = S_c$ at 1/2 V_{max}.

Generally, competitive inhibition does not change V_{max} but increases K_m, whereas noncompetitive inhibition does not change K_m but reduces V_{max}. Competitive enzyme inhibitors can be overcome by increasing the substrate concentration. An enzyme with a high K_m has a low affinity to its substrate (Engelking, 2015).

Kinetic parameters vary from enzymes to enzymes in wastewater. For instance, Li and Chrost (2005) reported varying kinetic parameters in the four enzymes studied. The values of V_{max} and K_m of the total alkaline phosphatase activity from activated sludge samples were approximately two to three times lower than those of the total β-glucosidase activity. It is also worthy of note that the source of wastewater can also affect the values of V_{max} and K_m because the source is a determinant of the types and concentration of the substrate present. This can be inferred from the results of Li and Chrost (2005) stating extremely high values of V_{max} and K_m for lipase and alkaline phosphatase in activated sludge and effluent samples of the dairy wastewater reactor, whereas the values of V_{max} and K_m of leucine aminopeptidase and β-glucosidase showed a slight decrease in the dairy wastewater reactor compared to the communal wastewater reactor.

Substrate Specificity This is the ability of an enzyme to choose the exact substrate from a group of similar chemical molecules. There are some enzymes that show extreme specificity to a particular substrate, whereas certain enzymes are relatively nonspecific and bind to many substrates. Most hydrolases (e.g., lipase) have broad substrate specificity, which is an advantage in wastewater treatment.

4 Modeling of Biological Wastewater Treatment Processes

Extracellular enzymes (mainly hydrolases) degrade complex organic matter (i.e., polymeric substrates such as proteins, lipids, and carbohydrates) into low-molecular-weight intermediates in activated sludge (Nybroe et al., 1992), which are assimilated by bacterial cells for metabolism to produce energy and carbon. This important process of depolymerization is central to several mathematical models that have been developed, explaining the biochemistry of biological wastewater treatment processes and predicting plant performance to assist wastewater treatment plant design engineers (Burgess & Pletschke, 2008). Such mathematical models have become an indispensable tool, especially for the simulation of complex biochemical processes involved in the activated sludge process, which requires a substantial amount of data related to wastewater and sludge characteristics, as well as kinetic and stoichiometric process parameters based upon the initial wastewater breakdown rate—i.e., hydrolysis by exoenzymes (Mu'azu et al., 2020). One of the internationally accepted models among the various dynamic and steady-state mathematical models describing the biological removal processes of organic matter and nitrogen, including nitrification and denitrification mechanisms, is the Activated Sludge Model (ASM) (Henze et al., 2000; Burgess & Pletschke, 2008; Mu'azu et al., 2020).

The Model was developed by the International Water Association (IWA) task group through early work on aerobic treatment. The first was Activated Sludge Model 1 (ASM1), which incorporates carbon oxidation, nitrification, and denitrification and has since been revised and expanded to create ASM2 and ASM2d, which include biological and chemical phosphorus removal (Henze et al., 1995, 1999), and later ASM3, in which biological substrate transport into cells and subsequent intracellular storage (i.e., bacterial membrane size restriction limitations) were proposed as the most important mechanisms of carbon and nitrogen utilization and removal from wastewater (Gujer et al., 2000; Henze et al., 2000). The models are based on the use of chemical oxygen demand (COD) to define carbonaceous material; ASM3 has a total organic carbon (TOC)-based version as well (Henze et al., 2000).

ASM1 has been considered the primary reference model because it triggered the universal recognition of Activated Sludge System modeling. With high potentials of providing a good depiction of the sludge production process, ASM1 primarily describes the removal of organic and nitrogenous compounds with concurrent NO_3 and O_2 consumption as acceptors of the electron. The important concepts adapted in the model were the *bisubstrate hypothesis* and the *death-regeneration* hypothesis. For the *bisubstrate hypothesis*, it was proposed that the biodegradable COD in the influent wastewater consisted of two fractions, which are readily (*Ss*) and slowly biodegradable COD (*Xs*). The readily biodegradable COD was presumed to consist of simple molecules able to pass through the cell membrane and can be immediately used in biosynthetic processes by the organisms. Moreover, the active biomass was divided into two types of organisms: heterotrophic biomass (*XH*) and autotrophic biomass (*XA*) according to the kind of substrate types they need for metabolism and process: autotrophic biomass produce nitrate (*SNO*) from ammonium ions (*SNH*) by the nitrification process and heterotrophic biomass use oxygen (*SO*) for the hydrolysis of the substrate (*SS*, *XS*). The slowly biodegradable COD (*XS*) which consists of larger complex molecules, entrapped by the sludge mass, is adsorbed and then required extracellular enzymatic breakdown before being transferred through the cell wall and used for metabolism. The essence of the introduction of the *death-regeneration* hypothesis was to concisely and elaborately describe the different reactions that occur when organisms die. The decayed cell material of the dead organisms was believed to be released through lysis. One fraction was considered nonbiodegradable and was to remain as an inert residue (*XI*), whereas the remaining fraction was considered to be slowly biodegradable (*XS*). This part could thus return to the process and be used by the remaining organisms as a substrate through hydrolysis (*SS*) (Henze et al., 2000; Szilveszter et al., 2010). There are a total of eight essential biochemical processes modeled in ASM1:

1. Aerobic growth of heterotrophic biomass.
2. Anoxic growth of heterotrophic biomass.
3. Aerobic growth of autotrophic biomass.
4. Heterotrophic biomass decay.
5. Autotrophic biomass decay.
6. Soluble organic nitrogen ammonification.
7. Hydrolysis of entrapped particulate organic matter.
8. Hydrolysis of entrapped organic nitrogen.

Activated Sludge Model 2 (ASM2) is an extension of ASM1 with additional biological processes included, primarily in order to deal with biological phosphorus removal (Szilveszter et al., 2010). Activated Sludge Model 3 (ASM3) was developed to correct some of the deficiencies of the earlier ASM1; the importance of storage polymers in heterotrophic conversions is recognized in the activated sludge processes of ASM3 (Gujer et al., 1999). It includes 12 biochemical processes and 13 components. Figure 1 shows the substrate flows for autotrophic and heterotrophic biomass in the ASM1 and ASM3 models.

5 Immobilization and Extraction of Microbial Enzymes

5.1 Immobilization of Enzymes

This is a process that converts an enzyme from its initial homogeneous form to an immobilized heterogeneous form to develop an immobilized catalyst (Zdarta et al., 2018). This immobilized enzyme can be used for the effective degradation of large volumes of wastewater (effluent) (Mugdha & Usha, 2012). Enzyme immobilization can be highly beneficial for large-scale applications as it helps boost the catalytic properties of enzymes compared to free homogeneous enzymes (Morsi et al., 2020). This was investigated in a research conducted by Morsi et al. (2020), which showed

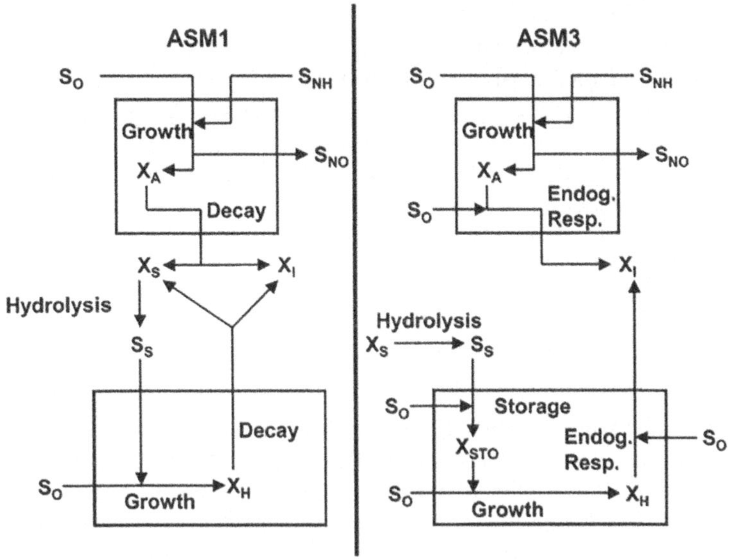

Fig. 1 Substrate flows for autotrophic and heterotrophic biomass in the ASM1 and ASM3 models. (Source: Szilveszter et al., 2010)

that soybean peroxidase (SBP) immobilization onto photocatalytic supports not only allowed for efficient recycling of the enzyme used but also created a potential hybrid catalyst, much more powerful than the free homogeneous enzyme.

The immobilization method reduces the loss of enzymes, thus increasing their reusability, and also minimizes the chances of loss of enzyme activity under harsh conditions. Use of immobilized enzymes in effluent treatment, as compared to free enzymes, results in multiple advantages like increased stability, reusability, ease of handling, and reduction in running cost. Immobilization of enzymes could be carried out conventionally by physical or chemical binding to an inert carrier such as zeolite, acrylic resin, polyacrylamide, and agarose; it could also be carrier-free immobilization, which is a recent technique and can be carried out by intermolecular cross-linking between adjacent enzyme molecules (Asgher et al., 2018). This alternative technique provides potential opportunities for research in the near future.

5.2 *Extraction of Microbial Enzymes from Wastewater and Sludge*

Enzymes are commercially produced for a large range of industrial uses and applications, as they can serve as a sustainable replacement to inorganic chemical catalysts. Sludge generated during wastewater treatment by the activated sludge system is a potential source of biomass for large-scale enzyme extraction (Liu &

Smith, 2019) and is also a cost-effective approach as compared to the synthetic culture media for commercial enzyme production (Tyagi & Lo, 2013). A vast community of microorganisms is established within the activated sludge system and is capable of degrading organic solid substrates in wastewater by producing substantial amounts of hydrolytic enzymes such as phosphatase, lipase, protease, cellulose, and dehydrogenase (Liu & Smith, 2021). Extraction of enzymes majorly involves the disruption of biomass of bacteria or different types of fungi including yeast and is formulated into solid or liquid products (Aberer et al., 2002).

5.3 *Methods of Enzyme Extraction*

According to the literature, different techniques have been adopted to extract crude microbial enzymes from activated sludge. One major technique is sonication, which involves the cellular disruption of biomass in order to shorten the hydrolytic phase of fermentation. This method helps improve the mineralization and stability of the fermented sludge (Zieliński et al., 2019). Other authors that have studied the extraction of enzymes using the sonication method include Balasundaram and Harrison (2006), who indicated that enzymes are released following the stepwise breakdown of the cell structure by sonication treatment. Nabarlatz et al. (2010) stated that ultrasonication allows more enzyme recovery, and Arun and Sivashanmugam (2017) observed that an increase in the duration of sonication treatment from 5 to 20 min significantly improved the extraction of microbial enzymes (amylase, protease, and lipase) from fruit wastes.

Temperature is one major factor that needs to be controlled during enzyme extraction by sonication treatment as high temperatures can have a detrimental effect on the protein structures of the enzymes. The characteristics of the solid content in sludge determine the effectiveness of the sonication treatment for cell disruption (Rubin et al., 2018). Sludge with higher solid contents greater than 3% Dissolved Solids were less susceptible to ultrasonic cell disruption due to the absorption of the sound energy, which reduced the disruption efficiency (Zhang et al., 2008).

Another method of enzyme extraction is the milling treatment, which involves the use of a bead mill consisting of a grinding chamber, filled with small beads and a rotating shaft (Liu & Smith, 2021). The milling treatment involves cell disruption to release cellular substances from sludge (Postma et al., 2015). The milling treatment for enzyme extraction possesses some advantages including high disruption efficiency, ease of loading sample biomass, and its applicability to varying sizes of biomass (Koubaa et al., 2020).

According to the experiment conducted by Nabarlatz et al. (2010), it was possible to recover the maximum concentration of protease and lipase after 1 h of extraction using the magnetic stirring method and a stirring velocity of 500 rpm was suitable for the extraction of enzymes. However, ultrasonication allows more enzyme recovery within a shorter time of extraction (10–20 min), which helps save time and has less power consumption as compared to the milling and stirring methods. In lieu of this, sonication is the more preferred treatment for the recovery of enzymes in industrial applications.

5.4 *Applications of Extracted Enzymes in Different Wastewaters*

Barber et al. (2020) have recently proposed that oxidoreductase enzymes cross-linked with flexible spacers (e.g., polyethylene glycol) could provide an effective approach to organic micropollutant degradation in municipal wastewater.

Protease extracted from activated sludge when mixed with milk wastewater at a volume of 1:1 yielded the degradation product (tyrosine) after 2 h. This implies that the extracted protease can be used for the treatment of diary wastewater (Jung et al., 2002).

Oxidases have been utilized in the removal of phenolic pollutants from wastewater (Mukherjee et al., 2013). This process is achieved by the hydroxylation of aromatic rings and, subsequently, the oxidation of monophenols or diphenols, which is usually oxygen-dependent (Mishra et al., 2013).

5.5 *Industrial Applications of Extracted Enzymes*

Microbial lipases have been utilized on a large scale in the bioremediation of petroleum contaminants, oil residues, effluents, and soil recovery (Hassan et al., 2018). Due to their reduced production cost, low energy input, high substrate specificity, and stability, they have been adopted for use in polymerization and in the pulp and paper and cosmetic industries (Arora et al., 2020; Gurung et al., 2013).

In the pharmaceutical industry, monooxygenases have been used as biocatalysts for the synthesis of pharmaceutical compounds. An example of such a monooxygenase is styrene monooxygenase, which is derived from *Pseudomonas* sp. (Arora et al., 2010). Oxygenases are relevant to the biodegradation of hydrocarbons that are environmental pollutants (Sun et al., 2018).

Microbial proteases are widely used in the food industry, leather industry, and also in the treatment of wastewater (Singh, 2003; Kumar & Sharma, 2019). They are highly important due to their low cost, high production rates, and activity (Kumar & Sharma, 2019). In the food industry, proteases are utilized in the manufacture of cheese; likewise, in the detergent industry, proteases are of immense importance. Proteases are also used in the pharmaceutical industry for the production of effective therapeutic agents (Rao et al., 1998).

Microbial peroxidases are utilized in the food industry, pharmaceutical industry, and paper and pulp industry and also actively in bioremediation (Karigar & Rao, 2011).

6 Conclusions

Microbial enzymes are essential in the biological treatment of wastewater to produce less toxic sludge that can be released into the environment. The understanding of the distribution and activities of these enzymes in wastewater and sludge aids in the construction and mathematical modeling of a more efficient biological wastewater treatment plant. Microbial enzymes in wastewater can be extracted from the biomass in activated sludge for other industrial applications.

References

Aberer, W., Hahn, M., Klade, M., Seebacher, U., Spök, A., Wallner, K., & Witzani, H. (2002). Collection of information on enzymes. *Office for Official Publications of the European Communities, Luxemburgo Acids Research, 37*(suppl_ 1), D261–D266.

Adams, M. A., & Jia, Z. (2005). Structural and biochemical evidence for an enzymatic quinone redox cycle in *Escherichia coli. Journal of Biological Chemistry, 280*, 8358–8363.

Al-Kassim, L., Nicell, J. A., Bewtra, J. K., & Taylor, K. E. (1993, March). Wastewater treatment by enzyme catalysed polymerization and precipitation. *Biodeterioration Abstracts, 7*, 1–8.

Arora, P. K., Kumar, M., Chauhan, A., Raghava, G. P., & Jain, R. K. (2009). OxDBase: A database of oxygenases involved in biodegradation. *BMC Research Notes, 2*(1), 1–8.

Arora, P. K., Srivastavs, A., & Singh, V. P. (2010). Application of monooxygenases in dehalogenation, desulphurization, denitrification and hydroxylation of aromatic compounds. *Journal of Bioremediation & Biodegradation, 1*, 112. https://doi.org/10.4172/2155-6199.100011

Arora, N. K., Mishra, J., & Mishra, V. (Eds.). (2020). *Microbial enzymes: Roles and applications in industries* (Vol. 11). Springer Nature.

Arun, C., & Sivashanmugam, P. (2017). Study on optimization of process parameters for enhancing the multi-hydrolytic enzyme activity in garbage enzyme produced from preconsumer organic waste. *Bioresource Technology, 226*, 200–210.

Asgher, M., Bashir, F., & Iqbal, H. M. (2018). Protease-based cross-linked enzyme aggregates with improved catalytic stability, silver removal, and dehairing potentials. *International Journal of Biological Macromolecules, 118*, 1247–1256.

Barber, E. A., Liu, Z., & Smith, S. R. (2020). Organic contaminant biodegradation by oxidoreductase enzymes in wastewater treatment. *Microorganisms, 8*(1), 122.

Balasundaram, B., & Harrison, S. T. L. (2006). Study of physical and biological factors involved in the disruption of E. coli by hydrodynamic cavitation. *Biotechnology Progress, 22*(3), 907–913.

Barjenbruch, M., & Kopplow, O. (2003). Enzymatic, mechanical and thermal pre-treatment of surplus sludge. *Advances in Environmental Research, 7*(3), 715–720.

Bhandari, S., Poudel, D. K., Marahatha, R., Dawadi, S., Khadayat, K., Phuyal, S., Parajuli, N., et al. (2021). Microbial enzymes used in bioremediation. *Journal of Chemistry, 2021*, 8849512.

Bhange, K., Chaturvedi, V., Bhatt, & R. (2016). Feather degradation potential of Stenotrophomonas maltophilia KB13 and feather protein hydrolysate (FPH) mediated reduction of hexavalent chromium. *3 Biotech, 6*(1), 42.

Bilal, M., Adeel, M., Rasheed, T., Zhao, Y., & Iqbal, H. M. (2019). Emerging contaminants of high concern and their enzyme-assisted biodegradation—A review. *Environment International, 124*, 336–353.

Burgess, J. E., & Pletschke, B. I. (2008). Hydrolytic enzymes in sewage sludge treatment: A minireview. *Water SA, 34*(3), 343–350.

Cadoret, A., Conrad, A., & Block, J. C. (2002). Availability of low and high molecular weight substrates to extracellular enzymes in whole and dispersed activated sludges. *Enzyme and Microbial Technology, 31*(1–2), 179–186.

Cairns, S. S., Cornish, A., & Cooper, R. A. (1996). Cloning, sequencing and expression in Escherichia coli of two rhizobium sp. genes encoding haloalkanoate dehalogenases of opposite stereospecificity. *European Journal of Biochemistry, 235*(3), 744–749.

Carlsson, C. G. H. (1979). *Aerobic, thermophilic degradation with enzyme addition*. U.S. Patent No. 4,132,638. U. S Patent and Trademark Office.

Czerska, B., Miksch, K., Matsché, N., & Franz, A. (1997). Application of enzymatic activity measurements to biological phosphate removal from wastewater. *Water Science and Technology, 36*(10), 87–95.

Demain, A. L., & Adrio, J. L. (2008). Contributions of microorganisms to industrial biology. *Molecular Biotechnology, 38*(1), 41–55.

Dircks, K., Henze, M., van Loosdrecht, M. C., Mosbæk, H., & Aspegren, H. (2001). Storage and degradation of poly-β-hydroxybutyrate in activated sludge under aerobic conditions. *Water Research, 35*(9), 2277–2285.

Engelking, L. R. (2015). *Textbook of veterinary physiological chemistry. Chapter 6* (pp. 32–35).

Flemming, H., & Wingender, J. (2010). The biofilm matrix. *Nature Reviews. Microbiology, 8*, 623–633.

Frølund, B., Keiding, K., & Nielsen, P. (1995). Enzymatic activity in the activated sludge flocs matrix. *Applied Microbiology and Biotechnology, 43*, 755–761.

Gaddad, S. M., & Rodgi, S. S. (1987). The effect of temperature on the growth and biochemical activities of Escherichia coli in sewage. *Environmental Pollution, 43*(4), 313–321.

Gescher, J., Ismail, W., Ölgeschläger, E., Eisenreich, W., Worth, J., & Fuchs, G. (2006). Aerobic benzoyl-coenzyme a (CoA) catabolic pathway in *Azoarcus evansii*: Conversion of ring cleavage product by 3,4-dehydroadipyl-CoA semialdehyde dehydrogenase. *Journal of Bacteriology, 188*(8), 2919–2927.

Gessesse, A., Dueholm, T., Petersen, S. B., & Nielsen, P. H. (2003). Lipase and protease extraction from activated sludge. *Water Research, 37*, 3652–3365.

Grosse, S., Laramee, L., Wendlandt, K. D., McDonald, I. R., Miguez, C. B., & Kleber, H. P. (1999). Purification and characterization of the soluble methane monooxygenase of the type II methanotrophic bacterium *Methylocystis* sp. strain WI 14. *Applied and Environmental Microbiology, 65*(9), 3929–3935.

Guellil, A., Thomas, F., Block, J. C., Bersillon, J. L., & Ginestet, P. (2001). Transfer of organic matter between wastewater and activated sludge flocs. *Water Research, 35*(1), 143–150.

Gujer, W., Henze, M., Mino, T., & Van Loosdrecht, M. (1999). Activated sludge model No. 3. *Water Science and Technology, 39*(1), 183–193.

Gujer, W., Henze, M., Mino, T., & Van, L. (2000). Activated sludge model No.3. In M. Henze, W. Gujer, T. Mino, & M. C. M. van Loosdrecht (Eds.), *Activated sludge models ASM1, ASM2, ASM2D and ASM3. IWA Scientific and Technical Report 9*. IWA Publishing, Caxton House.

Gurung, N., Ray, S., Bose, S., & Rai, V. (2013). A broader view: Microbial enzymes and their relevance in industries, medicine, and beyond. *BioMed Research International, 2013*, 329121.

Hassan, S. W., Abd El Latif, H. H., & Ali, S. M. (2018). Production of cold-active lipase by free and immobilized marine bacillus cereus HSS: Application in wastewater treatment. *Frontiers in Microbiology, 9*, 2377.

Henze, M., Gujer, W., Mino, T., Matsuo, T., Wentzel, M. C., & Marais, G. (1995). Wastewater and biomass characterization for the activated sludge model no. 2: Biological phosphorus removal. *Water Science and Technology, 31*(2), 13–23.

Henze, M., Gujer, W., Mino, T., Matsuo, T., Wentzel, M. C., Marais, G. V. R., & Van Loosdrecht, M. C. (1999). Activated sludge model no. 2d, ASM2d. *Water Science and Technology, 39*(1), 165–182.

Henze, M., Gujer, W., Mino, T., & Van Loosdrecht, M. C. M. (2000). *Activated sludge models: ASM1, ASM2, ASM2d and ASM3* (Vol. No. 9). Scientific and Technical Report.

Hiner, A. N., Ruiz, J. H., Lopez, J. N. R., Canovas, F. G., Brisset, N. C., Smith, A. T., Acosta, M., et al. (2002). Reactions of the class II peroxidases, lignin peroxidase and *Arthromyces ramosus* peroxidase, with hydrogen peroxide: Catalase-like activity, compound III formation, and enzyme inactivation. *Journal of Biological Chemistry, 277*(30), 26879–26885.

Johnson, E. A. (2013). Biotechnology of non-saccharomyces yeasts—The ascomycetes. *Applied Microbiology and Biotechnology, 97*(2), 503–517.

Johnsy, G., & Kaviyarasan, V. (2011). Nutrient composition on the effect of extracellular peroxidase production by Lentinus kauffmanii—Under submerged culture condition. *International Journal of ChemTech Research, 3*(3), 1563–1570.

Jung, J., Xing, X. H., & Matsumoto, K. (2002). Recoverability of protease released from disrupted excess sludge and its potential application to enhanced hydrolysis of proteins in wastewater. *Biochemical Engineering Journal, 10*(1), 67–72.

Karigar, C. S., & Rao, S. S. (2011). Role of microbial enzymes in the bioremediation of pollutants: A review. *Enzyme Research, 2011*, 1–11.

Kibret, M., Somitsch, W., & Robra, K. H. (2000). Characterization of a phenol degrading mixed population by enzyme assay. *Water Research, 34*(4), 1127–1134.

Koua, D., Cerutti, L., Falquet, L., Sigrist, C. J., Theiler, G., Hulo, N., & Dunand, C. (2009). PeroxiBase: A database with new tools for peroxidase family classification. *Nucleic Acids Research, 37*, D261–D266.

Koubaa, M., Imatoukene, N., Drévillon, L., & Vorobiev, E. (2020). Current insights in yeast cell disruption technologies for oil recovery: A review. *Chemical Engineering and Processing Process Intensification, 150*, 107868.

Kumar, A., & Sharma, S. (2019). *Microbes and enzymes in soil health and bioremediation* (pp. 353–366). Springer.

Li, Y., & Chróst, R. J. (2006). Microbial enzymatic activities in aerobic activated sludge model reactors. *Enzyme and Microbial Technology, 39*(4), 568–572.

Li, X. Y., & Yang, S. F. (2007). Influence of loosely bound extracellular polymeric substances (EPS) on the flocculation, sedimentation and dewaterability of activated sludge. *Water Research, 41*, 1022–1030.

Liang, Z., Li, G., & An, T. (2017). Purifying, cloning and characterizing a novel dehalogenase from bacillus sp. GZT to enhance the biodegradation of 2,4,6-tribromophenol in water. *Environmental Pollution, 225*, 104–111.

Liu, Z., & Smith, S. R. (2019). Enzyme activity of waste activated sludge extracts. *Water Science and Technology, 80*(10), 1861–1869.

Liu, Z., & Smith, S. R. (2021). Enzyme recovery from biological wastewater treatment. *Waste and Biomass Valorization, 12*(8), 4185–4211.

Metcalf and Eddy. (2003). *Wastewater Engineering: Treatment and reuse* (p. 384). McGraw Hill.

Mishra, B. B., Gautam, S., & Sharma, A. (2013). Free phenolics and polyphenol oxidase (PPO): The factors affecting post-cut browning in eggplant (*Solanum melongena*). *Food Chemistry, 139*(1–4), 105–114.

Morgenroth, E., Kommedal, R., & Harremoës, P. (2002). Processes and modeling of hydrolysis of particulate organic matter in aerobic wastewater treatment–A review. *Water Science and Technology, 45*(6), 25–40.

Morsi, R., Bilal, M., Iqbal, H. M., & Ashraf, S. S. (2020). Laccases and peroxidases: The smart, greener and futuristic biocatalytic tools to mitigate recalcitrant emerging pollutants. *Science of the Total Environment, 714*, 136572.

Mu'azu, N. D., Abubakar, I. R., & Blaisi, N. I. (2020). Public acceptability of treated wastewater reuse in Saudi Arabia: Implications for water management policy. *Science of the Total Environment, 721*, 137659.

Mugdha, A., & Usha, M. (2012). Enzymatic treatment of wastewater containing dyestuffs using different delivery systems. *Scientific Reviews and Chemical Communications, 2*(1), 31–40.

Mukherjee, S., Basak, B., Bhunia, B., Dey, A., & Mondal, B. (2013). Potential use of polyphenol oxidases (PPO) in the bioremediation of phenolic contaminants containing industrial wastewater. *Reviews in Environmental Science and Bio/Technology, 12*(1), 61–73.

Nabarlatz, D., Vondrysova, J., Jenicek, P., Stüber, F., Font, J., Fortuny, A., Bengoa, C., et al. (2010). Hydrolytic enzymes in activated sludge: Extraction of protease and lipase by stirring and ultrasonication. *Ultrasonics Sonochemistry, 17*(5), 923–931.

Nybroe, O., Jørgensen, P. E., & Henze, M. (1992). Enzyme activities in waste water and activated sludge. *Water Research, 26*(5), 579–584.

Pandey, K., Singh, B., Pandey, A. K., Badruddin, I. J., Pandey, S., Mishra, V. K., & Jain, P. A. (2017). Application of microbial enzymes in industrial waste water treatment. *International Journal of Current Microbiology and Applied Sciences, 6*(8), 1243–1254.

Park, Y. J., Yoo, C. B., Choi, S. Y., & Lee, H. B. (2006). Purifications and characterizations of a ferredoxin and its related 2-oxoacid: Ferredoxin oxidoreductase from the hyperthermophilic archaeon, Sulfolobus solfataricus P1. *BMB Reports, 39*(1), 46–54.

Parmar, N., Singh, A., & Ward, O. P. (2001). Enzyme treatment to reduce solids and improve settling of sewage sludge. *Journal of Industrial Microbiology and Biotechnology, 26*(6), 383–386.

Postma, P. R., Miron, T. L., Olivieri, G., Barbosa, M. J., Wijffels, R. H., & Eppink, M. H. M. (2015). Mild disintegration of the green microalgae Chlorella vulgaris using bead milling. *Bioresource Technology, 184*, 297–304.

Prasad, M. P., & Manjunath, K. (2011). Comparative study on biodegradation of lipid-rich wastewater using lipase producing bacterial species. *Indian Journal of Biotechnology, 10*(1), 121–124.

Price, N. C., & Stevens, L. (1999). *Fundamentals of enzymology: The cell and molecu lar biology of catalytic proteins*. Oxford University Press.

Ramesh, A., Lee, D. J., & Hong, S. G. (2006). Soluble microbial products (SMP) and extracellular polymeric substances (EPS) from wastewater sludge. *Applied Microbiology and Biotechnology, 73*, 219–225.

Ramesh, A., Lee, D. J., & Lai, J. Y. (2007). Membrane biofouling by extracellular polymeric substances or soluble microbial products from membrane bioreactor sludge. *Applied Microbiology and Biotechnology, 74*(3), 699–707.

Rao, M. B., Tanksale, A. M., Ghatge, M. S., & Deshpande, V. V. (1998). Molecular and biotechnological aspects of microbial proteases. *Microbiology and Molecular Biology Reviews, 62*(3), 597–635.

Rathner, R., Petz, S., Tasnádi, G., Koller, M., & Ribitsch, V. (2017). Monitoring the kinetics of biocatalytic removal of the endocrine disrupting compound 17 α -Ethinylestradiol from differently polluted wastewater bodies. *Journal of Environmental Chemical Engineering, 5*(2), 1920–1926.

Raunkjær, K., Hvitved-Jacobsen, T., & Nielsen, P. H. (1994). Measurement of pools of protein, carbohydrate and lipid in domestic wastewater. *Water Research, 28*(2), 251–262.

Razzaq, A., Shamsi, S., Ali, A., Ali, Q., Sajjad, M., Malik, A., & Ashraf, M. (2019). Microbial proteases applications. *Frontiers in Bioengineering and Biotechnology, 7*, 110.

Richards, S. R., Hastwell, C., & Davies, M. (1984). The comparative examination of 14-activatedsludge plants using enzymatic techniques. *Water Pollution Control (Maidstone), 83*(3), 300–313.

Robertson, L. R., LaZonby J. G., Krolezyk, J. J., & Melo, H. R. (1994). *Treatment of process waters to destroy filamentous bacteria*. U.S. Patent No. 5,324,432. U.S Patent and Trademark Office.

Rubin, D. M., Anderton, N., Smalberger, C., Polliack, J., Nathan, M., & Postema, M. (2018). On the behaviour of living cells under the influence of ultrasound. *Fluids, 3*(4), 82.

Rupley, J. A., Gratton, E., & Careri, G. (1983). Water and globular proteins. *Trends in Biochemical Sciences, 8*(1), 18–22.

Servos, M. R., Bennie, D. T., Burnison, B. K., Jurkovic, A., McInnis, R., Neheli, T., Ternes, T. A., et al. (2005). Distribution of estrogens, 17β-estradiol and estrone, in Canadian municipal wastewater treatment plants. *Science of the Total Environment, 336*(1–3), 155–170.

Shen, B., & Hutchinson, C. R. (1993). Tetracenomycin F1 monooxygenase: Oxidation of a naphthacenone to a naphthacenequinone in the biosynthesis of tetracenomycin C in *Streptomyces glaucescens*. *Biochemistry, 32*(26), 6656–6663.

Sheng, G. P., Yu, H. Q., & Li, X. Y. (2010). Extracellular polymeric substances (EPS) of microbial aggregates in biological wastewater treatment systems: A review. *Biotechnology Advances, 28*, 882–894.

Singh, C. J. (2003). Optimization of an extracellular protease of Chrysosporium keratinophilum and its potential in bioremediation of keratinic wastes. *Mycopathologia, 156*(3), 151–156.

Sponza, D. T. (2003). Investigation of extracellular polymer substances (EPS) and physicochemical properties of different activated sludge flocs under steady-state conditions. *Enzyme and Microbial Technology, 32*(3–4), 375–385.

Stevens, J. L., Northcott, G. L., Stern, G. A., Tomy, G. T., & Jones, K. C. (2003). PAHs, PCBs, PCNs, organochlorine pesticides, synthetic musks, and polychlorinated n-alkanes in UK sewage sludge: Survey results and implications. *Environmental Science & Technology, 37*(3), 462–467.

Sun, W. Q., Payne, G. F., Moas, M. S., Chu, J. H., & Wallace, K. K. (1992). Tyrosinase reaction/chitosan adsorption for removing phenols from wastewater. *Biotechnology Progress, 8*(3), 179–186.

Sun, H., Zhang, H., Ang, E. L., & Zhao, H. (2018). Biocatalysis for the synthesis of pharmaceuticals and pharmaceutical intermediates. *Bioorganic & Medicinal Chemistry, 26*(7), 1275–1284.

Szabolcs, S., Botond, R., Ildiko, M., Beata, A., Szabolcs, L., & Nicolae, D. R. (2009). *Enzymatic activity studies of biological wastewater treatment* (p. 113). Universitatis Babeş-Bolyai.

Szilveszter, S., Ráduly, B., Ábrahám, B., Lányi, S., & Niculae, D. R. (2010). Mathematical models for domestic biological wastewater treatment process. *Environmental Engineering and Management Journal, 9*(5), 629–636.

Tchobanoglous, G., Burton, F., & Stensel, H. (2003). In E. A. Jones (Ed.), *Wastewater engineering treatment and reuse*. Metcalf and Eddy Inc. New York: McGraw-Hill.

Tyagi, V. K., & Lo, S. L. (2013). Sludge: A waste or renewable source for energy and resources recovery? *Renewable and Sustainable Energy Reviews, 25*, 708–728.

Unuofin, J. O., Okoh, A. I., & Nwodo, U. U. (2019). Aptitude of oxidative enzymes for treatment of wastewater pollutants: A laccase perspective. *Molecules, 24*(11), 2064.

van Beilen, J. B., Duetz, W. A., Schmid, A., & Witholt, B. (2003). Practical issues in the application of oxygenases. *Trends in Biotechnology, 21*(4), 170–177.

Wang, Y., Wang, Q., & Liu, L. (2019). Crude oil degrading fingerprint and the overexpression of oxidase and invasive genes for n-hexadecane and crude oil degradation in the Acinetobacter pittii H9-3 strain. *International Journal of Environmental Research and Public Health, 16*(2), 188.

Whiteley, C. G., Heron, P., Pletschke, B., Rose, P. D., Tshivhunge, S., Van Jaarsveld, F. P., & Whittington-Jones, K. (2002). The enzymology of sludge solubilisation utilising sulphate reducing systems: Properties of proteases and phosphatases. *Enzyme and Microbial Technology, 31*(4), 419–424.

Whiteley, C. G., Burgess, J. E., Melamane, X., Pletschke, B., & Rose, P. D. (2003). The enzymology of sludge solubilisation utilising sulphate-reducing systems: The properties of lipases. *Water Research, 37*(2), 289–296.

Wong, D. W. (2009). Structure and action mechanism of ligninolytic enzymes. *Applied Biochemistry and Biotechnology, 157*(2), 174–209.

Ye, X., Peng, T., Feng, J., Yang, Q., Pratush, A., Xiong, G., Huang, T., & Hu, Z. (2019). A novel dehydrogenase 17β-HSDx from Rhodococcus sp. P14 with potential application in bioremediation of steroids contaminated environment. *Journal of Hazardous Materials, 362*, 170–177.

Yu, G. H., He, P. J., Shao, L. M., & Lee, D. J. (2007). Enzyme activities in activated sludge flocs. *Applied Microbiology and Biotechnology, 77*, 605–612.

Yu, G., He, P., Shao, L., & Lee, D. (2008). Extracellular enzymes in sludge flocs collected at 14 full-scale wastewater treatment plants. *Journal of Chemical Technology & Biotechnology: International Research in Process, Environmental & Clean Technology, 83*(12), 1717–1725.

Zdarta, J., Meyer, A. S., Jesionowski, T., & Pinelo, M. (2018). Developments in support materials for immobilization of oxidoreductases: A comprehensive review. *Advances in Colloid and Interface Science, 258*, 1–20.

Zhang, G., Zhang, P., Yang, J., & Liu, H. (2008). Energy-efficient sludge sonication: Power and sludge characteristics. *Bioresource Technology, 99*(18), 9029–9031.

Zieliński, M., Dębowski, M., Kisielewska, M., Nowicka, A., Rokicka, M., & Szwarc, K. (2019). Cavitation-based pretreatment strategies to enhance biogas production in a small-scale agricultural biogas plant. *Energy for Sustainable Development, 49*, 21–26.

Occurrence and Distribution of Microbial Enzymes in Freshwater

Femi Johnson Afolabi, Babafemi Raphael Babaniyi, and Monday Peter Obagunwa

Abstract This chapter examines the occurrence and distribution of microbial enzymes in freshwater. On the basis of solubility, molecular mass, chemical structure, and occurrence, organic matters in water can be broadly classified into four types, namely, bioorganic matter, dissolved organic matter, colloidal organic matter, and aggregate organic matter. However, their chemical structures vary widely at the molecular level. Therefore, a wide range of enzymes are needed to access the carbon and nutrients contained, but not immediately bioavailable, in plant, animal, and microbial detritus. The microorganisms that produce microbial enzymes in freshwaters are presumably superior competitors for the utilization of organic and inorganic nutrients as energy sources in aquatic environments. The quantity, type, and distribution of microbial enzymes in freshwater are directly related to the differences in the availability of nutrients in a particular environment and organic matter quantity, composition, and consumption in relation to microbial community diversity and growth. Freshwater microbial extracellular enzymes are either attached to the cell or dissolved in the water column. Moreover, the strategy of extracellular enzyme production also enables them to increase their growth and biomass production. However, a cursory look at the conditions in the freshwater system reveals that they are not supportive of the growth of microorganisms and the production of enzymes. This is partly due to low substrate concentration, insolubility of some substrates, diversity in substrates, and form of existence, i.e., they may be bound to humic substances, colloidal organic matter, and detritus. These conditions negatively affect the production of microbial enzymes in freshwater. Esterase and gelatinase have been found to have higher probability of being found in water than in soils and plants. Esterase activity has been proposed for use as a good index for estimating organic matter content in seas, lakes, and reservoirs. The key factors affecting the occurrence, distribution, and activity of enzymes in freshwater are temperature, pH, ionic strength, and the proper concentrations of essential components like substrates and enzymes.

F. J. Afolabi · B. R. Babaniyi (✉) · M. P. Obagunwa
Bioresources Development Centre, National Biotechnology Development Agency, Nigeria

N. R. Maddela et al. (eds.), *Ecological Interplays in Microbial Enzymology*, Environmental and Microbial Biotechnology,
https://doi.org/10.1007/978-981-19-0155-3_4

1 Introduction

Enzymes can be defined as biologically produced chemicals or molecules, which catalyze specific biochemical reactions. They are the biological counterparts of chemical catalysts, which help speed up biochemical and biological processes both within and outside the cell. Microorganisms (bacteria, fungi, molds, etc.) are surrounded by organic matter that is rich in carbon and the nutrients that are required for cell maintenance and growth. However, microbes cannot directly transport these macromolecules into the cytoplasm. Rather, they rely on the activities of the myriad of enzymes that they synthesize and release into their immediate environment (Jackson et al., 2013; Sala et al., 2019). These enzymes are the ones capable of breaking down polymers into monomers or smaller subunits that can be utilized by the cell (Maire et al., 2012). For example, enzymes such as cellobiohydrolase (CBH) and β-glucosidase are important for breaking down cellulose and work in synergy to catalyze the hydrolysis of cellulose to glucose (Jackson et al., 2013), which provides a utilizable carbon substrate for microbial uptake and assimilation. The enzyme phosphatase makes soluble inorganic phosphate groups available to microorganisms from organophosphates, essentially mineralizing phosphates (Dalal, 1977). Other enzymes, such as *N*-acetylglucosaminidases (NAGases), are important in chitin degradation and can make both carbon and nitrogen available for microbial nutrition (Sinsabaugh & Moorhead, 1995). In summary, they are vital to the sustenance of the ecosystem as the mineralization of nutrients, nutrient cycling, energy flow, and organic matter decomposition in the environment are all heavily dependent on them (Jackson et al., 2013). Furthermore, some of these enzymes are important degraders of contaminants both in water and in soil (Stottmeister et al., 2003). All these functions and activities of microbial enzymes occur both in the soil and in water bodies. This is especially true for heterotrophic microorganisms, which are responsible for controlling the metabolism of a particular ecosystem. The measurement of microbial activity in natural waters is highly important for understanding the dynamic aspects of the functioning of the whole ecosystem (Ryszard, 1990).

Apart from the physicochemical parameters of an aquatic environment, the composition and availability of organic matter are the other major factors that influence the development and activity of the microbial heterotrophic communities. Furthermore, microheterotrophs, particularly heterotrophic bacteria, are the major components of aquatic ecosystems that control the movement of the majority of organic compounds (Ryszard, 1990). They are the only biological populations capable of significantly altering both dissolved organic matter (DOM) and particulate organic matter (POM). Heterotrophic microorganisms are the most effective competitors for reduced carbon in all ecosystems (Williams, 1981). The metabolic activity of microheterotrophs is also important from another point of view. Small organisms, such as bacteria, have short generation times, high biogenic potentials, and one of the highest metabolic rates per unit of biomass. Microbial enzymes are the most important catalysts for a large number of biochemical transformations of organic and inorganic constituents in aquatic environments. Many of these transformations can only be mediated by heterotrophic bacteria because the enzyme systems required for these are not found in other organisms.

Microbial enzymes are key in the nutrition of microorganisms, especially heterotrophic bacteria, which are strategic in the nutrition of an aquatic ecosystem. A large chunk of the organic compounds produced in natural waters have a polymeric structure (Lochte & Ford, 1986) and they are too large to be readily assimilated. Here, microbial enzymes called permeases constantly facilitate the penetration of organic molecules across the microbial cell membranes. Only low-molecular-weight organic molecules (monomers or small polymers), which are products of the activities of microbial enzymes on polymers, in a process called enzymatic depolymerization, can therefore be taken up (Rogers, 1961).

For their nutrition through enzymatic depolymerization, microorganisms have essentially developed two strategies. Some microorganisms may engulf the polymer with the cytoplasmic membrane to form a vacuole within the cytoplasm. Enzymes are then released into this vacuole, and the polymers are broken down to subunits that are subsequently taken up by the organisms. This type of nutrition is known as pinocytosis. This method of nutrition is, however, restricted to eukaryotes with no cell wall, e.g., protozoa. For those prokaryotes and eukaryotes with cell walls, an alternative method for polymer assimilation has been adopted. Here, hydrolytic enzymes are secreted outside the cytoplasmic membrane where they hydrolyze macromolecules in close proximity to the cell. The resulting monomers or subunits are then transported across the cell membrane and used inside the cytoplasm (Ryszard, 1990). The hydrolysis of polymers by microbial enzymes is an acknowledged rate-limiting step in the utilization of organic matter by microorganisms in aquatic environments. The importance of microbial enzymatic activities in the mobilization, transformation, and turnover of organic and inorganic compounds in aquatic environments has been established by some authors.

On the basis of solubility, molecular mass, chemical structure, and occurrence, organic matters in water can be broadly classified into four types, namely, bioorganic matter, dissolved organic matter, colloidal organic matter, and aggregate organic matter (Piccolo, 2001; Cai et al., 2005; Kelleher & Simpson, 2006). However, their chemical structures vary widely at the molecular level. Therefore, a wide range of enzymes are needed to access the carbon and nutrients contained, but not immediately bioavailable, in plant, animal, and microbial detritus (Caldwell, 2005). The differences between their aggregation and sedimentary processes in water result in the diversity of organic preservation in sediments. Bioorganic matters can not only preserve organic matters themselves but also are the source of other types of organic matters in water. Dissolved organic matter and colloidal organic matter are widely distributed in water and have high chemical activities. They can aggregate with one another or with inorganic minerals, and, so, they play an important role in the process of organic matter aggregation and cycle. The formation of aggregated organic matter in water has a close relation with the bioorganic and physicochemical conditions (Cai et al., 2005).

2 Sources of Microbial Enzymes in Freshwater

The microorganisms that produce microbial enzymes in freshwaters are presumably superior competitors for the utilization of organic and inorganic nutrients as energy sources in aquatic environments. Most of the microorganisms in this category are also capable of breaking down a lot of polymeric compounds, which are the most prevalent in freshwaters. This is the secret of their survival when easily metabolizable sources of energy are inadequate or lacking (Chróst, 1992). Moreover, the strategy of extracellular enzyme production also enables them to increase their growth and biomass production. However, a cursory look at the conditions in the freshwater system reveals that they are not supportive of the growth of microorganisms and the production of enzymes. This is partly due to low substrate concentration, insolubility of some substrates, diversity in substrates, and form of existence, i.e., they may be bound to humic substances, colloidal organic matter, and detritus. These conditions negatively affect the production of microbial enzymes in freshwater. They do this by negatively impacting the coupling of an enzyme with its substrate and by increasing the rate of adsorption, which limits the amount of enzymes available for activity. As discussed earlier, a large chunk of the polymers in freshwater are converted to monomers, which can be utilized by microorganisms as a source of nutrition. This is usually achieved via the activities of microbial enzymes, which are either secreted into the environment or released as a result of lysis of microbial cells (Arnosti, 2011; Maire et al., 2012). At other times, microbial enzymes in freshwater may be inhibited by substances present in water bodies. Sometimes, microbial enzymes are denatured by physical and chemical factors in water bodies or may be hydrolyzed by proteases. For an extracellular enzyme to be of benefit to its producer microorganism, it is obvious that it must avoid destruction long enough to locate its substrate. Finally, the physical and chemical conditions of the reaction between an enzyme and its substrate after binding are dependent on factors that may not be suitable for catalysis. These factors include nonoptimal pH or temperature, the presence of inhibitors, the absence of activators, suboptimal ionic strength, etc. Generally, most microorganisms in freshwater produce extracellular enzymes that encounter many polymeric substrates and their growths are dependent on the success of their enzymes.

2.1 Types of Enzymes in Freshwater

The extent of the microbial diversity in nature can never be overstated. There exists a plethora of analyses of microbial communities in various environmental samples and systems (Vandenkoornhuyse et al., 2002). The way microorganisms have evolved with the peculiarities of their different habitats has culminated in their specific physiological and biochemical diversity, in which enzymes play a key role, especially with respect to adaptation (Lorenz & Schleper, 2002). The biochemical activities and the biotechnological potentials of the microorganisms in a particular ecosystem can only be understood if the microbial enzymes in such an environment are detected. Microbial enzymes are relatively more stable and have properties more diverse than do other enzymes derived from plants and animals (Mohapatra et al., 2003).

Alves et al. (2014) reported that esterase and gelatinase were more likely to be detected in water samples than in soils or plants. Wobus et al. (2003) also proposed the use of esterase activity as a good index for estimating organic matter content in seas, lakes, and reservoirs. It is generally accepted that in water samples, it is possible to find 100% of isolates having proteolytic and esterase activities, but amylolytic activity is seldom encountered. This is, however, in contrast to what has been found with yeasts. In a particular yeast screening study by Brandão et al. (2011), cellulolytic activity was shown to predominate in aquatic habitats. In the study by Alves et al. (2014), all isolates from a particular freshwater water sample produced esterase, and the majority of the isolates were able to produce gelatinase, esterase, and caseinase. This may be due to the fact that substrates for esterase production can be frequently found in this microenvironment, and it is also possible that this unique enzymatic profile is associated with natural selection acting on adapted microorganisms.

3 Distribution and Activity of Microbial Enzymes in Freshwater

The quantity, type, and distribution of microbial enzymes in freshwater are directly related to the differences in the availability of nutrients in a particular environment, organic matter quantity, composition, and consumption in relation to microbial community diversity and growth. The activities of such enzymes depend on their natural stability and the capacity of the environmental matrix to sorb and stabilize active enzymes through associations with particle surfaces and dissolved organic matter (DOM) (Nannipieri, 2006). It therefore means that the activity and turnover rate of enzymes may differ across systems; in fact, there may be multiple pools of active enzymes with different turnover rates in a particular system. Thus, enzyme pool sizes, turnover rates, and kinetics vary widely across systems and may require different methods of study. Factors affecting freshwater ecosystem microbial enzymes include, but are not limited to, small-scale gradients in solid surfaces, temperature, and pH. These will consequently affect the occurrence and distribution of microbial enzymes in that ecosystem.

Microbial enzymes in freshwater catalyze depolymerization of organic matter through either hydrolytic or oxidative reactions. Hydrolytic enzymes are known to be substrate-specific because their conformation allows them to catalyze reactions that cleave specific bonds (e.g., C–O and C–N bonds) that link monomers. This is in contrast to oxidative enzymes, which are known to act on broader classes of substrates that share similar bonds (e.g., C–C and C–O–C) and use either oxygen (oxygenases) or hydrogen peroxide (peroxidases) as electron acceptors (Wallenstein & Burns, 2011).

Generally, microbial enzymes in freshwater or stream sediments receiving wastewater are depressed (Kuhbier et al., 2002) and have been shown to reflect the nutrient status of wetland soils in the Florida Everglades (Wright & Reddy, 2001). Exogenous phosphatases are central to the regeneration of inorganic P concentration, particularly in P-limited freshwater ecosystems. The activity of phosphatase is usually repressed by high concentrations of dissolved reactive P (Cembella et al., 1984; Chróst, 1991). Moreover, an enzyme such as β-glucosidase, which is a part of the cellulase complex of enzymes involved in the regeneration of monosaccharides through hydrolysis of glycosides (Eivazi & Tabatabai, 1988), becomes available and is active as a result of the presence of the products of the hydrolysis of glycosides and can be correlated with detrital degradation rates (Sinsabaugh et al., 1994; McLatchey & Reddy, 1998).

4 Factors Influencing the Occurrence and Distribution of Microbial Enzymes in Freshwater

The main factors affecting the occurrence, distribution, and activity of enzymes in freshwater are temperature, pH, ionic strength, and the proper concentrations of essential components like substrates and enzymes.

4.1 Temperature

Raising the temperature generally speeds up a reaction, and lowering the temperature slows down a reaction. However, extreme high temperatures can cause an enzyme to lose its shape (denature) and stop working.

4.2 pH

Each enzyme has an optimum pH range. Changing the pH outside of this range will slow enzyme activity. Extreme pH values can cause enzymes to denature.

4.3 Enzyme Concentration

Increasing enzyme concentration will speed up a reaction, as long as there is a substrate available to bind to. Once all of the substrates are bound, the reaction will no longer speed up, since there will be nothing for additional enzymes to bind to.

4.4 Abiotic Factors

Surface Interactions as Controls on Enzymatic Activity

At micrometer scales, a freshwater ecosystem is a heterogeneous habitat, with living and nonliving components that are not evenly distributed (Sexstone et al., 1985). This heterogeneity even at the smallest level is largely determined by the availability of solid surfaces within a given area. This is depicted in the figure below. Enzymes and their activities are influenced by interactions with surfaces. In soils and sediments, mineral density, distribution, composition, and particle size strongly influence the activities of enzymes. This is because enzymes that are adsorbed may become less active or, in some instances, unavailable for activity. In some other cases, adsorption protects them from degradation and they may regain activity after desorption (Wetzel, 1993). In freshwater systems, the composition and density of particles vary in relation to factors like the source and proximity of terrestrial runoff, phytoplankton production, turbulence, and flocculation processes. DOM can form transient surfaces such as marine lake snow or river snow, which increase the available surface area and provide a reservoir of biogeochemical reactions and elevated enzyme activities (Ziervogel et al., 2010).

The chemical, physical, and biological properties of soil all affect enzyme diffusion, survival, and substrate turnover as well as the proportion of the product that is available to, and assimilated by, the producer cells.

For a holistic degradation of organic matters, microorganisms and their extracellular enzymes must be capable of detecting, migrating toward, and transforming organic debris to soluble monomers (or short oligomers) that are subsequently transported into the cytoplasm. Sometimes, the macromolecular components of living and dead plant, animal, and microbial tissues are often physically and chemically embedded with each other and sorb or entrap other low-molecular-weight organic compounds. This becomes a barrier to the microorganism and extracellular enzyme with respect to accessing the otherwise vulnerable soluble constituents of organic matter. In some cases, the substrates themselves may also be sequestered within soil components, thereby reducing or totally eliminating access by microbes or their enzymes (Jastrow et al., 2007).

Thus, the degradation of macromolecules by microbial enzymes in freshwater into more easily metabolized components requires not only enzyme production but also physical contact of enzymes with their target substrates and the sites of catalysis.

5 Synthesis and Secretion of Microbial Enzymes in Freshwater

Enzyme synthesis and secretion demands a lot of energy from the organism and requires nitrogen (Schimel & Weintraub, 2003) and may negatively impact the cell if a proportionate nutritional reward does not follow. Thus, the amount of energy and other cell resources available for enzyme synthesis must be balanced with the investment of the precious resources allocated to the production of enzymes with the energy and nutrients gained as a result of their activity. The key step in the control of the synthesis of extracellular enzymes is at the molecular level. This takes place at the transcription stage for both prokaryotes and eukaryotes. This phenomenon helps the organisms conserve energy that would have been wasted on transcripts that may never be translated. The only exemption to this is the synthesis of amylase and protease in *Bacillus amyloliquefaciens*, which regulates enzyme production at the translation stage. There are two known mechanisms for extracellular enzyme secretion, especially with respect to the "signal hypothesis." These are co-translational secretion, which is the main mode of exporting proteins from the cell, and post-translational secretion, which is a secondary mode. In the case of co-translational secretion, the signal peptide is released from the ribosome and subsequently connects with the inner surface of the cytoplasmic membrane of bacteria or with the endoplasmic reticulum of eukaryotic cells. As the polypeptide is elongated, it passes through the membrane, and a signal peptidase removes the signal peptide on the outer side of the membrane. In this case, the extracellular enzyme is activated outside of the membrane. In post-translational secretion, the extracellular enzyme is transported across the membrane after it has been completely synthesized. Post-translational secretion may occur in both eukaryotic and prokaryotic microorganisms.

Freshwater microbial extracellular enzymes are either attached to the cell or dissolved in the water column (Chróst, 1991; Baltar et al., 2010). Those enzymes that are cell-associated are bound to the cell surface or encamped in the periplasmic space (Reintjes et al., 2019). Due to the dilute nature of dissolved organic matter (DOM), they can help the cell preferentially access DOM (Chróst, 1991). However, the substrate must migrate into the cell wall or be in close proximity to the cell (Alison et al., 2012). However, some polysaccharide substrates could be directly taken up into the periplasm of a "selfish" organism without production of extracellular hydrolysis products (Reintjes et al., 2017, 2019; Hehemann et al., 2019) Dissolved enzymes, which belong to a kind of "living dead" realm (Baltar, 2018), may originate from active secretion by the cell (Alderkamp et al., 2007), bacterial starvation, and disruptions in the permeability of the cell (Albertson et al., 1990). Besides, they can also be produced in the process of grazing on bacterial communities and released during viral lysis (Bochdansky et al., 1995; Baltar, 2018). As they spread in the water body, these enzymes can hydrolyze substrates that are far away from the cell producing them. As a result of this phenomenon, the products of such hydrolysis may not be harvested by the parent cell. Dissolved enzymes, especially in deep waters, have a long lifespan and perform their important functions away from the producing cell (Baltar et al., 2013). Generally, dissolved extracellular enzymes in aquatic environment (EEA) could be as high as 100% of the total freshwater EEA (Baltar et al., 2010, 2016; Arnosti, 2010), which could indicate a disconnection between marine microbes and enzymatic activities (D'ambrosio et al., 2014).

6 Connecting Bacterial Communities and Enzyme Activity

The type of microorganisms in an ecosystem can determine the enzymes in such an environment. For instance, the marine ecosystem that harbors bacterial phyla such Bacteroidetes, Planctomycetes, Chloroflexi, and Proteobacteria, which are organisms capable of producing a wide range of extracellular enzymes, especially polysaccharases, will be a reservoir of enzymes produced by these same organisms. The bacterial community structure of seven stations were investigated and most were found in abundant. In the 16S rRNA gene clone libraries, marine *Roseobacter* lineages (OTU 3) within the class Alphaproteobacteria dominated. They are common members of coastal bacterioplankton and are often observed as particle colonizers. The next most abundant lineage was *Pseudoalteromonas* (OTU 8) within the class Gammaproteobacteria. This lineage can produce a broad range of hydrolases (e.g., alginate lyase, carrageenase, and peptidase) in response to available phytoplankton detritus. Other gammaproteobacterial genera (*Alteromonas* (OTU 5), *Vibrio* (OTU 149), *Pseudomonas* (OTU 91) *Psychrobacter* (COTU 614), and *Shewanella* (OTU 239)) that can secrete various enzymes to hydrolyze.

Tween 60, pullulan, and alginate were also found. Cultivating and molecular assays showed that some Chloroflexi populations can secrete xylanase, amylase, chitinase, esterase, galactosidase, and glucuronidase. Sphingobacteriia, Flavobacteriia, Cytophagia, and Bacteroidia within the phylum Bacteroidetes were observed in this study. They are candidates for the hydrolysis of complex HyMW carbohydrates and could assimilate phytoplankton phytodetritus quickly.

A lot of environmental factors come into play in the process of synthesis of enzymes by microorganisms in freshwaters. These factors consequently affect the molecular control of the synthesis. Therefore, the triggers for the expression of the right gene and the consequent production of the extracellular enzyme per time reach the cell from its surrounding. Based on the regulation of gene control, enzymes are classified as constitutive and inducible. Constitutive enzymes are those whose synthesis is constant, irrespective of the presence of a substrate in the environment, whereas inducible enzymes are those whose rate of synthesis is strongly dependent on the presence of their substrates (or derivatives thereof). It is already established, however, that inducible enzymes are usually synthesized at low basal levels in the absence of a substrate, whereas there is an astronomical increase in the production rate when a substrate is present. Synthesis continues at this amplified rate until the inducer is removed and/or the product of enzymatic catalysis accumulates. It then returns to the basal rate. A large chunk of the microbial enzymes in freshwater are inducible catabolic enzymes involved in the degradation of polymeric substrates that are not constantly available in the water. This makes the regular production of microbial enzymes in freshwaters needless as it will amount to the wastage of energy that would have been used for other important activities. This also gives an inkling of the evolutionary journey of most aquatic microorganisms and the advantages of induction in enzyme synthesis. So, generally, most microbial enzymes in freshwater are inducible and only a handful are constitutive (e.g., some amylases or proteases in bacteria). Sometimes, microorganisms in freshwater produce a little amount of their inducible enzymes, irrespective of the presence of their substrates, as a speculative sensing mechanism to detect substrates (Klonowska et al., 2002) or to depolymerize the available substrates to allow monomers and oligomers to diffuse into the cell to induce a higher production of the needed enzyme. Once concentrations of products are sufficient to meet the demand, enzyme production is repressed and returns to low constitutive levels (Chróst, 1991). This mechanism is called quorum sensing and has been well-described for many phytopathogens, e.g., *Erwinia carotovora*. Quorum sensing in the rhizosphere is believed to be an important controlling process for all sorts of catalytic activities (Pang et al., 2009).

Sometimes, the synthesis of some extracellular enzymes may occur only in the presence of a suitable substrate or some other inducer (Allison & Vitousek, 2005). However, at some other times, the inducer molecule may not need to enter the cell to stimulate the production of extracellular enzymes. Instead, they bind to cell wall receptor proteins called sensory kinases and initiate a process that signals the cell to produce the extracellular enzymes of interest.

As stated earlier, synthesis of extracellular enzymes by microorganisms in freshwater is carbon-, nitrogen-, and energy-intensive, and, so, microbes should produce only enzymes such as polysaccharases when nutrients and soluble C are scarce (Koch, 1985) or for the purpose of maintaining the stoichiometry of microbial biomass (Cleveland & Liptzin, 2007). Polysaccharase production is increased when there is plentiful supply of soluble nitrogen (Sinsabaugh et al., 2002), whereas excess carbon may increase protease synthesis. Another expression of this control is that extracellular enzyme secretion is usually inversely related to specific growth rate. When particular nutrients are scarce, on the other hand, microbes secrete enzymes to liberate those nutrients from organic matter (Harder & Dijkhuizen, 1983).

Furthermore, extracellular enzymes may be associated with the microbial cell's plasma membrane, contained within and attached to the walls of the periplasmic space, cell wall, and glycocalyx or released into the water body. Thus, the periplasm provides Gram-negative bacteria with a reservoir of activity that is retained until an external trigger for secretion is received. This is a strategic and rapid method to respond to the appearance of a potential substrate. Periplasmic enzymes may also enjoy the benefit of being adequately prepared for the hostile environment outside the cell. For example, glycosylation may take place in the periplasm (Feldman et al., 2005).

7 Evaluation of Microbial Enzymes in Water and Its Importance

Most of the enzymes used in various industries today have microbial origins. They are known to possess diverse properties and are more stable than their counterparts derived from plants and animals (Alves et al., 2014). In a report by Alves et al. (2014), there was a higher frequency in the production of esterase and gelatinase in freshwater samples. It therefore means that esterase activity can be used as an index to estimate the organic matter content of freshwater (Boschker & Cappenberg, 1998). Some works have reported 100% of isolates from water samples with proteolytic and esterase activities, but amylolytic activity is more rarely encountered (Ong et al., 2011). However, reports of yeast screening studies on freshwaters indicate a predominance of cellulolytic activity (Brandão et al., 2011). In the study conducted by Alves et al. (2014), it was reported that a large number of the isolates produced gelatinase, esterase, and caseinase. This may be due to the substrate enhancing esterase production, which is common in freshwater, or it may be that the organisms responsible for producing the enzymes have through the process of natural selection adapted to the freshwater environment. Also, the organic matter composition and the physicochemical conditions of the water bodies may have impacted on the activities and the microbial structure of that community conditions of the environment, controls the activity and structure of microbial communities (Chróst, 1992).

7.1 Enzyme Assays in Freshwater

Enzyme assays are usually carried out for two different reasons: (1) to identify a particular enzyme, especially with respect to its presence or absence in a sample like an organism, water, or tissue and (2) to determine the amount of the enzyme in the sample. For the first reason, which is usually qualitative, a positive or negative result with respect to the presence or absence will usually suffice. However, the second reason requires a qualitative approach, which presents the presence of the enzyme in concrete figures. One vital advantage of enzymes over functional proteins or nucleic acids is that they can be identified by their catalyzed reactions, unlike the latter that can only be detected by their direct detection. During the enzyme reaction, the product accumulates in amounts far exceeding the intrinsic enzyme concentration. However, assay procedures are usually adapted directly to the features of the individual enzyme and not to obey general standards. Enzymes are sensitive substances present in small amounts, and their activity in the cell can often be detected only at their optimum conditions. Various enzyme reactions require special conditions, e.g., if the thermodynamic equilibrium is unfavorable. Other enzymes, especially from extremophilic organisms, are only active under conditions completely different from the physiological range. For enzyme assays, it must be considered that enzyme reactions depend on more factors than pH, temperature, and ionic strength. Of great importance are the actual concentrations of all assay components. Further influences of compounds not directly involved in the action may occur, e.g., interactions of ions, especially metal ions, hydrophobic substances, or detergents with the protein surface.

Procedures for enzyme assays have been enumerated, detailed, documented, and cited in many standard books. These books include, but are not limited to, *Methods in Enzymology*; *Advances in Enzymology and Related Areas of Molecular Biology*; *Methods of Enzymatic Analysis* (Bergmeyer, 1983); *Springer Handbook of Enzymes* (Schomburg & Schomburg, 2009); *Practical Enzymology* (Bisswanger, 2011), and databases (ExPASy database; Brenda database). However, experiences have shown that strict adherence to these procedures does not necessarily guarantee accurate results. The same assays performed independently under obviously identical conditions may yield quite different results. In fact, the activity of enzymes depends on many factors and a general understanding of the particular features of enzymes is required, which cannot be described in detail due to the protocols for special enzyme assays.

Rapid enzyme assay techniques based on direct measurement of β-D-galactosidase (GALase) or β-D-glucuronidase (GLUase) activity without selective cultivation are used for rapid estimation of the level of coliform bacteria and *Escherichia coli* in water samples. Reported detection limits using fluorogenic substrates correspond to culturable target bacterial concentrations that can be appropriate within the present guidelines for recreational waters. Rapidity, that is, detection within 1 h, compromises the specificity of the assay; enzyme activity contributions from other than target bacteria need to be considered, particularly at low levels of target bacteria. Enzyme activities are more persistent than the culturability of target bacteria to environmental and disinfection stresses, and, thus, water samples may express enzyme activities of both culturable and viable nonculturable cells.

7.2 Determination of Microbial Enzyme Activity

Microbial community and the ecological makeup of the community determine the nature of enzymes and enzyme activities of a particular ecology in a freshwater network (Farris et al., 2016). For assessing water quality, the lowest number of each microbial family is considered; in this context, microbial concentration and enrichment are the key factors for detecting enzyme activities. There are various methods of assaying microbial enzymes. Amongst these procedures is the use of artificial *p*-nitrophenyl (pNP)-linked substrates; this approach was initially formulated to discover soil phosphatase activity. This method depends on the discovery of a colored end product of *p*-nitrophenol, which evolves when the artificial substrate is hydrolyzed by the appropriate enzyme. *p*-Nitrophenol can be further quantified using a colorimeter by measuring its absorbance at around 400–410 nm. This approach has been employed to discover more enzymes such as NAGase6 (Jackson et al., 2013).

Another approach for enzyme determination is the use of 4-methylumbelliferone (4-MUB)-linked substrates for glucosidase. The release of 4-methylumbelliferone as the end product is extremely fluorescent; this can be determined using a fluorometer between 360 and 460 nm of excitation. There are many types of MUB-linked artificial substrates, allowing the fluorometric determination of many enzymes such as β-glucosidase, cellobiohydrolase, NAGase, and phosphatase, which can be assayed by the use of the pNP substrate colorimetric method. More fervently, diverse microbial extracellular enzymes, such as the protein-degrading leucine aminopeptidase, can be evaluated with a fluorometer using 7-amino-4-methylcoumarin (COU)-linked substrates (Freeman et al., 1995). For a better understanding, a throughput illustration of assaying enzymes in freshwaters using the fluorescent MUB-linked substrate approach shall be made step by step.

Step 1. Preparation of the Substrates, Standards, and Buffer Solutions for Enzyme Assays Using Fluorometric Procedures

- Solutions of MUB-linked substrates (200 μM) such as 4-MUB–βglucopyranoside and 4-MUB-phosphate should be prepared by dissolving the appropriate substrate in distilled water, filled in 15 or 50 mL centrifuge tubes, and autoclaved. Enfold the tubes in aluminum foil to prevent light penetration and store in a refrigerator for a week so that the substrates can be stable.
- MUB standards will be prepared by making a stock solution of 100 μM 4-methylumbelliferone in distilled water aseptically in a bottle wrapped with foil or in an amber bottle. After autoclaving, store in a refrigerator. Dilute the 100 μM stock solution in 1/10 sterile water to make a working solution of 10 μM for enzyme assays.
- Prepare a stock solution of 100 mM bicarbonate buffer by dissolving 8.4 g of $NaHCO_3$ in 1 L of water and autoclave. Dilute this stock solution in 1/20 sterile water to make a working solution of 5 mM for enzyme assays.

Step 2. Preparation of Water Samples on a 96-Well Black Microplate

- Get a microplate for each enzyme. For adequate replication, standards, and controls, this method can analyze single enzymes for up to nine water samples on one 96-well black microplate.
- Measure exactly 5 mL of the first sample into a pipette reservoir and ensure that an eight-channel pipettor is used to pipet 200 μL into all of the wells in column 1 of the microplate(s). Spent pipet tips should be discarded, and the procedure should be repeated as required to fill each water sample in columns 1–9.

Step 3. Procedures for the Arrangement of the Samples, Standards, Quench, and Substrate Controls

- Controls are set up to monitor the activities of the samples, standards, substrates, and quenching on the same black microplate.

- The sample controls are composed of sample water and bicarbonate buffer, which does not take part in the calculations of activities but enhances the reading consistency throughout the experimental procedure. The quench controls are made up of a water sample and a standard fluorescent tag, which are used to quantify diffraction of fluorescence in sample water. Substrates and standard controls contain linked substrates or the standard fluorescent tag and the bicarbonate buffer.
- Introduce exactly 5 mL of 5 mM bicarbonate buffer in a clean pipette reservoir and pipet 50 μL of buffer into microplate wells 1–9 in Rows D and E; this will lead to formation of two replicate wells of sample controls per sample. Discarding the pipette tips and using new tips, transfer 200 μL of bicarbonate buffer to wells 10–12 in Rows A and H of the wells.
- Fluorescent standard is sensitive to light; therefore, turn off the light source to minimize ambient light effects.
- Pour exactly 5 mL of 10 μM 4-methylumbelliferone into a clean pipette reservoir and draw 50 μL into microplate wells 1–12 in Row H and also into wells 1–9 in Rows G and F to form three replicates of quench controls per sample and controls overall. The microplate is either placed in the dark or covered with an opaque lid to reduce light degradation of MUB.
- Turn on the fluorometer and set up any necessary software to be ready to read before adding the substrate. Note that some fluorometer bulbs may require a warm-up time of 3 min or more.
- Pour exactly 5 mL of the MUB-linked substrate into a clean pipette reservoir. A 12-channel pipettor is used to draw 50 μl into microplate wells 1–12 in Row A as well into wells 1–9 in Rows B and C to form three replicate assays for each sample and three substrate controls.

Step 4. Fluorescence Reading

- The initial fluorescence reading is taken immediately after the substrate is added to the microplate. Follow with the incubation of the microplate at RT (room temperature) (23 °C) covered with an opaque lid or in the dark to prevent light degradation of MUB.
- The incubation time for measuring enzyme activity in sample water relies on the concentration of enzymes within the sample. For accuracy purposes, readings are taken at intervals of 5 min within 1 h when assaying many enzymes; nonetheless, samples rich in enzymes may be identified before 10 min, courtesy of their peaks.

Step 5. Enzyme Activity Calculation per Volume of Water

- At intervals for each sample, calculate the initial mean of sample fluorescence (wells D and E), the final mean of sample fluorescence (wells A–C), the mean standard fluorescence (wells H10–11), and the mean quench control fluorescence (wells F–G).
- Moreover, at intervals of each time, calculate enzyme activity in nmoles/h/mL with the equation: enzyme activity = (mean sample fluorescence − mean initial

sample fluorescence) or ((mean standard fluorescence/0.5 mol) × (mean quench control fluorescence / mean standard fluorescence) × (0.2 mL) × (time in hours)).
- Check the values of activity calculated for each interval. Extrapolate the final activity from the interval with the highest activity. In the case in which activity values continue to increase, then later time steps may be required; if activity values decrease throughout the interval of the experiment, then repeat with a shorter time procedure. The final activity is in nmoles/h/mL of substrate consumed but can be scaled up to be expressed in µmoles/h/L (Jackson et al., 2013).

Quantifying enzyme activity in freshwater using a fluorometer has its limitations; MUB standards and MUB-linked substrates should not be exposed to light. Switching off the lights during pipetting and incubating the microplates in the dark is important to prevent light interference. This procedure requires the plates to be read multiple times as possible, leading to rapid switching between plates when analyzing many enzymes at the same time. More so, it is necessary to monitor the time it takes for the microplate reader to read the plate and stagger reading intervals when assaying multiple enzymes at a time. Turbid water samples or water with suspended particles should be stirred prior to pouring into the pipette reservoir and then withdrawn and ejected with the pipettor before loading onto the microplate even mixture. Higher quenching of the fluorescent signal is triggered by more numbers of particles in sample water. In addition, while assaying microbial enzyme activity in freshwater by adopting fluorometer techniques, it is important to note that the procedure measures microbial physiological processes, which have direct influence on transformations of carbon and nutrients of the ecosystem. The high-throughput microplate method permits simultaneous quantification of enzyme activity in larger numbers of samples than a single tube method. The microplate approach ensures evaluation of variation in enzyme activity in relation to the depth and environmental perturbations (Farris et al., 2016; Jackson et al., 2013).

8 Functions of Microbial Enzymes in Freshwater and Water Quality

Microbial enzyme activities of freshwater ecosystems per unit organic matter are usually higher compared to soil microbial enzymes due to the lack of stress in water; more so, nutrients are provided through runoff, urban, and industrial effluents. Enzyme activities in the hyporheic zone of a freshwater network depend on the types of microbes and the nature of the substrate consumed. In addition, enzymatic metabolic responses of most ecosystems of flowing and freshwater networks are attributed to biomass and biomass sediments. Notably, allochthonous inputs of plant litter account for a large fraction of organic matter contribution to many inland water networks. Microorganisms have served and will continue to serve as major sources of production of numerous enzymes in freshwater networks. Enzymes are helpful in so many applications as they work efficiently under mild conditions such as normal temperature and pH and under atmospheric conditions in which stress is minimal; therefore, it is needless to bother with the protection of substrate functional groups, as, in this state, they have a long half-life and, moreover, they work with natural substrates. Wastewaters which may be loaded with excess concentration of inorganic pollutants that can be effortlessly biodegraded, impacted the biological networks, either in Total Suspended Solids (TSS), Biochemical Oxygen Demand (BOD), or Chemical Oxygen Demand (COD), which might be in the tens of thousands mg/L. In the treatment of wastewater, the biological approach seems, by all accounts, to be a promising technology. Microbial enzymes are associated with playing a vital role as metabolic catalysts, bringing about their uses in freshwater applications (Kritika et al., 2017).

Microbial Oxygenases Oxygenases belong to the class of intracellular enzymes that enhance the biosynthesis and metabolism of microbes in freshwater; they have the potential to biodegrade hydrocarbons and their corresponding compounds of environmental pollutants. Oxygenases influence the regio-, stereo-, and enantioselective initiation of atomic oxygen into a helpful substrate, by transforming hydrophobic compounds, which are composed of endobiotic and xenobiotic origins, into more water-soluble and useful forms. In activated sludge, oxygenases are responsible for the cleavage of hydroxyl groups to carboniferous organic compounds and aromatic pollutants, thereby stimulating their oxidation. The triplet state of molecular oxygen ($3O_2$) makes it kinetically balanced because of the attachment of two unpaired electrons, which may lower the spontaneous oxidation of organic compounds. Oxygenases stimulate oxygen reactivity through O_2 activation, susceptibility of substrate attack by O_2 in freshwater, or treatment of a water network. Oxygenases, however, facilitate dehalogenation reactions of halogenated methanes, ethanes, and ethylenes in collaboration with multifunctional enzymes (John et al., 2019).

Microbial Monooxygenases The catalytic oxidation reactions of substrates from alkanes to complex molecules such as steroids and fatty acids are triggered by monooxygenase enzymes. For proper dissolvability of materials, molecular oxygen is needed for the enzyme's activities and the usage of substrate as a reducing agent. The enzymes direct the solitary reduction of atomic oxygen that subsequently appears on the addition of an individual hydroxyl group, which is often cofactor-dependent. Monooxygenases are efficient in catalyzing denitrification, desulfurization, hydroxylation, dehalogenation, ammonification, biotransformation, and biodegradation of different aromatic and aliphatic compounds in freshwater and in industrial water treatment plants (Kritika et al., 2017).

Microbial Laccases Laccases (*p*-diphenol: dioxygen oxidoreductase) belong to a group of multicopper oxidases secreted by certain plants, fungi, insects, and microorganisms. Laccase synthesis is responsive to the concentration of nitrogen in fungi because excess nitrogen is normally needed for greater amounts of laccase production. Microbial laccases trigger the catalysis oxidation of a wide range of reduced phenolic and aromatic substrates accompanying the reduction of subatomic oxygen to water. Aromatic compounds, comprising phenols and aromatic amines, are extremely toxic; many countries have a control approach toward it. They are found in the wastewaters of a wide assortment of industries including coal conversion, oil refining, resins and plastics, wood safeguarding, metal coating, colors and different synthetic substances, textiles, mining and dressing, and pulp and paper. Microbial laccases are efficient in oxidizing aromatic compounds and reducing, if not totally removing, their impact (Arora et al., 2010).

Microbial Cellulases Enzymatic hydrolysis of cellulose yields reducing sugars that can be fermented by yeasts or microorganisms to produce ethanol. The removal of cellulose microfibrils is due to cellulases, which are produced during washing, and the utilization of cotton-based materials. In the paper and pulp industry, cellulase is also used for the removal of ink when recycling paper. There has been an increase in interest in the enzymatic hydrolysis of cellulose for so many years. This interest is based on the benefits that such a procedure would provide, to be specific, the conversion of lignocellulosic and cellulosic waste into a useful energy source through the provision of sugars, ethanol, biogas, or other vigorous and useful end products (Sun & Jiayang, 2002).

Microbial Lipases Lipases are triacylglycerol acyl hydrolases that deed on carboxylic ester bonds. They belong to the class of serine hydrolases and do not need any cofactors. These enzymes are capable of degrading herbicides, detergents, and soaps, enhancing the removal of oil, and catalyzing numerous reactions in freshwater such as esterification, interesterification, alcoholysis, hydrolysis, and aminolysis. At the water interface, lipases rise and attain a neutral pH-dependent activity, thereby attaining equilibrium state. Microbial lipases that respond to acidic pH in freshwater enhance the neutrality and purification of the water network (Prem et al., 2020).

Microbial Peroxidases These enzymes are also tagged as heme-containing proteins. They are ubiquitous enzymes capable of catalyzing the oxidation of lignin and phenolic compounds and preventing oxidative damage to plant leaves. They influence the reduction of peroxides, such as hydrogen peroxide (H_2O_2), and the oxidation of many organic and inorganic compounds; peroxidases also mediate the detoxification of polluted water by cross-reaction with cosubstrates and phenolic or toxic compounds with harmless approach, leading to polymeric synthetic products like dimmers, trimmers, and oligomers. These products could be accumulated in the soil and waterways. Peroxidases are efficient in the treatment of wastewater contaminated with phenols and cresols and are also effective in the treatment of so many industrial effluents such as decolorization of textile dyes, removal of endocrine-disruptive chemicals, pesticide degradation, polychlorinated biphenyls, chlorinated alkanes, phenoxy alkanoic herbicides, chlorinated dioxins, and chlorinated insecticides (Neelam & Shamsher, 2013).

Microbial Hydrolytic Enzymes Freshwater pollutions by industrial, agricultural, and domestic wastes, coupled with hydrocarbon discharges, pose a difficult challenge to the water quality and the microbial community of the water networks. Microbial responses toward the pollutants result in the secretion of enzymes, which leads to hydrolysis of the pollutants. Furthermore, enzymes are needed for the degradation of conversion of organic polymers because only compounds with molecular mass less than 600 dalton can pass through the cell pores. The major chemical bonds in harmful molecules are hampered through enzyme hydrolysis, leading to reduced toxicity. This process can be demonstrated in the degradation of oil spills, organophosphates, carbamate insecticides, and heptachlor stability in water or in well-aerated soil but can readily degrade in anaerobic environments. Hydrolases also mediate other relevant reactions, including condensations and alcoholysis. Hydrolases are always available; this informs the most beneficial aspect of the enzymes together with lack of cofactor stereoselectivity, and they are tolerant to addition of water-miscible solutions (Chandrakant & Shwetha, 2011).

Microbial Dioxygenases These enzymes are synthesized by soil microbes and they partake in the transformation of aromatic precursors into aliphatic products. Dioxygenase enzymes have functional multicomponent systems that attach molecular oxygen to the substrate. These enzymes belong to a large family of Rieske nonheme iron oxygenases. Dioxygenases catalyze the oxygenation of different kinds of substrates, majorly the oxidation of aromatic compounds in water, and can be employed in environmental mediation. Most members of this family possess one or two electrons, which are responsible for moving proteins close to the preceding oxygenase components. The naphthalene dioxygenase crystal structure has affirmed the existence of a Rieske (2Fe–2S) cluster and mononuclear iron in each alpha subunit. Catechol dioxygenases have been one of the natural processes for aromatic molecule degradation in water and its environs (Chandrakant & Shwetha, 2011).

9 Conclusions

This chapter reviews the occurrence and distribution of microbial enzymes in freshwater. The microorganisms that produce microbial enzymes in freshwaters are presumably superior competitors for the utilization of organic and inorganic nutrients as energy sources in aquatic environments. Microbial enzyme activities of freshwater ecosystems per unit organic matter are usually higher compared to soil microbial enzymes due to lack of stress in water; more so, nutrients are provided through runoff, urban, and industrial effluents. Freshwater microbial extracellular enzymes are either attached to the cell or dissolved in the water column. Enzyme activities in the hyporheic zone of a freshwater network rely on the types of microbes and the nature of the substrate consumed. Microbial community and the ecological makeup of the community determine the nature of enzymes and enzyme activities of a particular ecology in a freshwater network. While assessing water quality, the lowest number of each microbial family is considered; in this context, microbial concentration and enrichment are the key factors toward detecting enzyme activities in freshwater. Quantifying enzyme activity in freshwater using a fluorometer has its limitations; MUB standards and MUB-linked substrates should not be exposed to light.

References

Albertson, N. H., Nyströn, T., & Kjelleberg, S. (1990). Exoprotease activity of two marine bacteria during starvation. *Applied and Environmental Microbiology, 56*, 218–223.

Alderkamp, A. C., Rijssel, M. V., & Bolhuis, H. (2007). Characterization of marine bacteria and the activity of their enzyme systems involved in degradation of the algal storage glucan laminarin. *FEMS Microbiology Ecology, 59*, 108–117. https://doi.org/10.1111/ja574F6941.2006.00219.x

Alison, S. D., Chao, Y., Farrara, J. D., Hatosy, S., & Martiny, A. C. (2012). Fine-scale temporal variation in marine extracellular enzymes of coastal southern California. *Frontiers in Microbiology, 3*, 301. https://doi.org/10.3389/fmicb.20132.00301

Allison, S. D., & Vitousek, P. M. (2005). Responses of extracellular enzymes to simple and complex nutrient inputs. *Soil Biology and Biochemistry, 37*, 937–944.

Alves, P. D. D., de Faria, F., Siqueira, S. F., Horta, C. C. R., Victória, J. M. N., & Kalapothakis, E. (2014). Survey of microbial enzymes in soil, water, and plant microenvironments. *The Open Microbiology Journal, 8*, 25–31.

Arnosti, C. (2010). Microbial extracelalar enzymes and the marine carbon cycle. *Annual Review of Marine Science, 3*, 401–425. https://doi.org/10.1146/annurevmarine-120709-142731

Arnosti, C. (2011). Microbial extracellular enzymes and the marine carbon cycle. *Annual Review of Marine Science, 3*, 401–425.

Arora, P. K., Srivastava, A., & Singh, V. P. (2010). Application of monooxygenases in dehalogenation, desulphurization, denitrification and hydroxylation of aromatic compounds. *Journal of Bioremediation & Biodegradation, 1*, 112. https://doi.org/10.4172/21556199.100011

Baltar, F. (2018). Watch out for the 'Living Dead': Cell-free enzymes and their fate. *Frontiers in Microbiology, 8*, 2438. https://doi.org/10.3389/fmicb.2017.02438

Baltar, F., Arnstegui, J., Gasol, J. M., Herndl, G. J., Sintes, E., & van Alken, H. M. (2010). High dissolved extracellular enzymatic activity in the deep central Atantic Ocean. *Aquatic Microbial Ecology, 58*, 287–302. https://doi.org/10.3354/ameo1377

Baltar, F., Aristegui, J., Gasol, J. M., Yokokawa, T., & Herndl, G. J. (2013). Bacterial versus archaeal origin of extracellular enzymatic activity in the Northeast Atlantic deep waters. *Microbial Ecology, 65*, 277–288. https://doi.org/10.1007/s00248-012-0126-7

Baltar, F., Legrand, C., & Pinhassi, J. (2016). Cell-free extracellular enzymatic activity is inked to seasonal temperature changes: A case study in the Baltic Sea. *Biogeosciences, 13*(2815), 2821. https://doi.org/10.5194/bg-13-2815-016

Bergmeyer, H. U. (1983). *Methods of enzymatic analysis* (3rd ed.). Verlag Chemie.

Bisswanger, H. (2011). *Practical enzymology* (2nd ed.). Wiley-Blackwell.

Bochdansky, A. B., Puskaric, S., & Herndl, G. J. (1995). Influence of zooplankton grazing on free dissolved enzymes in the sea. *Marine Ecology. Progress Series, 121*, 53–63. https://doi.org/10.3354/mepsi21053

Boschker, H. T. S., & Cappenberg, T. E. (1998). Patterns of extracellular enzyme activities in littoral sediments of Lake Gooimeer, The Netherlands. *FEMS Microbiology Ecology, 25*, 79–86.

Brandão, L. R., Libkind, D., Vaz, A. B. M., et al. (2011). Yeasts from an oligotrophic lake in Patagonia (Argentina): Diversity, distributionand synthesis of photoprotective compounds and extracellular enzymes. *FEMS Microbiology Ecology, 76*, 1–13.

Cai, J.-G., Guo, Z.-G., Li, C.-X., & Zhou, Z.-L. (2005). Types of organic matter in water and organic sedimentary processes. *Tongji Daxue Xuebao/Journal of Tongji University, 33*(9), 1213–1218.

Caldwell, B. A. (2005). Enzyme activities as a component of soil biodiversity: A review. *Pedobiologia, 49*, 637.

Cembella, A. D., Antia, N. J., & Harrison, P. J. (1984). The utilization of inolganic and organic phosphorus compounds as nutrients by eukaryotic microalgaeA multidisciplinary perspective. Part 1. *Critical Reviews in Microbiology, 10*, 317–391.

Chandrakant, S. K., & Shwetha, S. R. (2011). Review: Role of microbial enzymes in the bioremediation of pollutants. SAGE-Hindawi. *Enzyme Research, 805187*, 1–11. https://doi.org/10.4061/2011/805187

Chróst, R. J. (1991). Environmental control of the synthesis and activity of aquatic microbial ectoenzymes. In R. J. Chróst (Ed.), *Microbial enzymes in aquatic environments* (pp. 29–59). Springer-Verlag.

Chróst, R. J. (1992). Significance of bacterial ectoenzymes in aquatic environments. *Hydrobiologia, 243*(244), 61–70.

Cleveland, C. C., & Liptzin, D. (2007). C: N: P stoichiometry in soil: Is there a "Redfield ratio" for the microbial biomass? *Biogeochemistry, 85*, 235–252.

D'Ambrosio, L., Macgregor, B., Teske, A., & Arnosti, C. (2014). Composition and enzymatic function of particle associated and free living bacteria: a coastal/offshore comparison. *The ISME Journal, 8*, 2167–2179. https://doi.org/10.1038/1smej.2014.67

Dalal, R. C. (1977). Soil organic phosphorus. *Advances in Agronomy, 29*, 83–113.

Eivazi, F., & Tabatabai, M. A. (1988). Glucosidases and galactosidases soils. *Soil Biology and Biochemistry, 20*, 601–606.

Farris, M. H., Ford, K. A., & Doyle, R. C. (2016). Qualitative and quantitative assays for detection and characterization of protein antimicrobials. *Journal of Visualized Experiments, 110*, 53819. https://doi.org/10.3791/53819

Feldman, M. F., Wacker, M., Hernandez, M., Hitchen, P. G., Marolda, C. L., Kowarik, M., Morris, H. R., Dell, A., Valvano, M. A., & Aebi, M. (2005). Engineering N-linked protein glycosylation with diverse O antigen lipopolysaccharide structures in *Escherichia coli*. *Proceedings of the National Academy of Sciences of the United States of America, 102*, 3016–3021.

Freeman, C., Liska, G., Ostle, N. J., Jones, S. E., & Lock, M. A. (1995). The use of fluorogenic substrates for measuring enzyme activity in peatlands. *Plant and Soil, 175*, 147–152.

Harder, W., & Dijkhuizen, L. (1983). Physiological responses to nutrient limitation. *Annual Review of Microbiology, 37*, 1–23.

Hehemann, J.-H., Reintjes, G., Klassen, L., Smith, A. D., Ndeh, D., Arnosti, C., et al. (2019). Single cell fluorescence imaging of glycan uptake by intestinal bacteria. *The ISME Journal, 13*, 1883–1889. https://doi.org/10.1038/s41396-019-0406-2

Jackson, C. R., Tyler, H. L., & Millar, J. J. (2013). Determination of microbial extracellular enzyme activity in waters, soils, and sediments using high throughput microplate assays. *Journal of Visualized Experiments, 80*, e50399. https://doi.org/10.3791/50399

Jastrow, J. D., Amonette, J. E., & Bailey, V. L. (2007). Mechanisms controlling soil carbon turnover and their potential application for enhancing carbon sequestration. *Climatic Change, 80*, 5–23.

John, O. U., Okoh, A. I., & Nwodo, U. U. (2019). Review: Aptitude of oxidative enzymes for treatment of wastewater pollutants: A laccase perspective. *Molecules, 24*(2064), 1–36. https://doi.org/10.3390/molecules24112064

Kelleher, B. P., & Simpson, A. J. (2006). Humic substances in soils: Are they really chemically distinct? *Environmental Science and Technology, 40*, 4605–4611.

Klonowska, A., Gaudin, C., Fournel, A., Asso, M., Le Petit, J., Giorgi, M., & Tron, T. (2002). Characterization of a low redox potential laccase from the basidiomycete C30. *European Journal of Biochemistry, 269*, 6119–6125.

Koch, A. L. (1985). The macroeconomics of bacterial growth. In M. Fletcher & G. D. Floodgate (Eds.), *Bacteria in their natural environments* (pp. 1–42). Academic Press.
Kritika, P., Singh, B., Pandey, A. K., Badruddin, I. J., Pandey, S., Mishra, V. K., & Jain, P. A. (2017). Review: Application of microbial enzymes in industrial waste water treatment. *International Journal of Current Microbiology and Applied Sciences, 6*(8), 1243–1254. https://doi.org/10.20546/ijcmas.2017.608.151
Kuhbier, S., et al. (2002). Enzymatic response to pollution in sediments and aquatic systems. In R. G. Burns & R. P. Dick (Eds.), *Enzymes in the environment activity, ecology and applications* (pp. 433–446). Marcel Dekker.
Lochte, M. A., & Ford, T. E. (1986). Metabolism of dissolved organic matter by attached microorganisms in rivers. In F. Megusar & M. Gantar (Eds.), *Perspectives in microbial ecology* (pp. 367–374). Slovene Society of Microbiology.
Lorenz, P., & Schleper, C. (2002). Metagenome—a challenging source of enzyme discovery. *Journal of Molecular Catalysis B: Enzymatic, 19*, 13–19.
Maire, V., Alvarez, G., Colombet, J., Comby, A., Despinasse, R., Dubreucq, E., Joly, M., Lehours, A.-C., Perrier, V., Shahzad, T., & Fontaine, S. (2012). An unknown respiration pathway substantially contributes to soil CO2 emissions. *Biogeosciences Discussions, 6*, 8663–8691.
McLatchey, G. P., & Reddy, K. R. (1998). Regulation of organic matter decomposition and nutrient release in a wetland soil. *Journal of Environmental Quality, 27*, 1268–1274.
Mohapatra, B. R., Bapuji, M., & Sree, A. (2003). Production of industrial enzymes (amylase, carboxymethylcellulase and protease) by bacteria isolated from marine sedentary organisms. *Acta Biotechnologica, 23*, 75–84.
Nannipieri, P. (2006). Role of stabilised enzymes in microbial ecology and enzyme extraction from soil with potential applications in soil proteomics. *Nucleic Acids and Proteins in Soil, 8*, 75–94.
Neelam, B., & Shamsher, S. K. (2013). Review: Peroxidase(s) in environment protection. Hindawi. *The Scientific World Journal, 2013*, 714639. https://doi.org/10.1155/2013/714639
Ong, K. S., Chin, H. S., & Teo, K. C. (2011). Screening of antibiotic sensitivity, antibacterial and enzymatic activities of microbes isolated from ex-tin mining lake. *African Journal of Microbiology Research, 5*, 2460–2466.
Pang, Y., Liu, X., Ma, Y., Chernin, L., Berg, G., & Gao, K. (2009). Induction of systemic resistance, root colonisation and biocontrol activities of the rhizospheric strain of Serratia plymuthica are dependent on N-acyl homoserine lactones. *European Journal of Plant Pathology, 124*, 261–268.
Piccolo, A. (2001). The supramolecular structure of humic substances. *Soil Science, 166*, 810–832.
Prem, C., Enespa, R. S., & Arora, P. K. (2020). Review: Microbial lipases and their industrial applications: A comprehensive. *Microbial Cell Factories, 19*(169), 1–24. https://doi.org/10.1186/s12934-020-01428-8
Reintjes, G., Arnosti, C., Fuchs, B. M., & Amann, R. (2017). An alternative polysaccharide uptake mechanism of marine bacteria. *The ISME Journal, 11*, 1640–1650. https://doi.org/10.1038/ismej.2017.26
Reintjes, G., Arnosti, C., Fuchs, B., & Amann, R. (2019). Selfish, sharing and scavenging bacteria in the Atlantic Ocean: A biogeographical study of bacterial substrate utilisation. *The ISME Journal, 13*, 1119–1132. https://doi.org/10.1038/s41396-018-0326-3
Rogers, H. J. (1961). The dissimilation of high molecular weight organic substrates. In L. C. Gunsalus & R. Y. Stanier (Eds.), *The bacteria* (Vol. 2, pp. 261–318). Academic Press.
Ryszard, J. C. (1990). *Microbial ectoenzymes in aquatic environments*. Springer-Verlag.
Sala, M. M., Piontek, J., Endres, S., Romani, A. M., Dyhrman, S., & Steen, A. D. (2019). Editorial: Extracellular enzymes in aquatic environments: Exploring the link between genomic potential and biogeochemical consequences. *Frontiers in Microbiology, 10*, 1463. https://doi.org/10.3389/fmicb.2019.01463
Schimel, J. P., & Weintraub, M. N. (2003). The implications of exoenzyme activity on microbial carbon and nitrogen limitation in soil: A theoretical model. *Soil Biology and Biochemistry, 35*, 549–563.

Schomburg, D., & Schomburg, I. (Eds.). (2009). *Springer handbook of enzymes* (2nd ed.). Springer.

Sexstone, A. J., Revsbech, N. P., Parkin, T. B., & Tiedje, J. M. (1985). Direct measurement of oxygen profiles and denitrification rates in soil aggregates. *Soil Science Society of America Journal, 49*, 645–651.

Sinsabaugh, R. L., & Moorhead, D. L. (1995). Resource allocation to extracellular enzyme production: A model for nitrogen and phosphorus control of litter decomposition. *Soil Biology and Biochemistry, 26*, 1305–1311.

Sinsabaugh, R. L., Osgood, M. P., & Findlay, S. (1994). Enzymatic models for estimating decomposition rates of particulate detritus. *Journal of the North American Benthological Society, 13*, 160–169.

Sinsabaugh, R. L., Carreiro, M. M., & Repert, D. A. (2002). Allocation of extracellular enzymatic activity in relation to litter composition, N deposition, and mass loss. *Biogeochemistry, 60*, 1–24.

Stottmeister, U., Wießner, A., Kuschk, P., Kappelmeyer, U., Kästner, M., Bederski, O., Müller, R., & Moormann, H. (2003). Effects of plants and microorganisms in constructed wetlands for wastewater treatment. *Biotechnology Advances, 22*, 93–117.

Sun, Y., & Jiayang, C. (2002). Hydrolysis of lignocellulosic materials for ethanol production: A review. *Bioresource Technology, 83*(1), 1–11.

Vandenkoornhuyse, P., Baldauf, S. L., Leyval, C., Straczek, J., & Young, J. P. W. (2002). Extensive fungal diversity in plant roots. *Science, 295*, 2051.

Wallenstein, M. D., & Burns, R. G. (2011). *Ecology of extracellular enzyme activities and organic matter degradation in soil: A complex community—driven process*. Wiley.

Wetzel, R. (1993). Humic compounds from wetlands: Complexation, inactivation and reactivation of surface-bound and extracellular enzymes. *Verhandlungen der Internationalen Vereingang fur Liminologie, 25*, 122–128.

Williams, P. J. L. (1981). Incorporation of microheterotrophic processes into the classical paradigm of the planktonic food web. *Kieler Meeresforschungen, Sonderheft, 5*, 1–28.

Wobus, A., Bleul, C., Maassen, S., et al. (2003). Microbial diversity and functional characterization of sediments from reservoirs of different trophic states. *FEMS Microbiology Ecology, 46*, 331–347.

Wright, A. L., & Reddy, K. R. (2001). Phosphorus loading effects on extracellular enzyme activity in Everglades wetland soils. *Soil Science Society of America Journal, 65*, 588–595.

Ziervogel, K., Steen, A. D., & Arnosti, C. (2010). Changes in the spectrum and rates of extracellular enzyme activities in seawater following aggregate formation. *Biogeosciences, 7*, 1007–1017.

Marine Microbial Enzymes: An Overview

H. S. Auta, D. O. Aboyeji, J. D. Bala, O. P. Abioye, N. U. Adabara, Sesan Abiodun Aransiola, Auwalu Hassan, and A. Aziz

Abstract The modern world is now focusing on environmental-friendly products, and, hence, many chemical processes are being replaced by enzymatic methods. In recent years, enzymes have attracted huge attention due to their potential industrial, pharmaceutical, and cosmetic applications in everyday life. The marine environment has been identified as a reservoir of important microorganisms having the potential to generate multifarious enzyme systems with novel applications. Marine microbial enzymes, in particular, attract special interest due to their distinct habitat-related properties that enable them to be active in extreme environments. Hence, marine microbial enzymes including proteases, lipases, collagenases, agarases, celluloses, and other enzymes can offer novel biocatalysts with extraordinary properties. This chapter discusses marine microbial enzymes, their properties, and their applications in different fields of human endeavors.

H. S. Auta
Department of Microbiology, Federal University of Technology, Minna, Niger State, Nigeria

Africa Centre of Excellence for Mycotoxin and Food Safety (ACEMFS), Federal University of Technology, Minna, Niger State, Nigeria

D. O. Aboyeji (✉) · J. D. Bala · O. P. Abioye · N. U. Adabara
Department of Microbiology, Federal University of Technology, Minna, Niger State, Nigeria
e-mail: oyeyemi.aboyeji@st.futminna.edu.ng

S. A. Aransiola
Bioresources Development Centre, National Biotechnology Development Agency, Ogbomoso, Nigeria

A. Hassan
Department of Microbiology, Federal University of Technology, Minna, Niger State, Nigeria

Institute of Biological Sciences, Faculty of Science, University of Malaya, Kuala Lumpur, Malaysia

Department of Biological Sciences, Federal University, Kashere, Gombe, Nigeria

A. Aziz
Faculty of Marine Sciences, Lasbela University of Agriculture, Water and Marine Sciences, Uthal, Balochistan, Pakistan

N. R. Maddela et al. (eds.), *Ecological Interplays in Microbial Enzymology*, Environmental and Microbial Biotechnology,
https://doi.org/10.1007/978-981-19-0155-3_5

1 Introduction

Enzymes are biocatalysts that are involved in all phases of metabolism and biological reactions. They are biological materials or groups of biological macromolecules produced by living organisms that function as catalysts to speed up biological and biochemical reactions both inside and outside the cell. Generally, enzymes play a central role in mineralization and element cycling in various habitats as well as in biochemical reactions in living cells. As a result, every marine microbe should be a stable source of vital enzymes including proteases, amylases, lipases, chitinases, cellulases, ligninases, pectinases, xylanases, and nucleases (DNAses, RNAses, restriction enzymes). Therefore, there is need for efficient management of our rich marine microbial biodiversity toward deriving novel enzymes that could be recovered from marine microorganisms and used not only as cost-effective biocatalysts but also as an environmentally friendly reagent for various industrial processes.

Several industrial enzymes have been derived from terrestrial environments such as the savannahs (Noriler et al., 2018), forests (Pajares & Bohannan, 2016), deserts (Cui et al., 2018), the Artic (Malard & Pearce, 2018), and the Antarctic poles (Duarte et al., 2018), whereas, marine environments, which cover about 71% of Earth's surface and serve as a potential reservoir of useful enzymes, remain unexplored. Marine microbial enzymes have received unprecedented attention due to their wide industrial applications, but only few have been successfully isolated, purified, and characterized for their properties and applications.

The marine environment is considered one of the most significant sources of novel bioactive compounds in the world including enzymes. Marine microbial communities from marine environments are capable of producing an extensive spectrum of enzymes according to their habitats and their ecological roles. They constitute important ecological components of marine environments due to their role in biogeochemical processes and promising industrial applications (Barzkar, 2020). Marine microbial enzymes possess strong specificity, little reaction conditions, easy inactivation or control, and are eco-friendly when compared with the conventional chemical catalysts. Thus, they can meet the required market demand for novel biocatalysts with extraordinary properties suitable for various industrial processes.

Marine environment contains a myriad of potential microorganisms, fungi, plants, and animals, which are a rich source of biodiversity, with the ability to produce enzymes. A marine enzyme may be so unique that it is not found in any terrestrial organism or it may be a known enzyme from a terrestrial environment but with novel characteristics (Dadshahi et al., 2016; Homaei et al., 2016). Microbial enzymes are known to have some advantages over enzymes derived from plant or animal sources such as wide biodiversity, the ability to mass culture, ease of genetic exploration, high catalytic activity, cost-effectiveness, process efficiency, and sustainability (Beygmoradi & Homaei, 2017; Nguyen & Nguyen, 2017). Recently, the advancements in marine science, biotechnology, enzyme engineering, microbial fermentation technology, and other innovative technologies have necessitated for more studies by researchers to identify, characterize, and find useful applications for microbial enzymes. However, despite the huge benefits that marine microbial enzymes offer, they have not been fully explored as only few of these enzymes have been synthesized, characterized, and fully utilized. Therefore, this chapter reviews marine microbial enzymes, and their properties and applications are also summarized.

2 Classification of Enzymes

Different enzymes are secreted by bacteria and fungi in marine environments depending on their habitats and ecological activities. Marine microbial enzymes have sparked a lot of curiosity, and a few enzymes have been isolated from seawater and marine sediments, purified, and described for their properties and uses. The International Union of Biochemistry (IUB) initiated standards of enzyme nomenclature, which recommend that enzyme names must indicate both the substrate acted upon and the type of reaction catalyzed. According to the Enzyme Commission, enzymes are generally divided into six different classes, namely:

1. Oxidoreductases: Oxidation reactions involve the transfer of electrons from one molecule to another. In biological systems, there may be removal of hydrogen from the substrate. Typical enzymes in this class are called dehydrogenases, e.g., alcohol dehydrogenase catalyzes reactions of the type R–CH_2OH + AR–CHO + H_2A, where A is an acceptor molecule.
2. Transferases: This class of enzymes catalyzes the transfer of groups of atoms from one molecule to another. For instance, aminotransferases or transaminases promote the transfer of an amino group from an amino acid to an alpha-oxoacid.
3. Hydrolases: Hydrolases catalyze hydrolysis, the cleavage of substrates by water. They help break down larger molecules into smaller fragments. Their actions include the cleavage of peptide bonds in proteins, glycosidic bonds in carbohydrates, and ester bonds in lipids.
4. Lyases: Lyases catalyze the addition of groups to double bonds or the formation of double bonds through the removal of groups, e.g., pectate lyases split the glycosidic linkages by beta-elimination.
5. Isomerases: Isomerases catalyze the transfer of groups from one position to another in the same molecule. These enzymes change the structure of a substrate through the rearrangement of its atoms.
6. Ligases: Ligases join molecules together with covalent bonds. These enzymes participate in biosynthetic reactions in which new groups of bonds are formed. These types of reactions require the input of energy in the form of cofactors such as adenosine triphosphate (ATP) (Gurung et al., 2013).

3 Marine Microbial Enzymes

Microbial enzymes are enzymes derived from various microorganisms that are used in industries and on a commercial scale. Although plants and animals produce enzymes, microbes are preferable as a source of enzymes owing to their significant attributes of being cheap, exhibiting more predictable and controllable catalytic activities, generating high yield of enzymes, and being able to provide a consistent supply of culture due to the absence of seasonal fluctuations (Cheng et al., 2020). In addition, plant and animal tissues are considered more harmful compared to microbes.

Microbes inhabit various marine habitats including plankton, nekton, seston, epibiotic, endobiotic, pelagic, and benthic environments. These habitats house a diverse range of microbes such as archaebacteria, cyanobacteria, eubacteria, actinomycetes, yeasts, filamentous fungi, microalgae, algae, and protozoa. It is noteworthy that almost all of these groups are rich sources of useful enzymes that remain unexplored. Various researchers have identified marine bacteria as potential producers of a wide range of industrial enzymes (Zhang & Kim, 2010; Fulzele et al., 2011; El-Hassayeb & Abdel Aziz, 2016). These industrial enzymes, which include proteases (Fulzele et al., 2011; El-Hassayeb & Abdel Aziz, 2016), α-amylases, α-glucosidases, agarases, α-galactosidases, cellulases, chitinases, and lipases, are derived from marine bacteria-producing industrial enzymes such as *Aeromonas* sp., *Alteromonas* sp., *Arthrobacter* sp., *Chromobacterium* sp., *Clostridium* sp., *Cytophaga* sp., *Enterobacter* sp., *Flavobacterium* sp., *Klebsiella* sp., *Listonella* sp., *Moraxella* sp., *Pseudoalteromonas* sp., *Pseudomonas* sp., *Psychrobacter* sp., *Serratia* sp., *Streptomyces* sp., *Bacillus* sp. (El-Hassayeb & Abdel Aziz, 2016), *Vibrio* sp. (Zhang & Kim, 2010), and *Marinobacter* sp. (Fulzele et al., 2011).

Marine microbial enzymes have attracted special attention because of their stability, activity, and ability to withstand extreme conditions that most of the other enzymes cannot (Sana, 2013). Marine bacteria adapt successfully to diverse environmental parameters such as high salinity, acidic and alkaline pH, extreme temperature, extreme barometric pressure, and low nutrient availability. Thus, many marine microbial enzymes have been and are being utilized in biotechnology and other relevant fields. The recent advent of biotechnology has more growing interest and demand for enzymes with novel characteristics (Maddela & García, 2021; Maddela et al., 2021). Marine environments range from nutrient-rich regions to nutritionally low environments where only a few organisms can survive. However, the special adaptive properties possessed by marine microorganisms account for the significant differences observed between marine microbial enzymes and homologous enzymes from terrestrial microorganisms. This has, in recent years, resulted in the boost observed in marine microbial enzyme technology and the resulting remarkable products. These enzymes have been used as food additives, pharmaceuticals, and fine chemicals while some have yielded a considerable number of drug candidates (Zhang & Kim, 2010).

4 Properties and Applications of Some Marine Microbial Enzymes

4.1 Starch Hydrolyzing Enzymes

Starch hydrolyzing enzymes are enzymes involved in the conversion of starch to compounds with low molecular weight (such as glucose, maltose, and oligosaccharides). These enzymes include α-amylase, β-amylase, debranching enzymes (pullulanase), glucoamylase, and α-glucosidase. Because α-amylase, pullulanase, and α-glucosidase from archaea are all active in the same pH and high temperature range, they could be utilized in a one-step process for the industrial bioconversion of starch. This process is employed in order to significantly lower the cost of sugar syrup production through improvement of the starch-conversion process using new, efficient, and thermoactive enzymes.

4.2 α-Amylases

α-Amylase has a wide range of industrial applications due to the unique properties possessed by the enzyme. α-Amylase is used in the production of alcoholic drinks such as beer and alcohol, for animal feed preparation, in washing detergents, in the starch industry, in the textile industry as a desizing agent, in confectioneries, and in sugar syrup production. α-Amylase is one of the few major industrial enzymes in high demand. It is often produced by marine microorganisms such as thermophilic archaea *Pyrococcus woesei*, *Pyrococcus furiosus*, *Thermococcus celer*, *Fervidobacterium pennavorans*, *Desulfurococcus mucosus*, and *Thermotoga maritima* and psychrotrophic *Vibrio* isolated from deep-sea mud, *Vibrio gazogenes*, *Alteromonas rubra*, and *Mucor* sp.

4.3 α-Glucosidases

α-Glucosidase is mostly employed in the starch business alongside α-amylase. Two hyperthermophilic marine archaebacteria have been found to have a highly thermostable α-glucosidase, namely, *Pyrococcus furiosus* and *Pyrococcus woesei.*

4.4 Pullulanases (Debranching Enzymes)

The enzymes pullunases are also known as debranching enzymes. Similar to α-amylases and α-glucosidases, the starch industry primarily uses pullulanases.

Pullulanase type I is a bacterial enzyme that degrades branched oligosaccharides with 1,6 linkages. It is unable to attack 1,4 linkages.

4.5 Agarases

Agarase has been the subject of investigations for quite some time owing to its immediate applications in gene technology for the elution and isolation of deoxyribonucleic acid (DNA) fragments from agarose gels after electrophoresis; in the preparation of algal protoplasts such as from red alga *Gelidium robustum*; in seaweed polysaccharide characterization; in the production of simple sugars, including neoagarobiose, neoagarotetraose, and neoagarohexaose; in the degradation of agarose to oligosaccharides; in facilitating the liquefaction of agar and agarose gels; and in the defouling of fermentors and bioreactors.

Furthermore, the purified enzyme might be used to efficiently control red algae bloom contaminations, avoid biofouling of submerged marine surfaces or pipes by polluting complex polysaccharide layers, or remediate biofouled surfaces once they have been contaminated. Marine bacterial agarase depolymerizes complex polysaccharides, such as agar and agarose, with a high level of activity. Agarase is derived from a few numbers of bacteria. Agarase-0107, an endotype agarase that hydrolyzed the 1,4 linkage of agarose to generate neoagarotetraose and neoagarobiose at a pH of roughly 8, was isolated from *Vibrio* sp. JT0107, a marine salt-loving bacterium. *Pseudomonas stutzeri*, *Aeromonas* sp., and *Vibrio* sp., isolated from sea produce agarases, which have been characterized.

4.6 Cellulases

Cellulases are used for various industrial applications such as in the production of alcohol, food flavoring, maize gluten, silage, laundry and detergents, and wastewater treatment. Cellulase can be produced from a symbiotic bacterium found in the gland of Deshayes of a marine shipworm, *Aspergillus terreus* isolated from saltwater. Cellulose molecules are known to be strongly bound to each other; hence, cellulolysis is relatively difficult when compared to the breakdown of other polysaccharides (Brás et al., 2008). Different bacteria are able to produce cellulase, and these include *Cytophaga*, *Cellulomonas*, *Vibrio*, *Clostridium*, *Nocardia*, *and Streptomyces*. Moreover, some fungi such as *Trichoderma*, *Aspergillus*, *Fusarium*, *Chaetomium*, *Phoma*, *Sporotrichum*, *Penicillium*, etc. are also able to produce cellulase. Hemicellulase generally refers to hydrolase, that is, an enzyme that can hydrolyze polysaccharides, for example, xylanase, galactanase, and arabanase, among which xylanase is of higher economic value (Doi, 2008; Maki et al., 2009).

4.7 Proteases

Protease sales account for more than 60% of all industrial enzyme sales around the world. Proteases are commonly employed in modern life. They are employed in the detergent and leather industries as well as in pharmaceuticals such as digestive and anti-inflammatory medications (Zhu et al., 2019). Dane discovered alkaline protease from *Bacillus licheniformis* for the first time in 1960. So far, microbes have been discovered to be the most suitable resources for protease production. Nobou Kato identified a new type of alkaline protease from marine *Psychrobacter* in 1972, and many other proteases have been isolated from marine microbes since then.

An alkaline protease was tested as a cleansing addition after being obtained from a symbiotic bacterium found in the gland of Deshayes of a marine shipworm. Chi et al. (2007) identified a yeast strain (*Aureobasidium pullulans*) from a sea saltern in the China Yellow Sea with a high yield of alkaline protease, with a maximal synthesis of enzyme of 623.1 U/mg protein (7.2 U/mL). Haddar et al. (2009) discovered *Bacillus mojavensis* A21, which produces alkaline proteases from sea-water and purified two detergent-stable alkaline serine proteases (BM1 and BM2) from this strain. Both proteases were highly stable in the presence of nonionic surfactants.

4.8 Lipases

Lipases are enzymes that catalyze the breakdown of fats and oils, releasing free fatty acids, diacylglycerols, monoglycerols, and glycerol as a result. Lipases are also useful in a variety of processes, including esterification, transesterification, and aminolysis. Lipases have recently gotten a lot of attention, as indicated by the growing amount of information about them in the literature. Many microbial lipases are also commercially accessible, with the bulk of them being employed in detergents, paper manufacturing, cosmetic manufacturing, food flavoring, organic synthesis, and for other industrial uses (Sarmah et al., 2018).

In Europe, enzyme detergents now account for 90% of the market, whereas, in Japan, they account for roughly 80%. Because they work under mild circumstances and are highly stable in organic solvents, lipases are useful biocatalysts with broad substrate specificity (Zhu et al., 2019). Pelagic fishes have been the principal target of fisheries as a result of the exploration of marine resources: these species are resourceful and promising. However, because these species have a high fat content, humans must deal with particular challenges in terms of fish preservation, processing, and marketing.

5 The Role of Marine Microbial Enzymes

Marine microbial enzymes have been used in a variety of industrial applications. They play a vital role in the marine environment. Marine microorganisms take active part in the mineralization of complex organic matter through degradative pathways of their metabolism in marine environments. They play vital roles in major biogeochemical cycles, changes, and processes occurring in marine environments. Therefore, marine microorganisms are critically important to the environmental as well as human health. They also participate in the regulation of Earth's climate since they can release carbon products, particularly CO_2 and CH_4 (Sivaperumal et al., 2017). Various marine microbial enzymes form complexes used in the waste management system for remediation of toxic pollutants. Both domestic wastes and industrial sewages pose great threats and challenges to the ecosystem as well as human beings. Microbial enzymes in combinations (two or more enzymes together) or alone are used to minimize these hazardous materials containing compounds such as nitriles, phenols, and aromatic amines by the degradation of these noxious chemical compounds into harmless products (Pandey et al., 2011; Rubilar et al., 2008).

Furthermore, a number of enzymes are used for waste treatment such as amylases, amyloglucosidases, cellulases, lipases, amidases, glucoamylases, proteases, and pectinases (Karigar & Rao, 2011). Proteases, one of the major enzymes frequently isolated from marine microorganisms, have found useful applications in the detergent industry, tanneries, and the pharmaceutical industry, whereas some other enzymes such as agarases, amylases, cellulases, carrageenases, chitinases, lipases, and lignocelluloses, which are also isolated from marine microorganisms, are used in the production of bioethanol and for other purposes. This class of oxidoreductase enzymes, such as manganese peroxidase, lignin peroxidase, laccase, and tyrosinase, has been successfully used to eliminate industrial effluents of chlorinated phenolic compounds (Le Roes-Hill & Prins, 2016).

Different microbial enzymes with the marine organisms producing them, their properties, and their applications are summarized in Table 1.

6 Medicinal Use of Marine Microbial Enzymes

6.1 *Treatment of Damaged Tissues*

For the removal of dead skin from burns, a vast range of proteolytic enzymes of plant and bacterial origins have been explored. Clinical trials are currently underway for a variety of enzymes of improved grade and purity. The United States Food and Drug Administration (USFDA) approved a phase II clinical trial for Debrase Gel Dressing, which contains a blend of various enzymes isolated from pineapple, in 2002, for the treatment of partial-thickness and full-thickness burns. Vibrilase™, a proteolytic enzyme derived from *Vibrio proteolyticus*, has been demonstrated to be efficient in

Table 1 Industrial and commercial applications of numerous marine microbial enzymes

Enzyme	Microorganisms producing them	Reactions acted upon	Applications	References
Esterase	*Bacillus subtilis*, *Erythrobacter seohaensis* SW-13, *Thalassospira* sp. GB04J01, *Oleispira antarctica*, *Vibrio fischeri*, *Pseudoalteromonas arctica*, *Bacillus* sp., *Pseudonocardia antitumoralis* SCSIO 01299, and *Pseudoalteromas* sp. strain 643A	Hydrolyzes the ester bonds of water-soluble fatty acid esters with short-chain (≤8 carbon) acyl groups	Pharmaceutical industry, pulp and paper industry, degradation of man-made plastics, biodegradation of organophosphorus compounds, and production of short-chain flavor esters	Hao et al. (2014), Huo et al. (2017), De Santi et al. (2016), Lemak et al. (2012), Wang et al. (2016), Al Khudary et al. (2010), Cao et al. (2016), and Barzkar et al. (2021)
Protease	*Pseudoalteromonas* sp. 129–1, *Bacillus pumilus* MP27, *Yarrowia lipolytica* YlTun15, *Bacillus pumilus* TMS55, *Pseudomonas aeruginosa*, *Bacillus subtilis* NS, *Aspergillus oryzae*, *Psiloteredo healdi*, *Pseudoalteromonas arctica* PAMC 21717, *Micrococcus* NH54PC02, *Aureobasidium pullulans* 10, and *Bacillus* sp. SD11	Cleaves peptide bonds to produce small peptides or amino acids	Detergent industry, contact lens cleansing, leather industry, marine waste treatment, anti-biofilm and antifouling agents, fibrinolytic activity, and antibacterial activity	Wu et al. (2015), Yepuru (2018), Bessadok et al. (2017), Maruthiah et al. 2016a, b), Baweja et al. (2017), Karikalan and Mohankumar (2016), Park et al. (2018), Hou et al. (2017) and Barzkar (2020)
Agarase	*Vibrio* sp., *Pseudomonas stutzeri*, *Aeromonas* sp., *Cytophaga*, *Bacillus*, *Alteromonas*, *Pseudoalteromonas*, *Streptomyces*, *Alteromonas agarlyticus* GJ1B, *Thalassomonas* sp. JAMB-A33, *Catenovulum agarivorans*, *Alteromonas* sp., *Cytophaga* sp., *Agarivorans gilvus*, and *Pseudoalteromonas* sp.	Degrades agar, a highly heterogeneous polysaccharide	Control of red algae bloom, antibiofouling agents, food additives in food and moisturizing additives in cosmetics, food industry, gene technology, and agarose gel electrophoresis to separate DNA fragments	Hosoda et al. (2003), Ohta and Hatada (2006), Lee et al. (2013), Cui et al. (2014), Vijayaraghavan and Rajendran (2012), Hu et al. (2009), Chi et al. (2012) and Minegishi et al. (2013)

(continued)

Table 1 (continued)

Enzyme	Microorganisms producing them	Reactions acted upon	Applications	References
Lipase	*Aspergillus flavus*, *Penicillium oxalicum*, *Achromobacter* sp., *Alcaligenes* sp., *Arthrobacter* sp., *Chromobacterium* sp., *Staphylococcus* sp., *Pseudomonas* sp., *Aspergillus niger*, *Rhizopus delemar*, *Rhizopus japonicus*, *Rhizopus niveus*, *Rhizopus oryzae*, *Candida cylindracea*, *Humicola lanuginosa*, and *Mucor miehei*	Hydrolyzes esters of long-chain aliphatic acids from their Glycerol; breaks down triglycerides into fatty acids and glycerol	Biotechnology, especially dairy, detergents, drugs, chemicals, agricultural chemicals, and food, pharmaceutical, cosmetic, and agricultural industries	Renge et al. (2012), Do et al. (2013), Zhang and Kim (2010), Ananthi et al. (2014), Arora (2013), Charoenpanich et al. (2011), Ramani et al. (2013) and Beygmoradi and Homaei (2017)
Cellulase	*Cytophaga* (*C. diffluens*, *C. hutchinsonii*), *Marinobacter* sp. MS1032, *Cellulomonas*, *Vibrio*, *Paenibacillus* sp. BME-14, *Clostridium*, *Nocardia*, *Streptomyces*, *Trichoderma*, *Aspergillus*, *Fusarium*, *Chaetomium*, *Phoma*, *Sporotrichum*, and *Penicillium*	Breaks down cellulose and polysaccharides	Production of alcohol, flavoring in food, paper, corn gluten, detergents, sewage treatment, biotextile auxiliaries, and cotton, flax, and biofertilizer processing	Dong et al. (2010), Fu et al. (2010), Zhang and Kim (2010), Zhao et al. (2012), Shanmughapriya et al. (2010) and Beygmoradi and Homaei (2017)
Chitinase and chitosanase	*Vibrio parahaemolyticus*, *Listonella anguillarum*, *Aeromonas hydrophila*, *Aspergillus*, *Alteromonas*, *Penicillium*, *Rhizopus*, *Trichoderma*, *Myxobacter*, *Bacillus*, *Enterobacter*, *Klebsiella*, *Pseudomonas*, *Serratia*, *Chromobacterium*, *Clostridium*, *Flavobacterium*, *Arthrobacter*, and *Streptomyces*	Hydrolyzes chitin and chitosan	Anticancer drugs, inflammatory pain medications, antioxidants, treatment of genetic incompetence, wastewater treatment, tissue engineering, and food industry	Kavaz et al. (2010), Kumar et al. (2004), Madhumathi et al. (2010), Wei et al. (2010), Xia et al. (2008) and Beygmoradi and Homaei (2017)

Carrageenase	*Pseudomonas elongata*, *Cytophaga*, *Alteromonas atlantica*, and *Alteromonas carrageenovora*	Degrades carrageenan and carrageenin	Food industry, pharmaceuticals and cosmetics, antiviral medicines, and antibacterial, antitumor, and anticoagulant agent processing	Beygmoradi and Homaei (2017), Khambhaty et al. (2007) and Roberts et al. (2007)
Xylanase	*Streptomyces viridochromogenes*, *Streptomyces* M11, *Trichoderma longibrachiatum*, *Thermotoga maritima*, *Aureobasidium pullulans*, *Glaciecola mesophila*, and *Bacillus* sp.	Catalyzes the breakdown of xylan	Production of natural sweeteners, dough-softening ability, and the paper and pulp industry	Guo et al. (2013), Liu et al. (2014), Subramani and Aalbersberg (2012) and Yin et al. (2010)
Amylase	*Mucor* sp., *Thermotoga celer*, *Thermotoga maritime*, *Arxula adeninivorans*, *Pyrococcus furiosus*, *Lipomyces*, *Alteromonas haloplanktis*, *Saccharomycopsis*, *Schwanniomyces*, *Vibrio* sp., *Streptomyces* sp., *Candida japonica*, and *Filobasidium capsuligenum*	Breaks down starch into simple sugar-like compounds, such as glucose, maltose, and dextrin	Construction industry, alcoholic beverages such as beer, alcohol, animal feed, detergents for clothes, starch industry, confectioneries, and the textile industry	Bohlin et al. (2013), Liu et al. (2012), Manivasagan et al. (2015) and Venugopal (2016)
Collagenase	*Aspergillus oryzae*, *Clostridium histolyticum*, *Streptomyces parvulus*, and *Streptomyces* sp.	Degrades collagen	Food, medicinal products, cosmetics, and successful transplantation of specific organs	Beygmoradi and Homaei (2017), Chung et al. (2004), Kin et al. (2007) and Spök (2006)

(continued)

Table 1 (continued)

Enzyme	Microorganisms producing them	Reactions acted upon	Applications	References
Phytase	*Mucor*, *Rhizopus*, *Aspergillus flavus*, *Aspergillus ochraceus*, *Penicillium olsonii*, *Penicillium digitatum*, *Discostroma tricellulare*, and *Cladosporium gossypiicola*	Breaks down phytic acid to smaller parts	Food, dairy, and the nutrition industry	Chi et al. (2009), Noureddini and Dang (2009), Sadati and Barghi (2014) and Vats et al. (2009)
Tannase	*Phormidium valderianum* BDU140441	Breaks down glucose and tannic acid to produce gallic acid	Antioxidants, as ingredients of animal feed, treatment of fruit juices to reduce the bitterness and coffee-flavoured soft drinks, flavour improvement in grape wine, beer and application in food and pharmaceutical industries	Beena et al. (2010), Beygmoradi and Homaei (2017) and Chávez-González et al. (2012)

breaking down denatured proteins seen in burnt skin. Chondroitinases have been shown to aid in the regeneration of injured spinal cord by eliminating the glial scar and accumulating chondroitin sulfate, which inhibits axon development. Hyaluronidase has been reported to have a comparable hydrolytic action on chondroitin sulfate and may aid in nerve tissue regeneration (Gurung et al., 2013).

6.2 Treatment of Infectious Diseases

Lysozyme is a naturally occurring antibacterial agent that may break down carbohydrate chains in bacterial cell walls and is utilized in a variety of food and consumer products. Lysozyme has also been discovered to have anti-human immunodeficiency virus (HIV) activity, since RNase A and urinary RNase U present selective breakdown of viral RNA (Gurung et al., 2013), suggesting that it could be used to treat HIV infection. Another naturally occurring antibacterial agent is chitinase. Chitin is found in the cell walls of a variety of pathogenic species, including fungi, protozoa, and helminths, and is a good target for antimicrobials. The cell walls of *Streptococcus pneumoniae*, *Bacillus anthracis*, and *Clostridium perfringens* are targeted by a lytic enzyme generated from bacteriophage. Lytic bacteriophages can be utilized to treat a variety of infections and may be effective in the fight against new drug-resistant bacterial strains.

6.3 Treatment of Cancer

Enzyme therapies have been successfully used in cancer studies. The arginine-degrading enzyme (pegylated arginine deaminase) has been shown to suppress human melanoma and hepatocellular carcinomas in a recent research. Another pegylated enzyme, Oncaspar1 (pegaspargase), has demonstrated promising results in the treatment of children newly diagnosed with acute lymphoblastic leukemia and is already in clinical use. Normal cells can produce asparagine, but malignant cells cannot, and, hence, they perish when an asparagine-degrading enzyme is present. Polyethylene glycol (PEG) with asparaginase is a useful addition to regular chemotherapy. Proliferation is another crucial aspect of oncogenesis. It has been proven that chondroitinase AC and, to a lesser extent, chondroitinase B inhibit tumor growth, metastasis, and neovascularization by removing chondroitin sulfate proteoglycans.

Antibody-directed enzyme prodrug therapy (ADEPT) describes how enzymes can be used as therapeutic agents in cancer. An enzyme specific to cancer cells is carried by a monoclonal antibody, which activates a prodrug and kills cancer cells but not normal cells. Cancer therapies based on tumor-targeted enzymes that activate prodrugs are being discovered and developed using this method. This initiative will also make use of the targeted enzyme prodrug therapy (TEPT) platform, which involves enzymes with antibody-like targeting domains (Gurung et al., 2013).

6.4 General Therapeutic Applications of Enzymes

Therapeutic enzymes have a wide variety of specific uses such as oncolytics, thrombolytics, and anticoagulants and as replacements for metabolic deficiencies. Proteolytic enzymes serve as good anti-inflammatory agents. The list of enzymes that have the potential to become important therapeutic agents and their microbial sources are shown Table 2.

A number of limitations reduce the potential utility of microbial enzymes once we enter the medical area, including the enormous molecular size of biological catalysts, which hinders their distribution within somatic cells, and the immune system's response to the foreign enzyme protein after injection (Gurung et al., 2013). Therapeutically beneficial enzymes are required in smaller quantities than are industrial enzymes, but the degree of purity and specificity should be high in general. These enzymes have low K_m and high V_{max} kinetics, allowing them to be optimally effective even at low enzyme and substrate concentrations. The sources of such enzymes should be chosen with care to avoid any unwanted contamination by incompatible material and to allow for easy purification (Gurung et al., 2013).

The majority of therapeutic enzymes are sold as lyophilized pure formulations with biocompatible buffering salts and mannitol diluents. These enzymes are expensive, but their costs are equivalent to those of therapeutic drugs or treatments. Urokinase, for example, is made from human urine and is used to dissolve blood clots. One of the most common uses of therapeutic enzymes is in the treatment of cancer and other disorders. Asparaginase enzymes have shown promise in the treatment of acute lymphocytic leukemia. Their effectiveness is dependent on the

Table 2 List of some common enzymes found in different groups of microorganisms that produced them and their uses (Gurung et al., 2013)

Enzyme	Use	Source
Asparaginase	Leukemia	*Escherichia coli*
Collagenase	Skin ulcers	*Clostridium perfringens*
Glutaminase	Leukemia	*E. coli*
Lysozyme	Antibiotic	*Homo sapiens*
Ribonuclease	RNA hydrolysis	Yeast and bacteriophages
Streptokinase	Blood clots	*Bacillus subtilis*
Trypsin	Protein hydrolysis	*Streptococci* sp.
Uricase	Gout	*Aspergillus flavus*
Urokinase	Blood clots	*Bacillus subtilis*
β-Lactamase	Antibiotic resistance	*Citrobacter freundii*, *Serratia marcescens*, and *Klebsiella pneumonia*

absence of aspartate–ammonia ligase activity in tumor cells, which prevents the synthesis of the nonessential amino acid L-asparaginase.

As a result, they are collected from bodily fluids. Asparaginase has little effect on normal cells that can synthesize enough for their own needs; nevertheless, it lowers the free exogenous concentration, causing a condition of lethal hunger in sensitive cancer cells. The enzyme is administered intravenously and is solely efficient in lowering asparagine levels in the blood. It has a half-life of around a day. Using polyethylene glycol-modified asparaginase, this half-life can be enhanced by 20 times (Gurung et al., 2013).

7 Modern Applications of Enzymes: A Biotechnological Perspective

Enzymes are one of the most significant biomolecules, with a wide range of industrial and medicinal applications, as shown in Table 3. It is now one of the most essential compounds that have been used by humans since the dawn of civilization. Enzymes appear to be one of the most vital molecules that have a tremendous impact on every industry, whether it is dairy, industrial, agricultural, or medicine, with the growing population and rising demands.

Marine microorganisms have been recognized as a potential source of novel enzymes because they are relatively more stable than are the corresponding enzymes derived from plants and animals. Enzymes from marine environments also differ from homologous enzymes in terrestrial microorganisms based on salinity, pressure, temperature, and lighting conditions. Marine microbial enzymes can be used in diverse industrial applications (Gurung et al., 2013).

Biotechnology has the potential to increase the production of goods to suit a variety of human needs. Enzyme technology is a subfield of biotechnology in which new processes have been developed and are still being developed to manufacture both bulk and high value-added products using enzymes as biocatalysts to meet the needs in food (e.g., bread, cheese, beer, and vinegar), fine chemicals (e.g., amino acids, vitamins), agriculture (growth hormones), and pharmaceuticals (insulin) (Gurung et al., 2013).

Enzymes are also utilized to deliver services, such as in washing and environmental operations (particularly cleanup processes) as well as in analytical and diagnostic procedures. The development of new and better products, processes, and services to meet these needs, as well as the improvement of processes to produce the existing products from new raw materials, such as biomass, has been and will continue to be the driving force in the development of enzyme technology, both in academic research and in industry. The purpose of these methods is to create innovative goods and processes that are not only competitive but also sustainable and economically viable (Gurung et al., 2013).

Table 3 A broad-spectrum idea about using the applications of enzymes in different areas (Gurung et al., 2013)

Types of industries	Enzymes	Uses
Alcohol/ beverages	Amylases, glucanases, proteases, amyloglucosidases, pullulanases, and acetolactate decarboxylases	For degradation of starch and biodegradation of polycarbonates into simple sugars In addition, for degrading complex into sugars, thus to increase the fermentation efficiency For production of low-calorie beer
Fruit drinks	Cellulases and pectinases	To clarify fruit juice
Baby food	Trypsins	To predigest baby foods
Food processing	Amylases, proteases, and papains	For degradation of starch and complex proteins and for softening of meat
Diary	Rennins, lipases, and lactases	For hydrolyzing proteins, cheese production, and glucose production from lactose
Detergent	Protease amylases, lipases, cellulases, and mannanases	To remove proteins after staining, remove insoluble starch in dish washing, remove oils and fats, and to increase the effectiveness of detergents
Textile	Amylases, pectinases, cellulases, catalases, and proteases	To remove starch size, glue between the fiber core and waxes, fabric finishing in denims, for degradation of residual hydrogen peroxide after the bleaching of cotton, for wool treatment, and for the degumming of raw silk, also known as biopolishing
Paper and pulp	Amylases, xylanases, cellulases, hemicelluloses, ligninases, and esterases	To degrade starch to lower viscosity, aiding in sizing, deinking, and coating paper. Xylanases reduce bleach required for decolorizing; lipases reduce pitch and lignin-degrading enzymes that remove lignin to soften paper, for esterification
Animal feedstock	Phytases	To increase the total phosphorous content for growth and increase in phytic acid need
Rubber	Catalases	To generate oxygen from peroxide to convert latex into foam rubber
Oil and petroleum	Cellulases, ligninases, and mannanases	For formation of ethanol, forming gel breaker in oil drilling
Biopolymer/ plastic	Laccases, peroxidases, lipases, and transglutaminases	For forming cross-links in biopolymers to produce materials in suit by means of polymerization processes

(continued)

Table 3 (continued)

Types of industries	Enzymes	Uses
Pharmaceutical	Nitrile hydratases, D-amino acid oxidases, glutaric acid acylases, penicillin acylases, penicillin G acylases, ammonia lyases, and humulines	For producing water-soluble intermediates, semisynthetic antibiotics, intermediates for aspartame, and biosynthetic human insulin
Molecular biology	Restriction enzymes, DNA ligases, and polymerases	To manipulate DNA in genetic engineering, essential for restriction of digestion and the polymerase chain reaction, also important in forensic science

As can be seen from Table 3, enzymes and enzyme bioengineering are now widely used in practically every industry. These biomolecules, also known as biocatalysts, are currently playing an increasingly important role in modern industrial development, which is primarily focused on cost-effective, high-efficiency, and environmentally friendly production of various products and by-products. Enzymes are now a major focus of research on a variety of human disorders (Gurung et al., 2013).

Enzymes, like other proteins, are made inside cells by ribosomes, which form chains of amino acids. Although microbes manufacture the bulk of industrial enzymes, the enzymes are formed in the same manner as they are in human cells. The genetic instructions encoded in the deoxyribonucleic acid (DNA) found in the cell's chromosomes dictate the structure and properties of the enzymes generated by that cell. DNA uses a four-base code to facilitate the creation of certain enzymes: adenine (A), guanine (G), cytosine (C), and thymine (T). The double helix of DNA is made up of two complementary strands of these bases that are bound together by hydrogen bonds.

A is always paired with T, whereas C is always paired with G. The sequence of amino acids in the enzyme protein molecule is determined by the order in which these bases are formed in the DNA double helix. Each completely functional section of DNA – or gene – determines the structure of a certain protein, with each of the 20 amino acids having its own set of three bases.

"Enzyme engineering," also known as protein engineering, is a modern term for modifying an enzyme's structure and thus altering/improving its function by modifying the catalytic activity of isolated enzymes to produce new metabolites (Gurung et al., 2013) or to convert from one compound to another (a process known as biotransformation).

References

Al Khudary, R., Venkatachalam, R., Katzer, M., Elleuche, S., & Antranikian, G. (2010). A cold-adapted esterase of a novel marine isolate, *Pseudoalteromonas arctica*: Gene cloning, enzyme purification and characterization. *Extremophiles, 14*(3), 273–285.

Ananthi, S., Ramasubburayan, R., Palavesam, A., & Immanuel, G. (2014). Optimisation and purification of lipase through solid state fermentation by *Bacillus cereus* MSU as isolated from the gut of a marine fish *Sardinella longiceps*. *International Journal of Pharmacy and Pharmaceutical Sciences, 6*(5), 291–298.

Arora, P. K. (2013). *Staphylococcus lipolyticus* sp. nov., a new cold-adapted lipase producing marine species. *Annals of Microbiology, 63*(3), 913–922.

Barzkar, N. (2020). Marine microbial alkaline protease: A recent developments in biofilm an ideal choice for industrial application. *International Journal of Biological Macromolecules, 161*, 1–1618. https://doi.org/10.1016/j.ijbiomac.2020.06.072

Barzkar, N., Sohail, M., Jahromi, S. T., Gozari, M., Poormozaffar, S., Nahavandi, R., & Hafezieh, M. (2021). Marine bacterial esterases: Emerging biocatalysts for industrial applications. *Applied Biochemistry and Biotechnology, 193*, 1187–1214. https://doi.org/10.1007/s12010-020-03483-8

Baweja, M., Singh, P. K., Sadaf, A., Tiwari, R., Nain, L., Khare, S. K., & Shukla, P. (2017). Cost effective characterization process and molecular dynamic simulation of detergent compatible alkaline protease from *Bacillus pumilus* strain MP27. *Process Biochemistry, 58*, 199–203.

Beena, P. S., Soorej, M. B., Elyas, K. K., Sarita, G. B., & Chandrasekaran, M. (2010). Acidophilic tannase from marine *aspergillus awamori* BTMFW032. *Journal of Microbiology and Biotechnology, 20*(10), 1403–1414.

Bessadok, B., Masri, M., Breuck, T., & Sadok, S. (2017). Characterization of the crude alkaline extracellular protease of *Yarrowia lipolytica YlTun15*. *Journal of Fisheries Science.com, 11*(4), 019–024.

Beygmoradi, A., & Homaei, A. (2017). Marine microbes as a valuable resource for brand new industrial biocatalysts. *Biocatalysis and Agricultural Biotechnology, 11*, 131–152. https://doi.org/10.1016/j.bcab.2017.06.013

Bohlin, C., Praestgaard, E., Baumann, M. J., Borch, K., Praestgaard, J., Monrad, R. N., & Westh, P. (2013). A comparative study of hydrolysis and transglycosylation activities of fungal β-glucosidases. *Applied Microbiology and Biotechnology, 97*(1), 159–169.

Brás, N. F., Cerqueira, N. M. F. S. A., Fernandes, P. A., & Ramos, M. J. (2008). Carbohydrate binding modules from family 11: Understanding the binding mode of polysaccharides. *International Journal of Quantum Chemistry, 108*, 2030–2040.

Cao, Y., Deng, D., Sun, A., Zhang, Y., & Hu, Y. (2016). Functional characterization of a novel marine microbial esterase and its utilization in the enantioselective preparation of (R)-methyl 2-chloropropionate. *Applied Biochemistry and Biotechnology, 180*(2), 210–227.

Charoenpanich, J., Suktanarag, S., & Toobbucha, N. (2011). Production of a thermostable lipase by *Aeromonas* sp. EBB-1 isolated from marine sludge in Angsila, Thailand. *Science Asia, 37*, 105–114.

Chávez-González, M., Rodríguez-Durán, L. V., Balagurusamy, N., Prado-Barragán, A., Rodríguez, R., Contreras, J. C., & Aguilar, C. N. (2012). Biotechnological advances and challenges of tannase. *Food and Bioprocess Technology, 5*(2), 445–459.

Cheng, T. H., Ismail, N., Kamaruding, N., Saidin, J., & Danish-Daniel, M. J. B. R. (2020). Industrial enzymes producing marine bacteria from marine resources. *Biotechnology Reports, 27*, e00482.

Chi, Z. M., Ma, C., Wang, P., & Li, H. F. (2007). Optimization of medium and cultivation conditions for alkaline protease production by the marine yeast Aureobasidium pullulans. *Bioresource Technology, 98*, 534–538.

Chi, Z., Chi, Z., Zhang, T., Liu, G., Li, J., & Wang, X. (2009). Production, characterization and gene cloning of the extracellular enzymes from the marine-derived yeasts and their potential applications. *Biotechnology Advances, 27*(3), 236–255.

Chi, W. J., Chang, Y. K., & Hong, S. K. (2012). Agar degradation by microorganisms and agar-degrading enzymes. *Applied Microbiology and Biotechnology, 94*, 917–930.

Chung, L., Dinakarpandian, D., Yoshida, N., Lauer Fields, J. L., & Fields, G. B. (2004). Collagenase unwinds triple-helical collagen prior to peptide bond hydrolysis. *EMBO Journal, 23*, 3020–3030.

Cui, F., Dong, S., Shi, X., Zhao, X., & Zhang, X. H. (2014). Overexpression and characterization of a novel thermostable β-agarase YM01–3, from marine bacterium *Catenovulum agarivorans* YM01T. *Marine Drugs, 12*(5), 2731–2747.

Cui, Y., Fanga, L., Guo, X., Wang, X., Wang, Y., & Li, P. (2018). Responses of soil microbial communities to nutrient limitation in the desert-grassland ecological transition zone. *Science of The Total Environment, 642*, 45–55.

Dadshahi, Z., Homaei, A., Zeinali, F., Sajedi, R. H., & Khajeh, K. (2016). Extraction and purification of a highly thermostable alkaline caseinolytic protease from wastes *Penaeus vannamei* suitable for food and detergent industries. *Food Chemistry, 202*, 110–115.

De Santi, C., Leiros, H. K. S., Di Scala, A., de Pascale, D., Altermark, B., & Willassen, N. P. (2016). Biochemical characterization and structural analysis of a new cold-active and salt-tolerant esterase from the marine bacterium *Thalassospira* sp. *Extremophiles, 20*(3), 323–336.

Do, H., Lee, J. H., Kwon, M. H., Song, H. E., An, J. Y., Eom, S. H., & Kim, H. J. (2013). Purification, characterization and preliminary X-ray diffraction analysis of a cold-active lipase (CpsLip) from the psychrophilic bacterium *Colwellia psychrerythraea* 34H. *Structural Biology and Crystallization Communications, 69*(8), 920–924.

Doi, R. H. (2008). Cellulases of mesophilic microorganisms: Cellulosome and noncellulosome producers. *Annals of the New York Academy of Sciences, 1125*, 267–279.

Dong, H. P., Wang, Y. J., & Zheng, Y. G. (2010). Enantioselective hydrolysis of diethyl 3-hydroxyglutarate to ethyl (S)-3-hydroxyglutarate by immobilized Candida antarctica lipase B. *Journal of Molecular Catalysis B: Enzymatic, 66*(1–2), 90–94.

Duarte, A. W. F., dos Santos, J. A., Vianna, M. V., Vieira, J. M. F., Mallagutti, V. H., & Inforsato, F. J. (2018). Cold-adapted enzymes produced by fungi from terrestrial and marine Antarctic environments. *Critical Reviews in Biotechnology, 38*, 600–619.

El-Hassayeb, H. E. A., & Abdel Aziz, S. M. Z. (2016). Screening, production and industrial application of protease enzyme from marine bacteria. *International Journal of Current Microbiology and Applied Sciences, 5*, 863–874. https://doi.org/10.20546/ijcmas.2016.507.099

Fu, X., Liu, P., Lin, L., Hong, Y., Huang, X., Meng, X., & Liu, Z. (2010). A novel endoglucanase (Cel9P) from a marine bacterium Paenibacillus sp. BME-14. *Applied Biochemistry and Biotechnology, 160*(6), 1627–1636.

Fulzele, R., Desa, E., Yadav, A., Shouche, Y., & Bhadekar, R. (2011). Characterization of novel extracellular protease produced by marine bacterial isolate from the Indian ocean. *Brazilian Journal of Microbiology, 42*, 1364–1373. https://doi.org/10.1590/S1517-83822011000400018

Guo, B., Li, P. Y., Yue, Y. S., Zhao, H. L., Dong, S., & Song, X. Y. (2013). Gene cloning, expression and characterization of a novel xylanase from the marine bacterium, *Glaciecola mesophila* KMM241. *Marine Drugs, 11*, 1173–1187.

Gurung, N., Ray, S., Bose, S., & Rai, V. (2013). A broader view: Microbial enzymes and their relevance in industries, medicine, and beyond. *Biomedical Research International, 2013*, 329121.

Haddar, A., Agrebi, R., Bougatef, A., Hmidet, N., Sellami-Kamoun, A., & Nasri, M. (2009). Two detergent stable alkaline serine-proteases from Bacillus mojavensis A21: Purification, characterization and potential application as a laundry detergent additive. *Bioresource Technology, 100*(13), 3366–3373.

Hao, J., Liu, J., & Sun, M. (2014). Identification of a marine *Bacillus* strain C5 and parathion-methyl degradation characteristics of the extracellular esterase B1. *BioMed Research International, 2014*, 863094.

Homaei, A., Lavajoo, F., & Sariri, R. (2016). Development of marine biotechnology as a resource for novel proteases and their role in modern biotechnology. *International Journal of Biological Macromolecules, 88*, 542–552.

Hosoda, A., Sakai, M., & Kanazawa, S. (2003). Isolation and characterization of agar-degrading *Paenibacillus* spp. associated with the rhizosphere of spinach. *Bioscience, Biotechnology, and Biochemistry, 67*(5), 1048–1055.

Hou, E., Xia, T., Zhang, Z., & Mao, X. (2017). Purification and characterization of an alkaline protease from *micrococcus sp*. isolated from the South China Sea. *Journal of Ocean University of China, 16*(2), 319–325.

Hu, Z., Lin, B. K., Xu, Y., Zhong, M. Q., & Liu, G. M. (2009). Production and purification of agarase from a marine agarolytic bacterium *Agarivorans* sp. HZ105. *Journal of Applied Microbiology, 106*, 181–190.

Huo, Y. Y., Rong, Z., Jian, S. L., Xu, C. D., Li, J., & Xu, X. W. (2017). A novel halotolerant thermoalkaliphilic esterase from marine bacterium *Erythrobacter seohaensis SW-135*. *Frontiers in Microbiology, 8*, 2315.

Karigar, C. S., & Rao, S. S. (2011). Role of microbial enzyme in bioremediation of pollutants: A review. *Enzyme Research, 11*, 1–11.

Karikalan, S., & Mohankumar, A. (2016). Production, isolation and partially purification of alkaline protease from marine macro algae surface associated *Bacillus* spp. *International Journal of Biology Research, 1*(1), 63–66.

Kavaz, D., Odabas, S., Güven, E., Demirbilek, M., & Denkbas, E. (2010). Bleomycin loaded magnetic chitosan nanoparticles as multifunctional nanocarriers. *Journal of Bioactive and Compatible Polymers, 25*(3), 305–318.

Khambhaty, Y., Mody, K., & Jha, B. (2007). Purification and characterization of α-carrageenase from a novel α-Proteobacterium, *Pseudomonas elongata* (MTCC 5261) syn *Microbulbifer elongates* comb. nov. *Biotechnology and Bioprocess Engineering, 12*, 668–675.

Kin, T., Johnson, P. R., Shapiro, A. M., & Lakey, J. R. (2007). Factors influencing the collagenase digestion phase of human islet isolation. *Transplantation, 83*(1), 7–12.

Kumar, M. N. V. R., Muzarelli, R. A. A., Muzarelli, C., Sashiwa, H., & Domb, A. J. (2004). Chitosan chemistry and pharmaceutical perspectives. *Chemical Reviews, 104*(12), 6017–6084.

Le Roes-Hill, M., & Prins, A. (2016). *Biotechnological potential of oxidative enzymes from Actinobacteria*. Intech Open. https://doi.org/10.5772/61321

Lee, Y., Oh, C., De Zoysa, M., Kim, H., Wickramaarachchi, W. D., Whang, I., & Lee, J. (2013). Molecular cloning, overexpression and enzymatic characterization of glycosyl hydrolase family 16 β-agarase from marine bacterium *Saccharophagus* sp. AG21 in *Escherichia coli*. *Journal of Microbiology and Biotechnology, 23*(7), 913–922.

Lemak, S., Tchigvintsev, A., Petit, P., Flick, R., Singer, A. U., Brown, G., Evdokimova, E., Egorova, O., Gonzalez, C. F., & Chernikova, T. N. (2012). Structure and activity of the cold-active and anion-activated carboxyl esterase OLEI01171 from the oil-degrading marine bacterium *Oleispira Antarctica*. *Biochemical Journal, 445*(2), 193–203.

Liu, Y., Lei, Y., Zhang, Z., Gao, Y., Xiao, Y., & Peng, H. (2012). Identification and phylogenetic characterization of a new subfamily of α-amylase enzymes from marine microorganisms. *Marine Biotechnology, 14*(3), 253–260.

Liu, X., Huang, Z., Zhang, X., Shao, Z., & Liu, Z. (2014). Cloning, expression and characterization of a novel cold-active and halophilic xylanase from *Zunongwangia profunda*. *Extremophiles, 18*(2), 441–450.

Maddela, N. R., & García, L. C. (2021). *Innovations in biotechnology for a sustainable future* (p. VIII, 502). Springer Nature Switzerland AG. https://doi.org/10.1007/978-3-030-80108-3. ISBN: 978-3–030-80107-6. Retrieved from https://www.springer.com/gp/book/9783030801076

Maddela, N. R., Garcia, L. C., & Chakraborty, S. (2021). *Advances in the domain of environmental biotechnology* (p. XVIII, 717). Springer Nature Singapore. https://doi.org/10.1007/978-981-15-8999-7. ISBN: 978-981-15-8999-7. Retrieved from https://www.springer.com/gp/book/9789811589980.

Madhumathi, K., Sudheesh Kumar, P. T., Abilash, S., Sreeja, V., Tamura, H., & Manzoor, K. (2010). Development of novel chitin/nanosilver composite scaffolds for wound dressing applications. *Journal of Materials Science: Materials in Medicine, 21*(2), 807–813.

Maki, M., Leung, K. T., & Qin, W. (2009). The prospects of cellulase-producing bacteria for the bioconversion of lignocellulosic biomass. *International Journal of Biological Sciences, 5*, 500–516.

Malard, L. A., & Pearce, D. A. (2018). Microbial diversity and biogeography in Arctic soils. *Environmental Microbiology Reports, 10*, 611–625.

Manivasagan, P., Venkatesan, J., Kang, K. H., Sivakumar, K., Park, S. J., & Kim, S. K. (2015). Production of α-amylase for the biosynthesis of gold nanoparticles using *Streptomyces sp.* MBRC-82. *International Journal of Biological Macromolecules, 72*, 71–78.

Maruthiah, T., Immanuel, G., & Palavesam, A. (2016a). Purification and characterization of halophilic organic solvent tolerant protease from marine *bacillus sp.* APCMST-RS7 and its antioxidant potentials. *Proceedings of the National Academy of Sciences, India Section B: Biological Sciences, 87*(1), 207–216.

Maruthiah, T., Somanath, B., Jasmin, J. V., Immanuel, G., & Palavesam, A. (2016b). Production, purification and characterization of halophilic organic solvent tolerant protease from marine crustacean shell wastes and its efficacy on deproteinization. *3 Biotech, 6*(2), 157.

Minegishi, H., Shimane, Y., Echigo, A., Ohta, Y., Hatada, Y., & Kamekura, M. (2013). Thermophilic and halophilic β-agarase from a halophilic archae on *Halococcus* sp. 197A. *Extremophiles, 17*, 931–939.

Nguyen, T. H., & Nguyen, V. D. (2017). Characterization and applications of marine microbial enzymes in biotechnology and probiotics for animal health. *Advances in Food and Nutrition Research, 80*, 37–74.

Noriler, S. A., Savi, D. C., Aluizio, R., Palácio-Cortes, A. M., Possiede, Y. M., & Glienke, C. (2018). Bioprospecting and structure of fungal endophyte communities found in the Brazilian biomes, Pantanal, and Cerrado. *Frontiers in Microbiology, 9*, 1526.

Noureddini, H., & Dang, J. (2009). Degradation of phytates in distillers' grains and corn gluten feed by *aspergillus Niger* phytase. *Applied Biochemistry and Biotechnology, 159*(1), 11–23.

Ohta, Y., & Hatada, Y. (2006). A novel enzyme, lambda-carrageenase, isolated from a deep-sea bacterium. *Journal of Biochemistry, 140*(4), 475–481.

Pajares, S., & Bohannan, B. J. M. (2016). Ecology of nitrogen fixing, nitrifying, and denitrifying microorganisms in tropical forest soils. *Frontiers in Microbiology, 7*, 1045.

Pandey, D., Singh, R., & Chand, D. (2011). An improved bioprocess for synthesis of acetohydroxamic acids using DTT (dithiothreitol) treated resting cells of *Bacillus* sp. *APB-6*. *Bioresource Technology, 102*(11), 6579–6586.

Park, H. J., Han, S. J., Yim, J. H., & Kim, D. (2018). Characterization of an Antarctic alkaline protease, a cold-active enzyme for laundry detergents. *Korean Journal of Microbiology, 54*(1), 60–68.

Ramani, K., Saranya, P., Jain, S. C., & Sekaran, G. (2013). Lipase from marine strain using cooked sunflower oil waste: Production optimization and application for hydrolysis and thermodynamic studies. *Bioprocess and Biosystems Engineering, 36*(3), 301–315.

Renge, V. C., Khedkar, S. V., & Nandurkar, N. R. (2012). Enzyme synthesis by fermentation method. *Scientific Reviews & Chemical Communications, 2*, 585–590.

Roberts, J. N., Christopher, B. B., Cynthia, D. T., Rhonda, K., Marcelino, B., Peter, L. C., Douglas, R. L., & John, T. (2007). Genital transmission of HPV in a mouse model is potentiated by nonoxynol-9 and inhibited by carrageenan. *Nature Medicine, 13*, 857–861.

Rubilar, O., Diez, M. C., & Gianfreda, L. (2008). Transformation of chlorinated phenolic compounds by white rot fungi. *Critical Reviews in Environmental Science and Technology, 38*, 227–268.

Sadati, R., & Barghi, A. (2014). Urmia hypersaline lake: A potential source of actinomycets possessing. *Iranian Journal of Public Health, 43*(2), 149.

Sana, B. (2013). Marine microbial enzymes: Biotechnological and biomedical aspects. In S. K. Kim (Ed.), *Marine microbiology: Bioactive compounds and biotechnological applications*. Wiley-VCH Verlag GmbH & KGaA.

Sarmah, N., Revathi, D., Sheelu, G., Rani, K. Y., Sridhar, S., Mehtab, V., & Sumana, C. (2018). Recent advances on sources and industrial applications of lipases. *Biotechnology Progress, 34*, 5–28.

Shanmughapriya, S., Kiran, G. S., Selvin, J., Thomas, T. A., & Rani, C. (2010). Optimization, purification, and characterization of extracellular mesophilic alkaline cellulase from sponge-associated Marinobacter sp. MSI032. *Applied Biochemistry and Biotechnology, 162*(3), 625–640.

Sivaperumal, P., Kamala, K., & Rajaram, R. (2017). Waste through enzyme producing marine microorganisms. *Advances in Food and Nutrition Research, 80*, 165–179. https://doi.org/10.1016/bs.afnr.2016.10.006

Spök, A. (2006). Safety regulations of food enzymes. *Food Technology and Biotechnology, 44*, 197–209.

Subramani, R., & Aalbersberg, W. (2012). Marine actinomycetes: An ongoing source of novel bioactive metabolites. *Microbiological Research, 167*(10), 571–580.

Vats, P., Bhushan, B., & Banerjee, U. C. (2009). Studies on the dephosphorylation of phytic acid in livestock feed using phytase from *aspergillus Niger* van Teighem. *Bioresource Technology, 100*(1), 287–291.

Venugopal, V. (2016). Enzymes from seafood processing waste and their applications in seafood processing. *Advances in Food and Nutrition Research, 78*, 47–69.

Vijayaraghavan, R., & Rajendran, S. (2012). Identification of a novel agarolytic gamma -Proteobacterium *Microbulbifer maritimus* and characterization of its agarase. *Journal of Basic Microbiology, 52*(6), 705–712.

Wang, Y., Zhang, Y., Sun, A., & Hu, Y. (2016). Characterization of a novel marine microbial esterase and its use to make D-methyl lactate. *Chinese Journal of Catalysis, 37*(8), 1396–1402.

Wei, X. H., Niu, Y. P., Xu, Y. Y., Du, Y. Z., Hu, F. Q., & Yuan, H. (2010). Salicylic acid-grafted chitosan oligosaccharide nanoparticle for paclitaxel delivery. *Journal of Bioactive and Compatible Polymers, 25*(3), 319–335.

Wu, S., Liu, G., Zhang, D., Li, C., & Sun, C. (2015). Purification and biochemical characterization of an alkaline protease from marine bacteria *Pseudoalteromonas sp. 129-1. Journal of Basic Microbiology, 55*(12), 1427–1434.

Xia, W., Liu, P., & Liu, J. (2008). Advance in chitosan hydrolysis by non-specific cellulases. *Bioresource Technology, 15*, 6751–6762.

Yepuru, S. K. (2018). Production of alkaline protease from *aspergillus oryzae* isolated from seashore of bay of Bengal. *Journal of Applied and Natural Science, 10*(4), 1210–1215.

Yin, L. J., Lin, H. H., Chiang, Y. I., & Jiang, S. T. (2010). Bioproperties and purification of xylanase from bacillus sp. YJ6. *Journal of Agricultural and Food Chemistry, 58*(1), 557–562.

Zhang, C., & Kim, S. K. (2010). Research and application of marine microbial enzymes: Status and prospects. *Marine Drugs, 8*, 1920–1934. https://doi.org/10.3390/md8061920

Zhao, K., Guo, L. Z., & Lu, W. D. (2012). Extracellular production of novel halotolerant, thermostable, and alkali-stable carboxymethyl cellulase by marine bacterium Marinimicrobium sp. LS-A18. *Applied Biochemistry and Biotechnology, 168*(3), 550–567.

Zhu, D., Wu, Q., & Hua, L. (2019). Industrial enzymes. In M. Moo-Young (Ed.), *Comprehensive biotechnology* (3rd ed., pp. 1–13). Pergamon.

Hydrolytic Enzyme-Producing Bacteria from Algerian Hot Springs: Attractive Industrial Molecules

Mounia Arab, Hafida Baoune, and Idris Hannous

Abstract Indigenous thermophilic microorganisms in hot environments, such as terrestrial hot springs, have shown interesting adaptive capacities and thus constitute an important source of unconventional bioactive molecules of which enzymes are a part. Of these microorganisms, species belonging to the genus *Bacillus* have been well known as among the best producers of enzymes. Thermostable enzymes, also called thermozymes, isolated from thermophilic microorganisms, have a biotechnological potential and industrial significance due to their inherent stability under harsh industrial conditions, in addition to their performance at high temperatures. Hydrolases, especially thermostable hydrolases, are considered as compounds of great commercial importance in various industrial applications in medical, agricultural, and environmental processes. Algeria has more than 282 hydrothermal springs that can be important sources of novel microorganisms, genes, and molecules, which might be used in a large number of applications in various fields. Although there are intensive studies on the isolation of enzyme-producing bacteria from terrestrial hot springs around the world, research studies on thermophilic enzymes produced by the *Bacillus* species isolated from Algerian hot springs remain relatively rare. The results of the first investigations carried out in Debagh Hot Springs, the hottest hydrothermal source in Algeria, showed an extremely promising potential of enzyme production. This chapter considers the current advances in this topic, emphasizing on the thermozymes produced by the genus *Bacillus*, particularly isolated from Algerian hot springs, and their potential applications in different fields.

M. Arab (✉)
Laboratory of Applied Microbiology in Food, Biomedical and Environment, Department of Biology, Faculty of Nature Sciences and Life, University of Tlemcen, Tlemcen, Algeria

Faculty of Biological Sciences, University of Sciences and Technology Houari Boumediene, Algiers, Algeria

H. Baoune
Faculty of Natural and Life Sciences, Université Kasdi Merbah Ouragla, Ouragla, Algeria

I. Hannous
Service de diabétologie, Hôpital Central de l'Armée, Alger, Algeria

N. R. Maddela et al. (eds.), *Ecological Interplays in Microbial Enzymology*, Environmental and Microbial Biotechnology,
https://doi.org/10.1007/978-981-19-0155-3_6

1 Introduction

Many thermophilic and hyperthermophilic microorganisms and aerobic and anaerobic organisms have been isolated from various terrestrial environments, often extreme, such as hydrothermal springs. These microorganisms present great opportunities for fundamental research to know about their survival mechanisms under extreme conditions and to highlight their uses in many industrial applications, notably by thermostable enzymes.

The enzymes produced by these microorganisms are extremely thermoactive and rigorously thermostable, having even activities at temperatures exceeding the maximum of microorganisms' growth temperatures. They are often resistant to chemical denaturants such as detergents, organic solvents, and extreme pH values (Antranikian et al., 2005).

These enzymes are of great importance in today's biotechnology with a variety of applications ranging from food and fermentation to textile and paper industries (Maddela & García, 2021; Maddela et al., 2021). The development of microbiology has led to a better understanding of the enzyme synthesis pathways in harsh environments. The industrial production of enzymes has been oriented toward production processes, the main advantages of which are, in particular, independent production of seasonal and geographical constraints; the possibility of using inexpensive raw materials; and important production yields as well as improving microbial strain production through genetic engineering and optimizing fermentation conditions (Kambourova, 2018).

In addition, thermozymes usually exhibit better resistance to environmental factors such as pH, salinity, and high temperatures. These abiotic changes make thermozymes interesting products for the industry where it is common that conditions are not optimal for the use of mesophilic enzymes. A remarkable industrial increase of enzyme production has been observed for several years. Hydrolases were among the classes of enzymes that have been investigated in various sectors of the industry (Sharma et al., 2019).

Bacteria belonging to the *Bacillus* genus are by far the most commonly used bacteria for enzyme production because they are considered as good hosts for the industrial sector (Emanuel & Lorrence, 2009). The metabolism and physiology of bacteria belonging to the genus *Bacillus* is extremely well studied. Some species have been given the generally recognized as safe (GRAS) status. Those bacteria have good protein export systems and can produce certain extracellular enzymes such as amylases, xylanases, and proteases (Schallmey et al., 2004).

In this chapter, we will describe thermostable microbial hydrolase enzymes, particularly produced by the genus *Bacillus*, having an industrial interest, and especially those isolated from Algerian hot springs.

2 Thermozymes

Thermozymes have been named in this manner because of their optimum activity and stability above 60 °C. Enzymes with a maximum activity of above 80 °C are considered hyperthermophilic enzymes (Li et al., 2005). The production of thermostable enzymes has increased due to advances in the isolation of a large number of thermophilic microorganisms from different ecological niches (Bhalla et al., 2014). Although thermozymes have the same mechanisms of action as mesophilic enzymes, they are slightly different in terms of their structure, which leads to their improved stability (Zeikus et al., 1998). Various mechanisms are used, including the addition of salt bridges, hydrophobic interactions, and hydrogen links. The use of different amino acids, such as isoleucines, alanines, and prolines in greater quantities in hydrophobic centers, ensures a better organizational efficiency of the protein (Bruins et al., 2001). Better stability of thermozymes for a greater variety of pH and salinity was also observed. The major advantage of using thermozymes in industrial biotechnology is the possibility of retaining the desired enzymatic activity under thermophilic conditions, while preventing the growth of pathogenic bacteria in the reaction medium when the temperature is above 70 °C, by reducing the viscosity of the fluids. Thermophilic conditions may also allow the destruction of pathogens, and further increase the bioavailability and solubility of organic compounds, and increase the joint diffusion of substrates and higher reaction rates (Kumar & Nussinov, 2001).

Although thermophilic organisms are expected to produce more thermozymes than their mesophilic homologues, it has been shown that the thermophilic characteristic of the microorganisms is not the unique requisite condition for the production of thermostable enzymes (Ibrahim et al., 2014). Leite et al. (2007) studied the β-glucosidase of the mesophilic yeast *Aureobasidium pullulans* and the thermophilic fungus *Thermoascus aurantiacus* and concluded that the enzyme produced by the mesophilic strain was more thermostable than the one produced by the thermophilic fungus and that proteins showed a higher level of glycosylation. Extracellular enzymes in filamentous fungi usually have a high level of glycosylation, and this has been considered an associated thermostability factor (Martins et al., 2013). Despite the thermostability approved for enzymes produced by some mesophiles, thermophilic microorganisms remain the most natural and promising source for thermostable enzyme production, particularly for the many benefits, now known, which they confer on industrialists.

3 The Genus *Bacillus*: Generalities

The traditional genus *Bacillus* represents one of the most diverse genera in the class of Bacilli. The members of this genus show a wide range in DNA base compositions and major amino acid compositions in the cell walls (Fahmy et al., 1985; Priest,

1993). It includes aerobic Gram-positive bacteria and aero-anaerobic bacteria, in the form of rods, capable of forming spores (Prescott et al., 2002). Recently, analysis of the 16S rRNA gene sequences has revealed a high level of phylogenetic heterogeneity in this genus. The common important physiological characteristics for their survival include the production of a multilayer cell wall structure, the formation of stress-resistant endospores and antibiotic peptide secretion, peptide molecules, and extracellular enzymes (Gardener, 2004). The genus *Bacillus* is particularly heterogeneous, and this is reflected by the wide variety of ecological niches, which many species occupy, and by the extreme diversity of their taxonomic status.

The genus *Bacillus* is currently the largest genus of the Bacillaceae family, composed of at least 226 species (September, 2014). New strains are constantly being added as new species but are also reclassified into new genera. For example, in the last decade, 10 existing species were transferred to other genera and 39 new species were added to the genus only in 2014. The inferred phylogeny of the *Bacillus* species is often based on the 16S rRNA gene sequencing, but this still cannot distinguish between species. Therefore, the use of DNA–DNA hybridization or gene sequencing is recommended for better classification. This latter approach is even more important when studying the strains of *Bacillus* at the subspecies level (Wang et al., 2007; Stefanic et al., 2012). In general, the different species of the *Bacillus* genus show a small divergence of their 16 s rRNA genes, and this divergence is poorly correlated with their phenotypic characteristics. The members of the *Bacillus* genus are used for the synthesis of an extremely wide range of products for medical, agricultural, and pharmaceutical purposes and other interests.

4 Diversity of the *Bacillus* Species in Algerian Hot Springs

Hot springs are the hot spots for unusual life forms. In Algeria, more than 282 hot springs have been inventoried by the "Agence Nationale des Ressources Hydrauliques" (ANRH), concentrated in the northeastern part of the country, with temperatures ranging from 20 °C to 98 °C. These hot springs are mainly used in balneotherapy (Ouali et al., 2018).

Debagh Hot Springs is the most flourishing hot spring in Algeria, and its water is considered as the warmest. This source contains nine griffons with water temperatures between 90 and 98 °C (Fig. 1) (Boughlali, 2003). Algerian hot springs have been extensively studied in terms of their geological properties. However, few studies have conducted microbiological analysis of these hot springs (Kecha et al., 2007; Bouanane-Darenfed et al., 2011; Amarouche-Yala et al., 2015), and only two studies have shed light on their microbial diversity (Arab et al., 2018; Gomri et al., 2018a).

Our laboratory group investigated the diversity of aerobic bacilli from the Debagh source, using two approaches, cultural and molecular (Arab et al., 2018). In this work, and based on phenotypic characterization tests and genotypic identification, 41 aerobic, thermophilic, and halotolerant bacterial strains were isolated.

Fig. 1 Close-up photography of Debagh Hot Springs. (https://www.vitaminedz.com/fr/Guelma/Hammam-meskhoutine/28833/Articles/1.html. Consulted 16/01/2020 at 9:30 p.m.)

Sequencing of the 16S rRNA revealed that the recovered isolates belonged to two branches: Firmicutes, with 3 bacterial genera dominated by the genus *Bacillus*, represented by the species *Bacillus mojavensis* (16), *Bacillus licheniformis* (11), *Bacillus subtilis* (2), *Bacillus atrophaeus* (1), *Bacillus amyloliquefaciens* (1), and *Bacillus pumilus* (1); by the genus *Aeribacillus*, represented by the species *Aeribacillus pallidus* (3); and by the genus *Geobacillus*, represented by the species *Geobacillus toebii* (2). The Proteobacteria branch included the genus *Hydrogenophilus*, represented by the species *Hydrogenophilus hirschii* (4). The matrix-assisted laser desorption ionization time-of-flight (MALDI-TOF) mass spectroscopy analysis determined that the isolates belonged to the genus *Bacillus*, grouping the species *B. licheniformis* (12), *B. mojavensis* (6), *B. subtilis* (2), *B. atrophaeus* (1), and *B. pumilus* (1). Gomri et al. (2018b) found similar results regarding the diversity of this same hydrothermal source, by thermophilic bacilli. In all, 4 bacterial genera were found, with the predominance of the genus *Bacillus* with 10 strains belonging to the species *Bacillus paralicheniformis* and *B. licheniformis*, followed by the genus *Anoxybacillus* with the species *Anoxybacillus gonensis*, *Anoxybacillus flavithermus*, and *Anoxybacillus thermarum*, the genus *Geobacillus* with the species *Geobacillus thermoleovorans*, and the genus *Brevibacillus* with the species *Brevibacillus thermoruber*. *Bacillus* bacteria are well adapted to warm environments (Kawasaki et al., 2011). They also generally have simple nutritional needs, which could explain their almost exclusive presence in an environment such as thermal springs. Thus, specific nutrients for their growth are not required (Khiyami et al., 2012).

5 Hydrolytic Enzymes Produced by Thermophilic *Bacillus* sp.

The choice of an appropriate taxonomic group for enzyme use is determined both by the benefits of the respective group and by the industrially preferred hosts for enzyme expression. For example, thermophilic bacilli are the main industrial bacterial producers of thermostable extracellular enzymes (proteases, lipases, amylases, pullulanases, xylanses, etc.) (Elleuche et al., 2015). Strains of the genus *Bacillus* are well known to produce high-value enzymes (Meintanis et al., 2008). This is due to their secretome, which codes many enzymes that hydrolyze proteins, lipids, polysaccharides, and other nutrients (Berrada et al., 2012). From a wide range of amylase-producing microorganisms, bacterial strains are most appreciated because of their rapid growth rate, easy cultivation, and performance in the production of recombinant enzymes. The species of the genus *Bacillus*, including *B. amyloliquefaciens*, *B. licheniformis*, *Bacillus cereus*, *Bacillus coagulans*, *Bacillus subtilis*, *Bacillus polymyxa*, *Bacillus stearothermophilus*, *Bacillus mesentericus*, *Bacillus vulgaris*, *Bacillus megaterium*, *Bacillus halodurans*, etc., are described as powerful candidates of amylolytic enzyme production (Singh et al., 2019). Several enzymes have been reported in the literature to be secreted by different species of the genus *Bacillus*, isolated from hydrothermal springs: amylases by *Bacillus* sp. HUTBS71 (Al-Quadan et al., 2009; Fooladi & Sajjadian, 2010), proteases by *Bacillus* sp. MLA64 and Htjtbs71 (Akel et al., 2009; Lagzian & Asoodeha, 2012), cellulases by *Bacillus* sp. CH43 and HR68 (Mawadza et al., 2000), chitinases by *Bacillus* sp. 13.26 (Yuli et al., 2004), and lipases by *Bacillus thermoleovorans* CCR1 1 (Castro-Ochoa et al., 2005). As for the works on the enzymatic abilities of *Bacillus* sp. and related genera in Algerian hot springs, they remain relatively rare. In our study on aerobic bacilli isolated from Debagh Hot Springs (Arab et al., 2018), and out of five enzyme activities tested, the isolated strains demonstrated strong hydrolytic abilities coupled with their thermo-halotolerance/alkalitolerance, highly appreciated in the industry, especially in lipolytic activities where 92.68% of the strains hydrolyzed olive oil (Arab et al., 2018). Microorganisms are the most widely used source of lipases in biotechnology applications. Among bacteria, the species of the genus *Bacillus* are the best potential candidates for the production of lipases. In the *Bacillus* species, the most common lipase producers are *B. licheniformis*, *Bacillus alcalophilus*, *B. coagulans*, *B. subtilis*, *B. pumilus*, and *B. stearothermophilus* (Singh et al., 2019). Lipases are also known to be produced by other aerobic bacterial species such as the genus *Geobacillus* in which *G. thermoleovorans* exhibits extracellular lipolytic activities with high growth rates on substrates such as olive oil, soybean oil, tributyrin, and Tween 20, 40, and 80 (Lee et al., 1999). In addition, the isolated strains of the Debagh hydrothermal source had at least two extracellular enzymes of the five tested, as well as gelatinase, and 26.82% of these strains had all the extracellular enzymes tested. The hydrolytic activity on agar media is shown in Fig. 2.

With respect to our results, in their work regarding Debagh Hot Springs, Gomri et al. (2018b) reported that 21 isolated *Bacillus* exhibited a positive result for at least 1 of the 7 tested extracellular hydrolytic activities. On the other hand, amylases and proteases were the most commonly active produced enzymes, whereas lipolytic and cellulolytic activities were the least expressed enzyme activities. Furthermore, in the same hot spring, Gomri et al. (2018a) characterized an acid extracellular protease produced by the strain *Brevibacillus thermoruber* OA30. Protease 32-F38 had an optimal activity at 50 °C and high heat stability for 240 min. Its optimal pH was 6.0. This protease was highly stable in the presence of various detergents and solvents, as

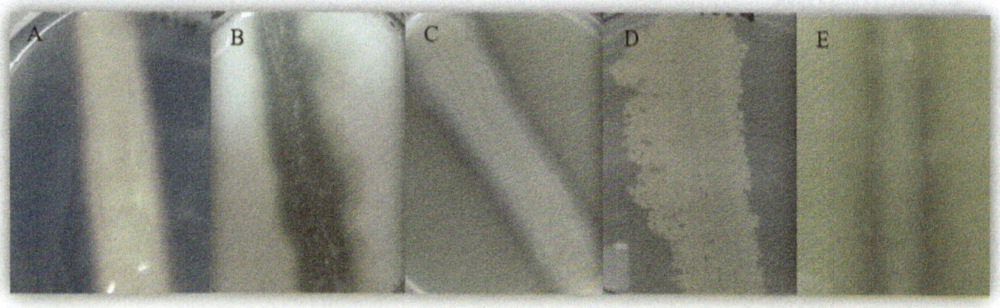

Fig. 2 Production of extracellular enzymes on different media (Arab et al., 2018. Data not shown) (**a**) Production of amylases on starch agar after revelation with lugol solution. (**b**) Production of proteinases on milk agar. (**c**) Production of lecithinases on egg yolk agar. (**d**) Production of lipases on Tween 80 agar. (**e**) Production of lipases on olive oil agar, with appearance of precipitate

it was inhibited by metalloprotease inhibitors. The environment offered to the microorganisms in Debagh Hot Springs remains nutrient-poor, clean, and away from human activity. However, the isolates appeared to have developed a genetic and physiological ability to use the available organic matter. This could contribute not only to the cycling of nutrients in nature (Rey et al., 2004) but also to the tendency of microbial societies to develop an adaptation to nutritional stress in order to survive with any available food (Derekova et al., 2007). Enzymatic activity is influenced, among other activities/factors, by the availability of nutrients in the direct environment of microorganisms and is a part of their adaptation to their natural environment (Cohen, 2011).

6 Some Applications of *Bacillus* Thermozymes

From a commercial point of view, the enzymes of thermophilic microorganisms, especially *Bacillus*, have had the greatest impact. They fall into several industrial sectors such as food, detergents, cosmetics, pulp and paper, pharmaceuticals, leather, biodiesel production, etc. (Sharma et al., 2019). For example, Yilmaz et al. (2016) purified an alkaline protease from the thermophilic strain *Bacillus licheniformis* A10 with an optimal activity at 80 °C and a pH of 9.0. Furthermore, the enzyme exhibited stability after 1 hour of incubation, which may make it a good candidate for the detergent industry. In fact, commercial lipases are mainly used for flavor development in the food processing industry, such as dairy products, meat, fruit, vegetables, etc. Bhosale et al. (2016) purified a hyperthermophilic alkaline lipase from *Bacillus sonorensis* 4R, and the enzyme showed an optimal activity at 80 °C. In this sense, it might be used in various industrial processes, including as additives in detergents, in the food industry, and in bioremediation.

7 Conclusions

The data reported in this chapter revealed that Algerian hot springs host a great diversity of thermophilic bacteria, not well explored yet, which could be promising sources of new thermostable enzymes.

The thermostable enzymes described in the few research studies have demonstrated high thermostability and activities in wide pH ranges, which are highly important properties for several industrial processes. Additionally, heat-stable enzymes are already in use in a number of industrial processes such as in the production and processing of foodstuffs, in production in the paper or textile industry, and in the production of biofuels; however, a great potential for other applications remains to be explored. In addition, it is necessary to develop laboratory technologies for the synthesis of these enzymes in optimized environments, thus allowing the transition to industrial-scale processes.

References

Akel, H., Al-Quadan, F., & Yousef, T. (2009). Characterization of a purified thermostable protease from hyperthermophilic *Bacillus* strain HUTBS7I. *European Journal of Scientific Research, 31*(2), 280–288.

Al-Quadan, F., Akel, H., & Natshi, R. (2009). Characteristics of a novel highly thermostable and extremely thermophilic aikalitolerant amylase from hyperthermophilic *Bacillus* strain 1-IUTBS7 1. *OnLine Journal of Biological Sciences, 9*(3), 67–74. https://doi.org/10.3844/ojbsci.2009.67.74

Amarouche-Yala, S., Benouadah, A., Bentabet, A. O., Moulla, A. S., Ouarezki, S. A., & Azbouche, A. (2015). Physicochemical, bacteriological, and radiochemical characterization of some Algerian thermal spring waters. *Water Quality Exposure and Health, 7*(2), 233–249. https://doi.org/10.1007/s12403-014-0130-x

Antranikian, G., Vorgias, G. C., & Bertoldo, C. (2005). Extreme environments as a resource for microorganisms and novel biocatalysts. *Advances in Biochemical Engineering and Biotechnology, 96*, 219–262. https://doi.org/10.1007/b135786

Arab, M., Bakour, S., Lalaoui, R., Aissaoui, N., Nas, F., Fournier, P. E., & Klouche-Khelil, N. (2018). Diversity of aerobic bacilli analysis using molecular and culture-based approaches in Debagh hot spring. *Geomicrobiology Journal, 36*(2), 137–147. https://doi.org/10.1080/01490451.2018.1520937

Berrada, I., Willems, A., De Vos, P., El Fahime, E., Swings, J., Bendaou, N., Melloul, M., & Amar, M. (2012). Diversity of culturable moderately halophilic and halotolerant bacteria in a marsh and two salterns a protected ecosystem of lower Loukkos (Morocco). *African Journal of Microbiology Research, 6*(10), 2419–2434. https://doi.org/10.5897/AJMR11.1490

Bhalla, A., Bansal, N., Kumar, S., Bischoff, K. M., & Sani, R. K. (2014). Bioresource technology improved lignocellulose conversion to biofuels with thermophilic bacteria and thermostable enzymes. *Bioresource Technology, 128*, 751–759. https://doi.org/10.1016/j.biortech.2012.10.145

Bhosale, H., Shaheen, U., & Kadam, T. (2016). Characterization of a hyperthermostable alkaline lipase from *Bacillus sonorensis* 4R. *Enzyme Research, 2016*, 4170684. https://doi.org/10.1155/2016/4170684

Bouanane-Darenfed, A., Fardeau, M. L., Grégoire, P., Joseph, M., Kebbouche-Gana, S., Benayad, T., Hacene, H., Cayol, J. L., & Ollivier, B. (2011). *Caldicoprobacter algeriensis* sp. nov. a new thermophilic anaerobic, xylanolytic bacterium isolated from an Algerian hot spring. *Current Microbiology, 62*(3), 826–832. https://doi.org/10.1007/s00284-010-9789-9

Boughlali, M. (2003). Thermalisme et thalassothérapie en Algérie. *La Presse Thermale et Climatique, 140*, 161–165.

Bruins, M. E., Janssen, A. E. M., & Boom, R. M. (2001). Thermozymes and their applications. In *Applied Biochemistry and Biotechnology* (pp. 90–155). https://doi.org/10.1385/ABAB:90:2:155

Castro-Ochoa, L., Rodriguez-Gomez, C., Valerio-Alfaro, G., & Ros, R. (2005). Screening, purification and characterization of the thermoalkalophilic lipase produced by *Bacillus thermoleovorans CCR1 1. Enzyme and Microbial Technology, 37*, 648–654. https://doi.org/10.1016/j.enzmictec.2005.06.003

Cohen, G. N. (2011). *Microbial biochemistry* (2nd ed., p. 545). Springer. https://doi.org/10.1007/978-94-017-8908-0

Derekova, A., Sjoholm, C., Mandeva, R., & Kambourova, M. (2007). *Anoxybacillus rupiensis* sp. Nov., a novel thermophilic bacterium isolated from Rupi basin (Bulgaria). *Extremophiles, 11*, 577–583. https://doi.org/10.1007/s00792-007-0071-4

Elleuche, S., Schäfers, C., Blank, S., Schröder, C., Katsifas, E., & Antranikian, G. (2015). Exploration of extremophiles for high temperature biotechnological processes. *Current Opinion in Microbiology, 25*, 113–119. https://doi.org/10.1016/j.mib.2015.05.011

Emanuel, G., & Lorrence, H. G. (2009). *Practical handbook of microbiology* (Vol. 13). CRC. https://doi.org/10.1201/9781420009330

Fahmy, F. J., Flossdorf, D., & Claus, R. (1985). The DNA base composition of the type strains of the genus *bacillus*. *Systematic and Applied Microbiology, 6*, 60–65. https://doi.org/10.1016/S0723-2020(85)80012-6

Fooladi, J., & Sajjadian, A. (2010). Screening the thermophilic and hyperthermophilic bacterial population of three Iranian hot-springs to detect the thermostable α-amylase producing strain. *Iranian Journal of Microbiology, 2*(1), 49–53. PMCID: PMC3279768.

Gardener, B. B. M. (2004). Ecology of *Bacillus* and *Paenibacillus* sp. in agricultural systems. *Phytopathology, 94*, 1252–1258. https://doi.org/10.1094/PHYTO.2004.94.11.1252

Gomri, M., Rico-Díaz, A., Escuder-Rodríguez, J., El Moulouk, K. T., González-Siso, M., & Kharroub, K. (2018a). Production and characterization of an extracellular acid protease from thermophilic *Brevibacillus* sp. OA30 isolated from an Algerian hot spring. *Microorganisms, 6*, 31. https://doi.org/10.3390/microorganisms6020031

Gomri, M., El Moulouk, K. T., & Kharroub, K. (2018b). Analysis of the diversity of aerobic, thermophilic endospore-forming bacteria in two Algerian hot springs using cultural and non-cultural methods. *Annales de Microbiologie, 68*(12), 915–929. https://doi.org/10.1007/s13213-018-1401-8

Ibrahim, D., Weloosamy, H., & Sheh-Hong, L. (2014). Potential use of nylon scouring pad cubes attachment method for pectinase production by *Aspergillus niger* HFD5A-1. *Process Biochemistry, 49*, 660–667. https://doi.org/10.1016/j.procbio.2014.01.012

Kambourova, M. (2018). Thermostable enzymes and polysaccharides produced by thermophilic bacteria isolated from Bulgarian hot springs. *Engineering in Life Sciences, 18*(11), 758–767. https://doi.org/10.1002/elsc.201800022

Kawasaki, Y., Aoki, M., Makino, Y., Sakai, H., Tsuboi, Y., Ueda, J., Watanabe, K., Yamamoto, S., & Kurosawa, N. (2011). Characterization of moderately thermophilic bacteria isolated from saline hot spring in Japan. *Microbiology Indonesia, 5*(2), 56–60. https://doi.org/10.5454/mi.5.2.2

Kecha, M., Benallaoua, S., Touzel, J. P., Bonaly, R., & Duchiron, F. (2007). Biochemical and phylogenetic characterization of a novel terrestrial hyperthermophilic archaeon pertaining to the genus *Pyrococcus* from an Algerian hydrothermal hot spring. *Extremophiles, 11*(1), 65–73. https://doi.org/10.1007/s00792-006-0010-9

Khiyami, M., Serour, E., & Shehata, M. (2012). Thermo-aerobic bacteria from geothermal springs in Saudi Arabia. *African Journal of Biotechnology, 11*(17), 4053–4062. https://doi.org/10.5897/AJB11.3339

Kumar, S., & Nussinov, R. (2001). How do thermophilic proteins deal with heat? *Cellular and Molecular Life Sciences, 58*(9), 1216–1233. https://doi.org/10.1007/PL00000935

Lagzian, M., & Asoodeha, A. (2012). An extremely thermotolerant, aikaliphilic subtilisin-like protease from hyperthermophilic *Bacillus* sp. MLA64. *International Journal of Biological Macromolecules, 51*, 960–967. https://doi.org/10.1016/j.ijbiomac.2012.08.009

Lee, D., Koh, Y., Kim, K., Kim, B., Choi, H., Kim, D., Suhartono, M. T., & Pyun, Y. (1999). Isolation and characterization of a thermophilic lipase from *bacillus thermoleovorans* ID-1. *FEMS Microbiology Letters, 179*, 393–400. https://doi.org/10.1111/j.1574-6968.1999.tb08754.x

Leite, R. S. R., Gomes, E., & da Silva, R. (2007). Characterization and comparison of thermostability of purified β-glucosidases from a mesophilic *Aureobasidium pullulans* and a thermophilic *Thermoascus aurantiacus*. *Process Biochemistry, 42*, 1101–1106. https://doi.org/10.1016/j.procbio.2007.05.003

Li, W. F., Zhou, X. X., & Lu, P. (2005). Structural features of thermozymes. *Biotechnology Advances, 23*(4), 271–281. https://doi.org/10.1016/j.biotechadv.2005.01.002

Maddela, N. R., & García, L. C. (2021). *Innovations in biotechnology for a sustainable future* (p. VIII, 502). Springer Nature. https://doi.org/10.1007/978-3-030-80108-3. ISBN: 978–3–030-80107-6, Retrieved from https://www.springer.com/gp/book/9783030801076

Maddela, N. R., Garcia, L. C., & Chakraborty, S. (2021). *Advances in the domain of environmental biotechnology* (p. XVIII, 717). Springer Nature. https://doi.org/10.1007/978-981-15-8999-7. ISBN: 978-981-15-8999-7, Retrieved from https://www.springer.com/gp/book/9789811589980

Martins, E. D. S., Leite, R. S. R., da Silva, R., & Gomes, E. (2013). Purification and properties of polygalacturonase produced by thermophilic fungus *Thermoascus aurantiacus* CBMAI-756 on solid-state fermentation. *Enzyme Research, 2013*, 7. https://doi.org/10.1155/2013/438645

Mawadza, C., Hatti-Kaul, R., Zvauya, R., & Mattiasson, B. (2000). Purification and characterization of cellulases produced by two *bacillus* strains. *Journal of Biotechnology, 83*, 177–187. https://doi.org/10.1016/S0168-1656(00)00305-9

Meintanis, C., Chalkou, K., Kormas, K., Lymperopoulou, D. S., Katsifas, E., Hatzinikolaou, D., & Karagouni, A. (2008). Application of *rpoB* sequence similarity analysis, REP-PCR and BOX-PCR for the differentiation of species within the genus *Geobacillus*. *Letters in Applied Microbiology, 46*(3), 395–401. https://doi.org/10.1111/j.1472-765X.2008.02328.x

Ouali, S., Hadjiat, M. M., Ait Ouali, A., Salhi, K., & Malek, A. (2018). Cartographie et caractérisation géométriques de l'Algérie. *Revue des Energies Renouvelables, 21*(1), 54–61. Retrieved from https://revue.cder.dz/index.php/rer/article/view/669

Prescott, L. M., Harley, J. P., & Klein, D. A. (2002). *Microbiology: Food and industrial microbiology* (5th ed.). McGraw-Hill.

Priest, F. G. (1993). Systematics and ecology of *bacillus*. In A. L. Sonenshein, J. A. Hoch, & R. Losick (Eds.), *Bacillus subtilis and other gram-positive bacteria: Biochemistry, physiology, and molecular genetics* (pp. 3–16). ASM. https://doi.org/10.1128/9781555818388.ch1

Rey, M. W., Ramaiya, P., Nelson, B. A., Brody-Karpin, S. D., Zaretsky, E. J., Tang, M., de Leon, A. L., Xiang, H., Gusti, V., Clausen, I. G., Olsen, P. B., Rasmussen, M. D., Andersen, J. T., Jørgensen, P. L., Larsen, T. S., Sorokin, A., Bolotin, A., Lapidus, A., Galleron, N., … Berka, R. M. (2004). Complete genome sequence of the industrial bacterium *Bacillus licheniformis* and comparisons with closely related *Bacillus* species. *Genome Biology, 5*(10), r77. https://doi.org/10.1186/gb-2004-5-10-r77

Schallmey, M., Singh, A., & Ward, O. P. (2004). Development in the use of *Bacillus* species for industrial production. *Canadian Journal of Microbiology, 50*, 1–17. https://doi.org/10.1139/w03-076

Sharma, S., Vaid, S., Bhat, B., Singh, S., & Bajaj, B. K. (2019). Thermostable enzymes for industrial biotechnology. In *Advances in enzyme technology* (pp. 469–495). https://doi.org/10.1016/b978-0-444-64114-4.00017-0

Singh, R. S., Singh, T., & Pandey, A. (2019). Microbial enzymes—An overview. Chapter 1. In *Advances in enzyme technology* (pp. 1–40). https://doi.org/10.1016/B978-0-444-64114-4.00001-7

Stefanic, P., Decorosi, F., Viti, C., Petito, J., Cohan, F. M., & Mandic-Mulec, I. (2012). The quorum sensing diversity within and between ecotypes of *Bacillus subtilis*. *Environmental Microbiology, 14*, 1378–1389. https://doi.org/10.1111/j.1462-2920.2012.02717.x

Wang, L. T., Lee, F. L., Tai, C. J., & Kasai, H. (2007). Comparison of *gyrB* gene sequences, 16S rRNA gene sequences and DNA-DNA hybridization in the *Bacillus subtilis* group. *International Journal of Systematic and Evolutionary Microbiology, 57*, 1846–1850. https://doi.org/10.1099/ijs.0.64685-0

Yilmaz, B., Baltaci, M. O., Sisecioglu, M., & Adiguzel, A. (2016). Thermotolerant alkaline protease enzyme from *bacillus licheniformis* A10: Purification, characterization, effects of surfactants and organic solvents. *Journal of Enzyme Inhibition and Medicinal Chemistry, 31*, 1241–1247. https://doi.org/10.3109/14756366.2015.1118687

Yuli, P., Suhartono, M., Rukayadi, Y., Hwang, J., & Pyun, Y. (2004). Characteristics of thermostable chitinase enzymes from the indonesian *bacillus* sp.13.26. *Enzyme and Microbial Technology, 2*(5), 174–153. https://doi.org/10.1016/j.enzmictec.2004.03.017

Zeikus, G. J., Vieille, C., & Savchenko, A. (1998). Thermozymes: Biotechnology and structure-function relationships. *Extremophiles, 2*(3), 179–183. https://doi.org/10.1007/s007920050058

Enzymology of Microbial Biofilms

Amira Bouaziz, Aicha Asma Houfani, and Hafida Baoune

Abstract In nature, a biofilm forms when microorganisms embed their cells within a matrix of extracellular polymeric substances (EPSs) and adhere to each other and/or surfaces. EPSs are comprised of exopolysaccharides, extracellular DNA, proteins, lipids, and other biomolecules. The biofilm growth cycle encloses bacterial adhesion starting with the initial physical attraction to a substrate and ending with the eventual release of cell clusters from the biofilm matrix. Biofilms can act as reservoirs of microbially produced enzymes, which play a crucial role in many physiological and biochemical processes in bacteria, including the formation of a characteristic biofilm architecture, nutrition acquisition, dispersion of biofilms, and the ability to survive in a variety of environmental settings. Comparative studies of enzyme activities in attached and planktonic cells have shown that enzymatic activities are increased in biofilms. This chapter will highlight the main secreted enzymes within the context of biofilms and their applications in the industry and medical world.

1 Introduction

To date, several studies have focused on isolating microorganisms from different habitats (Abusrewil et al., 2020; Miao et al., 2021; Rana, Kour, Yadav, Yadav, & Saxena, 2020). Evidence shows that bacteria and archaea present around half of all

---

A. Bouaziz (✉)
Laboratoire de Biotechnologie des Molécules Bioactives et de la Physiopathologie Cellulaire (LBMBPC), Faculté des Sciences de la Nature et de la Vie, Université Batna 2, Batna, Algeria

A. A. Houfani
Michael Smith Laboratories and Department of Biochemistry and Molecular Biology, University of British Columbia, Vancouver, BC, Canada
e-mail: aicha.asma.houfani@msl.ubc.ca

H. Baoune
Faculty of Natural and Life Sciences, Université Kasdi Merbah Ouragla, Ouargla, Algeria

N. R. Maddela et al. (eds.), *Ecological Interplays in Microbial Enzymology*,
Environmental and Microbial Biotechnology,
https://doi.org/10.1007/978-981-19-0155-3_7

extant life on Earth. Consequently, microorganisms have significant influence on our environment and our daily life. Indeed, studying these tiny ubiquitous life forms can generate huge leaps forward for the human society, whether in terms of improving and protecting the environment, public health, or their applications in various industries, such as energy, food, water treatment, and mining (Dzianach, Dykes, Strachan, Forbes, & Pérez-Reche, 2019). Overall, the most significant impact on microbiology was the recognition of the extent to which microbial growth and development occur on surfaces in complex communities (van Wolferen, Orell, & Albers, 2018).

In addition to carrying "bags of enzymes," microorganisms are taxonomically diverse and are represented within the three-domain system: Archaea, Bacteria, and Eukarya. Traditionally, microorganisms could grow in homogeneous planktonic populations or as multicellular associations attached to certain surfaces. Biofilms are one of the most successful forms of life on the planet, and they may be found in every ecosystem. In 1708, a microbial biofilm was first discovered by Antonie van Leeuwenhoek using his new, and at the time, highly powerful microscope to examine tissues invaded by microorganisms, i.e., "animalcules," in dental plaque from his own mouth (Römling et al., 2014). Nevertheless, the relative contributions of genetics (active response) and environmental circumstances (passive response) to study the formation of biofilm structure and development are still a point of contention among biofilm researchers (Hall-Stoodley, Costerton, & Stoodley, 2004).

A biofilm is a community of microorganisms that can grow on biotic or abiotic surfaces; they are encased in an extracellular polymeric substance (EPS) matrix. As a result of their interactions and supported by the EPS matrix, more than 99% of known bacteria can adhere to and create three-dimensional (3D) architectures and form biofilms (Bauman, 2012). In fact, the biofilm mode of development is considered as a natural microbial life form (Puiu et al., 2017; Romaní et al., 2008). Microbial biofilms have shown novel characteristics in terms of gene expression, protein synthesis, growth rate, and metabolic activities compared to planktonic cells (Arnaouteli, Bamford, Stanley-Wall, & Kovács, 2021). The EPS matrix protects extracellular enzymes and hydrolysis products involved in the recycling of organic molecules inside the biofilm (Romaní et al., 2008). Polysaccharides, proteins, extracellular DNA (eDNA), and lipids are the main components of EPSs. They offer mechanical stability, as they facilitate adherence to surfaces, and create a cohesive 3D polymer network that interconnects and transiently immobilizes biofilm cells. Furthermore, the biofilm matrix acts as an external digestive system by keeping extracellular enzymes near the cells. These enzymes are responsible for solubilizing colloidal and solid biopolymers, making them accessible (Bender, Buckley, & Stahl, 2019; Flemming et al., 2016). Additionally, the biofilm matrix's architectural design assures high gene expressions, phenotypic changes of colony morphology, transfer of plasmid-carrying antibiotic resistance genes, and production of large amounts of extracellular polymers. Improved access to nutrients, besides close proximity between cells, facilitates mutualistic or synergistic associations and protection (Puiu et al., 2017). Microbial biofilms are ubiquitous on various surfaces exposed to bulk liquid environments in natural, industrial, and medicinal contexts. Biofilms

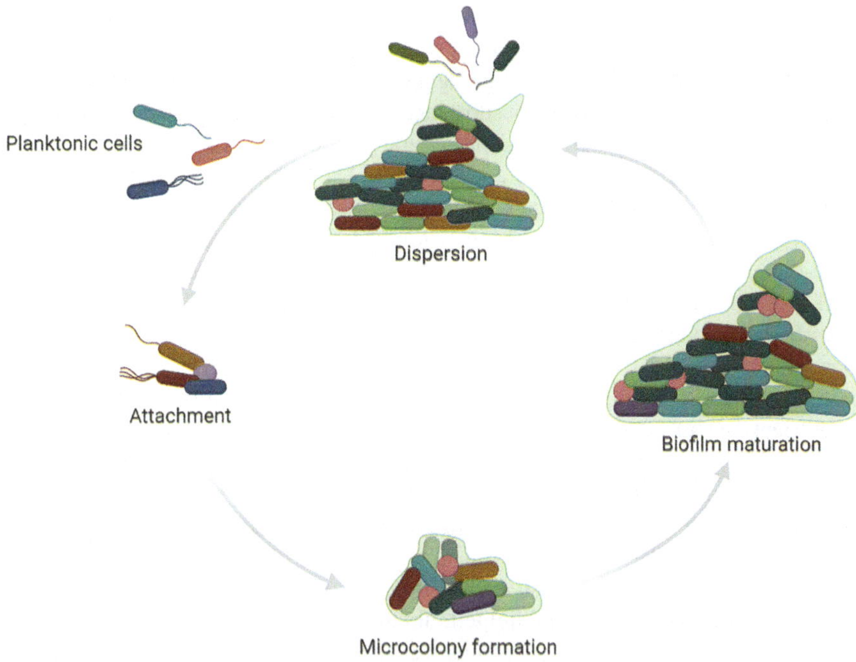

Fig. 1 Stages of biofilm formation. Attachment: reversible adhesion in which the bacteria attach to the surface. Microcolony formation: aggregation of bacterial cells. Biofilm maturation: production of extracellular polymeric substances and formation of a three-dimensional structure. Dispersion: Release of cells from the biofilm. Adapted from "Polymicrobial Biofilm", created by BioRender.com (2021). Retrieved from https://app.biorender.com/biorender-templates

have been found on a wide range of biotic and abiotic surfaces, such as river rocks, deep-sea vents, plant roots, water pipelines, food-processing surfaces, and medical implant devices (Berne, Ellison, Ducret, & Brun, 2018).

Some microorganisms are capable of forming biofilms in a variety of ways. Microscopic examinations revealed that the transition from a free-swimming to a surface-attached community-based lifestyle occurred in different phases, establishing a sophisticated cellular structural arrangement (Monds & O'Toole, 2009).

Various attractive and repulsive forces occur between the bacterial cell surface and environmental surfaces. In fact, negatively charged surfaces repel the attachment of negatively charged bacterial cells from a distance of 10–20 nm (Palmer, Flint, & Brooks, 2007). Nevertheless, the attractive van der Waals forces allow bacteria to get close to the material surface. Besides, fimbriae and flagella strengthen the attachment (Rabin et al., 2015).

A bacterial life cycle includes both planktonic and biofilm phases, as shown in Fig. 1. The first stage in the development of a biofilm is reversible attachment in which the bacteria attach to the surface. A variety of factors influence the attachment, including substratum surface roughness, surface conditions, and hydrophobic and electrostatic interactions, where at this stage, cells can detach. The interactions between the bacteria and the surface using fimbriae, pili, or lipopolysaccharides retain the irreversible attachment. Aggregation of bacterial cells leads to the production of extracellular polymeric substances (EPSs) (Ünal Turhan, Erginkaya, Korukluoğlu, & Konuray, 2019). Cells generally proceed to proliferate and generate small (5–200 μm wide) clusters (Stacy, McNally, Darch, Brown, & Whiteley, 2015). These aggregates, known as microcolonies, are often believed to be a transient phase of early biofilm development (Ch'ng, Chong, Lam, Wong, & Kline, 2018).The three-dimensional structure of adhering cells in the biofilm matrix comprises networks of channels for providing nutritional substances as well as a variety of microbial cell-to-cell communications (quorum sensing) that synchronize the activities of the microbial consortium (Facundo Rodriguez Ayala et al., 2017). As biofilms expand in size, cells in the deepest layers of the biofilm may lose access to nutrients or suffer from toxic waste product accumulations; as a result, their microenvironment may become unfavorable (Karatan & Watnick, 2009). Bacterial detachment from the biofilm is the final stage in the developing cycle. These cells that are lost from the biofilm colonize distant locations and repeat the growth cycle (Richards & Melander, 2009).

In this chapter, a comprehensive understanding of the most common enzymes of microbial biofilms will be highlighted. In the first part of this chapter, biofilm enzymes will be discussed. Subsequently, the second part will review biofilm enzymes applied in both the industrial and medical fields as supported by the literature.

2 Biofilm Enzymes

An essential element of biofilms is the presence and the action of enzymes, which control the development of many physiological and biochemical processes in bacteria (Garrett, Bhakoo, & Zhang, 2008). Biofilms involve a variety of extracellular enzymes including glycoside hydrolases (GHs), which break down the glycosidic bonds between carbohydrates and destroy the adhesive components of biofilms (Kaplan, 2010). Each cleaves a specific type of linkage, for instance, α-1,4 bond hydrolysis by α-amylase and β-1,4 bond hydrolysis by cellulase (Fleming & Rumbaugh, 2018).

There are several patterns of enzyme activity in biofilms. Exoenzymes are synthesized inside cells and then released into the extracellular space where they hydrolyze molecules or chemicals. Some of these extracellular enzymes break down large substrates, mainly complex organic compounds such polysaccharides that cannot be carried into cells (Engelkirk & Duben-Engelkirk, 2015; O'Flaherty, Collins, & Mahony, 2010). Examples of exoenzymes in biofilms are amylases, cellulases, lipases, and proteases. The opposing scenario occurs in the case of endoenzymes that are stored inside the cells and perform their function within the biofilm cells (Cowan & Smith, 2012; Talaro & Talaro, 2002). Based on the literature, this section will discuss the most common enzymes involved in biofilm metabolism from various microorganisms.

3 Amylases and Glycosyltransferases

The best known amylolytic enzyme is α-amylase (EC 3.2.1.1) (α-1,4-glucan-4-glucanohydrolase), found in a variety of GH families (GH13, GH57, GH119, and GH126) (Janeček, Svensson, & MacGregor, 2013). GH13, also known as an α-amylase family, is one of the largest GH families (Majzlová, Pukajová, & Janeček, 2013). α-Amylase is classified as an endoamylase, enabling the hydrolysis of the internal α-1,4 glycosidic bonds in starch into maltose/isomaltose, maltotriose, and larger oligosaccharides (Abeleda, Javier, Murillo, & Baculi, 2020; Bowen & Koo, 2011; Mandel, Gachons, des Plank, Alarcon, & Breslin, 2010).

Amylases that can promote biofilm growth include pancreatic amylases (PAs) and salivary amylases (SAs), which are structurally similar and in sequence (Jowiya et al., 2015). In the chromosome, the α-amylase genes cluster (Fig. 2) with the salivary amylase genes (*AMY1*), two pancreatic amylase genes (*AMY2A* and *AMY2B*), and a related pseudogene (Santos et al., 2012). *Campylobacter jejuni* (*C. jejuni*) can utilize pancreatic amylase as a signal to regulate growth and biofilm formation. The findings of the study by Jowiya et al. (2015) showed that *C. jejuni* secretes α-dextran as a biofilm component in response to PA and that its protease Cj0511 is required for this process.

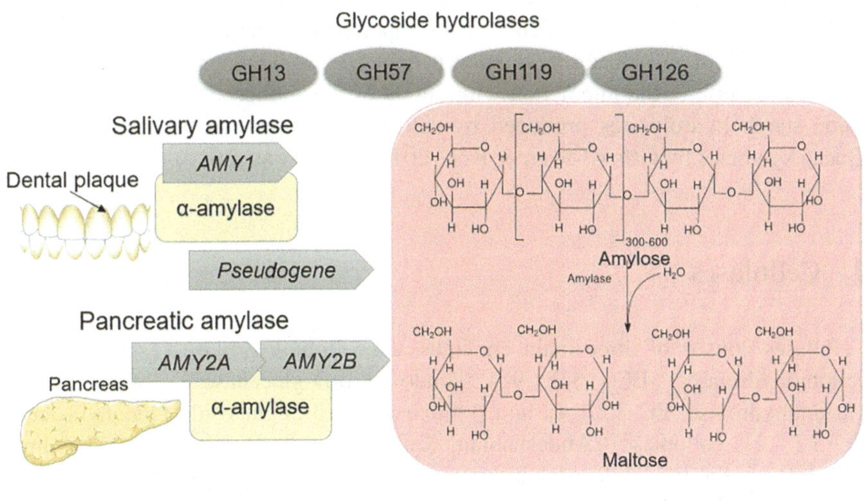

Fig. 2 Amylases in biofilms. Both the saliva and the pancreas contain biofilm amylases that are involved in amylase reactions. Amylases are represented by four GH families, and the genes of amylases are represented along with amylase reactions

Salivary α-amylase (sAA), also referred to as ptyalin, is a monomeric calcium-binding enzyme among the most abundant proteins in the human saliva of the oral cavity (Chaudhuri, Rojek, Vickerman, Tanzer, & Scannapieco, 2007; Nikitkova, Haase, & Scannapieco, 2013; Santos et al., 2012). sAA attaches to the amylase-binding protein, an adhesin of oral Streptococci that form dental plaque and are designated as α-amylase-binding streptococci (Lahiri et al., 2021; Nikitkova et al., 2013; Wilson, 2008). Communications and cell-to-cell signaling play important roles in biofilm communities, and they have also been demonstrated between oral Streptococci and *Veillonella atypica* (*V. atypica*), where *Streptococcus gordonii* (*S. gordonii*) increases the expression of an amylase gene named *amyB* after receiving a signal from *V. atypica* (Egland, Palmer, & Kolenbrander, 2004). The amylases in *S. gordonii* include amylase-binding protein A (AbpA) and amylase-binding protein B (AbpB) (Chaudhuri et al., 2007). On the surfaces of actively dividing cells, amylase appears to associate with receptors on the surface of *S. gordonii* that bind to it with high affinity (Rogers, Palmer, Kolenbrander, & Scannapieco, 2001). As a biofilm forms, pili provide attachments for cells to attach to surfaces; for humans, this happens when pathogens attach to cells and tissues (Pommerville, 2011). The implication of long filamentous pili to bind to α-amylase has also been demonstrated in *Streptococcus sanguinis* (*S. sanguinis*) to promote biofilm formation in the oral cavity. Furthermore, *S. sanguinis* produces a protease capable of breaking down sIgA (IgA protease), whereas *S. gordonii* can bind to salivary α-amylase, allowing them to degrade starch (Marsh, Martin, Lewis, & Williams, 2009).

A biofilm matrix is composed of α-glucan exopolymers synthesized by streptococcal glucosyltransferases (Gtfs) (Souza et al., 2020). Gtfs and α-amylases are both biofilm-promoting enzymes that are involved in bacterial colonization of the tooth surface (Kirsch et al., 2017). The Gtf activity of oral streptococci is pronounced in mutants lacking AbpA. Researchers have reported that the enzymatic activity of streptococcal Gtf and glucan production is reduced on saliva-coated hydroxyapatite surfaces because of the interaction with amylases (Chaudhuri et al., 2007). Indeed, some starch hydrolysates produced by salivary amylases can be integrated into glucan synthesis by Gtfs (Bowen & Koo, 2011; Klein et al., 2010).

4 Cellulases

Cellulases consist of three main enzymes: exocellobiohydrolase (1,4-β-D-glucan cellobiohydrolase, EC 3.2.1.91), endo-1,4-β-D-glucanase (1,4-β-D-glucan glucanohydrolase, EC 3.2.1.4), and β-glucosidase (β-D-glucoside glucohydrolase, EC 3.2.1.21; cellobiase) (Abdeshahian, Kadier, Rai, & da Silva, 2020; Krishna, 2007). Recently, lytic polysaccharide monooxygenases (LPMOs), a class of oxidative enzymes, have been added to this list of main cellulases (Ladevèze et al., 2017).

The participation of cellulose-degrading enzymes does not exclusively occur in biofilm inhibition, but it can also be part of biofilm formation (Scapin et al., 2017).

Numerous studies on biofilms and cellulosic biomass degradation demonstrate the critical role that cellulolytic biofilms play in cellulose utilization (Wang et al., 2013). The characteristics of biofilms, such as cell concentration and surface attachment, can result in an increased and multiplied cellulolytic activity. Furthermore, cellulose hydrolysis in biofilms supports the idea that microbial adhesion to cellulose facilitates substrate uptake (Lu, Zhang, & Lynd, 2006). Cellulolytic microbial cells adhere to their substrates, staying near areas of high enzyme concentrations and hydrolysis products; this may ease the uptake of solubilized growth substrates (Alonso, Pomposiello, & Leschine, 2008). Indeed, the cellulolytic biofilms allow a high level of enzyme activity to be concentrated on the surface of the solid substrate, and the hydrolysis products can be captured directly at the hydrolysis site, increasing their efficiency (Brethauer, Shahab, & Studer, 2020).

A wide variety of actinobacteria have been shown to have the cellulose degradation ability, and the advent of genomic sequencing has allowed the identification of cellulase producer genera other than *Streptomyces*, such as *Mycobacterium* (Van Wyk et al., 2017). Moreover, not only soil mycobacteria such as *Mycobacterium bovis* but also mammalian host pathogenic *Mycobacterium tuberculosis* (Mtb) strains are known to have cellulases (Varrot et al., 2005). According to a recent study, Mtb biofilms containing cellulose are relevant to human tuberculosis. Hence, the presence of cellulose as a matrix component in biofilms can act as a biomarker for detecting Mtb biofilms (Chakraborty, Bajeli, Kaushal, Radotra, & Kumar, 2021). Several cellulose-producing organisms showed higher production of cellulose in the presence of endoglucanase or carboxymethylcellulose. Thus, these enzyme-degrading celluloses may facilitate glycan maturation and/or transport through the cell wall in connection with cellulose formation by bacteria (Sheppard & Howell, 2016).

Bacterial extracellular polysaccharides such as celluloses play a key role in biofilm formation (Omadjela et al., 2013). It has been reported that the microfibril structure of cellulose plays a role in the attachment of the developing biofilm at an early stage of the growth process. Besides providing structural integrity to mature biofilms, cellulose is also an anchor and a link between the microcolonies inside the biofilms (Nahm et al., 2017). A biofilm of cellulolytic bacteria forms on cellulose fibers, and the close interaction of bacteria with cellulose particles is essential for the efficient solubilization of cellulose (O'Sullivan, Burrell, Pasmore, Clarke, & Blackall, 2009). *Salmonella* is among the bacteria that advance the knowledge of cellulose production regulation. It has also been found that biofilm formation and multicellular behavior (rdar morphotype) are associated with cellulose synthesis in *Escherichia coli* and that treatment with cellulase completely disperses the existing biofilms (Beloin, Roux, & Ghigo, 2008).

Caldicellulosiruptor sp. and *Clostridium thermocellum* (*C. thermocellum*) are thermophiles and anaerobes that can degrade cellulosic biomass by forming biofilms on the cellulose they consume (Wang et al., 2013). *C. thermocellum* is recognized as the model of cellulolytic bacterium that forms characteristically distinct and thin biofilms, often monolayers, without the polymeric matrix usually enclosed in biofilms (Dumitrache, Wolfaardt, Allen, Liss, & Lynd, 2013). A study has demonstrated that *C. thermocellum* and *Caldicellulosiruptor obsidiansis* exhibited four times greater cellulase activity than did planktonic cultures (Morrell-Falvey, Elkins, & Wang, 2015).

Similarly, cellulase production by *Aspergillus niger* biofilms, an important fungus in cellulase production, increased significantly over free-living submerged cultures. This increase in enzyme production is not the result of increased biomass concentrations but is rather the result of increased gene expression when fungal surfaces meet supports (Gutiérrez-Correa, Ludeña, Ramage, & Villena, 2012). Another fungal model system for studying biodegradation of plant biomass is *Aspergillus nidulans*; its adaptation to lignocellulose involves hydrophobins, according to the RNAseq study by Brown et al. (2016). This study showed that hydrophobins promote the growth of biofilms in sugarcane bagasse, improving the use of lignocellulose, where on the fungal biofilm formed on the lignocellulosic fibers, RodA was the major hydrophobin (Brown et al., 2016).

5 Lipases

Lipids are common components on the surface of a biofilm matrix and are involved in their interaction, attachment, and maintenance (Nahar, Mizan, Ha, & Ha, 2018). Several reports have demonstrated that lipases produced by bacterial biofilms enhance biofilm formation. *Staphylococcus aureus* (*S. aureus*) produces lipolytic enzymes that improve their colonization and their growth in lipid-rich environments such as the skin, leading to acne lesion. In addition, it was demonstrated that the deletion of lipase-coding genes decreases biofilm formation (Hu, Xiong, Zhang, Rayner, & Chen, 2012; Nguyen et al., 2018; Saising, Singdam, Ongsakul, & Voravuthikunchai, 2012).

6 Nucleases

Bacterial extracellular DNases play a crucial role in the physiology and the pathogenicity in several ways (Kiedrowski, Crosby, Hernandez, Malone, & Horswill, 2014). Within biofilm structures, eDNA acts as a structural scaffold inside EPSs, enabling bacterial adherence, aggregation, and horizontal gene transfers (Okshevsky & Meyer, 2015; Whitchurch, Tolker-Nielsen, Ragas, & Mattick, 2002). It is possible that the presence of eDNA in a biofilm matrix comes from active secretion or controlled cell lysis and it is sometimes linked to competence development. Through acid–base interactions, eDNA extended from the cell surface and then adsorbed on it, thus increasing the adhesion strength of abiotic surfaces (Okshevsky & Meyer, 2015). Some examples of common bacterial species producing nucleases are listed below.

Gonorrhea is one of the most frequently reported infectious diseases (Cornelissen, Fisher, & Harvey, 2013) caused by *Neisseria gonorrhoeae* (*N. gonorrhoeae*), an obligatory human pathogen Recently, *N. gonorrhoeae* has shown the ability to form biofilms on different types of surfaces (Falsetta et al., 2009; Greiner et al., 2005). On cervical epithelial cell surfaces, *N. gonorrhoeae* synthesizes a thermonuclease, Nuc, which may be secreted and may play a role in biofilm remodeling. eDNA is a key component of the gonococcal biofilm matrix. It seems that the secreted nuclease is capable of altering its own biofilm structure and/or that bacterial competition would be a key factor of essential virulence mechanism (Steichen, Cho, Shao, & Apicella, 2011).

For *Vibrio cholerae* (*V. cholerae*), the extracellular DNA is identified and characterized as a component of its biofilm matrix. It should be highlighted that *V. cholerae* inhabits soil and marine and freshwater habitats, which leads to developed mechanisms in order to use eDNA as a carbon, nitrogen, and phosphate source for its persistence in nutrient-poor ecosystems (Meibom et al., 2004). Furthermore, findings showed that two extracellular nucleases, Dns and Xds, regulate and control extracellular DNA. In fact, eDNA and extracellular nucleases are involved in a variety of activities, including the formation of a particular biofilm architecture, nutrition acquisition, separation from biofilms, and the colonization fitness of biofilm clumps after host ingestion (Seper et al., 2011).

S. aureus possesses a diverse set of virulence characteristics, as it is capable of forming biofilms (Torok, Moran, & Cooke, 2009). Interestingly, *S. aureus*-synthesized staphylococcal nuclease, encoded by the *nuc1* gene, encodes an essential virulence component. Consequently, biofilm formation may be blocked in staphylococcal nuclease-producing strains. However, the suppression of the *nuc1* gene significantly increases biofilm production. In a recent study, it has been demonstrated that both recombinant NUC 1 protein and staphylococcal nuclease have a visible effect on biofilm formation of other species, such as *Pseudomonas aeruginosa* (*P. Aeruginosa*), *Actinobacillus pleuropneumoniae*, and *Haemophilus parasuis* (Tang et al., 2011). Previous reports have shown that eDNA is an important component of a biofilm matrix. Kiedrowski et al. (2014) indicated that Nuc levels are closely related to biofilm development (Kiedrowski et al., 2011). During infections, a second nuclease (Nuc2) and a surface-attached and functional DNase are both expressed by *S. aureus*, having the same biochemical characteristics as those of the secreted Nuc enzyme (Kiedrowski et al., 2014).

7 Proteases

Studying the metabolic activities of bacterial biofilms and planktonic cells may help clarify the differences; for example, some secreted proteases are re-regulated in biofilms but not in planktonic cells, and, as a result, different proteolytic cleavage patterns occur at each growth stage (Lohse, Gulati, Johnson, & Nobile, 2017). In this part, we review some of the protease-producing biofilm bacteria.

Biofilms of *Candida albicans* (*C. albicans*) are developed on mucosal surfaces, epithelial cell linings, and implanted medical devices such as catheters, dentures, and heart valves (Lohse, Gulati, Craik, Johnson, & Nobile, 2020). A variety of methods, including biofilms, are employed by *C. albicans* to avoid host immune response. *C. albicans* secretes a family of aspartyl proteases (Sap), which have been linked to several aspects of pathogenesis, including invasion, hyphal cell aggregation, cell wall protein shedding, nutrition acquisition, immune evasion, and host inflammatory response activation (Kumar, Saraswat, Tati, & Edgerton, 2015; Naglik, Challacombe, & Hube, 2003; Pericolini et al., 2015; Schild et al., 2011). During biofilm formation, these proteases are highly expressed (Ganguly et al., 2011; Lohse et al., 2017).

The development of biofilms and pathogenicity has also been reported for *Enterococcus faecalis* (*E. faecalis*). Some virulence components in *E. faecalis*, such as serine protease, gelatinase, and collagen-binding protein (ace), enhance the bacterial cell adherence to dentin (Carniol & Gilmore, 2004; Wang et al., 2011). *E. faecalis* serine protease (SprE) contributes to biofilm formation and provides insights into understanding the development process regulation of these protease activities. On the contrary, SprE-deficient cells generate substantially more eDNA as a component of the biofilm matrix. In some reports, authors have suggested that the interaction of both secreted and coregulated proteases (GelE and SprE) controls the autolysis and the release of high-molecular-weight eDNA, required for the formation of *E. faecalis* biofilms (Thomas, Thurlow, Boyle, & Hancock, 2008; Waters, Antiporta, Murray, & Dunny, 2003).

The biofilm mode of life contributes considerably to the development and the persistence of *P. aeruginosa* under different environmental conditions, technological systems, and clinical context (Ma et al., 2009). Environmental and clinical *P. aeruginosa* isolates produce proteolytic enzymes, which play an important role in their ability to survive in several environmental settings (Tseng et al., 2018). Consequently, the type of extracellular enzyme and the expression level differ between strains, and these seem to be directly linked to the colonized habitat (Tielen et al., 2010).

In human oral cavity, *Streptococcus mutans* (*S. mutans*) is able to aggregate in a specific order on hard surfaces of the teeth, developing a protective EPS and producing oral biofilms, known as dental plaques (Ostadhossein et al., 2021). According to the findings of Lee, Li, and Bowden (1996), the surface proteins of a *S. mutans* monolayer biofilm were released by an indigenous enzymatic activity known as surface protein-releasing enzyme (SPRE) activity. In addition, the active release of surface proteins was used by *S. mutans* to detach from a colonized surface (Lee et al., 1996).

8 Industrial Biofilm Enzymes

The last few decades have witnessed the use of biofilms as a cost-effective process in industrial biotechnology, due to their resistance to toxic compounds, long-term activity, high biomass production, and stability. Biofilms have been used as bioinoculants and biocatalysts to transform a variety of agriculture compounds (LaPara, Konopka, Nakatsu, & Alleman, 2001; Qureshi, Annous, Ezeji, Karcher, & Maddox, 2005; Stewart & Franklin, 2008). In addition, they are involved in bioenergy production. For instance, *Zymomonas mobilis* biofilms have been known to be efficient in producing ethanol (Todhanakasem et al., 2019).

In nature, lignocellulose decomposition occurs over a long period of time, and biofilms allow enzyme concentrations from microorganisms to act together toward the goal of breaking down the most abundant recalcitrant polymers in the biosphere (Brethauer et al., 2020). In contrast to some of other fields, in lignocellulose degradation, biofilms are desired. Degradation of lignocellulose can result in the formation of biofilms, irrespective of whether they are fungal, bacterial, fungal–bacterial (Bomble et al., 2017), or algal biofilms (Zhang et al., 2019). As a biocatalyst, a biofilm is an efficient eco-friendly way to produce bioethanol from lignocellulosic materials, significantly reducing the operating costs and complexity of the process (Todhanakasem et al., 2019; Todhanakasem, Narkmit, Areerat, & Thanonkeo, 2015).

Extracellular enzymes in biofilms have great potential for biofuel production (Mitra, Sana, & Mukherjee, 2014). Biofilm technology has developed in the treatment of lignocellulose during several bioprocessing steps at once as summarized in Fig. 3, including delignification, saccharification, fermentation, and separation

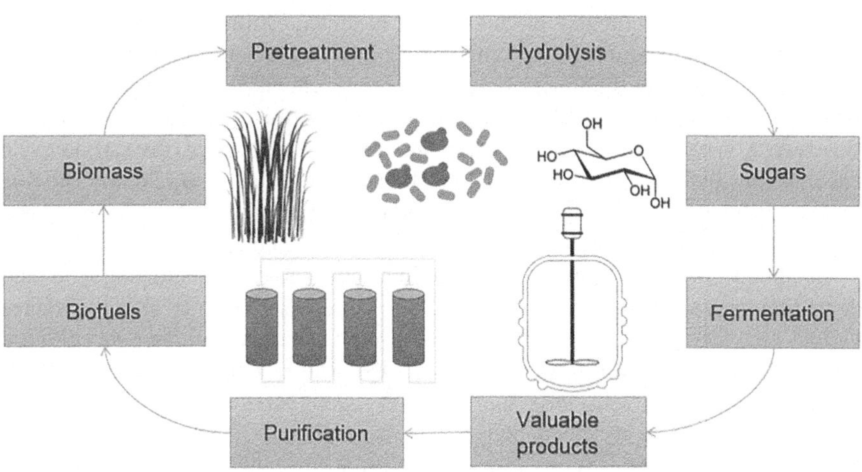

Fig. 3 The bioprocessing steps involved in valorizing lignocellulose while employing biofilms. Biorefineries can utilize various lignocellulosic co-products. After pretreatment and enzymatic hydrolysis, the biomass is fermented into biofuel

(Wang & Chen, 2009). Microbial biofilms have been used in the field of consolidated bioprocessing for direct conversion of plant biomass and have resulted in enhancing the overall enzymatic activities (Xiros & Studer, 2017; Xiros, Topakas, & Christakopoulos, 2013). Researchers have developed methods for consolidated bioprocessing of lignocellulose to ethanol. The biotechnological use of cellulases for the degradation of cellulose and conversion into glucose has become commonplace, through which valuable compounds can be biologically produced (Muffler et al., 2014). During a single multispecies biofilm membrane reaction, aerobic and anaerobic conditions are created to produce both fungal cellulolytic enzymes and yeast fermentation to convert cellulose-derived sugars into energy (Brethauer & Studer, 2014). Xiros, Shahab, and Studer (2019) found that a cellulolytic fungal biofilm enhanced the rumen microbiome's ability to convert cellulose into short-chain fatty acids. A fungal biofilm resulted in positive effects on the yields and productivity in the process due to the enhancement of cellulolytic activity compared to microbiome-only processes (Xiros et al., 2019).

Another type of energy from microbial lignocellulose degradation is bioelectrical power; it has been demonstrated that the *Geobacillus* species produce enzymes capable of hydrolyzing complex polymeric materials in biomass containing lignocellulosic materials. For instance, the *Geobacillus* sp. strain WSUCF1 has been demonstrated to use its unique lignocellulolytic activity, biofilm formation, and biochemical pathway to directly convert corn stover into electricity (Shrestha, Kognou, Zhang, & Qin, 2020). A procedure involving electrochemistry was also reported for enhancing the degradation of lignocellulose and the formation of a biofilm in the *Geobacillus* strain WSUCF1. As a result of microbial electrocatalysis, lignocellulosic materials can be utilized more quickly to become biofuels, thus showing promise for practical applications in the future (Rathinam, Gorky, Salem, & Sani, 2020).

It is interesting to note that the function of biofilms is infinite, as they are widely discussed in the literature. In fact, microbial biofilms play an important ecological role and are crucial to the functioning of the biogeochemical cycle where extracellular enzymes are responsible for up to 80% of all microbial activities (Battin, Besemer, Bengtsson, Romani, & Packmann, 2016). For instance, the carbon cycle encompasses the decomposition of lignocellulose (Brethauer et al., 2020). The role of extracellular enzymes in carbon, nitrogen, and phosphorus cycling has been widely studied in numerous ecosystems (Luo, Meng, & Gu, 2017).

In agriculture, soil biofilms play an interesting role in stabilizing soil and degrading organic matter (Ma et al., 2017). Moreover, biofilms in soil produce various enzymes able to fix nitrogen, solubilize phosphorus, and degrade cellulose (Pandit, Adholeya, Cahill, Brau, & Kochar, 2020; Rather, Gupta, & Mandal, 2021). It was reported that rhizobacterial biofilms of *Pseudomonas* enhance plant growth by catalyzing organic acids (Panichikkal & Krishnankutty, 2020). In the food and dietary industry, biofilms have been used to produce organic acids. In addition, a higher enzymatic activity of cellulases was recorded by bacterial biofilm cells during the food waste digestion process (Fung et al., 2021).

Bacterial biofilms, in an aquaculture system, play a significant role. The recirculation of water within a fish farm decreases its quality and increases the development of microbial pathogens, which lead to disease transmission. As pointed out, dispersed bacteria are rarely found in aquatic ecosystems, and their existence is more likely related as biofilms (Ríos-Castillo, Thompson, Adams, Mateo, & Rodríguez-Jerez, 2018). Aquatic biofilms produce extracellular enzymes that can remain stable for weeks (Zoppini & Marxsen, 2010). Iijima, Washio, Okahara, and Morikawa (2009) suggested that *Pseudoalteromonas* bacteria found in the biofilm community help eliminate excess proteinaceous matter from fish farm sediment sludge. *Pseudoalteromonas* were able to protect fish breeding due to their potent enzymes, including extracellular proteases (Iijima et al., 2009). In addition, biofilms were found to be more effective in the treatment of wastewater compared to dispersed bacterial cells (LaPara et al., 2001). Wastewater biofilms are able to transform or metabolize toxic compounds using different enzymes, such as oxido-reductases, transferases, hydrolases, lyases, isomerases, ligases (synthetases), and phosphatase (Yadav & Chandra, 2020).

9 Medical Biofilm Enzymes

Microbial biofilms show negative effects on medical tools, causing serious health problems, such as persistent infections, which may lead to death (Bryers, 2008). Infections induced by bacterial biofilms can be caused by biofilms formed on abiotic surfaces (medical devices) or by native biofilms of host tissues (Costerton, Montanaro, & Arciola, 2005). In fact, the most common infections occur in the urinary tract, caused by microbial biofilms grown on the surface of medical implants (Francolini & Donelli, 2010).

Over the last few decades, antibiotic resistance has been considered as a global health security issue occurring worldwide (Rodríguez-Baño, Gutiérrez-Gutiérrez, Machuca, & Pascual, 2018). Microbial biofilms have become a significant problem in health due to their resistance to antibiotics and host immune cells (Sharma, Misba, & Khan, 2019). One of the principal causes of chronic infections (approximately 80%), described by persistent inflammation and tissue damage, has been multidrug resistance of biofilm-growing bacteria (Høiby, Bjarnsholt, Givskov, Molin, & Ciofu, 2010).

A panel of studies highlighted that antibiotic resistance mechanisms in bacterial biofilms are different than planktonic ones. Several mechanisms are used by bacteria inside biofilms to resist and/or to tolerate antibiotics. The key factor of this phenomenon is the multicellular nature and the synergy between bacterial species of the biofilm (Hall & Mah, 2017; Pirlar et al., 2020; Sharma et al., 2019). Another critical factor of antibiotic resistance is that the diffusion of antibiotics through the biofilm matrix depends on bacterial species, the antimicrobial agent, cell density, and biofilm growth stage (Stewart, 2002). Both antibiotic-modifying enzymes and cell-modifying enzymes contribute to antibiotic resistance. Indeed, it was demonstrated that slow penetration of antibiotics such as β-lactams and aminoglycosides is caused by either their inactivation or their modification due to the action of these enzymes (Stewart, 2002). Consequently, genotypic and phenotypic adaptation responses result (Coenye, 2010). Handal, Olsen, Walker, and Caugant (2004) and Mah et al. (2003) pointed out that biofilms of *P. aeruginosa* synthesize membrane-bound NdvB glycosyltransferase located in the periplasm to sequester and prevent tobramycin from reaching its action site. In another study, the same strain showed a resistance to meropenem and ceftazidime in the presence of β-lactamases (Bowler, Zhanel, Ball, & Saward, 2012).

Although bacteria inside a biofilm may have antibiotic targets or produce antibiotic-degrading enzymes, studies have shown that multidrug resistance of bacterial biofilms might be due to the horizontal gene transmission between cells, in particular, traditional antibiotics, such as β-lactams, aminoglycosides, and fluoroquinolones (Høiby et al., 2010). Madsen, Burmølle, Hansen, and Sørensen (2012) have reported that high cell density and the abundance of mobile genetic elements increased the efficiency of conjugation (gene horizontal transfer) inside biofilms. In addition, EPSs were found to be a physical and chemical barrier, providing protection to biofilms through the diffusion–reaction phenomenon or enzymatically based reactions by blocking, trapping, or stopping antibiotics from achieving bacterial cells (Bi et al., 2021; Davenport, Call, & Beyenal, 2014). Khan et al. (2010) reported that EPSs provide tolerance to aminoglycosides in *Pseudomonas aeruginosa* biofilms. Besides, treatments with β-lactams showed only an effective response on individual cells of *Pseudomonas aeruginosa* compared to biofilms (Høiby et al., 2010). In this regard, the work of Perez et al. (2014) demonstrated that in mixed biofilms, the B-lactamase produced by *Moraxella catarrhalis* protects both *Streptococcus pneumonia* and *Haemophilus influenzae* from amoxicillin and ampicillin attacks. According to the literature, the effects of antibiotics rely on the biofilm's development stage, where mature biofilms are difficult to be treated due to the poor penetration of antibiotics, the effect of efflux pumps, and also the presence of persister cells (Bi et al., 2021; Monzón, Oteiza, Leiva, Lamata, & Amorena, 2002), which are a small subpopulation of bacteria in a slow-grow or starving mode, which is highly tolerant to killing by antibiotics (Lewis, 2006).

10 Conclusions

Biofilms are one of the interesting fields to study in microbiology. However, little is known about the factors that contribute to biofilm development in nature. As reviewed in this chapter, the composition of microbial communities can be highly different, providing a distinctive effect on biofilm enzymatic capacity. The long-term activities of biofilms would enable for continuous biofilms, where the synthesis of extracellular enzymes are greater than that of intracellular enzymes. Besides, due to the complexity and the heterogenicity of biofilm communities, the application of biofilm enzymes required for biotechnological interests is increasing. In fact, a better comprehension of the mechanisms involved in biofilms presents a challenge. Nowadays, researchers are contributing to building and developing fundamental methods in order to analyze biofilm systems. Besides, progress studies in this field are quite diverse. Taking everything together, we conclude this chapter highlighting the key findings and contributions of biofilm enzymes.

References

Abdeshahian, P., Kadier, A., Rai, P. K., & da Silva, S. S. (2020). Lignocellulose as a renewable carbon source for microbial synthesis of different enzymes. In *Lignocellulosic biorefining technologies* (pp. 185–202). Springer. https://doi.org/10.1002/9781119568858.CH9

Abeleda, H. E. P., Javier, A. P., Murillo, A. Q. M., & Baculi, R. Q. (2020). Alpha-amylase conjugated biogenic silver nanoparticles as innovative strategy against biofilm-forming multidrug resistant bacteria. *Biocatalysis and Agricultural Biotechnology, 29*, 101784. https://doi.org/10.1016/J.BCAB.2020.101784

Abusrewil, S., Alshanta, O. A., Albashaireh, K., Alqahtani, S., Nile, C. J., Scott, J. A., & McLean, W. (2020). Detection, treatment and prevention of endodontic biofilm infections: What's new in 2020? *Critical Reviews in Microbiology, 46*(2), 194–212. https://doi.org/10.1080/1040841X.2020.1739622

Alonso, A. N., Pomposiello, P. J., & Leschine, S. B. (2008). Biofilm formation in the life cycle of the cellulolytic actinomycete Thermobifida fusca. *Biofilms, 2008*, 1–11. https://doi.org/10.1017/S1479050508002238

Arnaouteli, S., Bamford, N. C., Stanley-Wall, N. R., & Kovács, Á. T. (2021). Bacillus subtilis biofilm formation and social interactions. *Nature Reviews Microbiology, 19*(9), 600–614. https://doi.org/10.1038/s41579-021-00540-9

Battin, T. J., Besemer, K., Bengtsson, M. M., Romani, A. M., & Packmann, A. I. (2016). The ecology and biogeochemistry of stream biofilms. *Nature Reviews Microbiology, 14*(4), 251–263. https://doi.org/10.1038/nrmicro.2016.15

Bauman, R. W. (2012). *Microbiology with diseases by body system* (3rd ed.). Pearson Education, Inc..

Beloin, C., Roux, A., & Ghigo, J.-M. (2008). Escherichia coli biofilms. *Current Topics in Microbiology and Immunology, 322*, 249.

Bender, M., Buckley, S., & Stahl. (2019). *Brock biology of microorganisms* (15th ed.). Pearson Education Limited.

Berne, C., Ellison, C. K., Ducret, A., & Brun, Y. V. (2018). Bacterial adhesion at the single-cell level. *Nature Reviews Microbiology, 16*(10), 616–627. https://doi.org/10.1038/s41579-018-0057-5

Bi, Y., Xia, G., Shi, C., Wan, J., Liu, L., Chen, Y., Wu, Y., Zhang, W., Zhou, M., He, H., & Liu, R. (2021). Therapeutic strategies against bacterial biofilms. *Fundamental Research, 1*(2), 193–212. https://doi.org/10.1016/J.FMRE.2021.02.003

Bomble, Y. J., Lin, C. Y., Amore, A., Wei, H., Holwerda, E. K., Ciesielski, P. N., Donohoe, B. S., Decker, S. R., Lynd, L. R., & Himmel, M. E. (2017). Lignocellulose deconstruction in the biosphere. *Current Opinion in Chemical Biology, 41*, 61–70. https://doi.org/10.1016/j.cbpa.2017.10.013

Bowen, W. H., & Koo, H. (2011). Biology of Streptococcus mutans-derived glucosyltransferases: Role in extracellular matrix formation of cariogenic biofilms. *Caries Research, 45*(1), 69–86. https://doi.org/10.1159/000324598

Bowler, L. L., Zhanel, G. G., Ball, T. B., & Saward, L. L. (2012). Mature Pseudomonas aeruginosa biofilms prevail compared to young biofilms in the presence of ceftazidime. *Antimicrobial Agents and Chemotherapy, 56*(9), 4976–4979. https://doi.org/10.1128/AAC.00650-12

Brethauer, S., Shahab, R. L., & Studer, M. H. (2020). Impacts of biofilms on the conversion of cellulose. *Applied Microbiology and Biotechnology, 104*(12), 5201–5212. https://doi.org/10.1007/S00253-020-10595-Y

Brethauer, S., & Studer, M. H. (2014). Consolidated bioprocessing of lignocellulose by a microbial consortium. *Energy & Environmental Science, 7*(4), 1446–1453. https://doi.org/10.1039/C3EE41753K

Brown, N. A., Ries, L. N. A., Reis, T. F., Rajendran, R., Corrêa dos Santos, R. A., Ramage, G., Riaño-Pachón, D. M., & Goldman, G. H. (2016). RNAseq reveals hydrophobins that are involved in the adaptation of Aspergillus nidulans to lignocellulose. *Biotechnology for Biofuels, 9*(1), 1–17. https://doi.org/10.1186/S13068-016-0558-2

Bryers, J. D. (2008). Medical biofilms. *Biotechnology and Bioengineering, 100*(1), 1–18. https://doi.org/10.1002/BIT.21838

Carniol, K., & Gilmore, M. S. (2004). Signal transduction, quorum-sensing, and extracellular protease activity in Enterococcus faecalis biofilm formation. *Journal of Bacteriology, 186*(24), 8161–8163. https://doi.org/10.1128/JB.186.24.8161-8163.2004

Ch'ng, J.-H., Chong, K. K. L., Lam, L. N., Wong, J. J., & Kline, K. A. (2018). Biofilm-associated infection by enterococci. *Nature Reviews Microbiology, 17*(2), 82–94. https://doi.org/10.1038/s41579-018-0107-z

Chakraborty, P., Bajeli, S., Kaushal, D., Radotra, B. D., & Kumar, A. (2021). Biofilm formation in the lung contributes to virulence and drug tolerance of Mycobacterium tuberculosis. *Nature Communications, 12*(1), 1–17. https://doi.org/10.1038/s41467-021-21748-6

Chaudhuri, B., Rojek, J., Vickerman, M. M., Tanzer, J. M., & Scannapieco, F. A. (2007). Interaction of salivary alpha-amylase and amylase-binding-protein A (AbpA) of Streptococcus gordonii with glucosyltransferase of S. gordonii and Streptococcus mutans. *BMC Microbiology, 7*(1), 1–9. https://doi.org/10.1186/1471-2180-7-60

Coenye, T. (2010). Response of sessile cells to stress: From changes in gene expression to phenotypic adaptation. *FEMS Immunology and Medical Microbiology, 59*(3), 239–252. https://doi.org/10.1111/J.1574-695X.2010.00682.X

Cornelissen, C. N., Fisher, B. D., & Harvey, R. A. (2013). *Lippincott illustrated reviews: Microbiology* (3rd ed.). Lippincott Williams & Wilkins.

Costerton, J. W., Montanaro, L., & Arciola, C. R. (2005). Biofilm in implant infections: Its production and regulation. *International Journal of Artificial Organs, 28*(11), 1062–1068. https://doi.org/10.1177/039139880502801103

Cowan, M. K., & Smith, H. (2012). *Microbiology: A systems approach* (3rd ed.). The McGraw-Hill Companies, Inc..

Davenport, E. K., Call, D. R., & Beyenal, H. (2014). Differential protection from tobramycin by extracellular polymeric substances from Acinetobacter baumannii and Staphylococcus aureus biofilms. *Antimicrobial Agents and Chemotherapy, 58*(8), 4755–4761. https://doi.org/10.1128/AAC.03071-14

Dumitrache, A., Wolfaardt, G. M., Allen, D. G., Liss, S. N., & Lynd, L. R. (2013). Tracking the cellulolytic activity of Clostridium thermocellum biofilms. *Biotechnology for Biofuels, 6*(1), 1–14. https://doi.org/10.1186/1754-6834-6-175

Dzianach, P. A., Dykes, G. A., Strachan, N. J. C., Forbes, K. J., & Pérez-Reche, F. J. (2019). Challenges of biofilm control and utilization: Lessons from mathematical modelling. *Journal of the Royal Society Interface, 16*, 155. https://doi.org/10.1098/RSIF.2019.0042

Egland, P. G., Palmer, R. J., & Kolenbrander, P. E. (2004). Interspecies communication in Streptococcus gordonii–Veillonella atypica biofilms: Signaling in flow conditions requires juxtaposition. *Proceedings of the National Academy of Sciences, 101*(48), 16917–16922. https://doi.org/10.1073/PNAS.0407457101

Engelkirk, P. G., & Duben-Engelkirk, J. L. (2015). *Burton's microbiology for the health sciences* (10th ed.). Wolters Kluwer Health.

Facundo Rodriguez Ayala, C. B., Bauman, C., Cogliati, S., Leñini, C., Bartolini, M., & Grau, R. (2017). Microbial flora, probiotics, *Bacillus subtilis* and the search for a long and healthy human longevity. *Microbial Cell, 4*(4), 133. https://doi.org/10.15698/MIC2017.04.569

Falsetta, M. L., Bair, T. B., Shan, C. K., Vanden Hoven, R. N., Steichen, C. T., McEwan, A. G., Jennings, M. P., & Apicella, M. A. (2009). Transcriptional profiling identifies the metabolic phenotype of gonococcal biofilms. *Infection and Immunity, 77*(9), 3522–3532. https://doi.org/10.1128/IAI.00036-09

Fleming, D., & Rumbaugh, K. (2018). The consequences of biofilm dispersal on the host. *Scientific Reports, 8*(1), 1–7. https://doi.org/10.1038/s41598-018-29121-2

Flemming, H.-C., Wingender, J., Szewzyk, U., Steinberg, P., Rice, S. A., & Kjelleberg, S. (2016). Biofilms: An emergent form of bacterial life. *Nature Reviews Microbiology, 14*(9), 563–575. https://doi.org/10.1038/nrmicro.2016.94

Francolini, I., & Donelli, G. (2010). Prevention and control of biofilm-based medical-device-related infections. *FEMS Immunology and Medical Microbiology, 59*(3), 227–238. https://doi.org/10.1111/J.1574-695X.2010.00665.X

Fung, A. H. Y., Rao, S., Ngan, W. Y., Sekoai, P. T., Touyon, L., Ho, T. M., Wong, K. P., & Habimana, O. (2021). Exploring the optimization of aerobic food waste digestion efficiency through the engineering of functional biofilm Bio-carriers. *Bioresource Technology, 341*, 125869. https://doi.org/10.1016/J.BIORTECH.2021.125869

Ganguly, S., Bishop, A. C., Xu, W., Ghosh, S., Nickerson, K. W., Lanni, F., Patton-Vogt, J., & Mitchell, A. P. (2011). Zap1 control of cell-cell signaling in candida albicans biofilms. *Eukaryotic Cell, 10*(11), 1448–1454. https://doi.org/10.1128/EC.05196-11

Garrett, T. R., Bhakoo, M., & Zhang, Z. (2008). Bacterial adhesion and biofilms on surfaces. *Progress in Natural Science, 18*(9), 1049–1056. https://doi.org/10.1016/J.PNSC.2008.04.001

Greiner, L. L., Edwards, J. L., Shao, J., Rabinak, C., Entz, D., & Apicella, M. A. (2005). Biofilm formation by Neisseria gonorrhoeae. *Infection and Immunity, 73*(4), 1964–1970. https://doi.org/10.1128/IAI.73.4.1964-1970.2005

Gutiérrez-Correa, M., Ludeña, Y., Ramage, G., & Villena, G. K. (2012). Recent advances on filamentous fungal biofilms for industrial uses. *Applied Biochemistry and Biotechnology, 167*(5), 1235–1253. https://doi.org/10.1007/S12010-012-9555-5

Hall, C. W., & Mah, T.-F. (2017). Molecular mechanisms of biofilm-based antibiotic resistance and tolerance in pathogenic bacteria. *FEMS Microbiology Reviews, 41*(3), 276–301. https://doi.org/10.1093/FEMSRE/FUX010

Hall-Stoodley, L., Costerton, J. W., & Stoodley, P. (2004). Bacterial biofilms: From the natural environment to infectious diseases. *Nature Reviews Microbiology, 2*(2), 95–108. https://doi.org/10.1038/nrmicro821

Handal, T., Olsen, I., Walker, C. B., & Caugant, D. A. (2004). β-Lactamase production and antimicrobial susceptibility of subgingival bacteria from refractory periodontitis. *Oral Microbiology and Immunology, 19*(5), 303–308. https://doi.org/10.1111/J.1399-302X.2004.00159.X

Høiby, N., Bjarnsholt, T., Givskov, M., Molin, S., & Ciofu, O. (2010). Antibiotic resistance of bacterial biofilms. *International Journal of Antimicrobial Agents, 35*(4), 322–332. https://doi.org/10.1016/J.IJANTIMICAG.2009.12.011

Hu, C., Xiong, N., Zhang, Y., Rayner, S., & Chen, S. (2012). Functional characterization of lipase in the pathogenesis of Staphylococcus aureus. *Biochemical and Biophysical Research Communications, 419*(4), 617–620. https://doi.org/10.1016/J.BBRC.2012.02.057

Iijima, S., Washio, K., Okahara, R., & Morikawa, M. (2009). Biofilm formation and proteolytic activities of Pseudoalteromonas bacteria that were isolated from fish farm sediments. *Microbial Biotechnology, 2*(3), 361–369. https://doi.org/10.1111/J.1751-7915.2009.00097.X

Janeček, Š., Svensson, B., & MacGregor, E. A. (2013). α-Amylase: An enzyme specificity found in various families of glycoside hydrolases. *Cellular and Molecular Life Sciences, 71*(7), 1149–1170. https://doi.org/10.1007/S00018-013-1388-Z

Jowiya, W., Brunner, K., Abouelhadid, S., Hussain, H. A., Nair, S. P., Sadiq, S., Williams, L. K., Trantham, E. K., Stephenson, H., Wren, B. W., Bajaj-Elliott, M., Cogan, T. A., Laws, A. P.,

Wade, J., Dorrell, N., & Allan, E. (2015). Pancreatic amylase is an environmental signal for regulation of biofilm formation and host interaction in Campylobacter jejuni. *Infection and Immunity, 83*(12), 4884–4895. https://doi.org/10.1128/IAI.01064-15

Kaplan, J. B. (2010). Biofilm dispersal: Mechanisms, clinical implications, and potential therapeutic uses. *Journal of Dental Research, 89*(3), 205. https://doi.org/10.1177/0022034509359403

Karatan, E., & Watnick, P. (2009). Signals, regulatory networks, and materials that build and break bacterial biofilms. *Microbiology and Molecular Biology Reviews, 73*(2), 310–347. https://doi.org/10.1128/MMBR.00041-08

Khan, W., Bernier, S. P., Kuchma, S. L., Hammond, J. H., Hasan, F., & O'Toole, G. A. (2010). Aminoglycoside resistance of Pseudomonas aeruginosa biofilms modulated by extracellular polysaccharide. *International Microbiology: The Official Journal of the Spanish Society for Microbiology, 13*(4), 207. https://doi.org/10.2436/20.1501.01.127

Kiedrowski, M. R., Crosby, H. A., Hernandez, F. J., Malone, C. L., & Horswill, A. R. (2014). Staphylococcus aureus Nuc2 is a functional, surface-attached extracellular nuclease. *PLoS ONE, 9*(4), e95574. https://doi.org/10.1371/JOURNAL.PONE.0095574

Kiedrowski, M. R., Kavanaugh, J. S., Malone, C. L., Mootz, J. M., Voyich, J. M., Smeltzer, M. S., Bayles, K. W., & Horswill, A. R. (2011). Nuclease modulates biofilm formation in community-associated methicillin-resistant Staphylococcus aureus. *PLoS ONE, 6*(11), e26714. https://doi.org/10.1371/JOURNAL.PONE.0026714

Kirsch, J., Hannig, C., Pötschke, S., Basche, S., Bowen, W. H., Rupf, S., Trautmann, S., Umanskaya, N., & Hannig, M. (2017). Enzymology and ultrastructure of the in situ pellicle in caries-active and caries-inactive patients. *Caries Research, 51*(2), 109–118. https://doi.org/10.1159/000452226

Klein, M. I., DeBaz, L., Agidi, S., Lee, H., Xie, G., Lin, A. H.-M., Hamaker, B. R., Lemos, J. A., & Koo, H. (2010). Dynamics of Streptococcus mutans transcriptome in response to starch and sucrose during biofilm development. *PLoS ONE, 5*(10), e13478. https://doi.org/10.1371/JOURNAL.PONE.0013478

Krishna, C. (2007). Fungal solid-state fermentation systems for bioconversion of lignocellulosic biomass: Process protocol and applications. In *Manual of environmental microbiology* (3rd ed., pp. 1107–1111). Wiley. https://doi.org/10.1128/9781555815882.CH89

Kumar, R., Saraswat, D., Tati, S., & Edgerton, M. (2015). Novel aggregation properties of Candida albicans secreted aspartyl proteinase sap6 mediate virulence in oral Candidiasis. *Infection and Immunity, 83*(7), 2614–2626. https://doi.org/10.1128/IAI.00282-15

Ladevèze, S., Haon, M., Villares, A., Cathala, B., Grisel, S., Herpoël-Gimbert, I., Henrissat, B., & Berrin, J.-G. (2017). The yeast Geotrichum candidum encodes functional lytic polysaccharide monooxygenases. *Biotechnology for Biofuels, 10*(1), 1–11. https://doi.org/10.1186/S13068-017-0903-0

Lahiri, D., Nag, M., Banerjee, R., Mukherjee, D., Garai, S., Sarkar, T., Dey, A., Sheikh, H. I., Pathak, S. K., Edinur, H. A., Pati, S., & Ray, R. R. (2021). Amylases: Biofilm inducer or biofilm inhibitor? *Frontiers in Cellular and Infection Microbiology, 10*, 355. https://doi.org/10.3389/FCIMB.2021.660048

LaPara, T. M., Konopka, A., Nakatsu, C. H., & Alleman, J. E. (2001). Thermophilic aerobic treatment of a synthetic wastewater in a membrane-coupled bioreactor. *Journal of Industrial Microbiology and Biotechnology, 26*(4), 203–209. https://doi.org/10.1038/SJ.JIM.7000110

Lee, S. F., Li, Y. H., & Bowden, G. H. (1996). Detachment of Streptococcus mutans biofilm cells by an endogenous enzymatic activity. *Infection and Immunity, 64*(3), 1035–1038. https://doi.org/10.1128/IAI.64.3.1035-1038.1996

Lewis, K. (2006). Persister cells, dormancy and infectious disease. *Nature Reviews Microbiology, 5*(1), 48–56. https://doi.org/10.1038/nrmicro1557

Lohse, M. B., Gulati, M., Craik, C. S., Johnson, A. D., & Nobile, C. J. (2020). Combination of antifungal drugs and protease inhibitors prevent Candida albicans biofilm formation and disrupt mature biofilms. *Frontiers in Microbiology, 10*, 1027. https://doi.org/10.3389/FMICB.2020.01027

Lohse, M. B., Gulati, M., Johnson, A. D., & Nobile, C. J. (2017). Development and regulation of single- and multi-species Candida albicans biofilms. *Nature Reviews Microbiology, 16*(1), 19–31. https://doi.org/10.1038/nrmicro.2017.107

Lu, Y., Zhang, Y.-H. P., & Lynd, L. R. (2006). Enzyme–microbe synergy during cellulose hydrolysis by Clostridium thermocellum. *Proceedings of the National Academy of Sciences, 103*(44), 16165–16169. https://doi.org/10.1073/PNAS.0605381103

Luo, L., Meng, H., & Gu, J. D. (2017). Microbial extracellular enzymes in biogeochemical cycling of ecosystems. *Journal of Environmental Management, 197*, 539–549. https://doi.org/10.1016/J.JENVMAN.2017.04.023

Ma, L., Conover, M., Lu, H., Parsek, M. R., Bayles, K., & Wozniak, D. J. (2009). Assembly and development of the Pseudomonas aeruginosa biofilm matrix. *PLoS Pathogens, 5*(3), e1000354. https://doi.org/10.1371/JOURNAL.PPAT.1000354

Ma, W., Peng, D., Walker, S. L., Cao, B., Gao, C.-H., Huang, Q., & Cai, P. (2017). Bacillus subtilis biofilm development in the presence of soil clay minerals and iron oxides. *Npj Biofilms and Microbiomes, 3*(1), 1–9. https://doi.org/10.1038/s41522-017-0013-6

Madsen, J. S., Burmølle, M., Hansen, L. H., & Sørensen, S. J. (2012). The interconnection between biofilm formation and horizontal gene transfer. *FEMS Immunology and Medical Microbiology, 65*(2), 183–195. https://doi.org/10.1111/J.1574-695X.2012.00960.X

Mah, T.-F., Pitts, B., Pellock, B., Walker, G. C., Stewart, P. S., & O'Toole, G. A. (2003). A genetic basis for Pseudomonas aeruginosa biofilm antibiotic resistance. *Nature, 426*(6964), 306–310. https://doi.org/10.1038/nature02122

Majzlová, K., Pukajová, Z., & Janeček, Š. (2013). Tracing the evolution of the α-amylase subfamily GH13_36 covering the amylolytic enzymes intermediate between oligo-1,6-glucosidases and neopullulanases. *Carbohydrate Research, 367*, 48–57. https://doi.org/10.1016/J.CARRES.2012.11.022

Mandel, A. L., Gachons, C. P., des Plank, K. L., Alarcon, S., & Breslin, P. A. S. (2010). Individual differences in AMY1 gene copy number, salivary α-amylase levels, and the perception of oral starch. *PLoS ONE, 5*(10), e13352. https://doi.org/10.1371/JOURNAL.PONE.0013352

Marsh, P. D., Martin, M. V., Lewis, M. A. O., & Williams, D. (2009). Oral microbiology. In *Oral microbiology* (5th ed.). Elsevier Limited.

Meibom, K. L., Li, X. B., Nielsen, A. T., Wu, C.-Y., Roseman, S., & Schoolnik, G. K. (2004). The Vibrio cholerae chitin utilization program. *Proceedings of the National Academy of Sciences, 101*(8), 2524–2529. https://doi.org/10.1073/PNAS.0308707101

Miao, L., Yu, Y., Adyel, T. M., Wang, C., Liu, Z., Liu, S., Huang, L., You, G., Meng, M., Qu, H., & Hou, J. (2021). Distinct microbial metabolic activities of biofilms colonizing microplastics in three freshwater ecosystems. *Journal of Hazardous Materials, 403*, 123577. https://doi.org/10.1016/J.JHAZMAT.2020.123577

Mitra, S., Sana, B., & Mukherjee, J. (2014). Ecological roles and biotechnological applications of marine and intertidal microbial biofilms. *Advances in Biochemical Engineering/Biotechnology, 146*, 163–205. https://doi.org/10.1007/10_2014_271

Monds, R. D., & O'Toole, G. A. (2009). The developmental model of microbial biofilms: ten years of a paradigm up for review. *Trends in Microbiology, 17*(2), 73–87. https://doi.org/10.1016/J.TIM.2008.11.001

Monzón, M., Oteiza, C., Leiva, J., Lamata, M., & Amorena, B. (2002). Biofilm testing of Staphylococcus epidermidis clinical isolates: Low performance of vancomycin in relation to other antibiotics. *Diagnostic Microbiology and Infectious Disease, 44*(4), 319–324. https://doi.org/10.1016/S0732-8893(02)00464-9

Morrell-Falvey, J. L., Elkins, J. G., & Wang, Z. W. (2015). Determination of the cellulase activity distribution in Clostridium thermocellum and Caldicellulosiruptor obsidiansis cultures using a fluorescent substrate. *Journal of Environmental Sciences, 34*, 212–218. https://doi.org/10.1016/J.JES.2015.03.009

Muffler, K., Lakatos, M., Schlegel, C., Strieth, D., Kuhne, S., & Ulber, R. (2014). Application of biofilm bioreactors in white biotechnology. *Advances in Biochemical Engineering/Biotechnology, 146*, 123–161. https://doi.org/10.1007/10_2013_267

Naglik, J. R., Challacombe, S. J., & Hube, B. (2003). Candida albicans secreted aspartyl proteinases in virulence and pathogenesis. *Microbiology and Molecular Biology Reviews, 67*(3), 400–428. https://doi.org/10.1128/MMBR.67.3.400-428.2003

Nahar, S., Mizan, M. F. R., Ha, A. J., & Ha, S.-D. (2018). Advances and future prospects of enzyme-based biofilm prevention approaches in the food industry. *Comprehensive Reviews in Food Science and Food Safety, 17*(6), 1484–1502. https://doi.org/10.1111/1541-4337.12382

Nahm, C. H., Lee, S., Lee, S. H., Lee, K., Lee, J., Kwon, H., Choo, K.-H., Lee, J.-K., Jang, J. Y., & Park, C.-H. L. (2017). Mitigation of membrane biofouling in MBR using a cellulolytic bacterium, Undibacterium sp. DM-1, isolated from activated sludge. *Journal of Microbiology and Biotechnology, 27*(3), 573–583. https://doi.org/10.4014/JMB.1610.10007

Nguyen, M. T., Luqman, A., Bitschar, K., Hertlein, T., Dick, J., Ohlsen, K., Bröker, B., Schittek, B., & Götz, F. (2018). Staphylococcal (phospho)lipases promote biofilm formation and host cell invasion. *International Journal of Medical Microbiology, 308*(6), 653–663. https://doi.org/10.1016/J.IJMM.2017.11.013

Nikitkova, A. E., Haase, E. M., & Scannapieco, F. A. (2013). Taking the starch out of oral biofilm formation: Molecular basis and functional significance of salivary α-amylase binding to oral streptococci. *Applied and Environmental Microbiology, 79*(2), 416–423. https://doi.org/10.1128/AEM.02581-12

O'Flaherty, V., Collins, G., & Mahony, T. (2010). Anaerobic digestion of agricultural residues. In *Environmental microbiology* (2nd ed., pp. 259–279). Elsevier. https://doi.org/10.1002/9780470495117.CH11

O'Sullivan, C., Burrell, P. C., Pasmore, M., Clarke, W. P., & Blackall, L. L. (2009). Application of flowcell technology for monitoring biofilm development and cellulose degradation in leachate and rumen systems. *Bioresource Technology, 100*(1), 492–496. https://doi.org/10.1016/J.BIORTECH.2008.06.028

Okshevsky, M., & Meyer, R. L. (2015). The role of extracellular DNA in the establishment, maintenance and perpetuation of bacterial biofilms. *Critical Reviews in Microbiology, 41*(3), 341–352. https://doi.org/10.3109/1040841X.2013.841639

Omadjela, O., Narahari, A., Strumillo, J., Mélida, H., Mazur, O., Bulone, V., & Zimmer, J. (2013). BcsA and BcsB form the catalytically active core of bacterial cellulose synthase sufficient for in vitro cellulose synthesis. *Proceedings of the National Academy of Sciences, 110*(44), 17856–17861. https://doi.org/10.1073/PNAS.1314063110

Ostadhossein, F., Moitra, P., Altun, E., Dutta, D., Sar, D., Tripathi, I., Hsiao, S.-H., Kravchuk, V., Nie, S., & Pan, D. (2021). Function-adaptive clustered nanoparticles reverse Streptococcus mutans dental biofilm and maintain microbiota balance. *Communications Biology, 4*(1), 1–16. https://doi.org/10.1038/s42003-021-02372-y

Palmer, J., Flint, S., & Brooks, J. (2007). Bacterial cell attachment, the beginning of a biofilm. *Journal of Industrial Microbiology and Biotechnology, 34*(9), 577–588. https://doi.org/10.1007/S10295-007-0234-4

Pandit, A., Adholeya, A., Cahill, D., Brau, L., & Kochar, M. (2020). Microbial biofilms in nature: unlocking their potential for agricultural applications. *Journal of Applied Microbiology, 129*(2), 199–211. https://doi.org/10.1111/JAM.14609

Panichikkal, J., & Krishnankutty, R. E. (2020). Rhizobacterial biofilm and plant growth promoting trait enhancement by organic acids and sugars. *Biofouling, 36*(8), 990–999. https://doi.org/10.1080/08927014.2020.1832219

Perez, A. C., Pang, B., King, L. B., Tan, L., Murrah, K. A., Reimche, J. L., Wren, J. T., Richardson, S. H., Ghandi, U., & Swords, W. E. (2014). Residence of Streptococcus pneumoniae and Moraxella catarrhalis within polymicrobial biofilm promotes antibiotic resistance and bacterial persistence in vivo. *Pathogens and Disease, 70*(3), 280–288. https://doi.org/10.1111/2049-632X.12129

Pericolini, E., Gabrielli, E., Amacker, M., Kasper, L., Roselletti, E., Luciano, E., Sabbatini, S., Kaeser, M., Moser, C., Hube, B., Vecchiarelli, A., & Cassone, A. (2015). Secretory aspartyl proteinases cause vaginitis and can mediate vaginitis caused by Candida albicans in mice. *MBio, 6*(3), 15. https://doi.org/10.1128/MBIO.00724-15

Pirlar, R. F., Emaneini, M., Beigverdi, R., Banar, M., Leeuwen, W. B., & Jabalameli, F. (2020). Combinatorial effects of antibiotics and enzymes against dual-species Staphylococcus aureus and Pseudomonas aeruginosa biofilms in the wound-like medium. *PLoS ONE, 15*(6), e0235093. https://doi.org/10.1371/JOURNAL.PONE.0235093

Pommerville, J. C. (2011). *Alcamo's fundamentals of microbiology* (9th ed.). Jones and Bartlett Publishers, LLC.

Puiu, R. A., Dolete, G., Ene, A. M., Nicoară, B., Vlăsceanu, G. M., Holban, A. M., Grumezescu, A. M., & Bolocan, A. (2017). Properties of biofilms developed on medical devices. In *Biofilms and implantable medical devices: Infection and control* (pp. 25–46). Elsevier. https://doi.org/10.1016/B978-0-08-100382-4.00002-2

Qureshi, N., Annous, B. A., Ezeji, T. C., Karcher, P., & Maddox, I. S. (2005). Biofilm reactors for industrial bioconversion processes: Employing potential of enhanced reaction rates. *Microbial Cell Factories, 4*(1), 1–21. https://doi.org/10.1186/1475-2859-4-24

Rabin, N., Zheng, Y., Opoku-Temeng, C., Du, Y., Bonsu, E., & Sintim, H. O. (2015). Biofilm formation mechanisms and targets for developing antibiofilm agents. *Future Medicinal Chemistry, 7*(4), 493–512. https://doi.org/10.4155/FMC.15.6

Rana, K. L., Kour, D., Yadav, A. N., Yadav, N., & Saxena, A. K. (2020). Agriculturally important microbial biofilms: Biodiversity, ecological significances, and biotechnological applications. In *New and future developments in microbial biotechnology and bioengineering: Microbial biofilms current research and future trends in microbial biofilms* (pp. 221–265). Elsevier. https://doi.org/10.1016/B978-0-444-64279-0.00016-5

Rather, M. A., Gupta, K., & Mandal, M. (2021). Impact of microbial biofilm on crop productivity and agricultural sustainability. In *Microbes in land use change management* (pp. 451–469). Elsevier. https://doi.org/10.1016/B978-0-12-824448-7.00023-1

Rathinam, N. K., Gorky, B., Salem, D. R., & Sani, R. K. (2020). Bioelectrochemical approach for enhancing lignocellulose degradation and biofilm formation in Geobacillus strain WSUCF1. *Bioresource Technology, 295*, 122271. https://doi.org/10.1016/J.BIORTECH.2019.122271

Richards, J. J., & Melander, C. (2009). Controlling bacterial biofilms. *ChemBioChem, 10*(14), 2287–2294. https://doi.org/10.1002/CBIC.200900317

Ríos-Castillo, A. G., Thompson, K. D., Adams, A., Mateo, M. M., & Rodríguez-Jerez, J. J. (2018). Biofilm formation of Flavobacterium psychrophilum on various substrates. *Aquaculture Research, 49*(12), 3830–3837. https://doi.org/10.1111/ARE.13849

Rodríguez-Baño, J., Gutiérrez-Gutiérrez, B., Machuca, I., & Pascual, A. (2018). Treatment of infections caused by extended-spectrum-beta-lactamase-, ampC-, and carbapenemase-producing enterobacteriaceae. *Clinical Microbiology Reviews, 31*(2), 17. https://doi.org/10.1128/CMR.00079-17

Rogers, J. D., Palmer, J., Kolenbrander, P. E., & Scannapieco, F. A. (2001). Role of Streptococcus gordonii amylase-binding protein A in adhesion to hydroxyapatite, starch metabolism, and biofilm formation. *Infection and Immunity, 69*(11), 7046–7056. https://doi.org/10.1128/IAI.69.11.7046-7056.2001

Romaní, A. M., Fund, K., Artigas, J., Schwartz, T., Sabater, S., & Obst, U. (2008). Relevance of polymeric matrix enzymes during biofilm formation. *Microbial Ecology, 56*(3), 427–436. https://doi.org/10.1007/S00248-007-9361-8

Römling, U., Kjelleberg, S., Normark, S., Nyman, L., Uhlin, B. E., & Åkerlund, B. (2014). Microbial biofilm formation: a need to act. *Journal of Internal Medicine, 276*(2), 98–110. https://doi.org/10.1111/JOIM.12242

Saising, J., Singdam, S., Ongsakul, M., & Voravuthikunchai, S. P. (2012). Lipase, protease, and biofilm as the major virulence factors in staphylococci isolated from acne lesions. *Bioscience Trends, 6*(4), 160–164. https://doi.org/10.5582/BST.2012.V6.4.160

Santos, J. L., Saus, E., Smalley, S. V., Cataldo, L. R., Alberti, G., Parada, J., Gratacòs, M., & Estivill, X. (2012). Copy number polymorphism of the salivary amylase gene: Implications in human nutrition research. *Lifestyle Genomics, 5*(3), 117–131. https://doi.org/10.1159/000339951

Scapin, S. M. N., Souza, F. H. M., Zanphorlin, L. M., Almeida, T. S., Sade, Y. B., Cardoso, A. M., Pinheiro, G. L., & Murakami, M. T. (2017). Structure and function of a novel GH8 endoglucanase from the bacterial cellulose synthase complex of Raoultella ornithinolytica. *PLoS ONE, 12*(4), e0176550. https://doi.org/10.1371/JOURNAL.PONE.0176550

Schild, L., Heyken, A., de Groot, P. W. J., Hiller, E., Mock, M., de Koster, C., Horn, U., Rupp, S., & Hube, B. (2011). Proteolytic cleavage of covalently linked cell wall proteins by Candida albicans Sap9 and Sap10. *Eukaryotic Cell, 10*(1), 98–109. https://doi.org/10.1128/EC.00210-10

Seper, A., Fengler, V. H. I., Roier, S., Wolinski, H., Kohlwein, S. D., Bishop, A. L., Camilli, A., Reidl, J., & Schild, S. (2011). Extracellular nucleases and extracellular DNA play important roles in Vibrio cholerae biofilm formation. *Molecular Microbiology, 82*(4), 1015–1037. https://doi.org/10.1111/J.1365-2958.2011.07867.X

Sharma, D., Misba, L., & Khan, A. U. (2019). Antibiotics versus biofilm: An emerging battleground in microbial communities. *Antimicrobial Resistance and Infection Control, 8*(1), 1–10. https://doi.org/10.1186/S13756-019-0533-3

Sheppard, D. C., & Howell, P. L. (2016). Biofilm exopolysaccharides of pathogenic fungi: Lessons from bacteria. *Journal of Biological Chemistry, 291*(24), 12529–12537. https://doi.org/10.1074/JBC.R116.720995

Shrestha, S., Kognou, A. L. M., Zhang, J., & Qin, W. (2020). Different facets of lignocellulosic biomass including pectin and its perspectives. In *Waste and biomass valorization* (Vol. 1, p. 3). Springer Science and Business Media. https://doi.org/10.1007/s12649-020-01305-w

Souza, J. G. S., Bertolini, M., Thompson, A., Mansfield, J. M., Grassmann, A. A., Maas, K., Caimano, M. J., Barao, V. A. R., Vickerman, M. M., & Dongari-Bagtzoglou, A. (2020). Role of glucosyltransferase R in biofilm interactions between Streptococcus oralis and Candida albicans. *The ISME Journal, 14*(5), 1207–1222. https://doi.org/10.1038/s41396-020-0608-4

Stacy, A., McNally, L., Darch, S. E., Brown, S. P., & Whiteley, M. (2015). The biogeography of polymicrobial infection. *Nature Reviews Microbiology, 14*(2), 93–105. https://doi.org/10.1038/nrmicro.2015.8

Steichen, C. T., Cho, C., Shao, J. Q., & Apicella, M. A. (2011). The Neisseria gonorrhoeae biofilm matrix contains DNA, and an endogenous nuclease controls its incorporation. *Infection and Immunity, 79*(4), 1504–1511. https://doi.org/10.1128/IAI.01162-10

Stewart, P. S. (2002). Mechanisms of antibiotic resistance in bacterial biofilms. *International Journal of Medical Microbiology, 292*(2), 107–113. https://doi.org/10.1078/1438-4221-00196

Stewart, P. S., & Franklin, M. J. (2008). Physiological heterogeneity in biofilms. *Nature Reviews Microbiology, 6*(3), 199–210. https://doi.org/10.1038/nrmicro1838

Talaro, K. P., & Talaro, A. (2002). *Foundations in microbiology* (4th ed.). The McGraw-Hill Companies.

Tang, J., Kang, M., Chen, H., Shi, X., Zhou, R., Chen, J., & Du, Y. (2011). The staphylococcal nuclease prevents biofilm formation in Staphylococcus aureus and other biofilm-forming bacteria. *Science China. Life Sciences, 54*(9), 863–869. https://doi.org/10.1007/S11427-011-4195-5

Thomas, V. C., Thurlow, L. R., Boyle, D., & Hancock, L. E. (2008). Regulation of autolysis-dependent extracellular DNA release by Enterococcus faecalis extracellular proteases influences biofilm development. *Journal of Bacteriology, 190*(16), 5690–5698. https://doi.org/10.1128/JB.00314-08

Tielen, P., Rosenau, F., Wilhelm, S., Jaeger, K.-E., Flemming, H.-C., & Wingender, J. (2010). Extracellular enzymes affect biofilm formation of mucoid Pseudomonas aeruginosa. *Microbiology, 156*(7), 2239–2252. https://doi.org/10.1099/MIC.0.037036-0

Todhanakasem, T., Narkmit, T., Areerat, K., & Thanonkeo, P. (2015). Fermentation of rice bran hydrolysate to ethanol using Zymomonas mobilis biofilm immobilization on DEAE-cellulose.

Electronic Journal of Biotechnology, 18(3), 196–201. https://doi.org/10.1016/J.EJBT.2015.03.007

Todhanakasem, T., Salangsing, O., Koomphongse, P., Kaewket, S., Kanokratana, P., & Champreda, V. (2019). Zymomonas mobilis biofilm reactor for ethanol production using rice straw hydrolysate under continuous and repeated batch processes. *Frontiers in Microbiology, 10*, 1777. https://doi.org/10.3389/FMICB.2019.01777

Torok, E., Moran, E., & Cooke, F. (2009). *Oxford handbook of infectious diseases and microbiology*. Oxford University Press. https://doi.org/10.1093/MED/9780198569251.001.0001

Tseng, B. S., Reichhardt, C., Merrihew, G. E., Araujo-Hernandez, S. A., Harrison, J. J., MacCoss, M. J., & Parsek, M. R. (2018). A biofilm matrix-associated protease inhibitor protects Pseudomonas aeruginosa from proteolytic attack. *MBio, 9*(2), 18. https://doi.org/10.1128/MBIO.00543-18

Ünal Turhan, E., Erginkaya, Z., Korukluoğlu, M., & Konuray, G. (2019). Beneficial biofilm applications in food and agricultural industry. In *Health and safety aspects of food processing technologies* (pp. 445–469). Springer. https://doi.org/10.1007/978-3-030-24903-8_15

van Wolferen, M., Orell, A., & Albers, S.-V. (2018). Archaeal biofilm formation. *Nature Reviews Microbiology, 16*(11), 699–713. https://doi.org/10.1038/s41579-018-0058-4

Van Wyk, N., Navarro, D., Blaise, M., Berrin, J.-G., Henrissat, B., Drancourt, M., & Kremer, L. (2017). Characterization of a mycobacterial cellulase and its impact on biofilm- and drug-induced cellulose production. *Glycobiology, 27*(5), 392–399. https://doi.org/10.1093/GLYCOB/CWX014

Varrot, A., Leydier, S., Pell, G., Macdonald, J. M., Stick, R. V., Henrissat, B., Gilbert, H. J., & Davies, G. J. (2005). Mycobacterium tuberculosis strains possess functional cellulases. *Journal of Biological Chemistry, 280*(21), 20181–20184. https://doi.org/10.1074/JBC.C500142200

Wang, L., Dong, M., Zheng, J., Song, Q., Yin, W., Li, J., & Niu, W. (2011). Relationship of biofilm formation and gelE gene expression in Enterococcus faecalis recovered from root canals in patients requiring endodontic retreatment. *Journal of Endodontics, 37*(5), 631–636. https://doi.org/10.1016/J.JOEN.2011.02.006

Wang, Z.-W., & Chen, S. (2009). Potential of biofilm-based biofuel production. *Applied Microbiology and Biotechnology, 83*(1), 1–18. https://doi.org/10.1007/S00253-009-1940-9

Wang, Z. W., Lee, S. H., Elkins, J. G., Li, Y., Hamilton-Brehm, S., & Morrell-Falvey, J. L. (2013). Continuous live cell imaging of cellulose attachment by microbes under anaerobic and thermophilic conditions using confocal microscopy. *Journal of Environmental Sciences, 25*(5), 849–856. https://doi.org/10.1016/S1001-0742(12)60104-1

Waters, C. M., Antiporta, M. H., Murray, B. E., & Dunny, G. M. (2003). Role of the Enterococcus faecalis GelE protease in determination of cellular chain length, supernatant pheromone levels, and degradation of fibrin and misfolded surface proteins. *Journal of Bacteriology, 185*(12), 3613–3623. https://doi.org/10.1128/JB.185.12.3613-3623.2003

Whitchurch, C. B., Tolker-Nielsen, T., Ragas, P. C., & Mattick, J. S. (2002). Extracellular DNA required for bacterial biofilm formation. *Science, 295*(5559), 1487. https://doi.org/10.1126/SCIENCE.295.5559.1487

Wilson, M. (2008). *Bacteriology of humans: An ecological perspective*. Blackwell Publishing Ltd..

Xiros, C., Shahab, R. L., & Studer, M. H.-P. (2019). A cellulolytic fungal biofilm enhances the consolidated bioconversion of cellulose to short chain fatty acids by the rumen microbiome. *Applied Microbiology and Biotechnology, 103*(8), 3355–3365. https://doi.org/10.1007/S00253-019-09706-1

Xiros, C., & Studer, M. H. (2017). A multispecies fungal biofilm approach to enhance the celluloyltic efficiency of membrane reactors for consolidated bioprocessing of plant biomass. *Frontiers in Microbiology, 10*, 1930. https://doi.org/10.3389/FMICB.2017.01930

Xiros, C., Topakas, E., & Christakopoulos, P. (2013). Hydrolysis and fermentation for cellulosic ethanol production. *Wiley Interdisciplinary Reviews: Energy and Environment, 2*(6), 633–654. https://doi.org/10.1002/wene.49

Yadav, S., & Chandra, R. (2020). Biofilm-mediated bioremediation of pollutants from the environment for sustainable development. In *New and future developments in microbial biotechnology and bioengineering: Microbial biofilms current research and future trends in microbial biofilms* (pp. 177–203). Elsevier. https://doi.org/10.1016/B978-0-444-64279-0.00014-1

Zhang, Q., Yu, Z., Jin, S., Zhu, L., Liu, C., Zheng, H., Zhou, T., Liu, Y., & Ruan, R. (2019). Lignocellulosic residue as bio-carrier for algal biofilm growth: Effects of carrier physicochemical proprieties and toxicity on algal biomass production and composition. *Bioresource Technology, 293*, 122091. https://doi.org/10.1016/J.BIORTECH.2019.122091

Zoppini, A., & Marxsen, J. (2010). Importance of extracellular enzymes for biogeochemical processes in temporary river sediments during fluctuating dry–wet conditions. In *Soil enzymology* (pp. 103–117). Springer. https://doi.org/10.1007/978-3-642-14225-3_6

Changes in the Attributes of the Oxisol "Arenito Caiuá" After the Use of the Crop–Livestock Integration System

Alex Eduardo Zaniboni, Higo Forlan Amaral, Hâmara Milaneze de Souza Zaniboni, Estyfany Kelle da Silva Kodaka Walichek, and José Ozinaldo Alves de Sena

Abstract Integrated systems such as crop-livestock integration (CLI) allow for the more remarkable preservation of soil and its biological characteristics as soil enzymes. CLI use can improve some soil characteristics and the recovery of degraded pastures, incredibly fragile soils for agriculture and livestock. The balance between nutrients and cycling could show how the agricultural model was adopted. Some CLI combinations and soil enzymes correlations with other soil attributes could demonstrate how sustainable these areas are. This study aimed to investigate the possible changes in an Oxisol's chemical and microbiological attributes when converting continuous pasture to CLI. This work was carried out in an agricultural area in the municipality of Amaporã in the state of Paraná, Brazil. Initially, there was continuous pasture cultivation with livestock in all six areas. At different times, the pasture was removed, soybeans and corn were cultivated, and then returned with the pasture and livestock with another type of forage. Friedman's test (a p-value of 0.05), correlation analysis, and principal component analysis were applied. Improved soil density and Ctot values are the macroeconomic indicators of the benefits of converting from continuous livestock (LS-c) to Crop-LS (1 and 2 years). The areas under Crop-LS at 1 and 2 years were the soil attributes that indicated a particular improvement the soil sustainability (mainly about carbon (C), microbial biomass of carbon (MBC), and soil enzyme —considering that this soil under continuous pasture is harmful to sustainable livestock. The crop-livestock system (CLS) is a sustainable production strategy that integrates agricultural and livestock

A. E. Zaniboni · H. M. de Souza Zaniboni · J. O. A. de Sena
Universidade Estadual de Maringá (UEM), Pós-Graduação em Agroecologia (PROFAGROEC), Maringá, Paraná, Brazil

H. F. Amaral (✉)
Universidade Estadual de Maringá (UEM), Pós-Graduação em Agroecologia (PROFAGROEC), Maringá, Paraná, Brazil

Departamento de Agronomia, Centro Universitário Filadélfia (UniFil), Londrina, Paraná, Brazil

E. K. da Silva Kodaka Walichek
Departamento de Agronomia, Centro Universitário Filadélfia (UniFil), Londrina, Paraná, Brazil

N. R. Maddela et al. (eds.), *Ecological Interplays in Microbial Enzymology*, Environmental and Microbial Biotechnology,
https://doi.org/10.1007/978-981-19-0155-3_8

activities carried out in the same area - considering this soil under continuous pasture is harmful to sustainable Livestock. The crop-livestock system (CLS) can be defined as a sustainable production strategy that integrates agricultural and livestock activities carried out in the same area. The conversion from continuous pasture to livestock crops, mainly soybeans and corn, caused positive conservation of soil carbon attributes and microbial activity. The studied soil conditions are favorable for improving agricultural and livestock activities and environmental quality and sustainability.

1 Introduction

The northwestern region of Paraná occupies 3.2 million hectares, representing 16% of this Brazilian state. The "Arenito Caiuá" (with a sandy clay loam texture) has a characteristic medium of sandy texture and, consequently, high erosion susceptibility and low water storage capacity (Fidalski, Tormena, & Alves, 2013). These soil and climate characteristics make the agrosystem areas exceptionally vulnerable to abiotic stresses, leading to plant and animal production restrictions, which have triggered the search for technologies to mitigate possible economic losses and enable greater productivity (Franchini, Balbinot Junior, Debiasi, & Conte, 2014; Nasielski et al., 2015).

Integrated systems such as crop–livestock integration (CLI) allow for more remarkable preservation of soil and its biological characteristics and result in increased productivity with favorable economic results in both medium and long terms (Ryschawy, Martin, Moraine, Duru, & Therond, 2017). CLI consolidates itself as an essential production system, which promotes technological, ecological, and socioeconomic components (Cordeiro, Vilela, Marchão, Kluthcouski, & Júnior, 2015), and also as an emerging way to mitigate greenhouse gases, sequester carbon, and promote the diversification of agricultural activities (de Moraes, Carvalho, Lustosa, Lang, & Deiss, 2014).

CLI is used for the improvement of the physical, chemical, and biological traits of soil and the recovery of degraded pastures; more efficient control of diseases and weeds; greater efficiency in the use of inputs and reduction in the use of pesticides; improvement of microclimatic conditions, increase in relative air humidity, and decrease in wind intensity; increase in animal welfare; and mitigation of the effects of greenhouse gas emissions by carbon sequestration and biodiversity promotion (Balbino, Barcellos, & Stone, 2011; Oliveira et al., 2018).

The balance between nutrients and their cycling and decomposition can determine the agricultural model adopted and new combinations. According to Navroski, Moreira, Colozzi-Filho, and Grange (2017), a crop–livestock rotation (CLR) system that integrates agricultural production with livestock increases the amount and activity of microbial biomass in the soil and contributes to the carbon stock. We hypothesized a plan for crop–livestock rotation (CLR) for soil recovery. Thus, the objective of this research was to evaluate the soil traits to measure how changes in these characteristics could improve soil recovery.

2 Materials and Methods

2.1 Description of the Location and Experimental Areas

The study was carried out at Fazenda Santa Helena located in the municipality of Amaporã, Paraná, Brazil (23° 13′ 19″ S and 52° 86′ 35″ W). The soil "Arenito Caiuá" is an oxisol with a sandy clay loam texture (Santos et al., 2018). The climate was characterized as "Cfa," with an average temperature of below 18 °C in the cold months and above 22 °C in the hot months (Peel, Finlayson, & McMahon, 2007). Rainfall indices show an average annual rainfall of 1200–1400 mm with higher concentrations in December, January, and February (Peel et al., 2007). Before 2015, the area was used as pasture without correcting soil acidity and nutrient balance (fertilizers). In 2015, six plots were subdivided as described in Fig. 1.

The area plots were managed with the characteristic soybeans, corn, and livestock. The historical de agriculture and Livestock was (below) regarding the management and use of the land at the time of soil sampling. In each area plot, eight samples composed of three replicates were collected along the largest transect (Table 1).

2.2 Soil Sampling and Density

In each field, a grid was established in the center for a quadrant of 50 m by 20 m, totaling 1000 m^2, defining 4 central points, and, at each end, 6 subsamples were

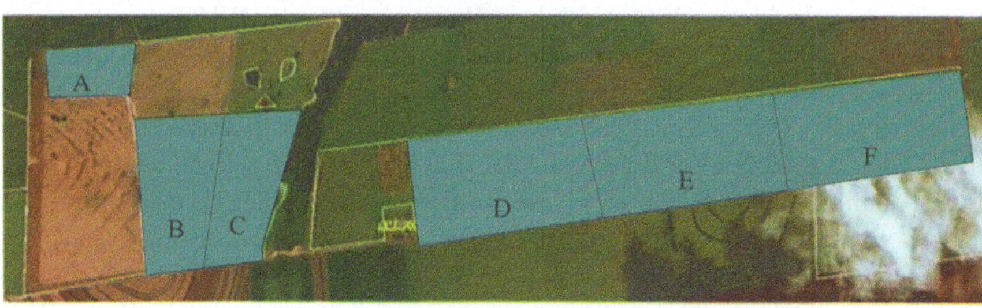

Note: in A, 6.98 ha of the area and 1.1 km of the perimeter; in B, 19.73 ha of the area and 1.9 km of the perimeter; in C, 17.5 ha of the area and 1.87 km of the perimeter; in D, 31,69 ha of the area and 2.35 km of the perimeter; in E, 32.52 ha of the area and 2.35 km of the perimeter; in F, 30.49 ha of the area and 2.36 km of the perimeter.

Fig. 1 A diagrammatic map of the experimental areas at different stages of the crop–livestock integration (CLI). "Arenito Caiuá" is an oxisol found in the municipality of Amaporã, Paraná, Southern Brazil. a = forest (For), b = livestock after 2-year crops (LS-2y), c = livestock after 1-year crop (LS-1y), d = continuous pasture (livestock) (LS-c), e = crops at 1 year after livestock integration (Crop-1y), and f = crops at 2 years after livestock integration (Crop-2y)

Table 1 Areas at different stages of the crop–livestock integration (CLI). "Arenito Caiuá" is an oxisol found in the municipality of Amaporã, Paraná, Southern Brazil

Code	Areas	Description
A	Forest (For)	There was a predominance of a tropical rainforest on the Third Plateau of Paraná
B	Livestock after 2-year crops (LS-Crop-2y)	Livestock with *Urochloa* (=Brachiaria) *brizantha* and animal production (4 unit ha^{-1}) after 2 years on soybeans and corn in the summer and winter
C	Livestock after 1-year crop (LS-Crop-1y)	Livestock with *Urochloa* (=Brachiaria) *brizantha* and animal production (2 unit ha^{-1}) after 1 year on soybeans and corn in the summer and winter
D	Continuous pasture (livestock) (LS-c)	>10 years with *Paspalum notatum* without animal density control and correction for acidity and soil nutrients. Animal production, 0.5 animals ha^{-1}
E	Crops at 1 year after livestock integration (Crop-1y)	Soybeans and corn in the first year of CLI; there was desiccation and harrowing of *Paspalum notatum* and correction of soil acidity and fertilization of these crops via mineral fertilization
F	Crops at 2 years after livestock integration (Crop-2y)	Soybeans and corn in the second year of CLI; soil pH and nutrients were monitored and corrected via mineral fertilization

collected, totaling 24 for each field. A Dutch-type tract was used at a depth of 0–10 cm. For chemical and microbiological analysis, the soil was sieved using a 2 mm mesh and stored at 5 °C.

Field-moist soil samples were sieved (4 mm) and stored in the dark at 6–8 °C until analysis less than 5 days later. Data were collected on a dry soil basis, and the soil gravimetric water content was determined by drying the subsamples in an oven for 48 h at 105 °C. Soil water holding capacity (WHC) was measured by repeatedly saturating soils (10 g fresh weight) with deionized water and draining in between 2.5 h in a funnel with an ash-free cellulose filter paper. The soil WHC of each sample was adjusted to 55–60% before performing the microbial activity and C microbial biomass analyses.

Soil density (DS), also at each subsampling point, was measured by the volumetric ring method (TEIXEIRA et al., 2017) at a depth from 0 to 10 cm.

A metallic cylinder or volumetric ring collected the samples. The sampling was handled with care, avoiding compaction of the soil inside the cylinder by excavating the soil around it to the extent that it was inserted into the ground. The dimensions of the cylinder containing the samples were measured and noted using a caliper. With these data, the volume of the cylinder was calculated. After collection, the samples from the cylinder were removed and transferred to a numbered container of known mass. Thereafter, they were dried in an oven at 105 °C for 48 h, and, after allowing them to cool in a desiccator, they were weighed.

The calculation used to measure the results is as follows: DS = (ma/*V*), where DS is the density of the soil (in g dm^{-3}, equivalent to g cm^{-3}), ma is the mass of the dry soil sample at 105 °C to constant weight (in g), and *V* is the cylinder volume (in cm^3).

2.3 Chemical Analyses

For the chemical analysis of the soil, the samples were air-dried, sieved (2 mm), and chemically analyzed according to the Paraná recommendations for the described procedures (Pavan, de Bloch, da Zempulski, Miyazawa, & Zocoler, 1992). The soil reaction (pH) was determined potentiometrically using a soil–$CaCl_2$ solution (0.01 mol L^{-1}) of 1:2.5. The single-buffer Shoemaker, McLean, and Pratt (SMP) method determined the potential acidity (H + Al) (Mclean, 2015). The total organic carbon (C) concentration in the soil was measured by the Walkley–Black sulfuric acid–potassium dichromate oxidation procedure (Nelson & Sommers, 2015). The method Mehlich-1 extracted phosphorus (P) and potassium (K^+), and their concentrations were determined colorimetrically using an ultraviolet (UV)-visible spectrophotometer and a flame photometer, respectively (Pavan et al., 1992). Calcium (Ca^{+2}) and magnesium (Mg^{+2}) were extracted with an unbuffered KCl solution (1.0 mol L^{-1}) and determined with an atomic absorption spectrophotometer (Pavan et al., 1992).

2.4 Microbiological Analyses

The microbial biomass of carbon (MBC) was evaluated by fumigation–extraction with chloroform (Vance, Brookes, & Jenkinson, 1987). Briefly, a subsample (20 g of moist soil) was fumigated with alcohol-free chloroform in closed desiccators for 24 h and extracted with K_2O_4 (0.5 mol L^{-1}). A second unfumigated set of soil subsamples was extracted under similar conditions. The amount of C in the extracts was measured with the potassium dichromate–sulfuric acid oxidation procedure (Nelson & Sommers, 2015). The microbial biomass of carbon (MBC) was calculated based on the differences between the C extracted from the fumigated and non-fumigated soil samples, with the conversion factors Kc = 0.33 for C of microbial biomass (Sparling & West, 1988), and the MBC: Ctot ratio was calculated, resulting in a metabolic quotient (qCO_2) (Anderson & Domsch, 1993).

Basal respiration (BR) was determined by incubating the soil samples for 7 days at 25 ± 2 °C in the dark in a closed flask with NaOH to trap CO_2, followed by flow injection analysis to measure electric conductivity (Doran & Zander, 2012). The following microbial indices were calculated: microbial quotient (qMIC) as BMC: Corg (Organic carbon = Corg) by dividing microbial biomass C by Corg and multiplying by 100 to express in percentage and the microbial metabolic quotient (=qCO_2) based on the relationship of BR (μg C–CO_2 g $soil^{-1}$ h^{-1}) per milligram of microbial biomass C.

The total microbial enzymatic activity was determined by the hydrolysis method of fluorescein diacetate (FDA), which, according to Schnurer and Rosswall (1996), reflects the soil microbial activity and is correlated with soil microbial biomass. Briefly, 5 g of soil was added into a 125 mL Erlenmeyer flask (3 repetitions). Adding

20 mL sodium phosphate buffer, 200 μl of the fluorescein diacetate solution, the vials were closed with aluminum foil, incubated, and the mixture placed on a shaker at 150 rpm at 24 °C for 20 min. After the incubation period, 20 mL of acetone was added to stop the reaction. The soil suspension was centrifuged at 4000 rpm for 10 min and filtered, collecting the supernatant aliquot for analysis using a spectrophotometer at 490 nm; the same procedure was performed for the control sample.

2.5 Statistical Approaches

Friedman's test (p-value of 0.05), correlation analysis, and principal component analysis were carried out by the PAST v.4 software.

3 Results and Discussion

3.1 Soil Density and Chemical Properties

The area (LS-c) had the highest standard deviation (SD) mean, and following from the largest to the smallest, it was Crop-2y > Forest > LS-1y > LS-2y (Table 2). All areas have high potential acidity under 5.9. The highest means of Ctot were under Forest, and the highest means of P-values were under LS-2y and LS-1y. Soils under LS-c and Crop-1y accumulated (in average, higher values of exchangeable bases), including $V\%$ of 50.8% under Crop-2y (Table 2).

Soil and water conservation requires that agricultural systems contribute not only to agricultural and livestock production but also to improving the conditions of the environment. Crop rotation, that is, alternating grasses, legumes, and other crops, which can be interspersed with fallow, and monitoring acidity are essential for conservationist agriculture (Telles, Righetto, da Costa, Volsi, & de Oliveira, 2019). Improved soil density and Ctot values are macroeconomic indicators of the benefits of converting from LS-c to Crop-LS. According to de Moraes et al. (2014), the integrated systems increase the values of organic C and its humus fraction. Phosphate fertilization can limit these CLR systems; after 2 years under crops (soybeans/corn), it reaches values close to the forest soil. The nutrients must be monitored to avoid degradation of soil and water sources in these types of soils; shown under Crop-1y and Crop-2y, the P-values decreased significantly, which weakens integrated crops as soybean and corn production integrates with livestock.

Table 2 Soil density and the chemical properties (0–10 cm layer) of the crop–livestock system (CLS) at different stages in Arenito Caiuá (an oxisol) found in Amaporã, Paraná state in Southern Brazil under subtropical climate conditions

Area descriptions	SD (g cm^{-3})	pH	C (g dm^{-3})	P (mg dm^{-3})	Ca (cmol$_c$ dm^{-3})	Mg (cmol$_c$ dm^{-3})	K (cmol$_c$ dm^{-3})	*V* (%)
Livestock after 2-year crops (crop-LS-2y)	0.042d[a]	5.50a	6.86b	12.06a	0.97b	0.58b	0.09c	42.45b
Livestock after one-year crop (crop-LS-1y)	0.062c	5.95a	7.07b	13.44a	1.25a	0.57b	0.10c	48.90a
Continuous pasture (livestock) (LS-c)	0.118a	5.20a	6.64b	4.24c	0.67d	0.65a	0.14b	40.00b
Crops at first year after livestock integration (Crop-1y)	0.045d	5.60a	7.67b	5.85c	1.06a	0.69a	0.30a	46.54a
Crop-2y after livestock integration (Crop-2y)	0.09b	5.80a	7.16b	7.42b	0.88c	0.63a	0.19b	50.80a
Forest	0.074c	4.60a	10.43a	8.82b	1.19a	0.44c	0.20b	33.03c
Coeffect of variation (%)	12.1	6.7	14.5	16.4	15.7	13.9	17.5	14.3

Here, pH is measured by $CaCl_2$ (0.01 mol L^{-1}), Ca and Mg by KCl (1 mol L^{-1}), *K* and *P* by Mehlich^{-1}, and *C* by the Walkley–Black procedure
[a] Means followed by the same letter do not differ by Friedman's test ($p < 0.005$)

Table 3 Soil microbial properties (0–10 cm layer) of the crop–livestock system (CLS) at different stages in Arenito Caiuá (an oxisol) found in Amaporã, Paraná state in Southern Brazil under subtropical climate conditions

Area descriptions	MBC (μg g $soil^{-1}$)	BR (μg C-CO_2 h^{-1} g $soil^{-1}$)	qCO_2 (μg C-CO_2 h^{-1} g $soil^{-1}$)	qMIC (%)	FDA (Enz-n)
Livestock after 2-year crops (LS-2y crop)	278.6a	0.67b	2.40c	2.75d	9.5b
Livestock after 1-year crop (LS-1y crop)	223.6a	0.57c	2.55c	3.20c	9.3b
Continuous pasture (livestock) (LS-c)	147.0c	1.08a	7.35a	4.52b	8.1c
Crops at first year after livestock integration (Crop-1y)	125.5c	0.48c	3.82b	5.47a	8.3c
Crop-2y after livestock integration (Crop-2y)	196.8b	0.72b	3.62b	3.55c	6.5d
Forest	275.1a	0.70b	2.54c	3.79c	16.5a
Coeffect of variation (%)	22.5	24.6	18.9	16.3	16.7

Means followed by the same letter do not differ by Friedman's test, $\alpha = 0.05$ of significance. *FDA* Enz-*n* per 5 g of soil + 20 mL sodium phosphate buffer and 200 μl of the fluorescein diacetate solution

3.2 *Microbial Biomass and Activity*

Soils under LS-2y, LS-1y, and Forest showed the highest BMC averages (Table 3). The highest average of BR (1.08) with a BMC value of 147 and a qCO_2 value of 7.35 indicates the most remarkable microbiological disturbance of the studied areas (Table 3). This qCO_2 value is (a minimum) double the sites at different moments of crop–livestock integration.

Ctot and BMC values to ideal (sustainable) values of qMIC close to soils under forest. Thus, Ctot there were no differences between cultivated areas in Livestock after two harvests (LS-2y) was the best condition for converting carbon into BMC and its activity - confirmed by the high FDA value(9.5).

In the soil under different crop–livestock systems, the values of enzymes by FDA were significantly ($p<0.005$) correlated (negatively and positively) with pH (−), Ctot (+), Mg (−), and V(−) (Fig. 2). Moreover, the FDA values had a positive association with both Ca and BMC.

The areas under Crop-LS at 1 and 2 years were the soil attributes indicating a certain degree of sustainability (mainly about C, MBC, and FDA) – considering that this soil under continuous pasture is harmful to sustainable livestock. The CLS can be defined as a sustainable production strategy that integrates agricultural and livestock activities carried out in the same area. The implementation of these systems is based on the principles of crop rotation and intercropping between grain and forage crops to produce, in the same area, grains and meat or milk throughout the year (Cordeiro et al., 2015). The diversity of activity generates income

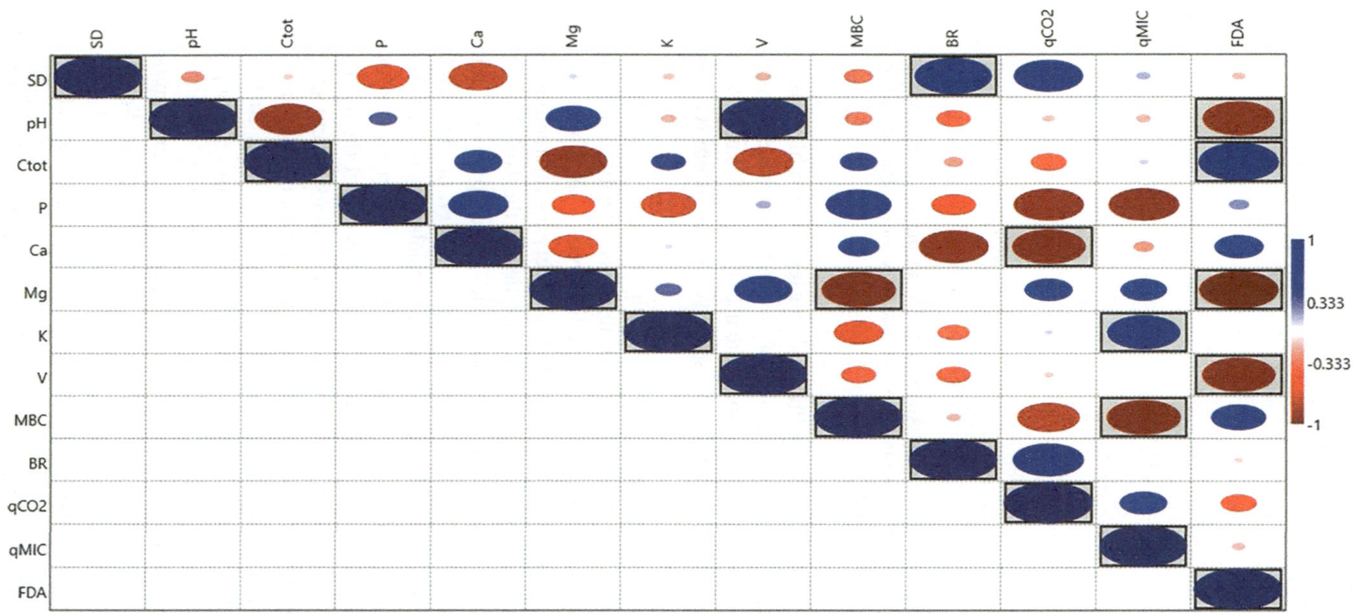

Note: blue spot indicates positive correlation (from 0 to 1), red spot negative correlation (from 0 to -1). Square around the spot means a significant correlation by the Spearman test ($p < 0.05$). When: SD = soil density, pH = hydrogen potential (H^+), Ctot = total of carbon, P = phosphorus, Ca = calcium, Mg = magnesium, K = potassium, V = base saturation, MBC = microbial biomass of carbon, BR = basal respiration, qCO2 =quotient metabolic, qMIC = microbial quotient and FDA = soil enzyme.

Fig. 2 Correlation of the soil attributes with the crop–livestock system (CLS) at different stages in Arenito Caiuá (an oxisol) found in Amaporã, Paraná state in Southern Brazil under subtropical climate conditions

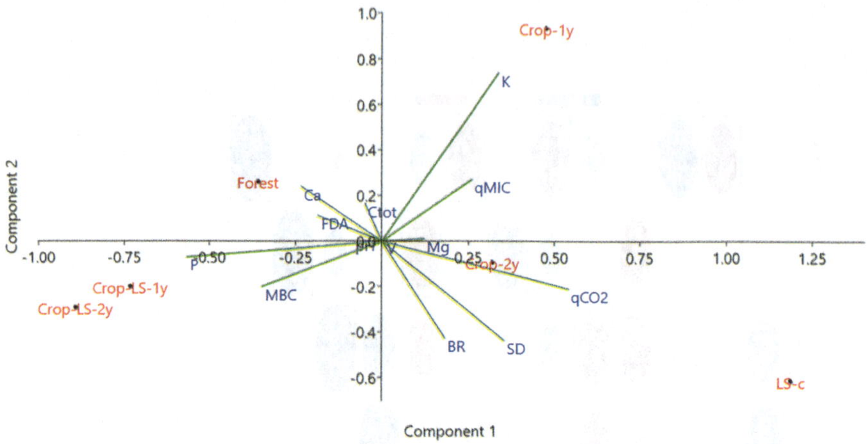

Fig. 3 Soil attributes Principal Component Analisys (PCA) in the six crop–livestock systems (CLSs) at different stages in Arenito Caiuá (an oxisol) found in Amaporã, Paraná state in Southern Brazil under subtropical climate conditions

diversification on the property, making the producer less vulnerable to bad weather and economic seasonality. Considering that agricultural activities have a higher commercial risk when compared to livestock, the use of the CLS can be an excellent alternative to promote greater cost-effective security in the system (Oliveira et al., 2018; Ryschawy et al., 2017).

Soil is a natural and vital resource for developing the ecosystem, and its phases (liquid, gaseous, and solid fractions) must be harmonious to enable equilibrium. Soil is composed of numerous living beings, including macro and microorganisms. Regardless of the genus and species, they live in society by interacting with each other, and many forms are highly active communities in the soil (Kallenbach & Grandy, 2011). These organisms, especially bacteria, make up the soil microbiota and play an essential role in decomposition of organic matter, nutrient cycling, and biological nitrogen fixation, among other aspects (Nasielski et al., 2015; Oliveira et al., 2018). Due to its significant participation in the soil dynamics, the microbiota stands out as an essential indicator of its quality since it is one of the most sensitive parameters about the changes in soil due to its use and management (Amaral, Ozinaldo Alves Sena, Regina Freitas Schwan-Estrada, Libório Balota, & Souza Andrade, 2011; Pontes, Amaral, & Igarashi, 2018; Schroeder et al., 2020).

According to Goedert et al. (2013), the significant increases in crop residues added by different cropping systems in response to P fertilizers increased soil C content, corroborating our results under LS-Crop 1 and 2 years and similar results under Forest sites (Table 2 and Fig. 3). Soil organic matter stocks are directly related to the amount of C added by the cropping system (Balota et al., 2014; Goedert et al., 2013), which increases microbial carbon biomass and soil enzyme activity (FDA) (Table 3 and Fig. 3).

4 Conclusions

Changes from continuous pasture to livestock crop, mainly soybeans and corn, caused positive conservation of soil carbon attributes and microbial activity. The studied soil conditions are favorable to improving agricultural and livestock activities and environmental quality and sustainability.

References

Amaral, H. F., Ozinaldo Alves Sena, J., Regina Freitas Schwan-Estrada, K., Libório Balota, E., & Souza Andrade, D. (2011). Soil chemical and microbial properties in vineyards under organic and conventional management in southern Brazil. *Revista Brasileira de Ciência do Solo, 35*, 1517–1526.

Anderson, T., & Domsch, K. (1993). The metabolic quotient for CO2 (qCO2) as a specific activity parameter to assess the effects of environmental conditions, such as ph, on the microbial biomass of forest soils. *Soil Biology and Biochemistry, 25*, 393–395. https://doi.org/10.1016/0038-0717(93)90140-7

Balbino, L. C., Barcellos, A. O., & Stone, L. F. (2011). *Reference document crop-livestock-forestry integration*. EMBRAPA.

Balota, E. L., Yada, I. F., Amaral, H., Nakatani, A. S., Dick, R. P., & Coyne, M. S. (2014). Long-term land use influences soil microbial biomass p and s, phosphatase and arylsulfatase activities, and s mineralization in a Brazilian oxisol. *Land Degradation and Development, 25*, 397–406. https://doi.org/10.1002/ldr.2242

Cordeiro, L. A. M., Vilela, L., Marchão, R. L., Kluthcouski, J., & Júnior, G. B. M. (2015). Integração lavoura-pecuária e integração lavoura-pecuária-floresta: Estratégias para intensificação sustentável do uso do solo Segundo Pretty (2008), o interesse pela sustentabilidade da agricultura das décadas de 1950 e 1960, apesar de existirem idei. *Cadernos de Ciência & Tecnologia, 32*, 15–43.

de Moraes, A., Carvalho, P. C., Lustosa, S. B., Lang, C. R., & Deiss, L. (2014). Research on integrated crop-livestock systems in Brazil. *Revista Ciência Agronômica, 45*, 1024–1031. https://doi.org/10.1590/s1806-66902014000500018

Doran, G., & Zander, A. (2012). An improved method for measuring soil microbial activity by gas phase flow injection analysis. *Revista Brasileira de Ciência do Solo, 36*, 349–357. https://doi.org/10.1590/S0100-06832012000200004

Fidalski, J., Tormena, C. A., & Alves, S. J. (2013). Intervalo hídrico ótimo de um latossolo vermelho distrófico, após o primeiro período de pastejo contínuo de Brachiaria ruziziensis, em sistema integração lavoura-pecuária. *Revista Brasileira de Ciência do Solo, 37*, 775–783. https://doi.org/10.1590/S0100-06832013000300024

Franchini, J. C., Balbinot Junior, A. A., Debiasi, H., & Conte, O. (2014). Soybean performance as affected by desiccation time of Urochloa ruziziensis and grazing pressures. *Revista Ciência Agronômica, 45*, 999–1005. https://doi.org/10.1590/S1806-66902014000500015

Goedert, W. J., dos Reis, B., Junior, F., Gomes de Sousa, D. M., de Carvalho, M. I., de Castro, A., Lopes, A., & Chaer, G. M. (2013). Interpretation of microbial soil indicators as a function of crop yield and organic carbon. *Soil Science Society of America Journal, 77*, 461. https://doi.org/10.2136/sssaj2012.0191

Kallenbach, C., & Grandy, A. S. (2011). Controls over soil microbial biomass responses to carbon amendments in agricultural systems: A meta-analysis. *Agriculture, Ecosystems and Environment, 144*, 241–252. https://doi.org/10.1016/j.agee.2011.08.020

Mclean, E. O. (2015). Soil pH and lime requirement. In *Methods of soil analysis: Part 2* (pp. 199–224). Springer.

Nasielski, J., Furze, J. R., Tan, J., Bargaz, A., Thevathasan, N. V., & Isaac, M. E. (2015). Agroforestry promotes soybean yield stability and N2-fixation under water stress. *Agronomy for Sustainable Development, 35*, 1541–1549. https://doi.org/10.1007/s13593-015-0330-1

Navroski, D., Moreira, A., Colozzi-Filho, A., & Grange, L. (2017). Atributos microbiologicos do solo em diferentes sistemas de manejo no oeste paranaense. *Agrotrópica (Itabuna), 29*, 127–134. https://doi.org/10.21757/0103-3816.2017v29n2p127-134

Nelson, D. W., & Sommers, L. E. (2015). Total carbon, organic carbon, and organic matter. In *Methods of soil analysis: Part 2* (pp. 539–579). Springer.

Oliveira, J. M., Madari, B. E., Carvalho, M. T., Assis, P. C., Silveira, A. L., de Leles Lima, M., Wruck, F. J., Medeiros, J. C., & Machado, P. L. (2018). Integrated farming systems for improving soil carbon balance in the southern Amazon of Brazil. *Regional Environmental Change, 18*, 105–116. https://doi.org/10.1007/s10113-017-1146-0

Pavan, M. A., de Bloch, M. F., da Zempulski, H. C., Miyazawa, M., & Zocoler, D. C. (1992). *Manual de análise química de solo e controle de qualidade*. IAPAR.

Peel, M. C., Finlayson, B. L., & McMahon, T. A. (2007). Updated world map of the Köppen-Geiger climate classification. *Hydrology and Earth System Sciences, 11*, 1633–1644. https://doi.org/10.5194/hess-11-1633-2007

Pontes, K. B., Amaral, H. F., & Igarashi, S. (2018). Qualidade do solo associada ao monitoramento e manejo da phakopsora pachyrhizie em áreas de diferentes sucessões da cultura da soja. *Revista Terra & Cultura: Cadernos de Ensino e Pesquisa, 34*, 333–359.

Ryschawy, J., Martin, G., Moraine, M., Duru, M., & Therond, O. (2017). Designing crop–livestock integration at different levels: Toward new agroecological models? *Nutrient Cycling in Agroecosystems, 108*, 5–20. https://doi.org/10.1007/s10705-016-9815-9

Santos, H., Jacomine, P., Anjos, L., Oliveira, V., Lumbreras, J., Coelho, M., Almeida, J., Araujo Filho, J., & Oliveira, J. (2018). *Sistema brasileiro de classificação de solos*. Embrapa.

Schnurer, J., & Rosswall, T. (1996). Fluorescein diacetate hydrolysis as a measure of total microbial activity in soil and litter. In *Applied and environmental microbiology* (pp. 47–67). Springer.

Schroeder, J., Jannoura, R., Beuschel, R., Pfeiffer, B., Dyckmans, J., Murugan, R., Chavannavar, S., Wachendorf, C., & Joergensen, R. G. (2020). Carbon use efficiency and microbial functional diversity in a temperate Luvisol and a tropical Nitisol after millet litter and N addition. *Biology and Fertility of Soils, 2020*, 1–12. https://doi.org/10.1007/s00374-020-01487-4

Sparling, G., & West, A. (1988). A direct extraction method to estimate soil microbial-C calibration in situ using microbial respiration and 14C-labeled cells. *Soil Biology and Biochemistry, 20*, 337–343.

Teixeira, P. C., Donagemma, G. K., Fontana, A., Teixeira, W. G. (2017). Manual de Métodos de Análise de Solo 3a edição revista e ampliada. Man. métodos análise solo - Capítulo 8 - Densidade partículas 1, 574

Telles, T. S., Righetto, A. J., da Costa, G. V., Volsi, B., & de Oliveira, J. F. (2019). Conservation agriculture practices adopted in southern Brazil. *International Journal of Agricultural Sustainability, 17*, 338–346. https://doi.org/10.1080/14735903.2019.1655863

Vance, E. D., Brookes, P. C., & Jenkinson, D. S. (1987). An extraction method for measuring soil microbial biomass C. *Soil Biology and Biochemistry, 19*, 703–707. https://doi.org/10.1016/0038-0717(87)90052-6

Part II
Microbial Enzymes: Role in the Environmental Sustainability

Microbial Enzymes: Role in Soil Fertility

Auwalu Hassan, Fauziah Shahul Hamid, Helen Shnada Auta, Agamuthu Pariatamby, Innocent C. Ossai, Jayanthi Barasarathi, and Aziz Ahmed

A. Hassan (✉)
Faculty of Science, Institute of Biological Sciences, University of Malaya, Kuala Lumpur, Malaysia

Faculty of Science, Center for Research in Waste Management, University of Malaya, Kuala Lumpur, Malaysia

Department of Biological Sciences, Faculty of Science, Federal University of Kashere, Kashere, Gombe, Nigeria

F. S. Hamid · I. C. Ossai
Faculty of Science, Institute of Biological Sciences, University of Malaya, Kuala Lumpur, Malaysia

Faculty of Science, Center for Research in Waste Management, University of Malaya, Kuala Lumpur, Malaysia
e-mail: fauziahsh@um.edu.my

H. S. Auta
Department of Microbiology, Federal University of Technology, Minna, Niger State, Nigeria

Africa Centre of Excellence for Mycotoxin and Food Safety (ACEMFS), Federal University of Technology, Minna, Niger State, Nigeria

A. Pariatamby
Jeffrey Sachs Center on Sustainable Development, Sunway University, Sunway, Malaysia
e-mail: agamutup@sunway.edu.my

J. Barasarathi
Faculty of Health and Life Science (FHLS), INTI International University Pesiaran Perdana BBN, Nilai, Negeri Sembilan, Malaysia
e-mail: jayanthi.baras@newinti.edu.my

A. Ahmed
Faculty of Science, Institute of Biological Sciences, University of Malaya, Kuala Lumpur, Malaysia

Faculty of Science, Center for Research in Waste Management, University of Malaya, Kuala Lumpur, Malaysia

Faculty of Marine Sciences, Lasbela University of Agriculture, Water and Marine Sciences, Uthal, Balochistan, Pakistan

N. R. Maddela et al. (eds.), *Ecological Interplays in Microbial Enzymology*, Environmental and Microbial Biotechnology,
https://doi.org/10.1007/978-981-19-0155-3_9

Abstract Chemical, biochemical, and physicochemical reactions are all involved in nutrient cycling in soil. Enzymes catalyze all biochemical processes in soil. Soil enzymes catalyze several biochemical processes that ensure the transformation of organic materials and the release of inorganic nutrients for plant growth and nutrient cycling. As a result of the significant role played by soil enzymes in improving the fertility of soil, an in-depth evaluation of the influence of soil microbial enzymes on the fertility of soil is essential for effective maintenance of soil fertility, utilization of soil resources, and enhancement of plant productivity. This chapter discusses (1) the detailed role of soil microbial enzymes in improving the fertility of soil, (2) the mechanisms of action of soil enzymes, and (3) the factors influencing the enzyme activity in soil.

1 Introduction

Soil is a nonrenewable, dynamic resource. It is an essential component of a terrestrial ecosystem, providing basic support to all living organisms on the planet. In various land-use scenarios, soil fertility is an important indicator of good agricultural productivity (Almeida, Naves, & Mota, 2015). The fertility of soil is referred to as its capability to function continuously within land-use boundaries and ecosystems, as a vital living system for biological productivity and animal, plant, and human well-being (Doran & Parkin, 1994). The status of soil has an impact on the ecosystem, food production, and global ecological equilibrium (Adetunji, Lewu, Mulidzi, & Ncube, 2017; Binkley & Fisher, 2012). Soil fertility is closely linked to biological properties that are extremely sensitive to changes in the environment. The soil microbiota and enzymes are closely related and play important roles in improving the fertility of soil (Joshi, Mohapatra, & Mishra, 2018; Pajares, Gallardo, & Masciandaro, 2011). Therefore, the sustainability of the soil ecosystem can be assessed using biologically based indicators (Adetunji et al., 2017; Piotrowska-Dlugosz & Charzynski, 2015). Recently, the biodegradability capacity of microorganisms has been assessed through the evaluation of the activity of soil enzymes (Fioretto, Papa, Curcio, Sorrentino, & Fuggi, 2000). The different microbes in soil establish a relationship with the other biological systems and release enzymes. In the soil environment, microorganisms release enzymes that break down organic substances into simple soluble molecules (Almeida et al., 2015). In soil, there are two types of enzymes. Constitutive enzymes are those that are always present in the body in a consistent amount for metabolic action. The addition of any substrate has no effect on these enzymes (Das & Varma, 2011). Phosphofructokinase, pyruvate kinase, hexokinase, phosphoglucose isomerase, and other enzymes involved in the glycolytic pathway, for example, are constitutive enzymes (Maitra & Lobo, 1971). Other enzymes in the soil such as urease and phosphatase are also constitutive enzymes (Kumar & Sharma, 2019; Margalef et al., 2017). Inducible or inductive enzymes are found in limited amounts and sometimes may be absent. The concentration of such enzymes may vary and increase with the presence of a substrate. For instance, cellulase (Kandeler, 2015) and amidase (Das & Varma, 2011) are some of the inductive enzymes found in soil.

Soil enzymes that are synthesized by soil-inhabiting microbes perform important functions in the cycling of nutrients and indicate soil fertility and microbial activity (Joshi et al., 2018). Extracellular and intracellular soil enzymes such as glucosidase (Almeida et al., 2015) and hydrolase (Bautista-Cruz & Ortiz-Hernandez, 2015) catalyze the breakdown of organic materials, whereas urease, amidase, and arylsulfatase are concerned with nutrient mineralization (Das & Varma, 2011; Kumar & Sharma, 2019). The decomposition of heavy metals in soil is also aided by catalase enzymes such as phosphatase and dehydrogenase. These are crucial in the remediation of heavy metal-impacted soils (Khan, Cao, Hesham, Xia, 2007). Soil enzymes catalyze and promote a variety of biochemical processes that result in soil organic matter transformation, breakdown of organic residues, mineralization of accessible nutrients for plant growth, and soil aggregation (Balezentiene, 2012). Enzymes are thus linked to the rate of breakdown. Enzymes are active facilitators of the degradation processes of soil mineral organic components. For instance, urease participates in nitrogen cycling and hydrolysis of urea to NH_3 and CO_2; sucrase catalyzes the hydrolysis of sucrose to release monosaccharides and improve the soil-soluble nutrients; and phosphatase hydrolyzes phosphate ester and participates in cycling and mineralization of phosphorus (Adamczyk, Kilpeläinen, Kitunen, & Smolander, 2014; Zhang et al., 2018). Earlier studies have revealed that the enzymes associated with the mineralization of N, C, and P are closely related to N: C: P stoichiometry of soil (Stock, Kster, Dippold, Nájera, & Kuzyakov, 2019; Xu et al., 2017). Thus, enzyme activity objectively reflects the fertility of soil. A decline in the activity of soil enzymes indicates a decrease in the quality of soil (Zhu, Wang, Chen, Li, & Wu, 2019). Therefore, these catalytic activities provide some vital information for the assessment of the rates of important reactions. Soil microbial enzyme activities (1) are largely connected to soil physical attributes, microbial biomass, and organic matter and (2) change more readily than do the other indicators, signaling changes in soil quality or health faster (Dick, 1994). The activity of soil enzymes can be used to assess soil productivity, microbial activity, and the inhibitory effects of soil pollutants. In comparison to other properties, the activity of soil enzymes responds quickly to management strategies such as crop rotations, amendments, and tillage systems (Lehman et al., 2015). Moreover, the responses of enzyme activity correlate with the other soil properties, which suggest that they can be utilized to differentiate how management practices may influence soil parameters such as pH, organic materials, and distribution of nutrients (Acosta-Martinez, Cano, & Johnson, 2018; Lehman et al., 2015). Therefore, an in-depth evaluation of the influence of soil microbial enzymes on the fertility of soil is essential for the effective maintenance of soil fertility, utilization of soil resources, and enhancement of plant productivity. This chapter discusses (1) the detailed role of soil microbial enzymes in improving the fertility of soil, (2) the mechanisms of action of soil enzymes, and (3) the factors influencing the enzyme activity in soil.

2 The Role of Microbial Enzymes in Improving Soil Fertility

Soil can be viewed as a biological entity. In other words, it is a living system where biochemical reactions take place, and those reactions are catalyzed by enzymes. It is believed that, without enzymes, the soil will be a lifeless and unaltered entity (Alkorta et al., 2003). Microbially aided reactions, which are catalyzed by enzymes, are the backbone of the performance of soil. Such performance includes the cycling of sulfur, phosphorus, nitrogen, and carbon in the soil. In addition, the processes also contribute to the cleanup of contaminated soil through degradation of contaminants such as hydrocarbons or immobilization, e.g., in the case of heavy metals. Enzymatic actions are also involved in the formation of soil structure (Nannipieri, Kandeler, & Ruggiero, 2002). All biochemical processes in soil are aided by enzymes; this makes enzymes suitable indicators of soil biological activity and health (Alkorta et al., 2003). Enzyme activities in soil have been regarded as important parameters that are utilized to biologically assess the function of soil (Alrumman, Standing, & Paton, 2015). Recently, a research has revealed the utilization of enzyme activities in soil as a measure of nutrient and carbon deficiencies, which can be employed to reveal the influence of regional anthropogenic stressors in soil (Pandey & Yadav, 2017).

Microbial enzymes that catalyze the numerous reactions in soil are necessary for the cycling of nutrients, decomposition of organic substances, formation of organic matter, life processes of soil microbes, and stabilization of soil structure (Burns et al., 2013). Some of the enzymes that catalyze the numerous reactions (N, S, C, and P cycling) in soil include ureases, dehydrogenases, phosphatases, catalases, cellulases, proteases, lipases, fluorescein diacetate, arylsulfatases, esterases, hydrolases, etc. (Fig. 1) (Banashree, Smrita, Nath, & Nirmali, 2017). The contributions of these enzymes to soil fertility and health are described in detail.

2.1 *Glucosidases*

Glucosidases are a group of hydrolytic enzymes that catalyze the breakdown of glycosides. They are extremely diverse, which is the result of the vast variety of glycosidic linkages and variations in substrates (Almeida et al., 2015). The four primary members of the glucosidases family are β- and α-galactosidase and β- and α-glucosidase. In soil, these enzymes are widely spread. The hydrolysis of α-D-glucopyranosides is catalyzed by α-glucosidase, whereas cellobiose and maltose are hydrolyzed by β-glucosidase (Utobo & Tewari, 2015). Glucosidase activity has been found in a variety of microorganisms, including *Flavobacterium johnsoniae* (Okamota, Nakano, Yatake, Kiso, & Kitahata, 2000), *Ceriporiopsis subvermispora* (Magalhaes, Ferraz, & Milagres, 2006), *Penicillium purpurogenum* (Dhake & Patil, 2005), *Lactobacillus plantarum* (Spano, Rinaldi, Ugliano, Beneduce, & Massa, 2005), and *Trichoderma harzianum* (An, Im, Yang, Yang, & Lee, 2005). In soil,

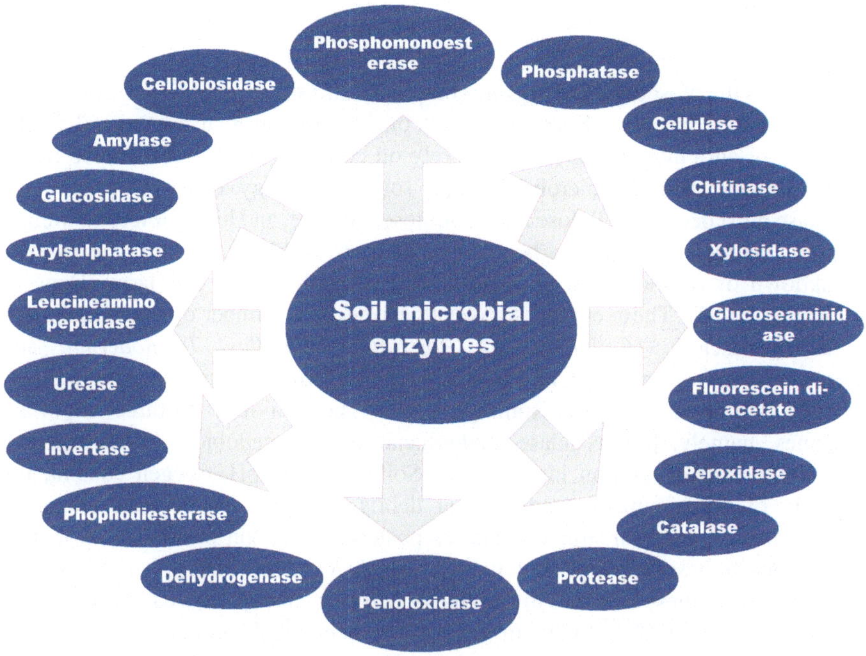

Fig. 1 Soil microbial enzymes

β-glucosidase plays a crucial role by facilitating the breakdown of different β-glucosides found in degrading plant materials (Veena, Poornima, Parvatham, & Sivapriyadharsini, 2011). Many soil bacteria use the end product of the breakdown (glucose) as a source of C for sustenance (Esen, 1993). β-Glucosidase is a crucial measure of soil health because it can stabilize organic materials in soils, represent historical biological activity in soils, and disclose the impacts of management activities on soils (Ndiaye, Sandeno, McGrath, & Dick, 2000). For this, it has been adopted for testing the quality of soil (Bandick & Dick, 1999). β-Glucosidase is highly sensitive to soil management practices and changes in pH (Madejon, Burgos, Lopez, & Cabrera, 2001). Such characteristics make it a useful indicator for assessing the ecological changes that result from the acidification of soil in situations that involve the activities of this enzyme (Das & Varma, 2010). Generally, the activity of β-glucosidase is closely associated with C cycling, organic matter, and biological activity and can provide an early signal of the alterations in organic carbon faster than can be determined using other methods. This forms the basis for its efficient applicability in agricultural practices (Adetunji et al., 2017).

2.2 Cellulases

Cellulose is the most common organic component in soil, accounting for over half of all synthesized biomass. Microbial development and survival are critical in the majority of agricultural soils, and they rely on cellulose as a carbon source in the soil, which serves as the microbe's primary source of energy. However, the enzyme cellulase must degrade cellulose into cellobiose, glucose, and high-molecular-weight oligosaccharides before the carbon can be made available to microorganisms. The breakdown of cellulose and polysaccharides is catalyzed by cellulases (Deng & Tabatabai, 1994). These enzymes are synthesized by a number of microorganisms including bacteria (*Cellulomonas*, *Clostridium*, *Bacillus*, *Trichoderma*, and *Thermomonospora*) and fungi such as *Aspergillus* (Kuhad, Gupta, & Singh, 2011; Micuţi, Bădulescu, & Israel-Roming, 2017). The activities of three important enzymes, namely, β-glucosidase, endoglucanase, and cellobiohydrolase, control the cellulose disintegration into glucose. Soil moisture, pH, oxygen content, the quantity of organic matter and/or plant debris, minerals and/or trace elements, organic matter chemical structure, and its position in the soil profile are all factors that influence these enzymes' activities (Arinze & Yubedee, 2000). Considering the sensitiveness of these enzymes toward these factors, their activities can be utilized as an early indication of the status of some physicochemical soil components, hence simplifying soil management in agriculture (Das & Varma, 2010).

2.3 Amylases

Amylase is a starch hydrolyzing enzyme and consists of α- and β-amylase. The enzyme is found abundantly in soil and is essential for the breakdown of starch, which is an important source of carbon for many soil-dwelling beneficial species. α-Amylase breaks down substrates that resemble starch into glucose and/or oligosaccharides, whereas β-amylase breaks down starch to maltose (Thoma, Spradlin, & Dygert, 1971). Amylase is synthesized by bacteria such as *Bacillus amyloliquefaciens*, *Bacillus stearothermophilus*, and *Bacillus licheniformis* and fungal species like *Penicillium expansum*, *Thermomyces lanuginose*, *Aspergillus niger*, and *Aspergillus oryzae* (Micuţi et al., 2017; Padma & Pallavi, 2016).

2.4 Phosphatases

Phosphatases belong to a group of enzymes that catalyze the hydrolysis of phosphoric acid anhydrides and esters (Condron, Turner, Cade-Menun, Sims, & Sharpley, 2005). Phosphatase enzymes are also produced by microbes in the soil. The phosphatase phosphomonoesterase is the most studied of the phosphatases found in soil. Phosphomonoesterase is a hydrolase enzyme that catalyzes the hydrolysis of phosphate monoester into free phosphate for biological absorption (Makoi & Ndakidemi, 2008). Polyphosphates, sugar phosphates, and nucleotides are among the low-molecular P-containing substances hydrolyzed by the enzyme (Dodor & Tabatabai, 2003). Phosphomonoesterase is active under alkaline and acidic conditions depending on its optimum pH. Alkaline phosphatase is active in the alkaline soil of pH 9–11, whereas acid phosphatase dominates in acidic soils with the pH range of 4–6 (Adetunji et al., 2017). The availability and content of phosphatase in soil vary depending on the extent of organic and mineral fertilizers, organic materials, microbial count, and agricultural practices (Banerjee, Sanyal, & Sen, 2012). These enzymes are believed to be important in the cycling of P in the soil environment. Phosphatases have been found to have a substantial relationship with plant development and P stress. Because plants only use inorganic P and a considerable amount of soil P is bonded to organic substances, the mineralization of this organically bound P will be critical because it will provide a valuable source of nutrients to the plants (Nannipieri, Giagnoni, & Landi, 2011). When there is a P shortfall in the soil, the soil microorganisms increase the production of this enzyme dramatically to improve the solubilization and remobilization of P. This has an impact on plants' ability to grow in P-stressed environments (Karthikeyan et al., 2002). As a result, the synthesis and activity of the phosphatase enzyme are directly linked to the requirement for P by microorganisms and plants (Condron et al., 2005). As a result, phosphatase activity can be used to determine the availability of inorganic P for microbes and plants (Piotrowska-Dlugosz & Charzynski, 2015).

2.5 Dehydrogenases

Dehydrogenase enzymes occur as an integral part of microbial cells. They are synthesized by bacteria such as *Pseudomonas entomophila*. The enzymes oxidize the soil organic matter through the transfer of electrons and protons from substrates to recipients. This activity forms part of the respiratory processes of soil microbes and is associated with the soil type and soil water–air conditions (Kandeler, 1996). The activity of dehydrogenase is mostly used to indicate the biological activity in soil. The fact is that the activity of dehydrogenase is part of the respiratory pathways of soil microbes; therefore, knowledge on the activity of dehydrogenase is highly essential as it will provide information on the soil potentials to support the biochemical processes that maintain the fertility of the soil. According to Brzezinska, Stepniewska, and Stepniewski (1998), temperature and the amount of water in soil affect dehydrogenase activity indirectly by changing the redox potentials of the soil. For example, during flooding, the available oxygen is quickly depleted, resulting in a shift in activity from aerobic to anaerobic. In the absence of oxygen, facultative anaerobic bacteria, for example, commence the metabolic processes employing dehydrogenase activity and Fe (III) as a terminal electron acceptor (Galstian & Awungian, 1974). This process may tamper with the availability of Fe to plants.

This type of redox transformation is closely related to microbial respiratory activities in soil. Therefore, the enzyme may serve as a measure of the microbial oxidative activities in soil. Dehydrogenases are often employed to gauge disruptions associated with trace metals, pesticides in soil, and in soil management practices (Frank & Malkomes, 1993; Hassan, Agamuthu, & Fauziah, 2020, 2021). They are also used to determine the type, extent, and significance of contamination in soil (Hassan et al., 2021). For instance, it has been reported that significant activity of dehydrogenase has been recorded in soil contaminated with effluents from the paper and pulp-making industry (McCarthy, Siddaramappa, Reight, Coddling, & Gao, 1994); meanwhile, in soil contaminated with fly ash, the activity was low (Pitchel & Hayes, 1990).

2.6 Peroxidases

Peroxidases are important in the breakdown of lignin, which is an essential component of the plant cell wall. The fact is that lignin constitutes a significant portion of the available polymers on Earth; therefore, its breakdown results in significant contribution to soil N and C pools and makes available nutrients to the soil microbes (Sinsabaugh, 2010). Peroxidase performs an important function in the decontamination of soil polluted with phenolics and toxic metals. It also helps lessen the negative impacts of reactive oxygen species in soil. Peroxidases are synthesized by the Ascomycota and Basidiomycota divisions of fungi as well as by various bacterial species (Micuţi et al., 2017; Sinsabaugh, Zak, Gallo, Lauber, & Amonette, 2004).

2.7 Chitinases

Chitinases are also called chitinolytic enzymes and catalyze the degradation or hydrolysis of chitin. They are considered an important part of fungal cell walls and serve as an effective defense system against pathogens. Chitinase is an agriculturally important enzyme that is synthesized by various microbes (Chet, 1987). The presence of chitinase in various forms has provided protection to cotton and beans against soil-borne diseases (Ordentlich, Elad, & Chet, 1988; Shapira, Ordentlich, Chet, & Oppenheim, 1989). One of the processes underlying the action that has been exhibited was the lysis activity by chitinase, which resulted in the degradation of the fungal pathogen (Singh, Shin, Park, & Chung, 1999). In the case of application in biological pest control, the enzyme was found to have significant applicability in terms of environmental friendliness, maintenance of soil health, and increasing plant growth and yields (Das & Varma, 2010).

2.8 Proteases

Proteases perform an important function in the mineralization of N in soil. This forms an essential process of regulating the available N for plant growth. Proteases are generally associated with organic and inorganic colloids. The level of activity of these enzymes indicates the biological capability of soil in terms of enzymatic conversion of substrates. The enzyme also serves an essential function in the ecology of microbes in the soil ecosystem (Burns, 1982).

2.9 Ureases

Urease catalyzes the hydrolysis of urea into NH_3 and CO_2, thus raising the pH of the soil in the process. This process results in rapid loss of N to the atmosphere via volatilization of NH_3 (Das & Varma, 2010). Urease also hydrolyzes other compounds such as dihydroxyurea, hydroxyurea, and semicarbazide using Ni as a cofactor (Alef & Nannipieri, 1995). Even though the enzyme is synthesized by various organisms, it is also synthesized by fungi, yeast, and bacteria (Machuca, Cuba-Díaz, & Córdova, 2015). Some of the bacteria that produce urease include *Helicobacter pylori*, *Bacillus pasteurii*, *Staphylococcus* sp., *Providencia* sp., *Klebsiella aerogenes*, and *Proteus mirabilis*, whereas the fungi that release urease include *Schizosaccharomyces pombe* and *Aspergillus* sp. (Krajewska, 2009). The enzyme is greatly distributed in nature. In soil, it occurs both as an intracellular and as an extracellular enzyme and its expression is mostly under the regulation of N (Mobley & Hausinger, 1989). The production of urease is normally stopped during microbial growth when NH_4^+ is used as the main source of N (Geisseler, Horwath, Joergensen, & Ludwig, 2010), whereas the synthesis of urease is initiated when urea or another accessible N source is present (Mobley, Island, & Hausinger, 1995). After urea fertilization, this step is critical for controlling N supply to plants. The activity of urease has gotten a lot of attention as a result of this function since it was first identified in 1935. Various factors influence the activity of urease in soil, including soil amendments, soil depth, heavy metal presence, organic matter concentration, cropping history, and environmental parameters such as temperature. The activity of urease generally increases with increase in temperature, and it was revealed that elevated temperature increases the coefficient of activity of this enzyme. The literature has shown that the activity of urease is easily hampered by elevated concentration of heavy metals (Yang, Liu, Zheng, & Feng, 2006). Therefore, because of its sensitivity and ability to provide information that connects the environmental factors and N cycling, the activity of soil urease has been of great importance and has been used as an index of soil quality. The activity of soil urease can also provide information on the management practices to be adopted, which can enhance the microbial metabolism, cycling of N, and soil fertility (Piotrowska-Dlugosz & Charzynski, 2015).

2.10 Arylsulfatases

These enzymes are widespread in soil. They are released into the external environment by bacteria (*Pseudomonas* sp., *Actinobacteria* sp., *Aerobacter* sp., *Klebsiella* sp., and *Raoultella* sp.) and fungi (*Eupenicillium* sp. and *Trichoderma* sp.) as a response to sulfur deficiency. Their presence in varying soil, in most cases, correlates with the rate of sulfur (S) immobilization and microbial biomass (Kertesz & Mirleau, 2004; Vong, Dedourge, Lasserre-Joulin, & Guckert, 2003). The enzyme induces the digestion of aromatic sulfate esters (R–O–SO_3), sulfate or sulfate sulfur (SO_4^{2-} or SO_4–S), or phenols (R–OH) (Banashree et al., 2017; Tabatabai, 1994a, b). Arylsulfatases are categorized according to the type of ester they hydrolyze. These categories are chondrosulfatases, glucosulfatases, steroid sulfatases, alkylsulfatases, and mycosulfatases (Tabatabai, 1982). Their presence in soil is related to the amount of organic carbon, the rate of microbial biomass, and the rate of S immobilization (Mirleau, Roy, Andrew, & Michael, 2005). Several factors, notably pH shifts, contaminants, and the type and amount of organic materials, influence the activity of these enzymes (Tyler, 1981). Their sensitivity toward these factors serves as an important criterion for using them as an index of soil quality.

3 Mechanisms of Action of Microbial Enzymes in Soil

Soil biota decomposes organic materials in the soil into nutrients that plants require and quickly absorb for optimum growth (Dotaniya et al., 2015; Meena et al., 2016). Soil microbes alter the nutrient kinetics in soil by accelerating the breakdown of compounds in the soil through the release of enzymes. The rhizosphere of roots supplies a significant number of low-molecular-weight organic acids that serve as sources of carbon for microorganisms and have a significant impact on soil enzyme synthesis. As a result, the synthesis of inorganic ions as plant nutrients is expedited (Dotaniya et al., 2014; Meena et al., 2017). In the root zone, the synthesized inorganic ions act as chelating agents, forming temporary complexes with the other plant nutrients. The created complexes then dissolve in the root zones, releasing the enzymes and plant nutrients. Some of the soil enzymes generated break down harmful molecules into innocuous substances, whereas others chelate poisonous ions like metals to prevent their uptake by plant roots (Gianfreda & Rao, 2014). Enzymes require substances to act upon in order to carry out their functions; these compounds are referred to as substrates (Das & Varma, 2011). β-Glucosidase, for example, degrades oligosaccharides with (1 $\rightarrow$ 4) glycosidic linkages, such as cellodextrins, cellobioses, and cellotrioses, to release glucose molecules. Every enzyme is specific to a substrate or a group of substrates that, under ideal conditions, fit into the active site of the enzyme, resulting in the creation of an enzyme–substrate complex (Das & Varma, 2011; Gianfreda & Rao, 2014). The enzyme catalyzes the reaction in the soil and separates from the products. The enzyme is then free to bind to the next substrate molecule and catalyze the reaction, resulting in new products. The enzyme undergoes many conformational changes from the initial complex to the ultimate release of the products. Enzymes are absorbed onto clay surfaces and remain active for a long time while being protected from environmental influences like photodegradation (Tietjen & Wetzel, 2003).

4 Factors that Influence the Activities of Soil Microbial Enzymes

4.1 Soil Factors

The activities of soil microbial enzymes are influenced by various factors (Fig. 2), such as changes in temperature. Temperature changes can modify the kinetics of microbial enzymes and the availability of nutrients in the soil (Chatterjee et al., 2019). The activity of soil enzymes increases with increase in temperature. Enzyme activity doubles for every 10 °C rise in temperature within the threshold limit. Above the threshold limit, the activity declines sharply and comes to a cease at extremely

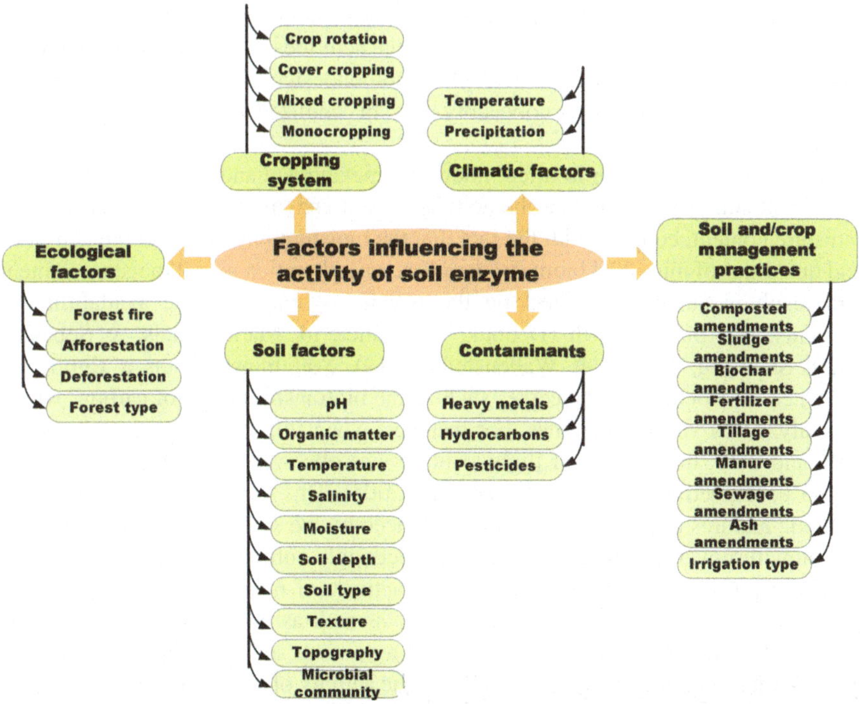

Fig. 2 Factors influencing the activity of soil microbial enzymes

high temperatures; this results in the inactivation of the enzymes (Dotaniya Aparna, Dotaniya, Singh, & Regar, 2019). However, in most cases, the thermal stability of enzymes varies depending on the source and type of enzyme. For instance, thermotolerant microorganisms release enzymes that can perform at a wider range of temperatures. Thermophilic microorganisms release enzymes that are specifically active at an elevated temperature and demonstrate less activity at a lower temperature (Dotaniya et al., 2019). The activities of β-glucosidase, fluorescein diacetate hydrolase, and dehydrogenase have been found to increase with increase in incubation temperature from the surrounding temperature; meanwhile, the activities of arylsulfatase and phosphomonoesterase decreased (Chatterjee et al., 2019). Fang et al. (2016) determined the warming effects on soil enzymes and realized that soil warming had no discernible impact on the activity of cellobiohydrolase, β-glucosidase, and *N*-acetylglucosaminidase; however, it increased the activity of oxidase and decreased the activity of acid phosphomonoesterase.

The salinization of soil, which is either caused by anthropogenic activities or natural factors, has been considered as a serious threat especially in arid and semiarid regions (Guangming et al., 2017; Wichelns & Qadir, 2014). Accumulation of salt has been known to have detrimental effects on the activity of soil microbial enzymes and biochemical processes (Karlen, Tomer, Neppel, & Cambardella, 2008). An increase in the salinity of soil has reportedly resulted in an exponential decrease in the activity of β-glucosidase (Rietz & Haynes, 2003). The activity of β-glucosidase in response to salinity can be utilized as a good indicator of soil quality. For instance, according to Boyrahmadi and Raiesi (2018), the activities of alkaline phosphomonoesterase, β-glucosidase, urease, acid phosphomonoesterase, L-glutaminase, invertase, and arylsulfatase were noticeably low in salinized soils in comparison to controls.

Soil moisture is known to affect microbial metabolism and hence the enzymatic activity in soil. The fact is that the activity of soil enzymes is strongly sensitive to moisture content, coupled with the fact that the moisture influences all the activities and quantities of microbial biomass; therefore, any alteration in soil moisture content will result in an adverse effect on the activity of enzymes, the availability of nutrients, and plant growth (Debouk, San Emeterio, Mari, Canals, & Sebastia, 2020; Steinweg, Dukes, & Wallenstein, 2012). Soil moisture has a significant influence on biochemical processes such as the biotransformation of carbon, which is catalyzed by various enzymes. For instance, when soil moisture was reduced by 10% and 21%, the activity of β-glucosidase was observed to fall by 10–80% and 35–83%, respectively (Sardans & Penuelas, 2005). This shows that the activity and catalytic features of β-glucosidase are influenced by the soil moisture, which results in a slow turnover of nutrients and lowers accessible nutrients to plants.

The depth of soil has an impact on the activity of microbial enzymes. This is closely related to organic matter availability as well as microbial activity. This is because the amount of organic matter in soil reduces as soil depth increases, and, as the amount of organic matter drops, so does the activity of soil microorganisms. This is due to the fact that soil enzyme activity is highly dependent on the availability of substrates and the microorganisms that synthesize the enzymes (Xiao-Chang & Qin,

2006). Many studies have found that the activity of soil enzymes reduces as the depth of the soil increases (Acosta-Martinez, Klose, & Zobeck, 2003; Xiao-Chang & Qin, 2006). The activity of enzymes in a vertical gradient is more pronounced in forest soil than in other ecosystems (Joshi et al., 2018).

The type and texture of soil have shown substantial influence on the activity of enzymes. According to Burns (1982), soil texture performs an important function in the stabilization of soil enzymes. The interactions with clay minerals and soil organic matter particularly affect the enzyme stability (Joshi et al., 2018). In an instance, lower activity of β-glucosidase has been reported in arable soils as compared to meadow and woodland soils (Bandick & Dick, 1999).

Soil enzymes are found to correlate with the abundance of individual microbial groups or microbial diversity (Kaiser, Koranda, Kitzler, Fuchslueger, & Schnecker, 2010). Because soil enzymes are mostly produced by microorganisms, any changes in the microbial community will have a major impact on soil enzyme synthesis (Xu et al., 2021). The enhanced activity of phosphatase in soil treated with mycorrhizal species has been observed in several investigations (Joner & Jakobsen, 1995; Van Aarle & Plassard, 2010). The link between phosphatase activity and mycorrhizal species supports phosphatase's degradative role in soil-bound phosphorus degradation (Van Aarle & Plassard, 2010). Mäder et al. (2011) discovered that amending soil with plant growth-promoting bacteria and arbuscular mycorrhizal fungi boosts the activities of dehydrogenase, urease, and phosphatase, resulting in improved soil quality. According to Wu, Wan, Wu, and Wong (2012), the presence of nitrogen-fixing bacteria increased the activities of phosphatase and urease. Furthermore, Xu et al. (2021) discovered a positive correlation between the soil bacterial community and the activity of β-1,4-glucosidase, which is engaged in C transformation, and a positive correlation between the fungi and the activity of oxidase, which is involved in C oxidation.

4.2 *Climatic Factors*

Precipitation and temperature are important climatic factors that influence the microbial communities and activities of enzymes in terrestrial ecosystems (Baldrian, Šnajdr, & Merhautová, 2013). Fluctuations in soil moisture and temperature occur seasonally, and the seasonal dynamics of microbial composition in soil are greatly related to the seasonal shifts in soil moisture and temperature (Rasche, Knapp, & Kaiser, 2010). Climatic factors are known to influence the microbial communities as well as the activities of their enzymes (Lanzen et al., 2016). This also results in affecting the fertility of the soil. In a study conducted by Sardans and Penuelas (2005), a decrease in precipitation had resulted in the reduction of β-glucosidase activity by 10–80%, protease by 15–66%, and urease by 10–67% while the further absence of moisture had resulted in a decline in 35–54%, 42–60%, 31–40%, and 35–83% of the activities of protease, urease, β-glucosidase, and acid phosphatase, respectively. Moreover, the activities of urease and protease were affected by drought (Sardans & Penuelas, 2005). On the other hand, about 33–80% reduction of laccase, peroxidase, and chitinase activities has been reported in soil samples collected during winter with a temperature of about 0 °C as compared to those collected during autumn when the temperature was around 15 °C. This shows that seasonal temperature can significantly influence the activity of microbial enzymes in soil ecosystems (Joshi et al., 2018). Zi, Hu, and Wang (2018) realized that short-term climatic changes can enhance the mineralization of plant nutrients and change the activity of soil enzymes in alpine meadow ecosystems. Some research found relatively higher activities of enzymes in the soil in colder environments (Jing, Wang, & Chung, 2014); meanwhile, other findings have reported substantial activities of soil enzymes during warmer periods (Baldrian et al., 2013; Jing et al., 2014). This implies that the relationships may differ in specific climatic zones (Luo He, Zeng, Li, & Yang, 2020).

4.3 Contaminants

The presence of contaminants in soil affects microbial metabolism, growth, and reproduction and eventually disrupts biochemical activities such as enzymatic activities. Contaminants can exert direct effects on enzyme activity, thereby destroying the spatial structure of enzyme active groups. For instance, the inhibition of invertase activity by contaminants has been revealed by many researchers, and most asserted that soil contaminants have a significant influence on microbial communities and soil respiration and have negative interactions with soil enzymes (Peyrot, Wilkinson, Desrosiers, & Sauvé, 2014; Tripathy, Bhattacharyya, Mohapatra, Som, & Chowdhury, 2014). The activities of enzymes can be altered by an elevated concentration of toxic metals (Duan et al., 2018). As the concentration of metal increases, the activities of most enzymes decrease drastically (Tripathy et al., 2014). The activity of dehydrogenase (38.9–18.1 g triphenylformazan/g soil/24 h), alkaline phosphatase (80.7–64.0 g p-nitrophenyl phosphate (PNP)/g soil/h), and acid phosphatase (73–55 g PNP/g soil/h) all decreased significantly as Pb concentrations increased from 0 to 300 mg/kg of soil (Dotaniya & Pipalde, 2018). Cao et al. (2020) revealed that the activities of urease, invertase, and cellulase decreased by 55.0–76.7%, 28.5–59%, and 17.3–34.1%, respectively, following an increase in the concentration of Cu. Hassan et al. (2020) revealed negative correlations between the concentrations of Cr, As, Cu, Mn, and Fe in landfill soil and the activities of urease and dehydrogenase. The application of pesticides to agricultural soils has resulted in several effects (positive and negative) on enzyme activities. The negative effects on enzymes such as oxidoreductases, hydrolases, and dehydrogenases have been broadly reported (Menon, Gopal, & Parsad, 2005; Monkiedje, Ilori, & Spiteller, 2002). The presence of high levels of crude oil and other heavy oil fractions can inhibit enzyme function by covering cell surfaces and organo-minerals, keeping soluble substrates away from enzyme molecules. According to Wang, Zhan, Zhou, and Lin (2010), the threshold level for activating or inhibiting the activities of dehydrogenase, phosphatase, and urease was 1000 mg/kg of mixed residual hydrocarbons. The toxic effects of some of the contaminants on the enzyme activities are depicted in Table 1.

4.4 Cropping System

The cropping system has been found to influence the activity of soil enzymes in different ways. For instance, phosphatase activity was found to be high under a crop rotation system involving meadow and oats, whereas under a monoculture system with soybean or corn, the activity was lower (Dodor & Tabatabai, 2003). In South African alluvial soil, Mukumbareza, Chiduza, and Muchaonyerwa (2015) discovered that rotating *Zea mays* with vetch and fertilized oat cover crop enhanced phosphatase activity and microbial biomass. The increased activity of phosphatase and microbial biomass in bicultures than in monocultures indicated the synergistic effects of the cover crops in the bicultures and can serve as a valuable avenue for enhancing the soil physicochemical properties and P cycling (Mukumbareza, Muchaonyerwa, & Chiduza, 2016). According to Chen, Guo, Guo, Tan, and Wang (2021), the increased duration of monocultures has reportedly decreased the activity of β-glucosidase, whereas, on the other hand, the activities of alkaline phosphatase and nitrate reductase increased nonlinearly. Extended monocultures of tea bush and tomato have reduced the microbial metabolic and enzymatic activities and resulted in shifts in the composition and structure of microbial communities (Fu et al., 2017; Li et al., 2017). Mganga, Razavi, and Kuzyakov (2016) realized that the fertility of the soil, the activity of the associated enzyme, and soil microbial biomass were enhanced in soil under traditional agroforestry systems than under monocropping with maize under the neutral to slightly acidic soil of tropical Africa.

4.5 Soil and/or Crop Management Practices

It is critical to understand the impact of various management strategies on the activity of enzymes in soil in order to improve soil quality and productivity. The activity of soil enzymes may be influenced by agricultural management practices. The activities of microbial enzymes and soil quality are altered by soil amendments under various management systems (Table 2). For example, when organic fertilizers such as sewage sludge, plant residues, compost, manure, and vermicompost were used, the activities of acid and alkaline phosphatase rose (Nannipieri et al., 2011; Piotrowska-Dlugosz & Wilczewski, 2014). Simultaneous addition of municipal solid waste or vermicompost and mineral N fertilizers has resulted in a higher activity of phosphatase than the application of individual fertilizers (Srivastava et al., 2012). Piotrowska-Dlugosz and Wilczewski (2014) revealed that the activity of phosphatase increased when soil containing low organic matter was supplemented

Table 1 Effects of contaminants on the activity of soil enzymes

Enzymes involved	Type of soil	Contaminant	Dosage of contaminant applied	Effect on soil enzymes	References
Dehydrogenase and phosphatase	Field soil	Kerosene and diesel	Kerosene (1% v/w) and diesel (1% v/w)	Inhibited the activities of the enzymes	(Alrumman et al., 2015)
Dehydrogenase, catalase, acid and alkaline phosphatase, and urease	Agricultural soil	Chlorothalonil	Chlorothalonil: 0.00, 1.660, and 16.60 mg/kg dry matter of soil	Inhibited the activity of dehydrogenase, catalase, and acid phosphatase	(Bacmaga, Wyszkowska, & Kucharski, 2018)
Invertase, urease, and cellulase	Paddy soil	Cu	Cu: 150 and 450 mg/kg soil	The activities of invertase, urease, and cellulase decreased by 28.5–59%, 55.0–76.7%, and 17.3–59%, respectively	(Cao et al., 2020)
Dehydrogenase, acid phosphatase, and alkaline phosphatase	Vertisol soil	Pb and Ni	0, 100, 150, and 300 mg/kg each for Pb and Ni	The activity of dehydrogenase decreased to 38.9, 32.1, 30.9, and 18.1 μg triphenylformazan/g soil/24 h at 0, 100, 150, and 300 mg/kg, respectively The activity of acid phosphatase decreased to 73, 61, 58, and 55 μg PNP/g soil/h at 0, 100, 150, and 300 mg/kg, respectively The activity of alkaline phosphatase decreased to 80.7, 69.4, 66.2, and 64.0 μg PNP/g soil/h at 0, 100, 150, and 300 mg/kg, respectively	(Dotaniya & Pipalde, 2018)

Invertase and β-glucosidase	Soil from uranium mining sites	As, Au, Cd, Cr, Cu, Mo, Ni, Pb, Sr, V, Zn, and U	Various treatments and dosages	The activities of the enzymes decreased with increase in metal concentrations and radiation	(Yang et al., 2018)
Fluorescein diacetate, β-glucosidase, and protease	Sediments	Cr, Cd, Cu, Ni, Pb, and Zn	Cd: 0.00–0.30 μg/g sediments Cu: 0–60 μg/g sediments Cr: 0–150 μg/g sediments Ni: 0–70 μg/g sediments Pb: 0–60 μg/g sediments Zn: 0–130 μg/g sediments	Enzyme activities negatively correlated with metal concentrations	(Jaiswal & Pandey, 2018)
Peroxidase and catalase	Field soil	Pesticides: dithiocarbamate, avermectin, organochlorine, and chlorothalonil	10 mL aqueous solution	The activities of the enzymes were affected negatively and positively	(Micuti, Badulescu, Burlacu, & Israel-Roming, 2018)
Protease, urease, dehydrogenase, and catalase	Agricultural soil	Pesticide: *cis*-nitromethylene neonicotinoid	20 mg/kg soil	Urease was initially inhibited	(Cai et al., 2016)

Table 2 Influence of different types of soil amendments on the activity of soil microbial enzymes

Enzymes involved	Type of soil amended	Material used for the amendment	Dosage/concentration applied	Effect of treatment on the activity of soil enzymes	References
Dehydrogenase, alkaline phosphatase, and urease	Agricultural soil	Sewage sludge	5, 10, 20, 30, and 50 t/ha	Increase the activity of the enzymes for up to 50–55%	(Dhanker, Chaudhary, Goyal, & Kumar, 2020)
β-glucosidase, β-D-cellobiosidase, β-xylosidase, *N*-acetyl-β-glucosaminidase, phosphatase, and leucine aminopeptidase	Aridisol soil	Biochar and dry manure	Biochar: 22.4 mg/ha and dry manure: 42 mg/ha	Biochar had no effect Dry manure had no effect Biochar + dry manure increased the activities of the enzymes	(Elzobair, Stromberger, Ippolito, & Lentz, 2016)
β-glucosaminidase, arylsulfatase, β-glucosidase, and acid phosphatase	Sandy soil	Biochar	5%, 10%, 0% v/v	Increased the activity of β-glucosaminidase by 5–30% and arylsulfatase by 12–46% Decreased the activity of β-glucosidase by 18–35% and acid phosphatase by 5–22%	(Frene et al., 2021)
Dehydrogenase, phosphatase, β-glucosidase, and urease	Agricultural soil	Poultry litter and biochar	Poultry litter: 70 kg N/ha biochar: 20 t/ha) poultry litter + biochar: 70 kg N/ha poultry litter and 20 t/ha biochar	Increased the activity of the enzymes	(Gao, Hoffman-Krull, & DeLuca, 2017)
Dehydrogenase, protease, β-glucosidase, and phosphomonoesterase	Agricultural soil	Organic pruning, manure, legume cover crop, and inorganic fertilizer	Organic pruning + manure: 0.63% N, 0.27% P_2O_5, 0.81% K_2O at a rate of 20 mg/ha Organic pruning + legume cover crop Inorganic fertilizer NPK: 8/4/12 at a rate of 250 kg/ha/year	Increased the activities of the enzymes	(Garcia-Orenes et al., 2016)

Dehydrogenase, β-glucosidase, fluorescein diacetate, cellulase, urease, protease, arylsulfatase, alkaline phosphatase, peroxidase, and phenoloxidase	Agricultural soil	Mineral fertilizers and organic fertilizers	Mineral fertilizers as used by farmers Recommended dose of mineral fertilizers: 50% inorganic + 50% organic 25% inorganic + 75% organic 75% organic + innovative 100% organic	Enzyme activities increased (50–75%) significantly in response to organic amendments	(Ghosh et al., 2020)
Urease, alkaline phosphatase, and catalase	Coastal saline soil	Hekang (soil modifier), chemical fertilizers, microbial inoculants, and organic fertilizers	Hekang: organic polymer Chemical fertilizers as NPK Organic fertilizer: organic matter (30%), NPK (8%) in the ratio of N: P: K = 3:2:3 Microbial inoculum: number of effective viable cells $\geq 10^{10}$/g	Enzyme activities improved by different amendments	(Guangming et al., 2017)
β-glucosidase, α-glucosidase, β-cellobiosidase, β-xylosidase, and *N*-acetyl-β-glucosaminidase	Paddy soil	Mineral fertilizers (NPK), organic fertilizers (livestock manure), and wheat straw	NPK: N, 351.75 kg/ha; P_2O_5, 75 kg/ha; K_2O, 84 kg/ha NPKM: N, 282.75 kg/ha; P_2O_5, 75 kg/ha; K_2O, 84 kg/ha) + livestock manure, 1500 kg/ha NPKS: N, 351.75 kg/ha; P_2O_5, 75 kg/ha; K_2O, 84 kg/ha + wheat straw, 3000 kg/ha	Activities of enzymes were enhanced	(Guo et al., 2018)

(continued)

Table 2 (continued)

Enzymes involved	Type of soil amended	Material used for the amendment	Dosage/concentration applied	Effect of treatment on the activity of soil enzymes	References
β-glucosidase, urease, and phosphodiesterase	Acidic tropical soil	Biochar and fertilizers	Palm kernel shell biochar (20 t/ha) Rice husk biochar: 20 t/ha Palm kernel shell biochar + fertilizer: 20 t/ha Rice husk biochar + fertilizer: 20 t/ha Fertilizer alone Control	All amendments significantly improved the enzyme activities	(Halmi, Hasenan, Simarani, & Abdullah, 2018)
β-glucosidase, urease, arylsulfatase, dehydrogenase, and acid and alkaline phosphatase	Agricultural soil	Inorganic fertilizers (NPK) and organic fertilizers	NPK: N, 2.4–2.7 kg/m/year; P, 0–12.6 kg/m/year; K, 0–0.6 kg/m/year Organic fertilizer: 9.4–10.5 kg/m/year	Organic fertilizers enhanced the activities of alkaline phosphatase, β-glucosidase, arylsulfatase, urease, and dehydrogenase by 41%, 26%, 47%, 39%, and 41%, respectively	(Igalavithana et al., 2017)
β-glucosidase, urease, acid phosphomonoesterase, and dehydrogenase	Agricultural soil	Olive pomace	Olive pomace: 50 mg/ha	The activities of enzymes in the treated plot were significant from those of the controls (without amendment)	(Innangi et al., 2017)
β-glucosidase, urease, acid and alkaline phosphatase, dehydrogenase, arylsulfatase, cellulase, and phenol oxidase	Agricultural soil	Biochar	0%, 15%, 20% w/w	Enhanced the enzyme resistance	(Jain et al., 2016)

Catalase, sucrase, and urease	Agricultural soil	Herbicide (acetochlor)	50%	The herbicide did not affect the activity of enzymes at first sampling	(Jiang et al., 2017)
β-glucosidase and urease	Agricultural soil	Biochar and fertilizers (NPK)	Biochar: 0 mg/ha + NPK: 32 kg/ha:14 kg/ha:16 kg/ha Biochar: 3.125 mg/ha + NPK: 32 kg/ha:14 kg/ha:16 kg/ha	The activities were significant than in controls	(Kaewpradit & Toomsan, 2019)

with P fertilizers, whereas there was no change in the activity of phosphatase when soil with high organic matter was also amended with P fertilizers. Other studies have revealed that amendments of soil with a combination of fertilizer treatments with vermicompost, compost, straw mulch, and municipal solid waste compost have resulted in increased activity of β-glucosidase than those without any compost and those supplemented with herbicides and synthetic fertilizers (Crecchio, Curci, Pizzigallo, Ricciuti, & Ruggiero, 2004; Meyer, Wooldridge, & Dames, 2015). Treatment of mining soil with biosolids in combination with a plant resulted in a substantial ($P < 0.05$) increase in β-glucosidase, alkaline phosphatase, and urease activities (Cele & Maboeta, 2016). Long-term irrigation with treated papermaking effluents, on the other hand, resulted in a considerable increase in urease, polyphenol oxidase, and invertase activities when compared to controls (Chen, Liang, Chen, Yang, & Ding, 2016). According to Pandey, Agrawal, and Bohra (2014), a reduction in the frequency of tillage in the no-tillage system had resulted in increased activity of β-glucosidase as compared to the conventional tillage system. In their study, Dominchin, Verdenelli, Aoki, and Meriles (2020) revealed that soil that was subjected to moderate water erosion had reduced microbial activity; meanwhile, the activity of dehydrogenase was increased. Furthermore, the activity of glucuronidase had reduced in the soil subjected to moderate water erosion.

4.6 Ecological Factors

Various ecological factors are known to affect the activity of soil enzymes. Soil enzymes and their relationships with ecological factors have received much attention in recent years (Ladwig, Sinsabaugh, Collins, & Thomey, 2015; McDaniel, Kaye, & Kaye, 2013; Zheng et al., 2018). Natural forest systems that have been transformed into agricultural fields have an impact not only on the plants but also on the soil's biological features. For instance, higher activities of β-glucosaminidase, β-glucosidase, phosphatases, arylamidase, arylsulfatase, and phosphodiesterase were observed in native grassland, rotation with other crops, and conservation reserves than with continuous cotton (Acosta-Martinez et al., 2003). According to Sicardi Garcia-Prechac, and Frioni (2004), conversion of natural grazed pastures to commercial plantations had significantly affected the activities of alkaline and acid phosphatase, dehydrogenase, soil respiration, and C mineralization. On the other hand, deforestation and afforestation are also found to influence the activity of soil enzymes. They are known to affect the quality of soil as compared to undisturbed soil. According to Bastida, Moreno, Hernandez, and Garcia (2006), the activities of protease and dehydrogenase were lower in deforested soil than in undisturbed soil. Furthermore, Izquierdo, Caravaca, Alguacil, Hernández, and Roldán (2005) claimed that removing vegetation had long-term negative consequences for soil microbial and metabolic activity. They further added that even after 15 years of deforestation, the soil quality has not improved. On the other hand, however, they realized that the activities of protease, urease, acid phosphatase, and β-glucosidase were higher in the soil after 4 years of revegetation (Izquierdo et al., 2005). Forest fire is regarded as a natural occurrence that causes numerous negative effects on soil ecosystems (Karaca, Cema, Turgay, & Kizilkaya, 2011). When there is a forest fire, most of the N found in soil and biomass escape into the atmosphere due to the low volatilization temperature of N. The effects of fire on the ecosystem are only differentiated by the activities of a few enzymes. The activities of various enzymes have been examined for differentiating the effects of fire-related stress on soil quality, and they have been found to increase or decrease (Karaca et al., 2011). For instance, the activities of protease and invertase were found to decline with burning, whereas the activities of peroxidase, polyphenol oxidase, and acid phosphatase increased (Zhang, Wu, Zhou, & Bao, 2005).

5 Conclusions

Soil microbial enzymes represent an important parameter for the quality of soil and plant well-being. Most of the degradative activities in soil are catalyzed by soil microbial enzymes. This provides essential sources of nutrients to the soil, thereby improving the fertility of the soil. The activities of soil enzymes are affected by various factors; this provides various signals regarding the status of the soil quality. It is therefore important that regular monitoring of the activity of soil enzymes should be put in place as it will give room for early correction of the soil condition. This will ensure effective maintenance of the quality and fertility of the soil.

References

Acosta-Martinez, V., Cano, A., & Johnson, J. (2018). Simultaneous determination of multiple soil enzyme activities for soil health-biogeochemical indices. *Applied Soil Ecology, 126*, 121–128. https://doi.org/10.1016/j.apsoil.2017.11.024

Acosta-Martinez, V., Klose, S., & Zobeck, T. (2003). Enzyme activities in semiarid soils under conservation reserve program, native rangeland, and cropland. *Journal of Plant Nutrition and Soil Science, 166*(6), 699–707. https://doi.org/10.1002/jpln.200321215

Adamczyk, B., Kilpeläinen, P., Kitunen, V. H., & Smolander, A. (2014). Potential activities of enzymes involved in N, C, P and S cycling in boreal forest soil under different tree species. *Pedobiologia, 57*(2), 97–102. https://doi.org/10.1016/j.pedobi.2013.12.003

Adetunji, A. T., Lewu, F. B., Mulidzi, R., & Ncube, B. (2017). The biological activities of β-glucosidase, phosphatase and urease as soil quality indicators: a review. *Journal of Soil Science and Plant Nutrition, 17*(3), 794–807. https://doi.org/10.4067/S0718-95162017000300018

Alef, K., & Nannipieri, P. (1995). *Methods in applied soil microbiology and biochemistry*. Academic Press.

Alkorta, I., Aizpurua, A., Riga, P., Albizu, I., Amézaga, I., & Garbisu, C. (2003). Soil enzyme activities as biological indicators of soil health. *Reviews on Environmental Health, 18*(1), 65–73. https://doi.org/10.1515/REVEH.2003.18.1.65

Almeida, R. F. D., Naves, E. R., & Mota, R. P. D. (2015). Soil quality: Enzymatic activity of soil β-glucosidase. *Global Journal of Agricultural Research and Reviews, 3*, 146–150.

Alrumman, S. A., Standing, D. B., & Paton, G. I. (2015). Effects of hydrocarbon contamination on soil microbial community and enzyme activity. *Journal of King Saud University - Science, 27*(1), 31–41. https://doi.org/10.1016/j.jksus.2014.10.001

An, D. S., Im, W. T., Yang, H. C., Yang, D. C., & Lee, S. T. (2005). *Dyella koreensis* sp. nov., a β-glucosidase producing bacterium. *International Journal of Systematic and Evolutionary Microbiology, 55*, 1625–1628. https://doi.org/10.1099/ijs.0.63695-0

Arinze, A. E., & Yubedee, A. G. (2000). Effect of fungicides on Fusarium grain rot and enzyme production in maize (Zea mays L.). *Global Journal of Pure and Applied Sciences, 6*, 629–634.

Bacmaga, M., Wyszkowska, J., & Kucharski, J. (2018). The influence of chlorothalonil on the activity of soil microorganisms and enzymes. *Ecotoxicology, 27*(9), 1188–1202. https://doi.org/10.1007/s10646-018-1968-7

Baldrian, P., Šnajdr, J., & Merhautová, V. (2013). Responses of the extracellular enzyme activities in hardwood forest to soil temperature and seasonality and the potential effects of climate change. *Soil Biology and Biochemistry, 56*, 60–68. https://doi.org/10.1016/j.soilbio.2012.01.020

Balezentiene, L. (2012). Hydrolases related to C and N cycles and soil fertility amendment: Responses to different management styles of agro-ecosystems. *Polish Journal of Environmental Studies, 21*(5), 1153–1159.

Banashree, S., Smrita, B., Nath, D. J., & Nirmali, G. (2017). Temporal responses of soil biological characteristics to organic inputs and mineral fertilizers under wheat cultivation in inceptisol. *Archives of Agronomy and Soil Science, 63*(1), 35–47. https://doi.org/10.1080/03650340.2016.1179385

Bandick, A. K., & Dick, R. P. (1999). Field management effects on soil enzyme activities. *Soil Biology and Biochemistry, 31*(11), 1471–1479. https://doi.org/10.1016/S0038-0717(99)00051-6

Banerjee, A., Sanyal, S., & Sen, S. (2012). Soil phosphatase activity of agricultural land: A possible index of soil fertility. *Agricultural Science Research Journal, 2*(7), 412–419.

Bastida, F., Moreno, J. L., Hernandez, T., & Garcia, C. (2006). Microbiological activity in a 15 years after its devegetation. *Soil Biology and Biochemistry, 38*(8), 2503–2507. https://doi.org/10.1016/j.soilbio.2006.02.022

Bautista-Cruz, A., & Ortiz-Hernandez, Y. (2015). Hydrolytic soil enzymes and their response to fertilization: A short review. *Comunicata Scientiae, 6*(3), 255. https://doi.org/10.14295/cs.v6i3.962

Binkley, D., & Fisher, R. (2012). *Ecology and management of forest soils*. John Wiley & Sons.

Boyrahmadi, M., & Raiesi, F. (2018). Plant roots and species moderate the salinity effect on microbial respiration, biomass, and enzyme activities in a sandy clay soil. *Biology and Fertility of Soils, 54*(4), 509–521. https://doi.org/10.1007/s00374-018-1277-6

Brzezinska, M., Stepniewska, Z., & Stepniewski, W. (1998). Soil oxygen status and dehydrogenase activity. *Soil Biology and Biochemistry, 30*(13), 1783–1790. https://doi.org/10.1016/S0038-0717(98)00043-1

Burns, R. G. (1982). Enzyme activity in soil: location and possible role in microbial ecology. *Soil Biology and Biochemistry, 14*(5), 423–427. https://doi.org/10.1016/0038-0717(82)90099-2

Burns, R. G., DeForest, J. L., Marxsen, J., Sinsabaugh, R. L., Stromberger, M. E., Wallenstein, M. D., Weintraub, M. N., & Zoppini, A. (2013). Soil enzymes in a changing environment: Current knowledge and future directions. *Soil Biology and Biochemistry, 58*, 216–234. https://doi.org/10.1016/j.soilbio.2012.11.009

Cai, Z. Q., Rong, Y., Chen, J., Wang, J., Ma, J. T., Zhang, W. J., & Zhao, X. Y. (2016). Effects of the novel cis-nitromethylene neonicotinoid insecticide Paichongding on enzyme activities and microorganisms in yellow loam and Huangshi soils. *Environmental Science and Pollution Research, 23*(8), 7786–7793. https://doi.org/10.1007/s11356-015-6036-8

Cao, Y. N., Ma, C. X., Chen, H. J., Chen, G. C., White, J. C., & Xing, B. S. (2020). Copper stress in flooded soil: Impact on enzyme activities, microbial community composition and diversity in the rhizosphere of Salix integra. *Science of the Total Environment, 704*, 135350. https://doi.org/10.1016/j.scitotenv.2019.135350

Cele, E. N., & Maboeta, M. (2016). Response of soil enzyme activities to synergistic effects of biosolids and plants in iron ore mine soils. *International journal of Environmental Science and Technology, 13*(9), 2117–2126. https://doi.org/10.1007/s13762-016-1043-y

Chatterjee, D., Kuotsu, R., Ao, M., Saha, S., Ray, S. K., & Ngachan, S. V. (2019). Does rise in temperature adversely affect soil fertility, carbon fractions, microbial biomass and enzyme activities under different land uses? *Current Science, 116*(12), 2044–2054. https://doi.org/10.18520/cs/v116/i12/2044-2054

Chen, A. H., Liang, H. X., Chen, T. M., Yang, W. J., & Ding, C. (2016). Influence of long-term irrigation with treated papermaking wastewater on soil ecosystem of a full-scale managed reed wetland. *Journal of Soils and Sediments, 16*(4), 1352–1359. https://doi.org/10.1007/s11368-015-1161-z

Chen, W. J., Guo, X. D., Guo, Q. N., Tan, X. L., & Wang, Z. G. (2021). Long-term chili monoculture alters environmental variables affecting the dominant microbial community in rhizosphere soil. *Frontiers in Microbiology, 12*, 681953. https://doi.org/10.3389/fmicb.2021.681953

Chet, I. (1987). *Trichoderma*-application, mode of action, and potential as biocontrol agent of soil borne pathogenic fungi. In I. Chet (Ed.), *Innovative approaches to plant disease control* (pp. 137–349). Wiley.

Condron, L. M., Turner, B. L., Cade-Menun, B. J., Sims, J., & Sharpley, A. (2005). Chemistry and dynamics of soil organic phosphorus. *Agriculture and Environment, 4*, 87–121. https://doi.org/10.2134/agronmonogr46.c4

Crecchio, C., Curci, M., Pizzigallo, M. D., Ricciuti, P., & Ruggiero, P. (2004). Effects of municipal solid waste compost amendments on soil enzyme activities and bacterial genetic diversity. *Soil Biology and Biochemistry, 36*(10), 1595–1605. https://doi.org/10.1016/j.soilbio.2004.07.016

Das, S. K., & Varma, A. (2010). Role of enzymes in maintaining soil health. In G. Shukla & A. Varma (Eds.), *Soil enzymology. Soil biology* (Vol. 22, pp. 25–42). Springer.

Das, S. K., & Varma, A. (2011). Role of enzymes in maintaining soil health. In G. Shukla & A. Varma (Eds.), *Soil enzymology* (pp. 43–60). Springer.

Debouk, H., San Emeterio, L., Mari, T., Canals, R. M., & Sebastia, M. T. (2020). Plant functional diversity, climate and grazer type regulate soil activity in natural grasslands. *Agronomy, 10*(9), 1291. https://doi.org/10.3390/agronomy10091291

Deng, S. P., & Tabatabai, M. A. (1994). Cellulase activity of soils. *Soil Biology and Biochemistry, 26*(10), 1347–1354. https://doi.org/10.1016/0038-0717(94)90216-X

Dhake, A. B., & Patil, M. B. (2005). Production of β-glucosidase by *Penicillium purpurogenum. Brazilian Journal of Microbiology, 36*(2), 170–176. https://doi.org/10.1590/S1517-83822005000200013

Dhanker, R., Chaudhary, S., Goyal, S., & Kumar, R. (2020). Soil microbial properties and functional diversity in response to sewage sludge amendments. *Archives of Agronomy and Soil Science*. https://doi.org/10.1080/03650340.2020.1855328

Dick, R. P. (1994). Soil Enzyme activities as indicators of soil quality. In J. W. Doran, D. C. Coleman, D. F. Bezdicek, & B. A. Stewart (Eds.), *Defining soil quality for a sustainable environment* (SSSA special publication no. 35) (p. 107). SSSA.

Dodor, D. E., & Tabatabai, M. A. (2003). Effect of cropping systems on phosphatases in soils. *Journal of Plant Nutrition and Soil Science, 166*(1), 7–13. https://doi.org/10.1002/jpln.200390016

Dominchin, M. F., Verdenelli, R. A., Aoki, A., & Meriles, J. M. (2020). Soil microbiological and biochemical changes as a consequence of land management and water erosion in a semiarid environment. *Archives of Agronomy and Soil Science, 66*(6), 763–777. https://doi.org/10.1080/03650340.2019.1638915

Doran, J., & Parkin, T. (1994). Defining and assessing soil quality. In J. W. Doran, D. C. Coleman, D. F. Bezdicek, & B. A. Stewart (Eds.), *Defining soil quality for a sustainable environment* (pp. 3–21). Soil Science Society of America.

Dotaniya, M., Aparna, K., Dotaniya, C., Singh, M., & Regar, K. (2019). Role of soil enzymes in sustainable crop production. In M. Kuddus (Ed.), *Enzymes in food biotechnology: Production, applications, and future prospects* (pp. 569–589). Elsevier.

Dotaniya, M., & Pipalde, J. (2018). Soil enzymatic activities as influenced by lead and nickel concentrations in a Vertisol of Central India. *Bulletin of Environmental Contamination and Toxicology, 101*(3), 380–385. https://doi.org/10.1007/s00128-018-2402-9

Dotaniya, M. L., Datta, S. C., Biswas, D. R., Meena, H. M., Rajendiran, S., & Meena, A. L. (2015). Phosphorus dynamics mediated by bagasse, press mud and rice straw in Inceptisol of North India. *Agrochimica, 59*(4), 358–369.

Dotaniya, M. L., Pingoliya, K. K., Lata, M., Verma, R., Regar, K. L., Deewan, P., & Dotaniya, C. K. (2014). Role of phosphorus in chickpea (*Cicer arietinum* L.) production. *African Journal of Agricultural Research, 9*(51), 3736–3743. https://doi.org/10.5897/AJAR2013.8427

Duan, C., Fang, L., Yang, C., Chen, W., Cui, Y., & Li, S. (2018). Reveal the response of enzyme activities to heavy metals through in situ zymography. *Ecotoxicology and Environmental Safety, 156*, 106–115. https://doi.org/10.1016/j.ecoenv.2018.03.015

Elzobair, K. A., Stromberger, M. E., Ippolito, J. A., & Lentz, R. D. (2016). Contrasting effects of biochar versus manure on soil microbial communities and enzyme activities in an Aridisol. *Chemosphere, 142*, 145–152. https://doi.org/10.1016/j.chemosphere.2015.06.044

Esen, A. (1993). β-glucosidases: Overview. In A. Esen (Ed.), *b-glucosidases and molecular biology* (pp. 9–17). American Chemical Society.

Fang, X., Zhou, G. Y., Li, Y. L., Liu, S. Z., Chu, G. W., Xu, Z. H., & Liu, J. X. (2016). Warming effects on biomass and composition of microbial communities and enzyme activities within soil aggregates in subtropical forest. *Biology and Fertility of Soils, 52*(3), 353–365. https://doi.org/10.1007/s00374-015-1081-5

Fioretto, A., Papa, S., Curcio, E., Sorrentino, G., & Fuggi, A. (2000). Enzyme dynamics on decomposing leaf litter of *Cistus incanus* and *Myrtus communis* in a mediterranean ecosystem. *Soil Biology and Biochemistry, 32*(13), 1847–1855. https://doi.org/10.1016/S0038-0717(00)00158-9

Frank, T., & Malkomes, H. P. (1993). Influence of temperature on microbial activities and their reaction to the herbicide Goltix in different soils under laboratory conditions. *Zentralblatt für Mikrobiologie, 148*(6), 403–412. https://doi.org/10.1016/S0232-4393(11)80305-7

Frene, J. P., Frazier, M., Liu, S., Clark, B., Parker, M., & Gardner, T. (2021). Early effect of pine biochar on peach-tree planting on microbial community composition and enzymatic activity. *Applied Sciences, 11*(4), 1473. https://doi.org/10.3390/app11041473

Fu, H., Zhang, G., Zhang, F., Sun, Z., Geng, G., & Li, T. (2017). Effects of continuous tomato monoculture on soil microbial properties and enzyme activities in a solar greenhouse. *Sustainability, 9*(2), 317. https://doi.org/10.3390/su9020317

Galstian, A. S., & Awungian, Z. S. (1974). Significance of the enzymes in oxidation of Fe and Mn oxides in soil. In *Trans. 10th international congress soil sci III* (pp. 130–135). Nauka Publishing House.

Gao, S., Hoffman-Krull, K., & DeLuca, T. H. (2017). Soil biochemical properties and crop productivity following application of locally produced biochar at organic farms on Waldron Island, WA. *Biogeochemistry, 136*(1), 31–46. https://doi.org/10.1007/s10533-017-0379-9

Garcia-Orenes, F., Roldan, A., Morugan-Coronado, A., Linares, C., Cerda, A., & Caravaca, F. (2016). Organic fertilization in traditional mediterranean grapevine orchards mediates changes in soil microbial community structure and enhances soil fertility. *Land Degradation and Development, 27*(6), 1622–1628. https://doi.org/10.1002/ldr.2496

Geisseler, D., Horwath, W. R., Joergensen, R. G., & Ludwig, B. (2010). Pathways of nitrogen utilization by soil microorganisms–a review. *Soil Biology and Biochemistry, 42*(12), 2058–2067. https://doi.org/10.1016/j.soilbio.2010.08.021

Ghosh, A., Singh, A., Kumar, R., Manna, M., Bhattacharyya, R., Rahman, M. M., Sharma, P., Rajput, P., & Misra, S. (2020). Soil enzymes and microbial elemental stoichiometry as bio-indicators of soil quality in diverse cropping systems and nutrient management practices of Indian Vertisols. *Applied Soil Ecology, 145*, 103304. https://doi.org/10.1016/j.apsoil.2019.06.007

Gianfreda, L., & Rao, M. A. (2014). *Enzymes in agricultural sciences*. OMICS International.

Guangming, L., Xuechen, Z., Xiuping, W., Hongbo, S., Jingsong, Y., & Xiangping, W. (2017). Soil enzymes as indicators of saline soil fertility under various soil amendments. *Agriculture, Ecosystems and Environment, 237*, 274–279. https://doi.org/10.1016/j.agee.2017.01.004

Guo, J. J., Liu, W. B., Zhu, C., Luo, G. W., Kong, Y. L., Ling, N., Wang, M., Dai, J. Y., Shen, Q. R., & Guo, S. W. (2018). Bacterial rather than fungal community composition is associated with microbial activities and nutrient-use efficiencies in a paddy soil with short-term organic amendments. *Plant and Soil, 424*(1-2), 335–349. https://doi.org/10.1007/s11104-017-3547-8

Halmi, M. F. A., Hasenan, S. N., Simarani, K., & Abdullah, R. (2018). Linking soil microbial properties with plant performance in acidic tropical soil amended with biochar. *Agronomy, 8*(11), 255. https://doi.org/10.3390/agronomy8110255

Hassan, A., Agamuthu, P., & Fauziah, S. H. (2021). Influence of bioaugmentation on the decontamination of heavy metal impacted landfill soil. *Open Access Journal of Waste Management & Xenobiotics, 4*(3), 000165. https://doi.org/10.23880/oajwx-16000165

Hassan, A., Agamuthu, P., Ossai, I. C., & Fauziah, S. H. (2020). Bioaugmentation assisted mycoremediation of heavy metal and/metalloid landfill contaminated soil using consortia of filamentous fungi. *Biochemical Engineering Journal, 157*, 107550. https://doi.org/10.1016/j.bej.2020.107550

Hassan, A., Pariatamby, A., Ossai, I. C., Ahmed, A., Muda, M. A., Wen, T. Z., & Hamid, F. S. (2021). Bioaugmentation-assisted bioremediation and kinetics modelling of heavy metal-polluted landfill soil. *International Journal of Environmental Science and Technology, 2021*, 1–26. https://doi.org/10.1007/s13762-021-03626-2

Igalavithana, A. D., Lee, S. S., Niazi, N. K., Lee, Y. H., Kim, K. H., Park, J. H., Moon, D. H., & Ok, Y. S. (2017). Assessment of soil health in urban agriculture: soil enzymes and microbial properties. *Sustainability, 9*(2), 310. https://doi.org/10.3390/su9020310

Innangi, M., Niro, E., D'Ascoli, R., Danise, T., Proietti, P., Nasini, L., Regni, L., Castaldi, S., & Fioretto, A. (2017). Effects of olive pomace amendment on soil enzyme activities. *Applied Soil Ecology, 119*, 242–249. https://doi.org/10.1016/j.apsoil.2017.06.015

Izquierdo, I., Caravaca, F., Alguacil, M. M., Hernández, G., & Roldán, A. (2005). Use of microbiological indicators for evaluating success in soil restoration after revegetation of a mining area under subtropical conditions. *Applied Soil Ecology, 30*(1), 3–10. https://doi.org/10.1016/j.apsoil.2005.02.004

Jain, S., Mishra, D., Khare, P., Yadav, V., Deshmukh, Y., & Meena, A. (2016). Impact of biochar amendment on enzymatic resilience properties of mine spoils. *Science of the Total Environment, 544*, 410–421. https://doi.org/10.1016/j.scitotenv.2015.11.011

Jaiswal, D., & Pandey, J. (2018). Impact of heavy metal on activity of some microbial enzymes in the riverbed sediments: Ecotoxicological implications in the Ganga River (India). *Ecotoxicology and Environmental Safety, 150*(2018), 104–115. https://doi.org/10.1016/j.ecoenv.2017.12.015

Jiang, Y. M., Lin, D., Guan, X. J., Wang, J. F., Cao, G. P., Zhu, D., & Peng, C. R. (2017). Effect of herbicide used with years (8+1) on soil enzymic activity and microbial population diversity. *Journal of Soils and Sediments, 17*(10), 2490–2499. https://doi.org/10.1007/s11368-017-1672-x

Jing, X., Wang, Y., & Chung, H. (2014). No temperature acclimation of soil extracellular enzymes to experimental warming in an alpine grassland ecosystem on the Tibetan Plateau. *Biogeochemistry, 117*, 39–54. https://doi.org/10.1007/s10533-013-9844-2

Joner, E., & Jakobsen, I. (1995). Growth and extracellular phosphatase activity of arbuscular mycorrhizal hyphae as influenced by soil organic matter. *Soil Biology and Biochemistry, 27*(9), 1153–1159. https://doi.org/10.1016/0038-0717(95)00047-I

Joshi, S., Mohapatra, B., & Mishra, J. (2018). Microbial soil enzymes: implications in the maintenance of rhizosphere ecosystem and soil health. In T. Adhya, B. Lal, B. Mohapatra, D. Paul, & S. Das (Eds.), *Advances in soil microbiology: Recent trends and future prospects* (Microorganisms for sustainability) (pp. 179–192). Springer.

Kaewpradit, W., & Toomsan, B. (2019). Impact of Eucalyptus biochar application to upland rice-sugarcane cropping systems on enzyme activities and nitrous oxide emissions of soil at sugarcane harvest under incubation experiment. *Journal of Plant Nutrition, 42*(4), 362–373. https://doi.org/10.1080/01904167.2018.1555849

Kaiser, C., Koranda, M., Kitzler, B., Fuchslueger, L., & Schnecker, J. (2010). Belowground carbon allocation by trees drives seasonal patterns of extracellular enzyme activities by altering microbial community composition in a beech forest soil. *The New Phytologist, 187*(4), 843–858. https://doi.org/10.1080/01904167.2018.1555849

Kandeler, E. (1996). Nitrate. In F. Schinner, R. Ohlinger, E. Kandeler, & R. Margesin (Eds.), *Methods in soil biology* (pp. 408–410). Springer.

Kandeler, E. (2015). Physiological and biochemical methods for studying soil biota and their functions. In E. A. Paul (Ed.), *Soil microbiology, ecology and biochemistry* (pp. 187–222). Elsevier.

Karaca, A., Cema, C. C., Turgay, O. C., & Kizilkaya, R. (2011). Soil enzymes as indicator of soil quality. In V. G. A. Shukla (Ed.), *Soil enzymology, soil biology* (pp. 119–148). Springer.

Karlen, D. L., Tomer, M. D., Neppel, J., & Cambardella, A. (2008). A preliminary watershed scale soil quality assessment in North Central Iowa, USA. *Soil & Tillage Research, 99*(2), 291–299. https://doi.org/10.1016/j.still.2008.03.002

Karthikeyan, A. S., Varadarajan, D. K., Mukatira, U. T., D'Urzo, M. P., Damaz, B., & Raghothama, K. G. (2002). Regulated expression of *Arabidopsis* phosphate transporters. *Plant Physiology, 130*(1), 221–233. https://doi.org/10.1104/pp.020007

Kertesz, M. A., & Mirleau, P. (2004). The role of soil microbes in plant sulphur nutrition. *Journal of Experimental Botany, 55*(404), 1939–1945. https://doi.org/10.1093/jxb/erh176

Khan, S., Cao, Q., Hesham, A. E. L., & Xia, Y. (2007). Soil enzymatic activities and microbial community structure with different application rates of Cd and Pb. *Environmental Sciences, 19*(7), 834–840. https://doi.org/10.1016/S1001-0742(07)60139-9

Krajewska, B. (2009). Ureases I. Functional, catalytic and kinetic properties: A review. *Journal of Molecular Catalysis B: Enzymatic, 59*(1-3), 9–21. https://doi.org/10.1016/j.molcatb.2009.01.003

Kuhad, R. C., Gupta, R., & Singh, A. (2011). Microbial cellulases and their industrial applications. *Enzyme Research, Sage-Hindawi Access Research, 2011*, 280696. https://doi.org/10.4061/2011/280696

Kumar, A., & Sharma, S. (2019). *Microbes and enzymes in soil health and bioremediation* (p. 401). Springer.

Ladwig, L. M., Sinsabaugh, R. L., Collins, S. L., & Thomey, M. L. (2015). Soil enzyme responses to varying rainfall regimes in Chihuahuan Desert soils. *Ecosphere, 6*(3), 1–10. https://doi.org/10.1890/ES14-00258.1

Lanzen, A., Epelde, L., Blanco, F., Martin, I., Artetxe, U., & Garbisu, C. (2016). Multi-targeted metagenetic analysis of the influence of climate and environmental parameters on soil microbial communities along an elevational gradient. *Scientific Reports, 6*, 28257. https://doi.org/10.1038/srep28257

Lehman, R. M., Cambardella, C. A., Stott, D. E., Acosta-Martinez, V., Manter, D. K., Buyer, J. S., Maul, J. E., Smith, J. L., Collins, H. P., Halvorson, J. J., Kremer, R. J., Lundgren, J. G., Ducey,

T. F., Jin, V. L., & Karlen, D. L. (2015). Understanding and enhancing soil biological health: The solution for reversing soil degradation. *Sustainability, 7*(1), 988–1027. https://doi.org/10.3390/su7010988
Li, Y., Li, Z., Arafat, Y., Lin, W., Jiang, Y., & Weng, B. (2017). Characterizing rhizosphere microbial communities in long-term monoculture tea orchards by fatty acid profiles and substrate utilization. *European Journal of Soil Biology, 81*, 48–54. https://doi.org/10.1016/j.ejsobi.2017.06.008
Luo, S. P., He, B. H., Zeng, Q. P., Li, N. J., & Yang, L. (2020). Effects of seasonal variation on soil microbial community structure and enzyme activity in a Masson pine forest in Southwest China. *Journal of Mountain Science, 17*, 1398–1409. https://doi.org/10.1007/s11629-019-5825-9
Machuca, A., Cuba-Díaz, M., & Córdova, C. (2015). Enzymes in the rhizosphere of plants growing in the vicinity of the polish arctowski antarctic station. *Journal of Soil Science and Plant Nutrition, 15*(4), 833–838. https://doi.org/10.4067/S0718-95162015005000057
Madejon, E., Burgos, P., Lopez, R., & Cabrera, F. (2001). Soil enzymatic response to addition of heavy metals with organic residues. *Biology and Fertility of Soils, 34*, 144–150. https://doi.org/10.1007/s003740100379
Mäder, P., Franziska, K., Adholeya, A., Singh, R., Uppal, H. S., Sharma, A. K., Srivastava, R., Sahai, V., Aragno, M., Wiemken, A., Johri, B. N., & Fried, P. M. (2011). Inoculation of root microorganisms for sustainable wheat-rice and wheat-black gram rotations in India. *Soil Biology and Biochemistry, 43*(3), 609–619. https://doi.org/10.1016/j.soilbio.2010.11.031
Magalhaes, P. O., Ferraz, A., & Milagres, A. F. M. (2006). Enzymatic properties of two β-glucosidases from *Ceriporiopsis subvermispora* produced in biopulping conditions. *Journal of Applied Microbiology, 101*(2), 480–486. https://doi.org/10.1111/j.1365-2672.2006.02946.x
Maitra, P., & Lobo, Z. (1971). A kinetic study of glycolytic enzyme synthesis in yeast. *The Journal of Biological Chemistry, 246*, 475–488.
Makoi, J. H., & Ndakidemi, P. A. (2008). Selected soil enzymes: Examples of their potential roles in the ecosystem. *African Journal of Biotechnology, 7*(3), 181–191.
Margalef, O., Sardans, J., Fernandez-Martinez, M., Molowny, H. R., Janssens, I. A., Ciais, P., Goll, D., Richter, A., Obersteiner, M., Asensio, D., & Penuelas, J. (2017). Global patterns of phosphatase activity in natural soils. *Scientific Reports, 7*, 1337. https://doi.org/10.1038/s41598-017-01418-8
McCarthy, G. W., Siddaramappa, R., Reight, R. J., Coddling, E. E., & Gao, G. (1994). Evaluation of coal combustion by products as soil liming materials: Their influence on soil pH and enzyme activities. *Biology and Fertility of Soils, 17*, 167–172. https://doi.org/10.1007/BF00336317
McDaniel, M. D., Kaye, J. P., & Kaye, M. W. (2013). Increased temperature and precipitation had limited effects on soil extracellular enzyme activities in a post-harvest forest. *Soil Biology and Biochemistry, 56*, 90–98. https://doi.org/10.1016/j.soilbio.2012.02.026
Meena, B. P., Shirale, A. O., Dotaniya, M. L., Jha, P., Meena, A. L., Biswas, A. K., & Patra, A. K. (2016). Conservation agriculture: A new paradigm for improving input use efficiency and crop productivity. In J. K. Bisht, V. S. Meena, P. K. Mishra, & A. Pattanayak (Eds.), *Conservation agriculture—Conservation agriculture—An approach to combat climate change in Indian Himalaya* (Vol. 1, pp. 39–69). Springer.
Meena, V. S., Meena, S. K., Verma, J. P., Kumrar, A., Aeron, A., Mishra, P. K., Bisht, J. K., Pattanayaka, A., Naveed, M., & Dotaniya, M. L. (2017). Plant beneficial rhizospheric microorganism (PBRM) strategies to improve nutrients use efficiency: A review. *Ecological Engineering, 107*, 8–32. https://doi.org/10.1016/j.ecoleng.2017.06.058
Menon, P., Gopal, M., & Parsad, R. (2005). Effects of chlorpyrifos and quinalphos on dehydrogenase activities and reduction of Fe3þ in the soils of two semiarid fields of tropical India. *Agriculture, Ecosystems and Environment, 108*(1), 73–83. https://doi.org/10.1016/j.agee.2004.12.008
Meyer, A. H., Wooldridge, J., & Dames, J. F. (2015). Variation in urease and β-glucosidase activities with soil depth and root density in a 'cripp's pink'/m7 apple orchard under

conventional and organic management. *South African Journal of Plant and Soil, 32*(4), 227–234. https://doi.org/10.1080/02571862.2015.1053155

Mganga, K. Z., Razavi, B. S., & Kuzyakov, Y. (2016). Land use affects soil biochemical properties in Mt. Kilimanjaro region. *Catena, 141*, 22–29. https://doi.org/10.1016/j.catena.2016.02.013

Micuti, M. M., Badulescu, L., Burlacu, A., & Israel-Roming, F. (2018). Activity of peroxidase and catalase in soils as influenced by some insecticides and fungicides. *AgroLife Scientific Journal, 7*(2), 99–104.

Micuţi, M. M., Bădulescu, L., & Israel-Roming, F. (2017). A review on the enzymatic indicators for monitoring soil quality. *Scientific Bulletin, 21*, 223–228.

Mirleau, P., Roy, W., Andrew, S., & Michael, A. K. (2005). *Importance of organosulfur utilization for survival of Pseudomonas putida in soil and rhizosphere, applied and environmental microbiology* (pp. 6571–6577). American Society for Microbiology.

Mobley, H., Island, M. D., & Hausinger, R. P. (1995). Molecular biology of microbial ureases. *Microbiological Reviews, 59*(3), 451–480. https://doi.org/10.1128/mr.59.3.451-480.1995

Mobley, H. L. T., & Hausinger, R. P. (1989). Microbial urease: significance, regulation and molecular characterization. *Microbiological Reviews, 53*(1), 85–108.

Monkiedje, A., Ilori, M. O., & Spiteller, M. (2002). Soil quality changes resulting from the application of the fungicides mefenoxam and metalaxyl to a sandy loam soil. *Soil Biology and Biochemistry, 34*(12), 1939–1948. https://doi.org/10.1016/S0038-0717(02)00211-0

Mukumbareza, C., Chiduza, C., & Muchaonyerwa, P. (2015). Effects of oats and grazing vetch cover crops and fertilisation on microbial biomass and activity after five years of rotation with maize. *South African Journal of Plant and Soil, 32*(4), 189–197. https://doi.org/10.1080/02571862.2015.1025446

Mukumbareza, C., Muchaonyerwa, P., & Chiduza, C. (2016). Bicultures of oat (*Avena sativa* l.) and grazing vetch (*Vicia dasycarpa* l.) cover crops increase contents of carbon pools and activities of selected enzymes in a loam soil under warm temperate conditions. *Soil Science & Plant Nutrition, 62*(5-6), 447–455. https://doi.org/10.1080/00380768.2016.1206833

Nannipieri, P., Giagnoni, L., & Landi, L. G. R. (2011). Role of phosphatase enzymes in soil. *Soil Biology & Biochemistry, 26*, 215–243. https://doi.org/10.1007/978-3-642-15271-9_9

Nannipieri, P., Kandeler, E., & Ruggiero, P. (2002). Enzyme activities and microbiological and biochemical processes in soil. In R. G. Burns & R. P. Dick (Eds.), *Enzymes in the environment* (pp. 1–33). Marcel Dekker.

Ndiaye, E. L., Sandeno, J. M., McGrath, D., & Dick, R. P. (2000). Integrative biological indicators for detecting change in soil quality. *American Journal of Alternative Agriculture, 15*(1), 26–36. https://doi.org/10.1017/S0889189300008432

Okamota, K., Nakano, H., Yatake, H., Kiso, T., & Kitahata, S. (2000). Purification and some properties of β-glucosidase from *Fusobacterium johnsonae. Bioscience, Biotechnology, and Biochemistry, 64*(2), 333–340. https://doi.org/10.1271/bbb.64.333

Ordentlich, A., Elad, Y., & Chet, I. (1988). The role of chitinase of *Serratia marcescens* in biocontrol of *Sclerotium rolfsii. Phytopathology, 78*, 84–87.

Padma, S., & Pallavi, K. (2016). Isolation and characterization of amylase producing *Bacillus* spp. from selected soil sample. *International Journal of Research in BioSciences, 5*(2), 24–29.

Pajares, S., Gallardo, J. F., & Masciandaro, G. (2011). Enzyme activity as an indicator of soil quality changes in degraded cultivated acrisols in the Mexican Transvolcanic Belt. *Land Degradation and Development, 22*(3), 373–381. https://doi.org/10.1002/ldr.992

Pandey, D., Agrawal, M., & Bohra, J. S. (2014). Effects of conventional tillage and no tillage permutations on extracellular soil enzyme activities and microbial biomass under rice cultivation. *Soil & Tillage Research, 136*, 51–60. https://doi.org/10.1016/j.still.2013.09.013

Pandey, J., & Yadav, A. (2017). Alternative alert system for Ganga river eutrophication using alkaline phosphatase as a level determinant. *Ecological Indicators, 82*, 327–343. https://doi.org/10.1016/j.ecolind.2017.06.061

Peyrot, C., Wilkinson, K. J., Desrosiers, M., & Sauvé, S. (2014). Effects of silver nanoparticles on soil enzyme activities with and without added organic matter. *Environmental Toxicology and Chemistry, 33*(1), 115–125. https://doi.org/10.1002/etc.2398

Piotrowska-Dlugosz, A., & Charzynski, P. (2015). The impact of the soil sealing degree on microbial biomass, enzymatic activity, and physicochemical properties in the ekranic technosols of toruń (Poland). *Journal of Soils and Sediments, 15*, 47–59. https://doi.org/10.1007/s11368-014-0963-8

Piotrowska-Dlugosz, A., & Wilczewski, E. (2014). Changes in enzyme activities as affected by greenmanure catch crops and mineral nitrogen fertilization. *Zemdirbyste, 101*(2), 139–146.

Pitchel, J. R., & Hayes, J. M. (1990). Influence of fly ash on soil microbial activity and populations. *Journal of Environmental Quality, 19*(3), 593–597. https://doi.org/10.2134/jeq1990.00472425001900030039x

Rasche, F., Knapp, D., & Kaiser, C. (2010). Seasonality and resource availability control bacterial and archaeal communities in soils of a temperate beech forest. *The ISME Journal, 5*, 389–402. https://doi.org/10.1038/ismej.2010.138

Rietz, D. N., & Haynes, R. J. (2003). Effects of irrigation-induced salinity and sodicity on soil microbial activity. *Soil Biology and Biochemistry, 35*(6), 845–854. https://doi.org/10.1016/S0038-0717(03)00125-1

Sardans, J., & Penuelas, J. (2005). Drought decreases soil enzyme activity in a mediterranean Quercus ilex l. *Soil Biology and Biochemistry, 37*(3), 455–461. https://doi.org/10.1016/j.soilbio.2004.08.004

Shapira, R., Ordentlich, A., Chet, I., & Oppenheim, A. B. (1989). Control of plant diseases by chitinase expressed from cloned DNA in *Escherichia coli*. *Phytopathology, 79*(11), 1246–1249. https://doi.org/10.1094/phyto-79-1246

Sicardi, M., Garcia-Prechac, F., & Frioni, L. (2004). Soil microbial indicators sensitive to land use conversion from pastures to commercial *Eucalyptus grandis* (Hill ex Maiden) plantations in Uruguay. *Applied Soil Ecology, 27*(2), 125–133. https://doi.org/10.1016/j.apsoil.2004.05.004

Singh, P. P., Shin, Y. C., Park, C. S., & Chung, Y. R. (1999). Biological control of *Fusarium* wilt of cucumber by chitinolytic bacteria. *Phytopathology, 89*(1), 92–99. https://doi.org/10.1094/PHYTO.1999.89.1.92

Sinsabaugh, R. L. (2010). Phenol oxidase, peroxidase and organic matter dynamics of soil. *Soil Biology and Biochemistry, 42*(3), 391–404. https://doi.org/10.1016/j.soilbio.2009.10.014

Sinsabaugh, R. L., Zak, D. R., Gallo, M. E., Lauber, C. L., & Amonette, R. (2004). Nitrogen deposition and dissolved organic matter production in northern temperate forests. *Soil Biology and Biochemistry, 36*(9), 1509–1515. https://doi.org/10.1016/j.soilbio.2004.04.026

Spano, G., Rinaldi, A., Ugliano, M., Beneduce, L., & Massa, S. (2005). A β-glucosidase producing gene isolated from wine *Lactobacillus plantarum* is regulated by abiotic stresses. *Journal of Applied Microbiology, 98*(4), 855–861. https://doi.org/10.1111/j.1365-2672.2004.02521.x

Srivastava, P. K., Gupta, M., Upadhyay, R. K., Sharma, S., Singh, N., Tewari, S. K., & Singh, B. (2012). Effects of combined application of vermicompost and mineral fertilizer on the growth of *Allium cepa* l. *Journal of Plant Nutrition and Soil Science, 175*(1), 101–107. https://doi.org/10.1002/jpln.201000390

Steinweg, J. M., Dukes, J. S., & Wallenstein, M. D. (2012). Modeling the effects of temperature and moisture on soil enzyme activity: linking laboratory assays to continuous field data. *Soil Biology and Biochemistry, 55*, 85–92. https://doi.org/10.1016/j.soilbio.2012.06.015

Stock, S. C., Kster, M., Dippold, M. A., Nájera, F., & Kuzyakov, Y. (2019). Environmental drivers and stoichiometric constraints on enzyme activities in soils from rhizosphere to continental scale. *Geoderma, 337*, 973–982. https://doi.org/10.1016/j.geoderma.2018.10.030

Tabatabai, M. A. (1982). Soil enzyme. In A. L. Page (Ed.), *Methods of soil analysis, part 2* (pp. 903–948). American Society of Agronomy.

Tabatabai, M. A. (1994a). Soil enzymes. In S. H. Mickelson (Ed.), *Methods of soil analysis, part 2. Microbiological and biochemical properties* (pp. 775–833). Soil Science Society of America.

Tabatabai, M. A. (1994b). Soil enzymes. In R. W. Weaver, J. S. Angle, & P. S. Bottomley (Eds.), *Methods of soil analysis, part 2. Microbiological and biochemical properties* (SSSA book series No. 5) (pp. 775–833). Soil Science Society of America.

Thoma, J. A., Spradlin, J. E., & Dygert, S. (1971). Plant and animal amylases. In P. D. Boyer (Ed.), *The enzymes* (5th ed., pp. 115–189). Academic.

Tietjen, T., & Wetzel, R. G. (2003). Extracellular enzyme-clay mineral complexes: Enzyme adsorption, alteration of enzyme activity, and protection from photodegradation. *Aquatic Ecology, 37*(4), 331–339. https://doi.org/10.1023/B:AECO.0000007044.52801.6b

Tripathy, S., Bhattacharyya, P., Mohapatra, R., Som, A., & Chowdhury, D. (2014). Influence of different fractions of heavy metals on microbial ecophysiological indicators and enzyme activities in century old municipal solid waste amended soil. *Ecological Engineering, 70*, 25–34. https://doi.org/10.1016/j.ecoleng.2014.04.013

Tyler, G. (1981). Heavy metals in soil biology and biochemistry. In E. A. Paul & J. N. Ladd (Eds.), *Soil biochemistry* (Vol. 5, pp. 371–414). Marcel Dekker.

Utobo, E., & Tewari, L. (2015). Soil enzymes as bioindicators of soil ecosystem status. *Applied Ecology and Environmental Research, 13*(1), 147–169. https://doi.org/10.15666/aeer/1301_147169

Van Aarle, I. M., & Plassard, C. (2010). Spatial distribution of phosphatase activity associated with ectomycorrhizal plants is related to soil type. *Soil Biology and Biochemistry, 42*(2), 324–330. https://doi.org/10.1016/j.soilbio.2009.11.011

Veena, V., Poornima, P., Parvatham, R., & Sivapriyadharsini, K. K. (2011). Isolation and characterization of β-glucosidase producing bacteria from different sources. *African Journal of Biotechnology, 10*(2), 14907–14912.

Vong, P. C., Dedourge, O., Lasserre-Joulin, F., & Guckert, A. (2003). Immobilized-S, microbial biomass-S and soil arylsulphatase activity in the rhizosphere soil of rape and barley as affected by labile substrate C and N additions. *Soil Biology and Biochemistry, 35*(12), 1651–1661. https://doi.org/10.1016/j.soilbio.2003.08.012

Wang, J., Zhan, X., Zhou, L., & Lin, Y. (2010). Biological indicators capable of assessing thermal treatment efficiency of hydrocarbon mixture-contaminated soil. *Chemosphere, 80*(8), 837–844. https://doi.org/10.1016/j.chemosphere.2010.06.009

Wichelns, D., & Qadir, M. (2014). Achieving sustainable irrigation requires effective management of salts, soil salinity, and shallow groundwater. *Agricultural Water Management, 157*, 31–38. https://doi.org/10.1016/j.agwat.2014.08.016

Wu, F., Wan, J. H. C., Wu, S., & Wong, M. (2012). Effects of earthworms and plant growth–promoting rhizobacteria (PGPR) on availability of nitrogen, phosphorus, and potassium in soil. *Journal of Plant Nutrition and Soil Science, 175*(3), 423–433. https://doi.org/10.1002/jpln.201100022

Xiao-Chang, W., & Qin, L. (2006). Beta-glucosidase activity in paddy soils of the Taihu lake region, China. *Pedosphere, 16*(1), 118–124. https://doi.org/10.1016/S1002-0160(06)60033-7

Xu, Z., Wang, S., Wang, Z., Dong, Y., Zhang, Y., Liu, S., & Li, J. (2021). Effect of drainage on microbial enzyme activities and communities dependent on depth in peatland soil. *Biogeochemistry, 155*, 1–19. https://doi.org/10.1007/s10533-021-00828-1

Xu, Z., Yu, G., Zhang, X., He, N., Wang, Q., Wang, S., Wang, R., Zhao, N., Jia, Y., & Wang, C. (2017). Soil enzyme activity and stoichiometry in forest ecosystems along the North-South transect in eastern China (NSTEC). *Soil Biology and Biochemistry, 104*(104), 152–163. https://doi.org/10.1016/j.soilbio.2016.10.020

Yang, G., Dong, F., Liu, M., Nie, X., Zong, M., Peng, C., Chen, H., Wei, H., Wang, P., & Zhang, W. (2018). Interactive effect of radioactive and heavy-metal contamination on soil enzyme activity in a former uranium mine. *Polish Journal of Environmental Studies, 27*(3), 1343–1351. https://doi.org/10.15244/pjoes/76182

Yang, Z. X., Liu, S. Q., Zheng, D. W., & Feng, S. D. (2006). Effects of cadium, zinc and lead on soil enzyme activities. *Journal of Environmental Sciences, 18*(6), 1135–1141. https://doi.org/10.1016/S1001-0742(06)60051-X

Zhang, W., Qiao, W., Gao, D., Dai, Y., Deng, J., Yang, G., Han, X., & Ren, G. (2018). Relationship between soil nutrient properties and biological activities along a restoration chronosequence of *Pinus tabulaeformis* plantation forests in the Ziwuling Mountains, China. *Catena, 161*, 85–95. https://doi.org/10.1016/j.catena.2017.10.021

Zhang, Y., Wu, N., Zhou, G., & Bao, W. (2005). Changes in enzyme activities of spruce (*Picea balfouriana*) forest soil as related to burning in the Eastern Qinghai-Tibetan Plateau. *Applied Soil Ecology, 30*(3), 215–225. https://doi.org/10.1016/j.apsoil.2005.01.005

Zheng, H., Liu, Y., Zhang, J., Chen, Y., Yang, L., Li, H., & Wang, L. (2018). Factors influencing soil enzyme activity in China's forest ecosystems. *Plant Ecology, 219*(1), 31–44. https://doi.org/10.1007/s11258-017-0775-1

Zhu, L., Wang, X., Chen, F., Li, C., & Wu, L. (2019). Effects of the successive planting of Eucalyptus urophylla on soil bacterial and fungal community structure, diversity, microbial biomass, and enzyme activity. *Land Degradation and Development, 30*(6), 1–11. https://doi.org/10.1002/ldr.3249

Zi, H. B., Hu, L., & Wang, C. T. (2018). Responses of soil bacterial community and enzyme activity to experimental warming of an alpine meadow. *European Journal of Soil Science, 69*(3), 429438. https://doi.org/10.1111/ejss.12547

Microbial Enzymes in the Recycling of Wastes

Oluwafemi Adebayo Oyewole, Abdullahi Dabban Idris, Aisha Bisola Bello, Japhet Gaius Yakubu, and Muhammed Muhammed Saidu

Abstract The increasing volume of different types of wastes from various sources is an important environmental problem due to the ever growing migration and successive urbanization. Enzymes are biological catalysts found in plants, animals, and microorganisms with numerous potential applications. Microbial enzymes have been used in the recycling and management of wastes through enzymatic degradation and remediation, resulting in less toxic useful products. Microbial enzymes are classified based on their mechanisms of action as oxidoreductases, transferases, hydrolases, lyases, isomerases, and ligases, with oxidoreductases and hydrolases being the most utilized in waste treatment and recycling. Microbial oxidoreductases are involved in catalyzing oxidation–reduction reactions in harmful biodegradable materials to produce nontoxic products. The oxidoreductases employed in waste degradation include oxygenases (monooxygenases and dioxygenases), laccases, and peroxidases. Microbial hydrolases catalyze the breakdown of waste biomass especially from the food, agricultural, chemical, and biomedical industries by addition of water molecules to the waste materials. Some microbial enzymes with hydrolytic properties include cellulases, hemicellulases, proteases, lipases, amylases, lactases, xylanases, and pullulanases. Compared to the conventional chemical methods, recycling of wastes using microbial enzymes has great significance in bioremediation as they are specific, fast, relatively cheap, can be applied across a wide variety of contaminants, and greatly reduce the waste and, at the same time, produce useful products. However, microbial enzymes are not devoid of limitations such as selection of the most suitable microbial enzymes for recycling and the ability of these enzymes to retain their active nature under normal conditions of operation for a prolonged period of time. Exploration of molecular studies will, in the near future, provide a clearer picture of the mechanisms of enzyme action either singly or in

O. A. Oyewole (✉) · J. G. Yakubu · M. M. Saidu
Department of Microbiology, Federal University of Technology Minna, Minna, Nigeria
e-mail: oa.oyewole@futminna.edu.ng; muhammad.m1603943@st.futminna.edu.ng

A. D. Idris · A. B. Bello
Department of Biological Sciences, The Federal Polytechnic Bida, Bida, Nigeria

N. R. Maddela et al. (eds.), *Ecological Interplays in Microbial Enzymology*,
Environmental and Microbial Biotechnology,
https://doi.org/10.1007/978-981-19-0155-3_10

consortium with other enzymes during biodegradation and recycling of wastes to produce more valuable products.

1 Introduction

The overall quality of the environment is inextricably linked to and highly dependent on the quality of life on planet Earth (Ilheme, Ukairo, Ibegbulem, Okorom, & Chibundu, 2017). The growing human population generates wastes of various kinds on a daily basis as a result of a wide range of different activities (Ebikapade & Jim, 2016). This leads to continuous accumulation in the volume and variety of wastes and thus poses a great threat to Earth's flora and fauna (Vergara & Tchobanoglous, 2012). The pollution of soil and water by industrial chemicals, petroleum hydrocarbons, polythene, plastics, metallic materials, and glass wastes is a serious problem of the modern world. Due to their extensive use, these are found as environmental contaminants in numerous aquatic and terrestrial ecosystems.

Recycling is described as the reprocessing of used materials into new products. This prevents or decreases utilization of raw materials and consumption of energy (Magram, 2011). The purpose of recycling is to convert waste products, which could be land-filled or waste streams, into feedstocks or raw materials for new and useful products (Dyson & Chang, 2005). Waste recycling poses a major environmental and economic challenge worldwide; it takes place out of sight and hence attracts less public concern and is the least priority for authorities (Dyson & Chang, 2005).

Naturally, wastes are spontaneously recycled by plants and microorganisms, especially bacteria and fungi, to maintain a healthy ecosystem (Fig. 1). Microbial enzymes are biological catalysts that are mostly proteins produced by

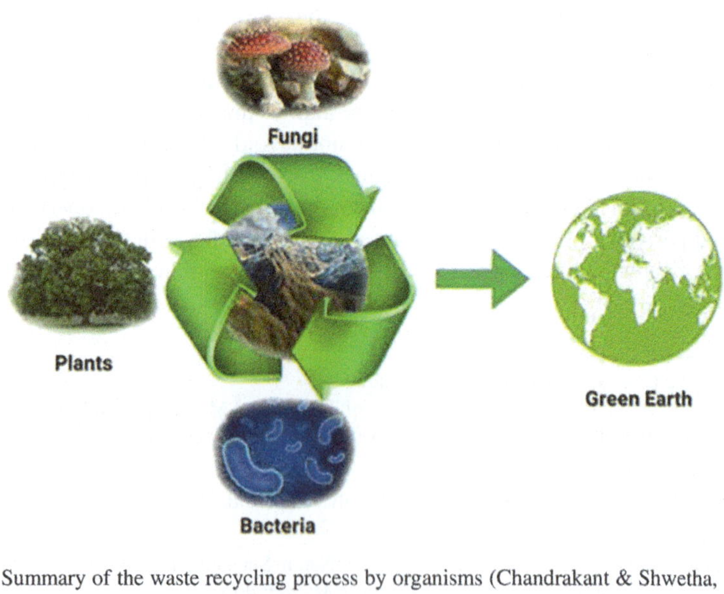

Fig. 1 Summary of the waste recycling process by organisms (Chandrakant & Shwetha, 2011)

microorganisms, especially bacteria, yeast, and molds, and have been significantly used in medicine, industries, and biotechnology (Maddela & García, 2021; Maddela, Garcia, & Chakraborty, 2021; Periasamy, Subash, Arzu, & Chaulagain, 2013). Compared to plant and animal enzymes, microbial enzymes are more recognized and preferred owing to their relatively high activity and stability coupled with their ease of production and recovery in large quantities (Shahid, Muhammad, Jabeen, & Mubeen, 2016).

Researchers have identified enzymatic treatment of wastes as an effective method of waste management compared to the conventional methods because microbial enzymes exhibit selective and specific activity; they are not inhibited by most toxic substances; they have low retention time; they function over a wide range of concentrations; they are less expensive and are produced in large quantities; and they provide a safe and economic alternative (Aitken, 1993; Karam & Nicell, 1997).

2 Wastes

Waste is any unwanted substance (solid, liquid, or gas) or material regarded as useless and to be disposed of as being broken, contaminated, or spoilt (Anifowose, Omole, & Akingbade, 2011; Ayilara, Olanrewaju, Babalola, & Odeyemi, 2020; Rajan, Robin, & Vandanarani, 2019). It is an unavoidable by-product of human activities whose continuous generation results in loss of resources (Cheremisinoff, 2003). Waste, now an important environmental problem, is a result of the increasing rate of development, urbanization, and migration to cities (Ayilara et al., 2020). Improper waste management is hazardous to the environment, humans, and animals alike. Wastes pollute the air when burnt and release gases that deplete the ozone layer such as carbon dioxide, hydrogen, and methane, thus causing climate change (Bhat, Dar, Dar, & Dar, 2018). When dumped in water, wastes affect aquatic lives, humans, soil organisms, and plants by lowering the pH and depositing metals, which increase the toxicity of water (Corral-Bobadilla, González-Marcos, Vergara-González, & Alba-Elías, 2019; Holanda & Johnson, 2020; Mani & Kumar, 2014; Sahay, Iqbal, Inam, Gupta, & Inam, 2019). Wastes harbor vectors of diseases such as mosquitoes and put refuse workers at risk of injuries and infections (Alam & Ahmade, 2013).

2.1 Classification of Wastes

Based on the state of matter, waste materials are classified as solid, liquid, and gaseous wastes. On the basis of biodegradability, they are categorized as nondegradable, partially degradable, and completely degradable. Completely degradable (biodegradable) solid wastes are wastes that undergo decomposition by microorganisms into their diverse components (Alam & Ahmade, 2013). Wastes from food, manure, and crop production can be completely decomposed (Lorenz,

Fischer, Schumacher, & Adler, 2013). Agricultural wastes from animal sources such as cow dung, poultry droppings, etc. are also classified as biodegradable wastes (Bhat et al., 2018). The purpose of biodegradation is to reduce the volume of the waste deposited and also reduce its harmful impact on the human health and environment by producing useful products with economic impact (Holm-Nielsen, Al Seadi, & Oleskowicz-Popiel, 2009).

Nonbiodegradable wastes are materials that cannot undergo biological or microbial decomposition or breakdown, and these include wastes from mines, mineral materials, polythene bags, leathers, plastics, glass, etc. (Baltrėnas, Jankaitis, & Raistenskis, 2005). Nonbiodegradable wastes can be grouped into recycled and nonrecycled wastes. Recycled wastes are sold to companies and converted into new products, whereas nonrecycled wastes are waste materials that are transported to dump sites and incinerated (Alam & Ahmade, 2013).

Solid waste materials are classified based on whether they can be incinerated, which are combustible and noncombustible, and the dangers they possess or are associated with, i.e., hazardous or nonhazardous (Demirbas, 2011). Hazardous solid wastes are a public health threat to humans, animals, and the environment, and these hazards include toxic gases, infectious diseases, and corrosive substances. Waste materials that do not pose potential hazardous or harmful threats are classified as nonhazardous (Buragohain, Nath, & Sharma, 2020).

2.2 Sources of Wastes

Wastes come from different sources, in different forms, and are dumped in different ways (Ahmed, 2013). The release of wastes into the environment affects the quality of life, and the impact on the environment is unquantifiable (Ahmed, 2013; Tulebayeva, Yergobek, Pestunova, Mottaeva, & Sapakova, 2020). Significant sources of waste generation include municipal, agricultural, industrial, biomedical, and electronic wastes (Amasuomo & Baird, 2016).

Municipal Wastes

Municipal solid wastes (MSWs), also known as garbage, are wastes collected from households, schools, markets, malls, gardens, streets, litter containers, etc. (Buragohain et al., 2020; Organisation for Economic Co-operation and Development, 2021). The increasing amount of MSW generation as a result of industrialization, migration, urbanization, and improper disposal of food waste poses a serious global challenge (Rajan et al., 2019). MSWs are generated from different sources where human activities take place. Developing countries generate about 55–80% of household wastes and 10–30% of commercial and market wastes, which consist of industries, streets, institutions, and many others (Nabegu, 2017). There are risks associated with the improper management of MSWs, which pose threats to the public health and environmental safety, from collection to reusable materials (WHO, 2018). Kaza, Yao, Bhada-Tata, and Van Woerden (2018) estimated that globally by the year 2050, the generation of MSWs will rise to 3.40 billion tons. The health risks associated with wastes are higher in low-income countries as a result of unpleasant methods of waste disposal such as uncontrolled dumping sites and burning of solid wastes (Ferronato & Torretta, 2019), with lower risks in high-income countries.

Controlled dumping sites, incineration, land filling, composite, anaerobic digestion, and recycling are some of the methods of waste treatment and disposal (Kaza et al., 2018; Vinti et al., 2021).

Industrial Wastes

Industrial wastes are produced as a result of industrial activities such as production of oil and gas, coal combustion, mining, and product manufacturing (Demirbas, 2011). By-products generated from manufacturing processes such as from mill mining and factories are also regarded as industrial waste materials. These wastes also include radioactive wastes, metals, paints, chemicals, sand papers, and paper products. Wastes from industrial sources are potentially toxic pollutants that necessitate thorough treatment before being discharged into the environment (Maczulak, 2010).

Biomedical Wastes

Biomedical wastes are wastes generated from healthcare institutions in the form of radioactive materials, blood, sharp and nonsharp objects, pharmaceutical products, and chemicals (Nwachukwu, Orji, & Ugbogu, 2013). About 85% of wastes generated in healthcare are nonhazardous, and the remaining 15% are hazardous, which may cause infections, toxicity to the environment, or are poisonous (radioactive) (WHO, 2018). WHO (2018) reported about 16 million injections being administered per year globally, resulting in the improper disposal of needles and syringes after use. Wastes generated from healthcare centers expose patients, care givers, and waste handlers to potential infections, injuries, and toxic materials while at the same time polluting the environment; such wastes include radioactive materials, pharmaceutical wastes, nonhazardous wastes, pathological wastes, and toxic wastes (Nwachukwu et al., 2013). To adequately manage healthcare wastes, separation, appropriate treatment, and safe disposal are important to enable proper recycling and disposal (Nwachukwu et al., 2013). Incineration of these wastes may result in the release of toxic chemicals and particles, causing pollution. Therefore, proper actions should be taken to make sure environmental safety and health management are put in place to prevent serious health and environmental impacts such as accidental release of chemicals and biological hazards including drug-resistant microorganisms (WHO, 2018).

Agricultural Wastes

The role of agriculture cannot be overemphasized in human and economic development with the growing human population, technological advancement toward the Green Revolution, and expansion of soil for agricultural production, resulting in increased waste generation, which may constitute serious public health challenges through pollution (Adejumo, Adebiyi, & Olufemi, 2020). Agricultural wastes are wastes generated from the growing and processing of raw farm products resulting in by-products that may be beneficial but have less economic value and high cost of management. Agro-wastes consist of animal and food wastes and harmful and toxic agricultural wastes (pesticides, herbicides, insecticides). The intensity of agriculture in developing countries may contribute to the increased generation of agro-wastes globally, with about 998 million tons of agro-wastes generated yearly (Agamuthu, 2009; Obi, Ugwuishiwu, & Nwakaire, 2016). Agricultural wastes are not properly managed because very little is known about the potential risks and benefits associated with proper management (Adejumo et al., 2020).

Agricultural wastes can be utilized through the absence of oxygen (anaerobic) digestion, fertilizer application, as absorbents in the removal of heavy metals, and in pyrolysis, animal feed, and direct combustion. Management of agro-wastes requires the need to consider wastes as potential resources rather than as undesirable to avoid water, air, and land contamination. Improper management of these wastes may also result in breeding sites for insects with the ability to transmit diseases, poor soil quality, and degradation such as phosphorous loading, emission of gases, and foul odors such as ammonia and methane (Obi et al., 2016).

Electronic Wastes (E-Wastes)

Electronic wastes (e-wastes) denote discarded electrical or electronic devices. E-wastes are among the types of wastes plaguing the world currently. Used electronics, which are destined for reuse, refurbishment, salvaging, and recycling through material recovery, disposal, or abandonment, are also considered e-wastes. The informal processing of e-wastes in developing nations can result in adverse effects on human health and lead to environmental pollution. Scrap components of electronics, such as central processing units (CPUs), contain potential harmful materials such as cadmium, lead, beryllium, and brominated flame retardants (Buragohain et al., 2020). The recycling and disposal of e-wastes can pose significant risks to the health of workers and communities in developed and developing countries (Abalansa et al., 2021).

3 Mechanisms of Enzyme Degradation of Wastes

There are six groups in which all known enzymes are classified based on their mechanisms of action, and they include oxidoreductases, transferases, hydrolases, lyases, isomerases, and ligases. Of these six classes, oxidoreductases and hydrolases are the most employed in the recycling of wastes.

3.1 Oxidoreductases

These are groups of enzymes that aid the transfer of oxygen, hydrogen, electrons, and protons from a donor to an acceptor. Oxidoreductases from microbial sources such as bacteria and fungi detoxify organic pollutants using oxidation–reduction reactions (Chandrakant & Shwetha, 2011). Microorganisms extract energy from biochemical exothermic reactions through the breakdown of chemical bonds by these oxidoreductases. These enzymes also utilize the energy for electron transfer from a donor organic compound to another accepting chemical compound. The donor is reduced, whereas the acceptor is oxidized. Safer substances are generated from the pollutants after the redox reaction process (Cirino & Arnold, 2002).

3.2 Transferases

Functional groups, e.g., acyl, alkyl, formal, glycosyl, hydroxymethyl, methyl, sulfate, and phosphate groups, are transferred using a nucleophillic substitution reaction from a donor to an acceptor by this class of enzymes (Pandeeti, Veeraiah, & Routhou, 2019).

3.3 Hydrolases

This class of enzymes mediates the breakdown of carbon to carbon, carbon to oxygen, and carbon to nitrogen bonds using water molecules (Chandrakant & Shwetha, 2011). The toxicity of pollutants is decreased through microbial hydrolase disruption of the main chemical bonds (Vasileva-Tonkova & Galabova, 2003). Alcoholysis and condensation reactions are also mediated by this class of enzymes. Constant availability, ability to withstand addition of solvents, and the absence of the need to select cofactors are some of the merits of this class of enzymes (Schmidt, 2006). In biomedical sciences, chemicals, and the food and feed industries, hydrolases have been greatly used for their wide potential applications (Sanchez-Porro, Mart, Mellado, & Ventosa, 2003).

3.4 Lyases

This enzyme class is responsible for catalyzing addition and elimination reactions. Lyases cleave the bonds between carbon atoms (C–C) and those between carbon and other atoms such as oxygen (C–O), nitrogen (C–N), etc. by elimination. Lyases break down double bonds in chemical pollutants and, subsequently, mediate the insertion of other chemical groups at the cleaved double bond (Pandeeti et al., 2019).

3.5 Isomerases

The structural rearrangement of molecules, i.e., isomerization is facilitated by microbial isomerases.

3.6 Ligases (Synthetases)

Also known as synthetases, ligases are enzymes with the catalyzing ability of joining two large molecules together, resulting in a new chemical bond. They are also generally associated with small chemical groups and linking of compounds. This class of enzymes establishes carbon–oxygen, phosphoric–ester, carbon–sulfur, carbon–nitrogen, nitrogen–metal, and carbon–carbon bonds (Lehninger & Nelson, 2004; Manjunuth, Lavanya, Pathakoti, & Kjell, 2018).

In many waste recycling techniques, two or more enzyme mechanisms are usually utilized.

4 Microbial Enzymes in Waste Degradation/Recycling

Microbial enzymes are continuously being used in the degradation, remediation, and recycling of different wastes (Buragohain et al., 2020).

4.1 Microbial Oxidoreductases

Oxidoreductases are a class of enzymes that catalyzes oxidation–reduction reactions. They are involved in the biodegradation of harmful waste compounds such as radioactive metals, halogenated compounds, and phenolic and other related aromatic and aliphatic hydrocarbons (Park, Park, & Kim, 2006; Vidali, 2001). Oxidoreductases from microorganisms have been employed for decolorization, degradation, and remediation of azo and other related synthetic dyes (Husain, 2006; Leung, 2004). The microbial oxidoreductases most studied in waste bioremediation due to their high efficacy in degrading harmful substances in the environment include oxygenases (monooxygenases and dioxygenases), laccases, and peroxidases (Arora, Srivastava, & Singh, 2010; Chandrakant & Shwetha, 2011).

Microbial Oxygenases

This family of oxidoreductases is responsible for the biodegradation of a broad range of waste materials by increasing their solubility, reactivity, and breakdown of aromatic rings present in the waste materials. The cleavage of the aromatic rings in the toxic wastes by oxygenases is achieved by introducing atoms of oxygen into the organic compounds (Arora, Kumar, Chauhan, Raghava, & Jain, 2009; Fetzner, 2003). The bacterial sources of oxygenases are the most researched in biodegradation and remediation of toxic waste materials (Chandrakant & Shwetha, 2011). A large majority of oxygenases oxidize reduced toxic substrates using flavin adenine dinucleotide (NAD), reduced nicotinamide adenine dinucleotide (NADH), or reduced nicotinamide adenine dinucleotide phosphate (NADPH) as cosubstrates (Fetzner, 2003).

The widespread use of insecticides, fungicides, herbicides, and other chemicals containing high concentrations of halogens makes them a major environmental pollutant. Specific microbial oxygenases are being used in the breakdown of these toxic contaminants. Oxygenases have also being utilized in consortium with other multifunctional enzymes to catalyze the removal of these halogens from methane-, ethane-, and ethylene-containing compounds (Chandrakant & Shwetha, 2011). In the process of pulp bleaching, chlorinated phenolic wastes from the paper and pulp industry are generated in abundance from incomplete breakdown of lignin (Rubilar, Diez, & Gianfreda, 2008). Oxygenases from fungal sources are extracellular and are released into nearby environments from the mycelium of fungi. Hence, due to this advantage, numerous suitable species of fungi are being used for biodegradation and bioremediation of environments contaminated with chlorinated phenolic compounds (Rubilar et al., 2008). Oxygenases are further grouped into monooxygenases and dioxygenases on the basis of the number of oxygen atoms used for oxygenation.

Microbial Monooxygenases

This enzyme group consists of a vast superfamily involved in the catalysis of a variety of simple (e.g., alkanes) to complex substrates (fatty acids and steroids) through oxidation reactions. Monooxygenases operate by integrating an atom from an oxygen molecule into their substrates. Their relatively high stereoselectivity on a broad variety of substrates makes them a useful tool in biodegradation and remediation processes. On the basis of the presence of cofactors, monooxygenases are divided into P450-dependent and flavin-dependent monooxygenases. P450 monooxygenases usually contain iron, and they are found in prokaryotic as well as eukaryotic organisms. Flavin-dependent monooxygenases, on the other hand, contain flavin as their prosthetic group, and they generally require NADPH or NADP as a coenzyme (Arora et al., 2010). Most monooxygenases require a cofactor. Although some members of this enzyme group can function properly without a cofactor, they require molecular oxygen for their action and use the substrate as a reducing agent (Cirino & Arnold, 2002).

Monooxygenases have been employed in the biodegradation and biotransformation of a wide variety of aliphatic and aromatic contaminants through the removal of sulfur and halogens as well as the addition of ammonia and hydroxyl groups. These properties have been explored in recent years for important applications in the recycling of recalcitrant wastes (Chandrakant & Shwetha, 2011).

Microbial Dioxygenases

This group utilizes a multicomponent system to catalyze in an enantiospecific manner by introducing molecular oxygen into their substrates. In other words, dioxygenases break down complex waste compounds by introduction of two atoms of oxygen into their substrates to produce simpler products. They primarily oxidize aromatic compounds and hence have been used in the remediation of pollutants in the environment. Dioxygenases usually contain proteins used in electron transport, which precedes their oxygenase components (Dua, Singh, Sethunathan, & Johri, 2002). Among other mechanisms that the nature employs to break down aromatic compounds in the environment is the utilization of catechol dioxygenases. The breakdown and subsequent biotransformation of aromatic compounds to produce simpler aliphatic compounds is catalyzed by catechol dioxygenases manufactured by many soil bacteria. The extradiol-degrading enzyme makes use of Fe(II) and sometimes Mn (II), whereas the intradiol-degrading enzyme uses only Fe(III) (Chandrakant & Shwetha, 2011).

Microbial Dehalogenases

Microbial dehalogenases have significant applications in bioremediation of halogenated organic compounds. Dehalogenase enzymes degrade a wide range of halogenated compounds by breaking the alkyl–halide bonds (Wang, Feng, Cao, Liu, & Xue, 2018) through three mechanisms: hydrolytic, reductive, and oxygenolytic methods. Dehalogenation is performed by replacing an atom of halogen with a hydroxyl group from water molecules (Wang et al., 2018). *Bacillus* sp. with the intrinsic ability to concurrently carry out debromination and mineralization of tribromophenol (TBP) has been reported (Zu, Li, An, & Wong, 2012). The bacteria utilize two pathways in the debromination step of which reductive bromination and methyl bromination are the major and minor pathways, respectively, producing CO_2 as the by-product of the mineralization (Zu et al., 2012). Other bacterial species such as *Pseudomonas umsongensis* YCIT161213 (Xue, Ya, Tong, Xiu, & Huang, 2018), the *Ancylobacter aquaticus* strain UV514, and *Rhizobium* sp. synthesize enzymes with the ability to transform a variety of halogenated pollutants (Kumar, Dhar, Vijay, Vaida, & Akolkar, 2016).

Microbial Laccases

These are multicopper-containing extracellular enzymes found in bacterial and fungal species and consist of mono-, di-, and tetrameric glycol proteins. Microbial laccases are produced by different microorganisms. The laccases from *Streptomyces* sp. are well identified, characterized, and the most studied. Various species of *Streptomyces* have been found to produce laccases. They include *Streptomyces ipomoea*, *Streptomyces cyaneus*, *Streptomyces bikiniensis*, and *Streptomyces coelicolor*. Of all the species, *S. coelicolor* is the most broadly characterized (Guan, Luo, Wang, Chen, & Liao, 2018). The presence of lignin and other phenolic compounds in a wide variety of agricultural waste materials (e.g., banana peels, rice bran, maize husk, saw dust, and other lignin-rich materials) elicits the production of laccases by these organisms (Muthukumarasamy, Jackson, & Joseph, 2015). Laccases are capable of oxidizing phenolic compounds, aromatic amines, and derivatives of these compounds, which tend to have varying functional groups. The oxidation is catalyzed through the formation of two molecules of water with the loss of electron from a single molecule of oxygen. It also catalyzes the oxidation of non-phenolic substrates that are less soluble and more stable (Janusz et al., 2020). Xenobiotic substances can be removed by microbial laccases, and they produce polymeric products used for bioremediation processes.

Polycyclic aromatic hydrocarbons (PAHs) are compounds with benzene rings arranged linearly. They are among the major contaminants of the environment (Abdel-Shafy and Mansour, 2016). Owing to their persistent, carcinogenic, mutagenic, and toxic nature, these pollutants and their derivatives pose severe threats to both the flora and fauna (Abdel-Shafy and Mansour, 2016). They are formed as a result of incomplete combustion of industrial wastes and fossil fuels. Due to poor degradation rate and low water solubility, they are regarded as xenobiotics (Ilheme et al., 2017). Polycyclic aromatic hydrocarbons are converted into quinone form by microbial laccases and are subsequently degraded to carbon dioxide (Patel et al., 2020). Textile dyes and phenols produced by the textile industry can also be detoxified and removed by laccases (Akram et al., 2022). Some of the applications of laccases reported include decolorization, degradation, and detoxification of various components of distillery effluents as well as wastes from the paper and pulp industry (Chandra & Chowdhary, 2015).

Microbial Peroxidases

Produced by plants and microorganisms, peroxidases are ubiquitous enzymes that catalyze the oxidation of lignin and other phenolic compounds at the expense of hydrogen peroxide (H_2O_2) in the presence of a mediator. Their activity greatly depends on the presence of peroxides, e.g., hydrogen peroxide, manganese peroxide, lignin peroxides, and other peroxidases from diverse sources. The enzyme is first oxidized by the peroxides, and, subsequently, oxidation of the substrate is catalyzed by the oxidized enzyme. In the treatment of aqueous aromatic pollutants, peroxidases from various sources have been greatly used (Karam & Nicell, 1997). Peroxidases are classified as heme and nonheme proteins (Koua, Cerutti, & Falquet, 2009).

Heme peroxidases are found in animals, plants, fungi, and prokaryotes. They are further subdivided into Classes I, II, and III on the basis of sequence comparison. Class I includes the ascorbate, cytochrome, and catalase peroxidases, which are all intracellular enzymes. Class II includes the manganese (MnP) and lignin (LiP) peroxidases. They are produced by certain fungal species, and their main function is the breakdown of plant lignin. Class III includes horseradish peroxidases (HRPs) from plant sources such as horseradish, soybean, or barley, and they catalyze the biosynthesis of plant cell walls and lignification reactions (Hiner, Ruiz, & Rodri, 2002).

Nonheme peroxidases, on the other hand, are grouped into five nonrelated independent families. They include alkyl hydroperoxidases, manganese catalase peroxidases, NADH peroxidases, nonheme haloperoxidases, and thiol peroxidases. Thiol peroxidases are the largest group with two subfamilies: peroxiredoxins and glutathione peroxidases (Koua et al., 2009). On account of their activity, enzyme source, and potential to naturally degrade toxic pollutants, peroxidases are also categorized into versatile peroxidase (VP), manganese-dependent peroxidase (MnP), and lignin peroxidase (LiP).

Microbial Versatile Peroxidases

Due to their wide substrate specificity and oxidation in the absence of manganese, members of this group have been used in bioremediation of recalcitrant wastes and other industrial processes (Tsukihara, Honda, Sakai, Watanabe, & Watanabe, 2006; Wong, 2009). Similar to manganese, lithium, and horseradish peroxidases, versatile peroxidases catalyze the oxidation of substrates such as Mn^{2+}, phenolic aromatic compounds, phenolic and non-phenolic lignin dimers, and methoxybenzene (Ruiz-Duenas, Morales, & Perez-Boada, 2007).

Microbial Manganese Peroxidases

The production of manganese peroxidase is stimulated by Mn^{2+}, thereby acting as a substrate for the enzyme. Produced extracellularly by the Basidiomycetes class of fungi, manganese peroxidases are heme enzymes that catalyze the oxidation of Mn^{2+} to Mn^{3+} in a series of reactions. The oxidant Mn^{3+} serves as an intermediary for phenolic compound oxidation. Due to its small size, the Mn^{3+} chelate oxalate diffuses into regions inaccessible to enzymes. Typical examples include lignin and xenobiotic pollutants buried deep in the soil and inaccessible to enzymes (Ten Have & Teunissen, 2001).

Microbial Lignin Peroxidases

These enzyme groups play an important role in the breakdown of lignin, a plant cell wall component. Produced majorly as a secondary metabolite by the white rot fungi, lignin peroxidases are heme proteins that catalyze the degradation of lignin and other phenolic compounds in the presence hydrogen peroxide (H_2O_2) (cosubstrate) and veratryl (mediator). These enzymes also catalyze the oxidation of aromatic compounds, but the mechanism of action is not well known (Piontek, Smith, & Blodig, 2001).

Microbial Dehydrogenases

Microbial dehydrogenases are oxidoreductases found majorly in bacteria and yeast. Microbial alcohol dehydrogenases catalyze the transformation of alcohols to yield aldehydes or ketones. They are grouped as nicotinamide adenine dinucleotide (NAD^+)- or nicotinamide adenine dinucleotide phosphate ($NADP^+$)-dependent dehydrogenases and NAD^+- or $NAD(P)^+$-independent dehydrogenases. NAD^+- and $NADP^+$-independent dehydrogenases use pyrroloquinoline, quinone, heme, or F420 as a cofactor (Chandrakant & Shwetha, 2011). In the same vein, aldehyde dehydrogenase catalyzes the $NADP^+$-dependent transformation of aldehyde to carboxylic acid (Nickolas & Vasiliou, 2003). Polyethylene glycol dehydrogenase from cell-free extracts was found to degrade polyethylene glycol and xenobiotics emitted from industries (Kawai & Yamanaka, 1989). Secretion of NAD^+-dependent polypropylene glycol dehydrogenase (PPG-DH) by *Stenotrophomonas maltophilia* oxidizes hydrophobic polymers with medium-chain secondary alcohols, dipropylene glycols, tripropylene glycols, and polypropylene glycols (Tachibana, Naka, Kawai, & Yasuda, 2008).

4.2 Microbial Hydrolytic Enzymes

The significance of this category of enzymes is due to their uses in the breakdown of waste biomass (Schmidt, 2006), especially from the food, agricultural, chemical, and biomedical industries. Microbial enzymes with hydrolytic activity employed in different waste treatments and recycling include cellulases, hemicellulases, proteases, lipases, amylases, lactases, xylanases, and pullulanases (Sanchez-Porro et al., 2003).

Microbial Cellulases

Over the years, the treatment of agricultural wastes rich in cellulose, lignocellulose, and related biomaterials using microbial enzymes has been continuously gaining attention (Chandrakant & Shwetha, 2011). The production of high-value products such as bioethanol, biogas, enzymes, and sugars from the conversion of agricultural and municipal wastes rich in cellulose and lignocellulose using microbial cellulases has increased the interests for application in industries (Sun & Cheng, 2002). Microbial cellulases, glucanases, cellobiohydrolases, glucosidases, carbohydrates, and cellobiases are continuously being used in the degradation and recycling of agricultural and municipal wastes. In order to meet the demands of the ever growing population, cellulases are continuously being used in food and feed production from the conversion of cellulosic waste resources (Bennet, Wunch, & Faison, 2002).

Cellulases produced by microorganisms could be intracellular or extracellular. Bacteria and fungi have been reported to liberate cellulases and related enzymes extracellularly but at extremely low levels (Adriano-Anaya, Salvador-Figueroa, Ocampo, & Garcıa-Romera, 2005). Species of *Bacillus* have been found to produce alkaline cellulases, whereas fungi such as *Trichoderma* and *Humicola* are known to produce neutral and acidic cellulases. More often, cellulases consist of a combination of a number of enzymes especially from microbial sources. In the process of hydrolysis, three main groups are implicated: the endoglucanase, exoglucanase, and β-glucosidase. In the cellulose fiber, the sites of low crystallinity are first acted upon by the endoglucanase creating free chain ends. Thereafter, the cellulose molecule is further broken down by exoglucanase (cellobiohydrolase) through cellobiose molecule elimination from the free chain ends. Finally, hydrolysis of cellobiose to glucose units is catalyzed by β-glucosidase (Fig. 2). In the enzymatic cellulose hydrolysis by cellulases to reducing sugars, which are fermented by bacteria and yeast to produce ethanol, the presence of some secondary enzymes along with the key enzymes has been reported (Sun & Cheng, 2002).

Cellulases have found various applications especially in the textile, brewing, paper, and pulp industries. In the textile industry, cellulases have been used for the brightening of colors and softening of materials. Since cellulases catalyze the removal of cellulose microfibrils formed during washing, these enzymes are also used in the manufacture of detergents. In the brewing industry, cellulases are

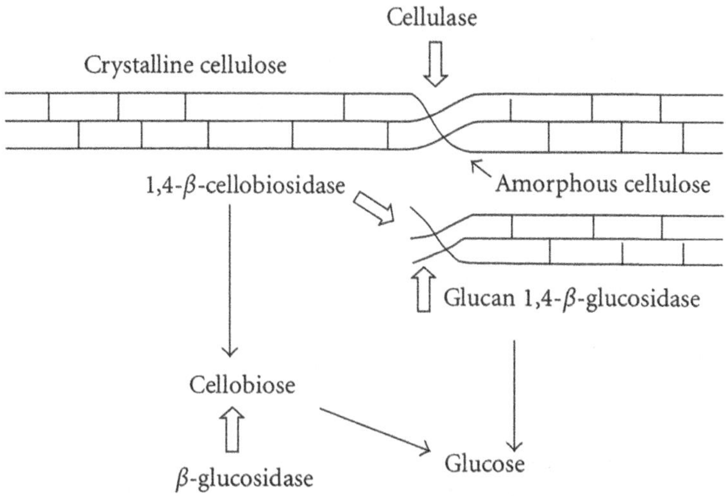

Fig. 2 Mechanism of cellulose hydrolysis by microbial cellulase (Schmidt, 2006)

employed in treatment of cellulose-rich biomass to produce ethanol. Similarly, addition of the enzyme in fruit pulp increases the liberation of the juice. Cellulases have also been used in the paper and pulp industry for elimination of ink during the paper recycling process.

Microbial Lipases

Lipases are ubiquitous and can be isolated from a wide group of plants, animals, and microorganisms; they are known to break down lipids to their corresponding monomers (free fatty acids and glycerol). The enzyme group has been reported to have a close relationship with soil organic pollutants. Thus, their activity was found to be accountable for the tremendous reduction in hydrocarbon pollutants present in soil. In industries, lipases from microbial sources are most commonly used than are other sources of the enzyme. Some of the reactions catalyzed by these enzymes include alcoholysis, esterification, hydrolysis, and aminolysis (Prasad & Manjunath, 2011).

Lipolytic activities of lipases occur in a two-phase system, i.e., the lipid–water interface (Fig. 3), where the substrates of lipases appear to be stable between three states, namely, the monomeric, micellar, and emulsification states (Prasad & Manjunath, 2011). Hence, two major groups of lipases have been reported based on the following factors: (a) improvement of lipase activity following the immediate emulsification of triacylglycerides and (b) lipases with their active site containing a protein-covering (lid) loop (Sharma, Sharma, & Shukla, 2011).

Natural fats and oils contain triglycerides as the major component, and the triglycerides can undergo hydrolysis to produce monoacylglycerols, diacylglycerols, free fatty acids, and glycerol. These resulting products of hydrolysis are used for different purposes. In cosmetics and in the pharmaceutical and food industries,

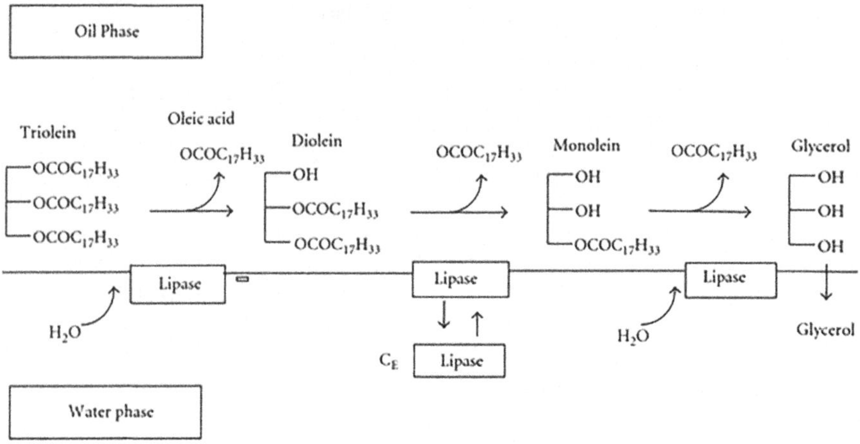

Fig. 3 A two-phase oil–water system of triolein hydrolysis by lipase isolated from *Candida rugosa* (Hermansyah, Wijanarko, & Gozan, 2007)

monoacylglycerols are greatly utilized as emulsifying agents. The activity of lipase is employed as a major valuable marker for assessing the degradation of hydrocarbons in polluted soil (Margesin, Zimmerbauer, & Schinner, 1999).

The prospects of lipases in the manufacturing, food, cosmetic, detergent, paper/pulp, and chemical industries are enormous coupled with their potentials as indicators in biodegradation and bioremediation of contaminated soil. However, application of these enzymes is limited to the industry due to the high cost associated with the enzyme production (Sharma et al., 2011).

Microbial Proteases

Their ability to hydrolyze peptide bonds of polymeric proteins in an aqueous environment into different amino acid monomers (Fig. 4) makes microbial proteases an important enzyme in the food, leather, tannery, and pharmaceutical industries (Singh, 2003). Proteases hydrolyze proteinaceous substances, which are introduced into the environment due to death of animals, shedding, and molting of body parts of animals, and protein-rich by-product production from the fishery, tannery, dairy, poultry, and related industries. Depending on the nature of peptide chain catalysis, proteases are categorized as exopeptidases and endopeptidases (Beena & Geevarghese, 2010).

Endopeptidases These are subdivided into metalloendopeptidases, cysteine endopeptidases, aspartic endopeptidases, and serine endopeptidases depending on the location of the active site. The action of endopeptidases on the peptide chain is usually in the inner regions of the polymer. There is a negative impact on enzyme activity as a result of the free carboxyl and amino terminals from cleavage of the peptide bonds (Beena & Geevarghese, 2010).

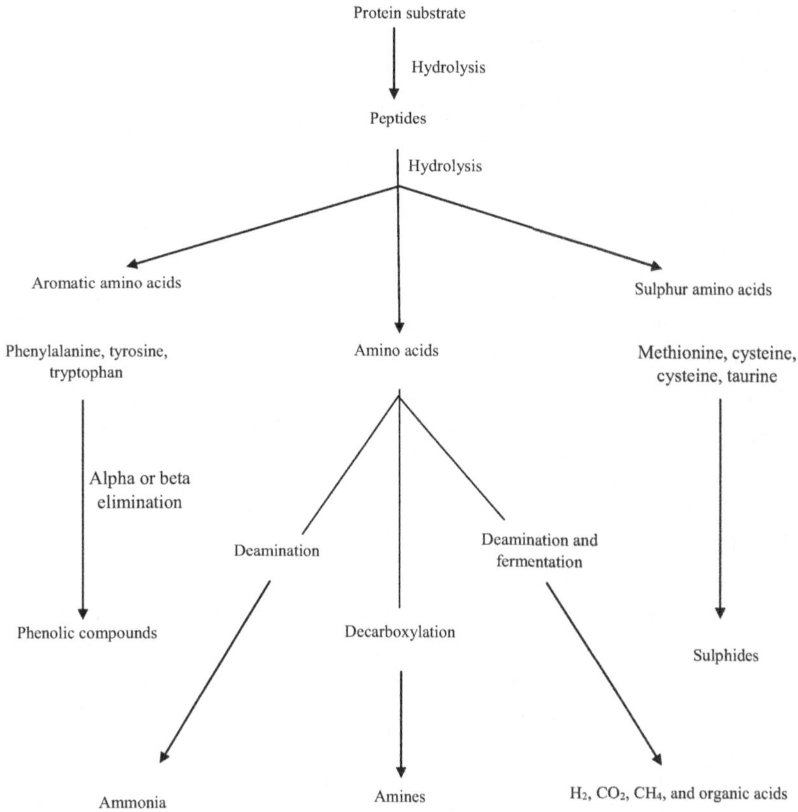

Fig. 4 Protease hydrolysis pathway (Chandrakant & Shwetha, 2011)

Exopeptidases The activity of this group is close to the terminal carboxylic or amino sites of the chain. The aminopeptidases and the carboxypeptidases then act on the free amino and carboxyl terminals, respectively (Singh, 2003).

Proteases have found applications in the detergent, leather, pharmaceutical, and food production/processing industries. In the leather industry, alkaline proteases are used in animal skin processing for removal of hairs and other parts of the skins. In addition, in dipeptide aspartame production (an artificial noncaloric sweetener), the use of protease has been reported (Rao, Tanksale, Ghatge, & Deshpande, 1998).

Microbial Amylases

Microbial amylases are hydrolases of polysaccharides, and they have found use in instantaneous saccharification, fermentation of starch, and, ultimately, treatment of food wastes rich in starch (Karam & Nicell, 1997; Shoemaker, 1986). Amylases have also been employed in the production of alcohol using rice processing wastewaters as substrates (Karam & Nicell, 1997; Shoemaker, 1986). The enzyme has also been found to improve the treatment of activated sludge wastewaters through the decrease in treatment time. Another interesting application of this enzyme is in the consortium of α-amylase and glucoamylase to convert starch-rich wastes in potato or cheese whey from the food processing industry to produce biodegradable and photodegradable plastics (Coleman, 1990; Karam & Nicell, 1997). Utilization of α-amylase to cleave long molecules of starch into smaller fragments is initially carried out. Glucoamylase is further used to attack these small fragments, producing glucose through saccharification of more than 90% of the starch. Lactic acid bacteria subsequently act on the resulting glucose, producing lactic acid. Finally, recovery, purification, and successive use of the lactic acid in production of environmentally friendly plastics are carried out. Proper combinations of the isomers of lactic acid alongside other compounds usually control the rate of decomposition of the plastics (Coleman, 1990; Karam & Nicell, 1997).

4.3 Other Enzymes

Many other enzymes from microbial sources have found use in waste recycling. Pectin lyase and pectinesterase from *Clostridium beijerinckii* and *Clostridium thermosulfurogenes*, respectively, have been used in pectin degradation. The processing of food wastes such as apple pomace has been utilized to produce butanol (Blasheck, 1992). *Candida norvegensis*, a yeast found to produce L-galactonolactone oxidase, an enzyme employed in the manufacture of L-ascorbic acid by biotransformation of excess galactose, is a product of lactose hydrolysis of whey (Shoemaker, 1986). Lactases have also been utilized in the recycling of dairy wastes rich in lactose and whey proteins to produce value-added products (Blasheck, 1992; Karam & Nicell, 1997).

Chitinase isolated from *Serratia marcescens* with the capability to degrade chitin has been reported. An alternative disposal of high chitin content contained in shellfish waste through bioconversion to single-cell proteins was proposed. The method requires pretreatment of shrimp waste by first reducing the size and then eliminating proteins and minerals to give rise to a chitin substance, which can easily be transformed by the action of chitinase on *N*-acetyl glucosamine, a substrate for production of a single-cell protein (Cosio, Fisher, & Carroad, 1982). Table 1 presents a summary of microbial enzymes, their sources, and their applications.

5 Significance of Microbial Enzymes in Waste Recycling

The role that microbial enzymes play in the recycling of different varieties of wastes cannot be overemphasized. Enzymatic bioconversion of wastes has the dual benefit of decreasing the quantity of otherwise worthless materials to be disposed and,

Table 1 Summary of some enzymes and their functions in waste recycling

Enzyme	Microbial source	Function
Alkylsulfatases	Bacteria	Surfactant degradation
Amylases, e.g., α- and β-amylases and glucoamylases	Bacteria and fungi	Starch hydrolysis and production of glucose
Cellulolytic enzymes, e.g., cellobiohydrolases, cellobiases, cellulases, and exo-1,4-b-D-glucosidases	Bacteria and fungi	Sugar, alcohol, and bioenergy production by hydrolysis of cellulose-rich sludge from paper, pulp, and municipal solid wastes
Chitinases	*Serratia marcescens*	*N*-acetyl glucosamine production from bioconversion of shellfish waste
Chloroperoxidases	*Caldariomyces fumago*	Phenolic compound oxidation
Cyanidases	*Alcaligenes denitrificans*	Decomposition of cyanide
Cyanide hydratases	Fungi, e.g., *Gloeocercospora sorghi* and *Stemphylium loti*	Cyanide hydrolysis
Dehalogenases	Bacteria	Bioremediation and transformation of halogenated organic compounds and debromination and mineralization of tribromophenol
L-galactonolactone oxidases	*Candida norvegensis*	L-ascorbic acid production from hydrolysis of galactose present in whey
Laccases	Several fungi and bacteria	Binding of aromatic amines and phenols to humus, elimination of phenols, and decolorization of effluents from kraft bleaching
β-Galactosidases	Bacteria and fungi	Processing of dairy wastes and subsequent production of high-value products
Lignin peroxidases	*Phanerochaete chrysosporium*	Decolorization of effluents from kraft bleaching industries and elimination of phenols and other aromatic waste constituents
Lipases	Bacteria and fungi	Enhanced dewatering of sludge
Lysozymes	Bacteria	Enhanced dewatering of sludge
Manganese peroxidases	*P. chrysosporium*	Oxidization of aromatic dyes and monoaromatic phenols
Oxygenases Monooxygenases Dioxygenases	Bacteria and fungi	Desulfurization, dehalogenation, hydroxylation, denitrification, ammonification, biotransformation, bioremediation, and biodegradation of various aliphatic and aromatic compounds

(continued)

Table 1 (continued)

Enzyme	Microbial source	Function
Parathion hydrolases	*Pseudomonas* sp. *Flavobacterium* *Streptomyces*	Hydrolysis of organophosphates in pesticides
Pectin lyases	*Clostridium beijerinckii*	Pectin degradation
Pectinesterases	*Clostridium thermosulfurogenes*	Pectin degradation
Peroxidases	Bacteria	Removal of phenols and aromatic amines, dewatering of sludge, and decolorization of effluents from kraft bleaching
Phosphatases	*Citrobacter* sp.	Heavy metal removal
Proteases	Bacteria and fungi	Sludge improvement, digestion of meat and fish wastes

Source: Karam and Nicell (1997) and Chandrakant and Shwetha (2011)

subsequently, creating products of significant value such as food, feed, biofuels, or other bioproducts. Conventional chemical and biological processes of efficient treatment, reduction, or removal of these waste materials from the environment have proved hard to achieve. Hence, enzymatic processing of wastes, which falls between these two traditional classes, has shown significant bioremediation potentials due to the following merits: faster and cheaper operation through a broad variety of factors such as pH, temperature, salinity, etc.; application to a wide variety of wastes; function at low and high concentrations of contaminants; nonexistence of impediments to biomass adaptation; nonexistence of shock load effects; reduction in sludge quantity; and also the ease in the control of the processes involved in waste treatment.

6 Limitations of Microbial Enzymes in Waste Recycling

Regardless of the immense potentials that microbial enzymes offer, some limitations still linger. Some of them are cost of enzymatic treatment/recycling of wastes; selection of the most suitable microbial enzyme or, in some cases, group of enzymes, which is also a function of the enzyme specificities; the need for cofactors of some enzymes; the ability of enzymes to retain their active nature under normal conditions of operation for a prolonged period of time; and difficulty in assessing the toxicity and subsequent disposal of enzymatic reaction by-products (Karam & Nicell, 1997).

7 Future Prospects

Environmental pollutants have serious health hazards on humans, animals, plants, and other life forms in nature, with various destructive effects such as respiratory disorders, cardiovascular disorders, cancer, allergic reactions, mental disorders, perinatal disorders, and even mortality, and, as such, recycling of these wastes is immensely significant. The acknowledgment of the wide potentials of microbial enzymes is exhibited in the recycling of the different sources and forms of wastes. Nonetheless, synergies between microbial enzymes are currently being recognized as an effective strategy for bioproduct development from waste biomass. In order to economically and sustainably recycle wastes to produce useful bioproducts, more extensive use of the microbial enzyme omics technologies, such as genomics, metabolomics, transcriptomics, proteomics, and interactomics, should be encouraged. Application of these molecular studies in the efficient enzymatic breakdown of waste materials will go a long way in providing a better understanding of the individual and interactive roles of the vast amount of microbial enzymes in degradation, biotransformation, and, ultimately, by-product creation and valorization.

References

Abdel-Shafy, H., & Mansour, M. S. M. (2016). A review on polycyclic aromatic hydrocarbons: Source, environmental impact, effect on human health and remediation. *Egyptian Journal of Petroleum, 25*(1), 107–123. https://doi.org/10.1016/j.ejpe.2015.03.011

Abalansa, S., El Mahrad, B., Icely, J., & Newton, A. (2021). Electronic waste, an environmental problem exported to developing countries: The GOOD, the BAD and the UGLY. *Sustainability, 13*(9), 5302. https://doi.org/10.3390/su13095302

Adejumo, O., Adebiyi, I., & Olufemi, A. (2020). Agricultural solid wastes: Causes, effects, and effective management. In H. M. Saleh (Ed.), *Strategies of sustainable solid waste management*. IntechOpen.

Adriano-Anaya, M., Salvador-Figueroa, M., Ocampo, J. A., & Garcıa-Romera, I. (2005). Plant cell-wall degrading hydrolytic enzymes of Gluconacetobacter diazotrophicus. *Symbiosis, 40*(3), 151–156.

Agamuthu, P. (2009). *Challenges and opportunities in agro-waste management: An Asian perspective*. Inaugural meeting of first regional 3R Forum in Asia, Tokyo, Japan.

Ahmed, E. H. (2013). *In the economy of green cities* (pp. 45–56). Springer.

Aitken, M. D. (1993). Waste treatment applications of enzymes: Opportunities and obstacles. *The Chemical Engineering Journal, 52*, B49–B58.

Akram, F., Ashraf, S., Haq, I. U., Shah, F. I., & Aqeel, A. (2022). Eminent industrial and biotechnological applications of Laccases from bacterial source: A current overview. *Applied Biochemistry and Biotechnology*. https://doi.org/10.1007/s12010-021-03781-9

Alam, P., & Ahmade, K. (2013). Impact of solid waste on health and the environment. *International Journal of Sustainable Development in Green Economy, 2*, 165–168.

Amasuomo, E., & Baird, J. (2016). Solid waste management trends in Nigeria. *Journal of Management and Sustainability, 6*(4), 35–44.

Anifowose, Y. B., Omole, K. E., & Akingbade, O. (2011). Waste disposal site selection using remote sensing and GIS: A study of Akure and its environs, South West Nigeria. In *Proceedings*

of environmental management conference, Federal University of Agriculture Abeokuta Nigeria (pp. 2–9). Akure.
Arora, P. K., Kumar, M., Chauhan, A., Raghava, G. P., & Jain, R. K. (2009). OxDBase: A database of oxygenases involved in biodegradation. *BMC Research Notes, 2*, 67.
Arora, P. K., Srivastava, A., & Singh, V. P. (2010). Application of monooxygenases in dehalogenation, desulphurization, denitrification and hydroxylation of aromatic compounds. *Journal of Bioremediation & Biodegradation, 1*, 1–8.
Ayilara, M. S., Olanrewaju, O., Babalola, O. O., & Odeyemi, O. (2020). Review: Waste management through composting: Challenges and potentials. *Sustainability, 12*(4456), 1–23.
Baltrėnas, P., Jankaitis, A., & Raistenskis, E. (2005). Experimental investigation of biodegradation processes in food waste. *Journal of Environmental Engineering and Landscape Management, 13*(4), 167–176.
Beena, A., & Geevarghese, P. I. (2010). A solvent tolerant thermostable protease from a psychrotrophic isolate obtained from pasteurized milk. *Developmental Microbiology and Molecular Biology, 1*, 113–119.
Bennet, J. W., Wunch, K. G., & Faison, B. D. (2002). *Use of fungi biodegradation*. ASM Press.
Bhat, R., Dar, S., Dar, D., & Dar, G. (2018). Municipal solid waste generation and current scenario of its management in India. *Internal Journal of Advanced Research in Science and Engineering, 7*, 419–431.
Blasheck, H. P. (1992). Approaches to making the food processing industry more environmentally friendly. *Trends in Food Science and Technology, 3*, 107–110.
Buragohain, P., Nath, V., & Sharma, H. K. (2020). Microbial degradation of waste: A review. *Current Trends in Pharmaceutical Research, 7*(1), 107–125.
Chandra, R., & Chowdhary, P. (2015). Properties of bacterial laccases and their application in bioremediation of industrial wastes. *Environmental Science: Processes & Impacts, 17*(2), 326–342.
Chandrakant, S. K., & Shwetha, S. R. (2011). Role of microbial enzymes in the bioremediation of pollutants: A review. *Enzyme Research, 2011*, 1–9. https://doi.org/10.4061/2011/805187
Cheremisinoff, N. P. (2003). *Handbook of solid waste management and waste minimization technologies*. Butterworth-Heinemann.
Cirino, P. C., & Arnold, F. H. (2002). Protein engineering of oxygenases for biocatalysis. *Current Opinion in Chemical Biology, 6*(2), 130–135.
Coleman, R. (1990). Biodegradable plastics from potato waste double savings to environment. *Agricultural Engineering, 71*(6), 20–22.
Corral-Bobadilla, M., González-Marcos, A., Vergara-González, E., & Alba-Elías, F. (2019). Bioremediation of waste water to remove heavy metals using the spent mushroom substrate of Agaricus bisporus. *Watermark, 11*, 454.
Cosio, I. G., Fisher, R. A., & Carroad, P. A. (1982). Bioconversion of shellfish chitin waste: Waste pretreatment, enzyme production, process design and economic analysis. *Journal of Food Science, 47*, 901–905.
Demirbas, A. (2011). Waste management, waste resource facilities and waste conversion processes. *Energy Conversion and Management, 52*(2), 1280–1287.
Dua, M., Singh, A., Sethunathan, N., & Johri, A. (2002). Biotechnology and bioremediation: Successes and limitations. *Applied Microbiology and Biotechnology, 59*(2-3), 143–152.
Dyson, B., & Chang, N. B. (2005). Forcasting municipal solid waste generatio in fast growing urban region with system dynamics modelling. *Waste Management, 5*, 669–679.
Ebikapade, A., & Jim, B. (2016). The concept of waste and waste management. *Journal of Management and Sustainability, 6*(4), 88.
Ferronato, N., & Torretta, V. (2019). Waste mismanagement in developing countries: A review of global issues. *International Journal of Environmental Research and Public Health, 16*, 1060.
Fetzner, S. (2003). Oxygenases without requirement for cofactors or metal ions. *Applied Microbiology and Biotechnology, 60*(3), 243–257.

Guan, Z. B., Luo, Q., Wang, H. R., Chen, Y., & Liao, X. R. (2018). Bacterial laccases: Promising biological green tools for industrial applications. *Cellular and Molecular Life Sciences, 75*(19), 3569–3592.

Hermansyah, H., Wijanarko, A., & Gozan, A. (2007). Consecutive reaction model for triglyceride hydrolysis using lipase. *Jurnal Teknologi, 2*, 151–157.

Hiner, A. N., Ruiz, J. H., & Rodri, J. N. (2002). Reactions of the class II peroxidases, lignin peroxidase and Arthromyce ramosus peroxidase, with hydrogen peroxide: Catalase-like activity, compound III formation, and enzyme inactivation. *The Journal of Biological Chemistry, 277*(30), 26879–26885.

Holanda, R., & Johnson, D. (2020). Removal of zinc from circum-neutral pH mine-impacted waters using a novel "hybrid" low pH sulfidogenic bioreactor. *Frontier Environmental Sciences, 8*, 22.

Holm-Nielsen, J. B., Al Seadi, T., & Oleskowicz-Popiel, P. (2009). The future of anaerobic digestion and biogas utilization. *Bioresource Technology, 100*, 5478–5484.

Husain, Q. (2006). Potential applications of the oxidoreductive enzymes in the decolorization and detoxification of textile and other synthetic dyes from polluted water: A review. *Critical Reviews in Biotechnology, 26*(4), 201–221.

Ilheme, C., Ukairo, D. I., Ibegbulem, C. O., Okorom, O. O., & Chibundu, K. (2017). Analysis of enzymes activities on domestic waste dump sites. *Journal of Bioremediation & Biodegradation, 8*(4), 1–6.

Janusz, G., Pawlik, A., Swiderska-Burek, U., Polak, J., Sulej, J., Anna Jarosz-Wilkołazka, A., & Andrzej Paszczynski, A. (2020). Laccase properties, physiological functions, and evolution. *International Journal of Molecular Sciences Review, 21*, 966. https://doi.org/10.3390/ijms21030966

Karam, J., & Nicell, J. A. (1997). Potential applications of enzymes in waste treatment. *Journal Chemistry, Technology and Biotechnology, 69*, 141–153.

Kawai, F., & Yamanaka, H. (1989). Inducible or constitutive polyethylene glycol dehydrogenase involved in the aerobic metabolism of polyethylene glycol. *Journal of Fermentation and Bioengineering, 67*(4), 300–302.

Kaza, S., Yao, L., Bhada-Tata, P., & Van Woerden, F. (2018). *What a waste: A global snapshot of solid waste management to 2050*. The World Bank.

Koua, D., Cerutti, L., & Falquet, L. (2009). Peroxidase: A database with new tools for peroxidase family classification. *Nucleic Acids Research, 37*(1), 261–266.

Kumar, S., Dhar, H., Vijay, V. B. J., Vaida, A. N., & Akolkar, A. B. (2016). Characterization of municipal solid waste in high-altitude sub-tropical region. *Environmental Technology, 37*(20), 2627–2637.

Lehninger, A. L., & Nelson, D. L. (2004). *Lehninger's principles of biochemistry* (4th ed.). W.H. Freeman.

Leung, M. (2004). Bioremediation: Techniques for cleaning up a mess. *Journal of Biotechnology, 2*, 18–22.

Lorenz, H., Fischer, P., Schumacher, B., & Adler, P. (2013). Current EU-27 technical potential of organic waste streams for biogas and energy production. *Waste Management, 33*, 2434–2448.

Maczulak, A. E. (2010). *Pollution: Treating environment toxins* (p. 120). Info Base Publishing.

Maddela, N. R., & García, L. C. (2021). *Innovations in biotechnology for a sustainable future*. Springer.

Maddela, N. R., Garcia, L. C., & Chakraborty, S. (2021). *Advances in the domain of environmental biotechnology*. Springer.

Magram, S. F. (2011). Worldwide solid waste recycling strategies: A review. *Indian Journal of Science and Technology, 4*(6), 692–702.

Mani, D., & Kumar, C. (2014). Biotechnological advances in bioremediation of heavy metals contaminated ecosystems: An overview with special reference to phytoremediation. *International Journal of Environmental Science and Technology, 11*, 843–872.

Manjunuth, M., Lavanya, G., Pathakoti, K., & Kjell, M. (2018). Enzymes as direct decontaminating agent - Mycotoxins. In S. N. Carlos & K. Vikas (Eds.), *Enzymes in human and animal nutrition: Principles and perspective* (pp. 313–330). Academic Press.

Margesin, R., Zimmerbauer, A., & Schinner, F. (1999). Soil lipase activity—A useful indicator of oil biodegradation. *Biotechnology Techniques, 13*(12), 859–863.

Muthukumarasamy, N. P., Jackson, B., & Joseph, R. (2015). A production of extracellular laccase from Bacillus subtilis MTCC 2414 using agroresidues as a potential substrate. *Biochemistry Research International, 2015*, 1–9.

Nabegu, A. B. (2017). An analysis of municipal solid waste in kano metropolis, Nigeria. *Journal of Human Ecology, 31*(2), 111–119.

Nickolas, A., & Vasiliou, V. (2003). Aldehyde dehydrogenase gene superfamily: The 2002 update. *Chemico-Biological Interactions, 143*, 5–22.

Nwachukwu, N. C., Orji, A. F., & Ugbogu, O. C. (2013). Health care waste management – Public health benefits, and the need for effective environmental regulatory surveillance in Federal Republic of Nigeria. In *Current topics in public health*. Intech Open.

Obi, F. O., Ugwuishiwu, B. O., & Nwakaire, J. N. (2016). Agricultural waste concept, generation, utilization and management. *Nigerian Journal of Technology, 35*(4), 957–964.

Organisation for Economic Co-operation and Development. (2021). *Municipal waste (indicator)*. OECD.

Pandeeti, E. V., Veeraiah, S., & Routhou, G. D. (2019). Emerging trends in the industrial production of chemical products by microorgansisms. In B. Viswanath (Ed.), *Recent development in applied microbiology and biochemistry* (pp. 107–125). Academic Press.

Park, J. W., Park, B. K., & Kim, J. E. (2006). Remediation of soil contaminated with 2,4-Dichlorophenol by treatment of minced Shepherd's purse roots. *Archives of Environmental Contamination and Toxicology, 50*(2), 191–195.

Patel, A. B., Shaikh, S., Jain, K. R., Desai, C., & Madamwar, D. (2020). Polycyclic aromatic hydrocarbons: Sources, toxicity, and remediation approaches. *Frontiers in Microbiology, 11*, 562813. https://doi.org/10.3389/fmicb.2020.562813

Periasamy, A., Subash, G. C., Arzu, C., & Chaulagain, P. B. (2013). Microbial enzymes and their applications in industries and medicine. *BioMed Research International, 2013*, 7–8.

Piontek, K., Smith, A. T., & Blodig, W. (2001). Lignin peroxidase structure and function. *Biochemical Society Transactions, 29*(2), 111–116.

Prasad, M. P., & Manjunath, K. (2011). Comparative study on biodegradation of lipid-rich wastewater using lipase producing bacterial species. *Indian Journal of Biotechnology, 10*(1), 121–124.

Rajan, R., Robin, D., & Vandanarani, M. (2019). Biomedical waste management in ayurveda hospitals–Current practices and future prospective. *Journal of Ayurveda Integrated Medicine, 2019*, 214–221.

Rao, M. B., Tanksale, A., Ghatge, M. S., & Deshpande, V. V. (1998). Molecular and biotechnological aspects of microbial proteases. *Microbiology and Molecular Biology Reviews, 62*(3), 597–635.

Rubilar, O., Diez, M. C., & Gianfreda, L. (2008). Transformation of chlorinated phenolic compounds by white rot fungi. *Critical Reviews in Environmental Science and Technology, 38*(4), 227–268.

Ruiz-Duenas, F. J., Morales, M., & Perez-Boada, M. (2007). Manganese oxidation site in Pleurotus eryngii versatile peroxidase: A site-directed mutagenesis, kinetic and crystallographic study. *Biochemistry, 46*(1), 66–77.

Sahay, S., Iqbal, S., Inam, A., Gupta, M., & Inam, A. (2019). Waste water irrigation in the regulation of soil properties, growth determinants and heavy metal accumulation in different brassica species. *Environmental Monitoring and Assessment, 191*, 107.

Sanchez-Porro, C., Mart, S., Mellado, E., & Ventosa, A. (2003). Diversity of moderately halophilic bacteria producing extracellular hydrolytic enzymes. *Journal of Applied Microbiology, 94*(2), 295–300.

Schmidt, O. (2006). *Wood and tree fungi*. Springer.

Shahid, R., Muhammad, W. S., Jabeen, S., & Mubeen, H. (2016). Applications and uses of enzymes. *International Journal of Medical and Health Research, 2*(3), 58–61.

Sharma, D., Sharma, B., & Shukla, A. K. (2011). Biotechnological approach of microbial lipase: A review. *Biotechnology, 10*(1), 23–40.

Shoemaker, S. (1986). The use of enzymes for water management in the food industry. In *Biotechnology in Food Processing* (pp. 259–267). Nayes Publication.

Singh, C. J. (2003). Optimization of an extracellular protease of chrysosporium keratinophilum and its potential in bioremediation of keratinic wastes. *Mycopathologia, 156*(3), 151–156.

Sun, Y., & Cheng, J. (2002). Hydrolysis of lignocellulosic materials for ethanol production: A review. *Bioresource Technology, 83*(1), 1–11.

Tachibana, S., Naka, N., Kawai, F., & Yasuda, M. (2008). Purification and characterization of cytoplasmic NAD+-dependent polypropylene glycol dehydrogenase from Stenotrophomonas maltophilia. *FEMS Microbiology Letters, 288*(2), 266–272.

Ten Have, R., & Teunissen, P. J. (2001). Oxidative mechanisms involved in lignin degradation by white-rot fungi. *Chemical Reviews, 101*(11), 3397–3413.

Tsukihara, T., Honda, Y., Sakai, R., Watanabe, T., & Watanabe, T. (2006). Exclusive overproduction of recombinant versatile peroxidase MnP2 by genetically modified white rot fungus, Pleurotus ostreatus. *Journal of Biotechnology, 126*(4), 431–439.

Tulebayeva, N., Yergobek, D., Pestunova, G., Mottaeva, A., & Sapakova, Z. (2020). Green economy: Waste management and recycling methods. *E3S Web of Conferences, 159*, 01012.

Vasileva-Tonkova, E., & Galabova, D. (2003). Hydrolytic enzymes and surfactants of bacterial isolates from lubricant contaminated wastewater. *Zeitschrift für Naturforschung, 58*(1-2), 87–92.

Vergara, S. E., & Tchobanoglous, G. (2012). Municipal solid waste and the environment: A global perspective. *Environment and Resources, 37*(37), 277–309.

Vidali, M. (2001). Bioremediation. An overview. *Pure and Applied Chemistry, 73*(7), 1163–1172.

Vinti, G., Bauza, V., Clasen, T., Medlicott, K., Tudor, T., & ans Zurbrügg C. (2021). Municipal solid waste management and adverse health outcomes: A systematic review. *International Journal of Environmental Research and Public Health, 18*, 4331.

Wang, Y., Feng, Y., Cao, X., Liu, Y., & Xue, S. (2018). Insights into the molecular mechanism of dehalogenation catalyzed by D-2-haloacid dehalogenase from crystal structures. *Scientific Reports, 8*(1), 1454.

WHO. (2018). *Fact sheets on detail health care waste*. WHO.

Wong, D. W. (2009). Structure and action mechanism of ligninolytic enzymes. *Applied Biochemistry and Biotechnology, 157*(2), 174–209.

Xue, F., Ya, X., Tong, Q., Xiu, Y., & Huang, H. (2018). Heterologous overexpression of Pseudomonas umsongensis halohydrin dehalogenase in Escherichia coli and its application in epoxide asymmetric ring opening reactions. *Process Biochemistry, 75*, 139–145.

Zu, L., Li, G., An, T., & Wong, K. (2012). Biodegradation kinetics and mechanism of 2,4,6-tribromophenol by Bacillus sp. GZT: A phenomenon of xenobiotic methylation during debromination. *Bioresource Technology, 110*, 153–159.

Soil Microbial Enzymes and Mitigation of Heavy Metal Uptake by Plants

Olaniran Victor Olagoke, Gabriel Gbenga Babaniyi, Femi J. Afolabi, and Babafemi Raphael Babaniyi

Abstract Heavy metals are contaminants that cause immense environmental problems as they are harmful to humans, animals, plant health, and the environment at large. The activities of hydrolytic enzymes and ligninolytic oxidases and peroxidases directly affect the rates of conversion of soil biopolymers into compounds that are accessible to microorganisms and plants. The role of plants and microorganisms in the biotransformation of heavy metals into nontoxic forms is well recognized, and understanding the molecular mechanisms of metal accumulation has numerous biotechnological implications in bioremediation of metal-contaminated sites. The process of bioremediation uses various agents such as bacteria, yeasts, fungi, algae, and higher plants as key tools in treating oil spills and heavy metals present in the environment. As a result of increasing metal concentrations in the soil due to either natural or anthropogenic contamination, it has been found that soil enzyme activities are influenced by different metals in diverse ways, depending on the type of metal and the metal salt. However, soil characteristics such as pH, clay content, and soil organic matter (OM) can change the negative or positive impacts of heavy metals on soil enzymes. Therefore, monitoring changes in soil metal content, an assessment of changes in soil enzyme activities, would be a useful tool for monitoring soil quality and fertility under heavy metal pollution. This absolutely depends on the enzyme, the metal, and its concentration.

O. V. Olagoke · G. G. Babaniyi
Department of Agricultural Development and Management, Agricultural and Rural Management Training Institute (ARMTI), Ilorin, Nigeria

F. J. Afolabi · B. R. Babaniyi (✉)
Bioresources Development Centre, National Biotechnology Development Agency, Nigeria

N. R. Maddela et al. (eds.), *Ecological Interplays in Microbial Enzymology*, Environmental and Microbial Biotechnology,
https://doi.org/10.1007/978-981-19-0155-3_11

1 Introduction

Soil microbial enzymes are major mechanisms of biological soil processes, such as the degradation of organic compounds, their mineralization, and the release or recycling of nutrients including nitrogen (N), phosphorus, sulfur, and other essential metals. The activities of hydrolytic enzymes and ligninolytic oxidases and peroxidases directly affect the rates of transformation of soil biopolymers into compounds that are accessible to microorganisms and plants. The study of enzymatic activities in environmental samples (soil, litter, lignocellulose, or other matrices) is a useful tool for assessing the functional diversity of soil microbial communities or soil organic mass turnover (Kandeler et al., 1999). Measuring microbial enzyme activities in soil has a long tradition in connection with evaluating soil fertility and quantifying processes in natural and seminatural ecosystems with a high turnover of organic compounds, such as in forest and grassland soils. The soil microbial activity indicates its quality and can be measured by microbial carbon (C) (Silva et al., 2010) and microbial N (Gama-Rodrigues et al., 2010). This activity may also be evaluated by enzymatic reactions, e.g., β-glycosidase, urease, acid phosphatase, and aryl sulfatase by the respiratory activity and microbial biomass of the soil (Lisboa et al., 2019) or by the most probable number (MPN) of bacteria and fungi in the soil (Silveira et al., 2006). The estimate of some microbial groups can probably indicate how biochemical actions take place in the soil. This is because the nutrient cycles in the soil are directly dependent on the microbial action (Silveira et al., 2006). The quantification of enzymes has become an effective soil bioindicator, since enzymatic activity can be used to evaluate the activity of the microbiota. However, the most studied enzymes are β-glycosidase, arylsulfatase, acid phosphatase, and urease (de Araújo and Monteiro, 2007). The β-glycosidase enzyme hydrolyzes the residues of cellobiose, acting in the final process of the cellulose decomposition (Eivazi and Tabatabai, 1988); therefore, changes in the activity of this enzyme can indicate the soil quality.

1.1 Enzymes with Special Characteristics in Biotechnology

Proteases

Although hydrolytic enzymes belong to the largest group of enzymes and are the most commercially applicable ones, among the enzymes within this group, microbial proteases have been extensively studied (Chudasama, Jani, Jajda, & Pate, 2010). Proteases prepared from microbial systems are of three types: acidic, neutral, and alkaline. Alkaline proteases are efficient under alkaline pH conditions and consist of a serine residue at their active site (Gupta, Beg, & Lorenz, 2002). Alkaline serine proteases have the largest applications in bioindustry. Alkaline proteases are of particular interest being more suitable for a wide range of applications, since these possess high activity and stability in abnormal conditions of extreme physiological parameters. Alkaline proteases have shown their capability to work under high pH, temperature, and in the presence of inhibitory compounds (Gupta, Joseph, Mani, & Thomas, 2008). Vijayalakshmi, Venkat Kumar, and Thankamani (2011) have optimized and characterized the cultural conditions for the production of an alkalophilic as well as a thermophilic extracellular protease enzyme from *Bacillus*. This bacterium named *Bacillus* RV. B2.90 was found to be capable of producing an enzyme preparation possessing special characteristics such as being highly alkalophilic, moderately halophilic and thermophilic, and exhibiting the quality of a thermostable protease enzyme. Alkaline proteases possess the property of great stability in their enzyme activity when used in detergents (Hadj-Ali et al., 2007). The alkaline protease produced from Bacilli and proteases from other microorganisms have found more applications overall in bioindustries such as the detergent, tannery, food, leather processing, and pharmaceuticals and for studies in molecular biology and in peptide synthesis (Chirumamilla, Muralidhar, Marchant, & Nigam, 2001).

Keratinases

Keratin is an insoluble and fibrous structural protein that is a constituent of feathers and wool. The protein is abundantly available as a by-product of keratinous wastes, representing a valuable source of proteins and amino acids that could be useful for animal feed production or as a source of nitrogen for plants (Gushterova, Vasileva-Tonkova, Dimova, Nedkov, & Haertle, 2005). However, keratin-containing substrates and materials have high mechanical stability and hence are difficult to be degraded by common proteases. Keratinases are specific proteolytic enzymes, which are capable of degrading insoluble keratins. The importance of these enzymes is being increasingly recognized in fields as diverse as animal feed production, textile processing, detergent formulation, leather manufacture, and medicine. Proteolytic enzymes with specialized keratinase activity are required to degrade keratins, and, for this purpose, keratinases have been isolated and purified from certain bacteria, actinomycetes, and fungi (Brandelli, Daroit, & Riffel, 2010). Keratinases have been classified as serine- or metalloproteases. Cloning and expression of keratinase genes in a variety of expression systems have also been reported (Gupta, Sharma, & Beg, 2013). A higher operation temperature is required in the degradation of materials like feathers and wool, which would be possible using a thermostable keratinase. This aspect is of added advantage in achieving a higher reactivity due to lower diffusional restrictions, and, hence, a higher reaction rate would be established. The enhanced stability of keratinase would increase the overall process yield due to the increased solubility of keratin and favorable equilibrium displacement in endothermic reactions.

Baihong et al., (2013) reported the enhanced thermostability of a preparation of keratinase by computational design and empirical mutation. The quadruple mutant of *Bacillus subtilis* has been characterized to exhibit synergistic and additive effects at 60 °C with an increase of 8.6-fold in the $t_{1/2}$ value. The N122Y substitution also led to an approximately 5.6-fold increase in catalytic efficiency compared to that of the wild-type keratinase. An alkalophilic strain of *Streptomyces albidoflavus* has been reported to produce extracellular proteases (Indhuja, Shiburaj, Pradeep, Thankamani, & Abraham, 2012). This particular type of protease is capable of hydrolyzing keratin. The biosynthesis of this specific enzyme was optimized in submerged batch cultures at a highly alkaline pH of 10.5, and the enzyme yield was stimulated using an inducer substrate containing keratin in the form of white chicken feathers. An enhanced (sixfold) protease production could be achieved with a modified composition of the culture medium containing the inducer at a concentration of 0.8% in the fermentation medium. The novelty of this crude enzyme has been reported to be its activity and stability in neutral and alkaline conditions. The maximum activity was obtained at a pH of 9.0 in the temperature range of 60–70 °C. This type of protease (keratinase hydrolyzing keratins) is of particular significance for its application in industries since the crude enzyme showed its tolerance to the detergents and solvents tested (Indhuja et al., 2012). Liu et al., (2013) studied the expression of extreme alkaline, oxidation-resistant keratinase from *Bacillus licheniformis* into the recombinant *Bacillus subtilis* WB600 expression system. The alkaline keratinase was characterized for its application in the processing of wool fibers.

Amylases

Amylases are significant enzymes for their specific use in the industrial starch conversion process. Amylolytic enzymes act on starch and related oligo- and polysaccharides (Pandey et al., 2000). The global research on starch hydrolyzing enzymes based on DNA sequence, structural analysis, and catalytic mechanism has led to the concept of one enzyme family—α-amylase. Amylolytic and related enzymes are classified as glycoside hydrolases. These enzymes are produced by a wide range of microorganisms and substrates (Sivaramakrishnan, Gangadharan, Nampoothiri, Soccol, & Pandey, 2006) and are categorized as exoenzymes, endoenzymes, debranching, and cyclodextrin-producing enzymes. The application of these enzymes has been established in starch liquefaction and in the paper, food, sugar, and pharmaceutical industries. In the food industry, amylolytic enzymes have a large scale of applications, such as the production of glucose syrups, high-fructose corn syrups, and maltose syrups, in the reduction of viscosity of sugar syrups and reduction of turbidity to produce clarified fruit juice for longer shelf-life, and solubilization and saccharification of starch in the brewing industry (Pandey et al., 2000). The baking industry uses amylases to delay the staling of bread and other baked products; the paper industry uses amylases for the reduction of starch viscosity to achieve the appropriate coating of paper. Amylase enzymes are used in the textile industry for warp sizing of textile fibers and as digestive aids in the pharmaceutical industry. Li, Niu, Zhang, Wang, and Shi (2013) have recently isolated, characterized, and cloned a thermotolerant isoamylase. For this purpose, the enzyme was biosynthesized using a thermophilic bacterium *Bacillus* sp. This novel enzyme has been reported to display its optimal activity at a remarkably high temperature of 70 °C as well as being active in the alkaline range. This thermophilic enzyme has also been found to be thermostable between 30 and 70 °C, and its activity has been reported to be stable within a pH range of 5.5–9.0.

Gurumurthy et al., (2012) completed the molecular characterization of an extremely thermostable α-amylase for industrial applications. This novel enzyme was produced by a bacterium *Geobacillus* sp., which was isolated from the thermal water of a geothermal spring. This isolated bacterium showed the characteristics of thermotolerance and alkali resistance. A purified preparation of amylase suitable for application was obtained using a diethylaminoethyl (DEAE)-cellulose column and Sephadex G-150 gel filtration chromatography. The enzyme is a novel α-amylase due to its optimum activity at an extremely high temperature of 90 °C and an alkaline pH of 8.0. However, this purified enzyme preparation was found to be stable only for 10 min at 90 °C.

Xylanases

Hemicellulose is one of the main constituents of agricultural residues and plants along with cellulose, lignin, and pectin (Polizeli et al., 2005). Xylan is the major component of hemicellulose consisting of β-1,4-linked D-xylopyranosyl residues. The hydrolysis of xylan in plant materials is achieved by the use of a mixture of hydrolytic enzymes including endo-β-1,4-xylanase and β-D-xylosidase (Polizeli et al., 2005). The importance of xylanase has tremendously increased due to its biotechnological applications in pentose production, fruit juice clarification, improving rumen digestion, and bioconversion of lignocellulosic agricultural residues to fuels and chemicals (Nigam & Pandey, 2009). Collins, Gerday, and Feller (2005) studied the xylanase enzyme and its families as well as the special xylanases possessing extremophilic characteristics. Xylanases have established their uses in the food, pulp, paper, and textile industries, agri-industrial residue utilization, and ethanol and animal feed production (Garg, Roberts, & McCarthy, 1998). The enzyme used for the purpose of biobleaching of wood pulp should be active in conditions of alkaline pH and high temperature, and, at the same time, it is desirable that this enzyme be stable at high reaction temperatures. Xylanase preparations used for wood processing in the paper industry should be free of cellulose activity. Cellulase-free xylanase preparations have applications in the paper industry to provide brightness to the paper due to their preferential solubilization of xylans in plant materials and selective removal of hemicelluloses from the kraft pulp. Kohli, Nigam, Singh, and Chaudhary (2001) studied the production of a cellulase-free extracellular endo-1,4-β-xylanase at a higher temperature of 50 °C and a pH of 8.5 employing a selected microorganism, *Thermoactinomyces thalophilus*. The enzyme preparation was found to be thermostable at 65 °C, retaining its activity at 50% after 125 min of incubation at 65 °C. The crude enzyme preparation showed no cellulase activity, and the optimum temperature and pH for maximum xylanase activity was found to be 65 °C and 8.5–9.0, respectively. A thermotolerant and alkalotolerant xylanase has been reported to be produced by *Bacillus* sp. (Marques, Alves, Ribeiro, Girio, & Amaralcollaco, 1998). To make the applications of xylanase viable on commercial scales, heterologous systems of *Escherichia coli*, *Pichia pastoris*, and *Bacillus* sp. have been used to express xylanase activity (Jhamb & Sahoo, 2012). The thermophilic microorganism *Humicola* spp. has been studied for its capability of biosynthesizing an alkali-tolerant β-mannase xylanase (Luo et al., 2012). Acidophilic xylanases stable under acidic conditions of reaction are reported to be produced by an acidophilic fungus *Bispora* (Luo et al., 2009); in contrast, a xylanase active under conditions of alkaline pH has been studied by Mamo, Thunnissen, Hatti-Kaul, and Mattiasson (2009) for the mechanism of their high pH catalytic adaptation. Recently, three novel xylanases, thermophilic in nature (XynA,B,C), have been characterized by Shuyun et al., (2013); these were produced by *Humicola* sp. for their potential applications in the brewing industry. One xylanase gene, *XynA*, has been found to adapt to alkaline conditions and stability at higher temperatures. This *XynA* also possessed higher catalytic efficiency and specificity for a range of substrates. Shuyun et al., (2013) reported the application of three xylanases, XynA-C, under simulated mashing conditions in the brewing industry and found a better performance of 37% on filtration acceleration and 13% reduction in the viscosity of the substrate in comparison to the performance of a commercial trade enzyme, Ultraflo, a product from Novozymes.

Laccases/Ligninases

Ligninolytic enzymes are applicable in the hydrolysis of lignocellulosic agricultural residues, particularly for the degradation of the complex and recalcitrant constituent lignin. This group of enzymes is a mixture of synergistic enzymes; hence, they are highly versatile in nature and can be used in a range of industrial processes (Dahiya, Singh, & Nigam, 1998). The complex enzyme system consists of three oxidative enzymes: lignin peroxidase (LiP), manganese peroxidase (MnP), and laccase. These enzymes have established their applications in bioremediation, pollution control, and in the treatment of industrial effluents containing recalcitrant and hazardous chemicals such as textile dyes, phenols, and other xenobiotics (Robinson & Nigam, 2008). The paper and pulp industry requires a step of separation and degradation of lignin from the plant material, where the pretreatment of wood pulp using ligninolytic enzymes is important for a milder and cleaner strategy of lignin removal compared to chemical bleaching. Bleach enhancement of mixed wood pulp has been achieved using coculture strategies, through the combined activity of xylanase and laccase (Dwivedi, Vivikanand, Pareek, Sharma, & Singh, 2010). The ligninolytic enzyme system is used in the biobleaching of kraft pulp and in other industries such as for the stabilization of wine and fruit juices, denim washing (Dahiya et al., 1998), the cosmetic industry, and biosensors (Pandey et al., 1999). Fungi are the most potent producers of lignin-degrading enzymes. White rot fungi for the production of these enzymes have been specifically studied by Robinson et al., (2001). For economical production of ligninolytic enzymes, agricultural residues have been used as the substrate in microbial production of lignin-degrading enzymes. Thermophilic laccase enzymes are of particular use in the pulping industry. Recently, Gali and Kotteazeth (2013) have reported the biophysical characterization of thermophilic laccase isoforms. These were initially isolated from the xerophytic plant species *Cereus pterogonus* and *Opuntia vulgaris* and showed thermophilic properties (Uthandi, Saad, Humbard, & Maupin-Furlow, 2010). In order to prepare laccase enzymes with special characteristics, several studies have been conducted to provide a scientific basis for the employment of laccases in biotechnological processes (Mishra and Kumar 2009). Forms of laccase with unusual properties have been isolated from the Basidiomycetes culture of *Steccherinum ochraceum*, *Polyporus versicolor*, and a microbial consortium (Wongwilaiwalin et al., 2010).

Cellulases

Cellulase enzymes are the third most important enzyme for industrial uses: worldwide research has been focused on the commercial potential of cellulolytic enzymes for the commercial production of glucose feedstock from agricultural cellulosic materials (Pandey et al., 1999). The significance of cellulose hydrolyzing thermophilic enzymes in various industries includes the production of bioethanol and value-added organic compounds from renewable agricultural residues (Hardiman, Gibbs, Reeves, & Bergquist, 2010). Cellulose is the most abundant natural resource available globally for bioconversion into numerous products in the bioindustry on a commercial scale. For efficient bioconversion, a strategy of efficient saccharification using cellulolytic enzymes is required. Hardiman et al., (2010) used the approach of thermophilic-directed evolution of a thermophilic β-glucosidase. Cellulase is a complex of three important enzymes, which work synergistically owing to the crystalline and amorphous complex structure of cellulose. These enzymes, acting synergistically, hydrolyze cellulose to cellobiose, glucose, and oligosaccharides. The endoglucanase enzyme is the first to act on amorphous cellulose fibers, randomly attacking the glucose–polymer chain, which releases small fibers consisting of free reducing and nonreducing ends. The free ends of the chain are then exposed to the activity of the exoglucanase enzyme, which produces cellobiose. The third component of cellulase is β-glucosidase, which hydrolyzes the cellobiose, producing glucose as the final product of cellulose saccharification. Thermostability is an important technical property for cellulases: since the saccharification of cellulose is faster at higher temperatures, the stability of enzyme activity is necessary to be maintained for the completion of the process. Although enzymes have been prepared using thermophilic microorganisms, these enzyme preparations are not necessarily heat-stable. The activity profile of the thermal activation and the stability of cellulases derived from two Basidiomycetes cultures were studied by Nigam and Prabhu (1988). The results proved that the prior heat treatment of enzyme preparations caused activation of exo- and endoglucanase activities and improved the stability of enzymes over a period of reaction time. Therefore, the efficiency of cellulolytic enzymes may be increased by heat treatment, by incubating buffered enzyme preparations without cellulose or substrates prior to the saccharification process. Cellulolytic enzymes have been produced by a range of microorganisms including bacteria and fungi. Studies have been performed for the biosynthesis of a high-activity preparation with high yields (Nigam & Prabhu, 1991).

1.2 Miscellaneous Enzymes in Biotechnology

Various enzymes other than those described above have a significant place in the list of microbial enzymes, which have established their applications in bioindustries. Lipases have been widely studied for their properties and utilization in many industries (Reddivari, Chirumamilla, & Nigam, 2002). Pectinases have established their role in the fruit and juice industries (Sunnotel & Nigam, 2002). Certain enzymes are specifically required in the pharmaceutical industry for diagnostic kits and analytical assays (Zhou, Nigam, Marchant, & Jones, 1995). Bornscheuer et al. (2012) mentioned that in all the research and developments so far in the field of biocatalysis, researchers have contributed to three waves of outcomes. These innovations have played an important role in the establishment of the current commercially successful level of bioindustries. As a result, the recent bioprocess technology has been capable of meeting the future challenges and requirements of conventional and modern industries, for example, Trincone (2013) reviewed the options for unique enzymatic preparation of glycosides. Earlier enzymatic processes were performed within the limitations of an enzyme, whereas, currently, with the knowledge of modern techniques, an enzyme can be engineered to be a suitable biocatalyst to meet the process requirement. Riva (2013) identified the scope of a long-term research in biocatalysis, since there are underlying problems in the shift from classical processes to bio-based processes for the commercial market. There is a tremendous scope for research and development to meet the challenges of third-generation biorefeneries (Riva, 2013), for the production of numerous chemicals and bioproducts from renewable biomasses, new glycoside hydrolases (Trincone, 2013), or new enzymes found in marine environments (Trincone, 2010). Although the research on hemicellulases as important biorefining enzymes has not been well established, biocatalysis for xylan processing is slowly progressing and a wide range of hemicellulases have been isolated and characterized (Dumon, Songa, Bozonneta, Fauréa, & O'Donohue, 2012). Specifically, regarding bio-based glycosynthesis, Trincone (2013) mentioned that the new prospects are open for the use of pentose sugars as the main building blocks for engineered pentosides to be used as nonionic surfactants or as ingredients for prebiotic food and feed preparations.

2 Quantification of Microbial Enzymes in Soil

2.1 *Enzyme Factors/Enzyme Sensitivity*

Shen et al., (2005) investigated the interactions of polycyclic aromatic hydrocarbons (PAHs) (phenanthrene, fluoranthene, benzo(a)pyrene) and heavy metals (cadmium, zinc, and lead) with soil enzymes (urease and dehydrogenase). The results showed that dehydrogenase was more sensitive to combined pollution than was urease. Similarly, Maliszewska-Kordybach and Smreczak (2003) demonstrated that dehydrogenase activity is most sensitive to the combined effects of pollutants (heavy metals and PAHs). Shen et al., (2005) reported that urease and dehydrogenase could be suitable indicators of combined pollution (heavy metals and PAHs), particularly at the early stages of pollution (Bååth, 1989; Yang and Han, 2000). Renella et al., (2003) reported that alkaline phosphatase was more susceptible in acidic soil, whereas acid phosphatase was more susceptible in alkaline soil. Żyszkowska and Paszkiewicz-Jasińska (2011) found that the metal sensitivities of enzymes followed the order: dehydrogenase > urease > alkaline phosphatase > acid phosphatase.

2.2 *Structural Inhibition of Enzymes*

Enzyme reactions are inhibited by heavy metals in three different ways: (1) complexation of the substrate; (2) combination with protein-active groups on the enzyme; and (3) reaction with the enzyme–substrate complex (Tejada et al., 2008a, b; Megharaj et al., 2003). D'Ascoli et al., (2006) reported that heavy metals inhibited enzyme activity in several ways:

1. By masking catalytically active groups
2. By denaturing the protein conformation
3. By competing with metal ions that are needed to form enzyme–substrate complexes (Gianfreda et al., 1999)

Khan et al., (2007) reported that extracellular enzymes were inactivated by heavy metals. The mechanisms involved the metals binding to some of the amino acids in the enzymes and indirectly reducing the number of microorganisms responsible for producing the enzymes (Kuperman and Carreiro, 1997; Bandick and Dick, 1999; Kunito et al., 2001). Geiger et al. (1998a, b) reported that the interaction of a metal cation with an enzyme is largely dependent on the amino acid composition of the protein. It is assumed that the catalytic reactions of cellulases involve a hydrolysis reaction that proceeds via an acid–base mechanism involving aspartic and glutamic acids. There are two components to this mechanism:

1. Acting as a catalyst (aspartic acid)
2. Acting as a nucleophile (glutamic acid)

Cellulose binds to cellulase in the region of the cellulose-binding domain (Esterbauer et al., 1991). Cellulose-binding domains contain plenty of glycine and cysteine, which are stabilized by two or three disulfide bridges (Wood and Garcia-Campayo, 1990). In other words, the shape of the active site of cellulase is mainly provided by amino acids (glycine and cysteine) and the bonds between them (disulfide bridges). The cellulose-binding domain also contains tryptophan residues (Pregitzer et al., 1995). Copper can form complexes with tryptophan residues in the cellulose-binding domain, resulting in the inhibition of cellulase. Khan et al., (2007) stated that "it is well documented that heavy metals react with sulfhydryl groups of enzymes and inhibit and/or inactivate the enzymatic activities." Lorenz et al., (2006) reported that enzyme activities decreased due to the binding of Cd^{2+} to sulfhydryl groups (Sanadi, 1982). Hemida et al., (1997) reported that Juma and Tabatabai (1977) stated that "there was a marked decrease in urease activity with increasing trace element ion concentrations due to the reaction of –SH groups on urease (which are involved in urease activity) with the trace element ions." Kundu et al., (2007) specified that As (arsenic) ions inactivate enzymes by reacting with sulfhydryl groups resulting from the formation of arsenic sulfide. They also reported that As decreases enzyme activity in three ways:

1. By interacting with the enzyme–substrate complex
2. By denaturing the enzyme protein
3. By interacting with the active protein groups (Dick, 1997)

Hemida et al., (1997) indicated that the amidase activity in soil to which Cu^{2+} and Zn^{2+} had been added was not strongly inhibited compared to the activities of urease and nitrate reductase and explained this by citing the different functional groups at the active sites of amidase.

Weinstein (1974) stated that thiol groups had no direct effect on the catalytic activity of amidase, but they were necessary to stabilize the active amidase conformation. Frankenberger and Tabatabai (1980) suggested that α-amino groups may be effective at catalyzing amidase function and that these groups do not react with metal ions. Kundu et al., (2007) reported that phosphatase activity was negatively influenced by high phosphorus content in the soil because of the structural similarity of phosphate and arsenate (Juma and Tabatabai, 1977; Speir et al., 1999). Arsenic is a highly inhibitory heavy metal, even at low concentrations, due to its chemical properties (uncharged at neutral pH, it can diffuse across the cell membrane). When arsenic reaches the inside of the cytoplasm, it cross-links with sulfhydryl groups and permanently inactivates the enzyme (Dick, 1997).

2.3 Seasonal Effects of Enzymes

Soil enzymes are season-dependent macromolecules because they derive from living organisms. Microorganisms, plants, and animals show seasonal fluctuations in activity. Zhang et al., (2008) found that there was a seasonal difference in the effects of heavy metals on soil enzymes—the effects of the heavy metals were more obvious in spring and summer than in autumn.

2.4 Soil Factors

Soil pH

Effron et al., (2004) revealed that enzyme activities are sensitive to changes in pH. Metals in the soil can effectively alter soil pH, which often results in acidification. Increasing pH influences Cd sorption, reducing the concentration of Cd in soil solution and making less Cd available in soil (Vig et al., 2003). Geiger et al. (1998a, b) found that the effect of copper on the enzymatic decomposition of cellulose by cellulase and β-glucosidase in suspensions of montmorillonite and aluminum-treated montmorillonite was the strongest in the pH range 5.0–5.5. Copper lowered the pH values corresponding to the optimal activities of cellulase and β-glucosidase. Generally, amino acids of enzymes are deprotonated at high pH involved in metal interaction. Geiger et al. (1998a, b) reported that in the presence of kaolinite, the optimal pH for clay-absorbed enzyme activity was shifted by one or two pH units toward alkaline values (Pflug, 1982). Campbell (1988) suggested that almond β-glucosidase had a catalytic function involving two key groups, i.e., aspartic and glutamic carboxyl groups, at the enzyme's active site, when they were in the appropriate protonation state. Campbell's model assumes that enzyme activity can be lost in two ways:

1. By deprotonation of the aspartic carboxyl group
2. By protonation of the glutamic carboxyl group

Geiger et al., (1998a, b) found that the effect of copper was the strongest in the pH range 5.0–5.5, in which case 200 mM Cu reduced the enzyme activities (of cellulase and β-glucosidase) by 25% or more. However, when the pH was close to 4, the enzyme activities were reduced by only 5% by the same level of copper.

Different enzymes can respond differently at the same pH values and metal levels. Under conditions of pH 5.5 and 600 mM copper, β-glucosidase activity was reduced by 90%, whereas cellulase activity dropped by 60%.

Soil Organic Matter

D'Ascoli et al., (2006) investigated the effects of heavy metal contamination on the biological and biochemical properties (fluorescein diacetate (FDA) hydrolase, dehydrogenase, β-glucosidase, urease, arylsulfatase, and acid phosphatase) of a soil onto which a river contaminated with Cr (III) and Cu overflowed. The results showed negative correlations between the activities of dehydrogenase, arylsulfatase, and acid phosphatase and Cr fractions (soluble, exchangeable, and carbonate-bound).

Although Cu pollution negatively influenced the biological and biochemical properties of the soil, the soil organic matter was able to mask these negative impacts of Cu on the microbial community. Similarly, many other studies have shown that organic amendments (with municipal wastes, composts, biosolid composts, leonardites, gyttjas, and litters) reduce the toxicities of heavy metals to soil enzymes (Karaca et al., 2006). Karaca et al., (2010) indicated that many of the effects of Cd were reduced by sewage sludge and phosphate fertilizer amendments. Therefore, reducing the amount of fertilizer added to a contaminated agricultural site will result in an increase in the availability of Cd at that site. A positive way of reducing the impact of Cd contamination is therefore to continue phosphate and sewage sludge/organic matter amendments, which are low in pollutants, on a limited basis. For example, if 80% of the Cd added to the soil remains in the topsoil each year (Taylor, 1997), the addition of phosphate or organic matter resulting in a <20% increase in the soil Cd content will eventually result in a reduction of Cd in the soil. This will also reduce the availability of Cd, resulting in less toxic soil and less Cd being sequestered by crop biomass. Tejada et al., (2008a, b) found that increasing Ni levels reduced soil enzyme activities and that soil amendments with organic wastes (crushed cotton gin composts, poultry manures) reduced the toxicity of nickel to soil enzyme activities (urease, Biochimica et Biophysica Acta (BBA) protease, alkaline phosphatase, β-glucosidase, and arylsulfatase). Organic amendments enhance soil enzyme activities for the following reasons: (1) intra- and extracellular enzymes stimulate microbial activity in the added materials and (2) carboxyl, phenolic, alcohol, and carbonyl functional groups in humic substances react with toxic ions, forming metal–humate complexes (metal chelation) and stabilizing them (Nannipieri, 1994; Dick, 1997; Pascual et al., 1998). Tejada et al., (2008a, b) summarized the following results from different studies. Carboxyl groups play an important role in stabilizing toxic ions in humic acids (McKnight et al., 2001). Although fulvic acids contain more carboxyl groups than do humic acids (Stevenson, 1994), studies show that metal chelation by humic acids is more effective than that by fulvic acids since humic acids provide more binding sites due to their larger molecules and more complex nature (Lobartini et al., 1994). Moreover, humic substances have more strongly acidic groups than do fulvic acids. Tejada et al. (2008a, b) concluded that soil microbial biomass and soil enzyme activities are greater in humic acid-amended (crushed cotton gin composts) than those in fulvic acid-amended (poultry manures) soils and that the addition of these organic materials may be considered a good strategy for heavy metal-polluted soil remediation and also that the addition of organic materials with a higher humic acid concentration than fulvic acid concentration is more advisable.

Clay Minerals

Zeng et al., (2007) studied the effect of lead treatment on soil enzyme activities in a soil–lead–rice system in a greenhouse pot experiment. High inhibition was observed in sandy soil with low organic matter content. Similarly, Renella et al., (2003) found that enzyme inhibition was greater in sandy than in fine-textured soils because the clay fraction protects soil enzyme activity. Geiger et al., (1998a, b) investigated the effect of copper on the enzymatic decomposition of cellulose by cellulase and β-glucosidase in suspensions of montmorillonite and aluminum-treated montmorillonite. The results showed that montmorillonite and Al montmorillonite reduced the activities of cellulase and β-glucosidase. Moreover, the use of montmorillonite resulted in the largest reduction in enzyme activity due to its larger specific surface and higher surface area. Gianfreda et al., (1991) indicated that the specific surface areas of montmorillonite and Al montmorillonite when fully dispersed were approximately 700 and 450 $m^2 g^{-1}$, respectively. There are various reasons for the different specific surface areas of these clay minerals: (1) the adsorption of enzyme molecules on both external and internal surfaces by montmorillonite (Fusi et al., 1989) and (2) the larger net negative charge of montmorillonite (87 mEq 100 g^{-1}) compared to Al montmorillonite (15 mEq 100 g^{-1}) (Lothenbach et al., 1998). Montmorillonite and Al montmorillonite did not reduce the toxic effects of the metal. However, Geiger et al., (1998a, b) cited a higher affinity of copper to cellulase and β-glucosidase than to montmorillonite or Al montmorillonite and the synergetic effects of clay minerals and copper on the inhibition of enzyme activity. Geiger et al., (1998a, b) proposed that clay surfaces interact with both enzymes and metals and ultimately reduce the toxicity of metals. Clay minerals can strongly affect extracellular enzyme activity in soil (Geiger et al., 1998a, b). The adsorption of enzymes on clay surfaces caused two different responses:

1. The inactivation of enzymes due to conformational changes (Burns, 1978; Boyd and Mortland, 1990; Geiger et al., 1998a, b)
2. Enzyme activity enhancement caused by increased concentrations of the enzyme and substrate at the solid–water interface (Burns, 1978)

Tietjen and Wetzel (2003) investigated the effect of clay adsorption on enzyme activities (alkaline phosphatase, glucosidase, protease, and xylosidase). Montmorillonite clay (M) and clay extracted from Elledge Lake (EL) were used in enzyme–clay solutions in an adsorption experiment. While adsorption onto the EL clay decreased alkaline phosphatase activity, adsorption onto the M clay decreased the activities of all of the studied enzymes. They also found that the adsorption of the enzyme onto clay protects the enzyme from photodegradation. Żyszkowska and Paszkiewicz-Jasińska (2011) investigated the effects of copper on soil enzymes (dehydrogenase, urease, acid phosphatase, and alkaline phosphatase) and its interactions with other heavy metals (Zn, Ni, Pb, Cd, Cr). They found that the activity of dehydrogenase was greater in heavy loamy sand, whereas the activities of other enzymes were higher in light silty clay. In another words, enzyme inhibition due to heavy metals was greater in heavy loamy sand than in light silty clay (except in the case of dehydrogenase).

3 Heavy Metals

Heavy metals are the most insidious pollutants because of their nonbiodegradable nature and properties that affect all forms of the ecological system. Despite the toxicity potentials of heavy metals, some are essential for normal healthy growth and reproduction at low, but critical, concentrations. Among the heavy metals identified in the polluted environment, cadmium, lead, and mercury are more toxic than are other heavy metals because of their potentials to cause harmful effects even at low concentrations. Over the years, conventional or additional treatment methods of metal recovery have been used to ameliorate the effects of heavy metal pollution, but the traditional cleanup methods are highly expensive and efficiently low. The World Health Organization (WHO) has estimated that environmental exposures contribute to 19% of cancer incidence worldwide (Vineis and Xun, 2009). Heavy metals pose a critical concern to human health and the environment due to their common occurrence as contaminants, their low solubility in biota, and the classification of several heavy metals as carcinogenic and mutagenic (Diels et al., 2002). Industrial and agricultural activities have led to a considerable increase in heavy metals in different environmental compartments, especially in soils, over the course of recent decades. In recent years, increasing attention has been paid to the remediation of polluted soils, among which the use of plants and microbes to remove hazardous metals ions is particularly emphasized.

Heavy metals are naturally occurring elements and are present in varying concentrations in all ecosystems. There is huge number of heavy metals. They are found in elemental form and in a variety of other chemical compounds. Those that are volatile and those that become attached to fine particles can be widely transported on extremely large scales. Each form or compound has different properties, which also affect what happens to it in the food web and how toxic it is. Human activities have drastically changed the biochemical cycles and the balance of some heavy metals. Between 1850 and 1990, production of copper, lead, and zinc has increased 10-fold (Nriagu 1996; CACAR 2003). The main anthropogenic sources of heavy metals are various industrial processes, mining, foundries, smelters, combustion of fossil fuel and gasoline, and waste incinerators. The major heavy metals of concern to European Monitoring and Evaluation Programme (EMEP) are Hg, Cd, and Pb because they are the most toxic and have known serious effects on, e.g., human health.

Heavy metals are natural elements in the environment. However, anthropogenic releases, including industrial and domestic effluents, urban storms, water runoffs, landfill leachates, atmospheric sources, and dumping of sewage sludge, can give rise to higher concentrations of the metals relative to the normal background values. The term "heavy metal" refers to a metal or metalloid with a density exceeding 5 g cm^{-3} and is usually associated with pollution and toxicity, although some of these elements (essential metals) are actually required by organisms at low concentrations (Adriano, 2001). Several heavy metals, such as copper, zinc, and iron, are essential for the physiological functioning of living organisms, but they all become toxic at high concentrations. The toxicity of a metal depends on the metal itself, its total concentration, the availability of the metal to the organism, and the organism itself. Depending on the organism and the metal, different modes of action are recognized, namely, binding to macromolecules (proteins, DNA, RNA), disruption of enzymatic functions, catalysis of radical formation, etc. For example, zinc (Zn) is a component found in a variety of enzymes (dehydrogenases, proteinases, peptidases), but it is also involved in the metabolism of carbohydrates, proteins, phosphates, auxins, and in RNA and ribosome formation in plants (Kabata-pendias and Pendias 2001; Mengel and Kirkby, 1982).

3.1 Sources of Heavy Metals in Soil

The two main sources of heavy metals in soil are natural and anthropogenic/human.

The natural factors include soil erosion, volcanic activities, urban runoffs, and aerosol particulate, whereas the human factors include metal finishing and electroplating processes, mining extraction operations, textile industries, and nuclear power. The main natural sources of heavy metal pollutants in the soil are volcanic activities, soil erosion, urban runoffs, and aerosol particles. It is reported that volcanic eruptions produce hazardous impacts on the environment, climate, and health of exposed individuals. Apart from the deterioration of social and chemical conditions and the gases (carbon dioxide, sulfur dioxide, carbon monoxide, hydrogen sulfide) released during eruptions, various organic compounds and heavy metals, such as mercury, lead, and gold, are also released. The presence of these heavy metals in soil and water bodies is known to significantly deteriorate the quality of such soils and waters. Several rocks and volatiles of volcanic origins are indicated to be responsible for the presence of metals in soils and waters. This is because the diffusion of acidic volcanic gases through water-permeable rocks contributes to the hydrological material transfer in the volcanic strata. The activities from volcanoes are reported to be responsible for the release of metals such as arsenic, mercury, aluminum, rubidium, lead, magnesium, copper, zinc, and a host of others (Amarlal et al., 2006). Soil erosion is also indicated to be a source of heavy metal pollution in soils. The two main agents of soil erosion are wind and water. During rainfall, sediment-bound heavy metals are distributed in the soil. Water containing agrochemicals with toxic metal concentrations drop these sediment-bound metals in the soil even as they cause erosion.

In addition, some aerosol (fine colloidal particles or water droplets in the air; in some cases, they can be gas) particles may carry different kinds of the contaminant, like smoke cloud and heavy metals. These heavy metal-containing aerosols usually accumulate on leaf surfaces in the form of fine particulates and can enter the leaves via the stomata (Sardar et al., 2013). Some of the human sources of heavy metals in soil are metal finishing and electroplating, mining and extraction operations, textile activities, and nuclear power. Metal finishing and electroplating involve the deposition of thin protective layers onto the prepared surfaces of the metal using electrochemical processes. When this happens, toxic metals may be released into wastewater effluents. This may be either through rinsing of the product or through spillage and dumping of process baths. It is also indicated that the cleaning of process tanks and treatment of wastewater can generate substantial quantities of wet sludge containing high levels of toxic metals (Cushnie, 1985).

Similarly, mining activities can release toxic metals into the environment. Metal mining and smelting activities are regarded as the major sources of heavy metals in the environment. In environments where these activities take place, it is indicated that a large amount of toxic metal deposits are found in their water, soil, crops, and vegetables (Wei et al., 2008).

Additionally, textile industries are indicated to be the major sources of heavy metal pollutants in soil and water. This is said to mostly originate from the dyeing process, which is a major process in such industries. The compounds used for these dyeing processes (coloration) include copper, chromium, nickel, and lead, which are highly toxic and carcinogenic. In some cases, nuclear-generating facilities have also been described as the source of discharge of heavy metals like copper and zinc to surface soil and water. In nuclear plants, because a large amount of water is consumed for operation, after the operation, the nuclear effluents containing heavy metals are discharged into surface and groundwater bodies, which can pollute soil and aquatic systems (Hagberg et al., 2007; Wuana and Okieimen 2011).

Heavy metals occur naturally in the environment from pedogenetic processes of weathering of parent materials and also through anthropogenic sources. The most significant natural sources are weathering of minerals, erosion, and volcanic activity, whereas the anthropogenic sources depend upon human activities such as mining, smelting, electroplating, use of pesticides, and phosphate fertilizer discharge as well as biosolids (e.g., livestock manures, composts, and municipal sewage sludge), atmospheric deposition, etc. (Modaihsh et al., 2004, Sabiha-Javied et al., 2009). The disturbance of nature's slowly occurring geochemical cycle of metals by humans results in accumulation of one or more of heavy metals in soils and waters, and, above defined levels, this is enough to cause risks to human health, plants, animals, and aquatic biota (Summer, 2002). Heavy metals essentially become contaminants in the soil and water environments because of their excessive generation by natural and man-made activities, transfer from mines to other locations where higher exposure to humans occurs, discharge of high concentration of metal wastes through industries, and greater bioavailability.

4 Effects of Heavy Metals on Plants and Microbial Enzymes

4.1 Metal Accumulator Plants

Wang et al., (2008) defined metal accumulator plants as those that can grow in heavy metal-contaminated soils and have evolved mechanisms to tolerate high levels of heavy metals from the soil inside their cells (Tang and Yu 1999; Song et al., 2004). Mining sites, in particular, contain high heavy metal concentrations in soil and metal-tolerant plants. *Elsholtzia splendens* is a Cu-tolerant plant that is widely found at Cu mining sites and is used as a Cu mine indicator (Wang et al., 2008). Such plants can be used in the phytoremediation of heavy metal soils because they accumulate the metals and thus reduce metal levels in the soil. Wang et al., (2008) investigated the acid phosphatase activity in the rhizosphere of a copper accumulator (*Elsholtzia splendens*) and a nonaccumulator plant (*Trifolium repens*) upon different Cu treatments (0, 200, 500, 1000 mg kg^{-1}). Studies show that enzyme inhibition was strong in the unplanted and nonaccumulator plant rhizospheres and weak in the rhizosphere of the Cu accumulator plant. Wang and Qu 2007; Wang et al., (2007) studied the effect of heavy metal pollution on enzyme activity near a copper smelter. They found a strong inhibition of alkaline phosphatase activity near the copper smelter ($<$200 m).

Plant Community Effect

Yang et al. (2007a, b) investigated the effects of coexisting plant species on soil microbes and soil enzymes in lead-contaminated soils. In a mesocosm experiment carried out in a greenhouse, four different plant species (*Festuca arundinacea*: FA, *Kummerowia striata*: KS, *Echinochloa crus-galli*: EC, and *Solidago canadensis*: SC), three different species mixtures (one: FA, two: FA + KS, and four: FA + KS + EC + SC), and three different lead application rates (0, 300, and 600 mg kg^{-1}) were used. Urease activity was significantly affected by plant species and Pb concentration. It was significantly greater for the four-species mixture than for the one- or two-species mixtures. Alkaline phosphatase activity was not significantly impacted.

4.2 Effects of Heavy Metals on Microbial Enzymes

Heavy metals affect soil enzyme activities, thereby thwarting plant growth, most especially Pb concentration. Acid phosphatase and dehydrogenase were not significantly influenced by either species mixture or Pb concentration.

Special Inhibition Parameters

Ecological Dose

The effects of heavy metals on soil enzyme activities can be quantified by determining the ED 50 (ecological dose) parameter, which is the concentration of heavy metals at which the enzyme activity, or some other biological activity, is reduced to 50% of its uninhibited value (Tejada et al., 2008a, b). Tejada et al., (2008a, b) reported that ED 50 values may be more suitable indicators of the sensitivity of an ecosystem to stress because a 50% reduction in the basic ecological process may be too extreme for its continued functioning (Babich et al., 1983).

Understanding the Inhibition of Soil Enzymes by Heavy Metals

Combined Effects

Heavy metals exert inhibitory effects on soil enzymes, but these effects depend on many factors in the soil.

Combined Effects of Two Metals

Khan et al., (2007) investigated soil enzyme activities (catalase, alkaline phosphatase, and dehydrogenase) when various levels of Cd and/or Pb were applied to the soil. This work thus provides a good example of the combined effects of heavy metals on soil enzyme activities. Strong inhibition was observed at high heavy metal concentrations in both the single-metal and dual-metal systems; however, the inhibition was greater in the dual-metal system than in the single-metal system; in other words, a "synergistic effect" was observed. However, some combinations of metals exhibit this synergism, whereas others do not. Żyszkowska and Paszkiewicz-Jasińska (2011) concluded that treatment with copper alone was inhibitory toward soil enzyme activity than copper applied in conjunction with other heavy metals (Cu with Zn, Ni, Pb, Cd, and Cr).

Combined Effects of Three Metals

Yang et al., (2006) investigated the combined effects of Cd, Zn, and Pb on catalase, urease, invertase, and alkaline phosphatase in soil. The results showed that Cd significantly inhibited the activities of all of the enzymes studied, Zn only inhibited those of urease and catalase, whereas Pb was not significantly inhibitory compared to the other heavy metals toward the studied enzymes and actually had a protective influence on catalase activity when all of the metals were present (Cd, Zn, and Pb). Cd was the most effective enzyme inhibitor, followed by Zn. The order of the effect of Cd, Zn, and Pb was Cd > Zn > Pb. There was a negative synergistic inhibitory effect of Cd and Zn on urease and catalase activity in the presence of Cd, Zn, and Pb, which can be explained by the similar ionic properties of Zn and Cd. Urease activity was enhanced by Cd and Pb at low concentrations; however, it was inhibited at higher concentrations of Cd and Pb. Urease activity was reduced by 20–40% in the Cd–Zn–Pb-combined metal system. Therefore, three-metal treatments had a greater inhibitory effect than did single heavy metal treatments because of a synergistic effect of the metals on enzyme activity. In this study, the enzymes showed different sensitivities to the single- and three-metal treatments. Urease was the most sensitive of the enzymes to combined pollution (Cd, Zn, and Pb). Yang et al., (2006) reported that the magnitude of enzyme inhibition or activation depends on (1) the heavy metal ion, its concentration, and the type of enzyme assayed, (2) the interaction between the heavy metals, (3) the reactions between the heavy metals in solution and the functional groups of the enzymes, and (4) the chemical and physical properties of the soil (pH, organic matter content, and the type and amount of clay).

Combined Effects of pH, Organic Matter (OM), Clay, and Four Metals

Irha et al., (2003) studied the effect of heavy metals and PAHs on dehydrogenase in soil. Decreasing the organic matter, clay, and pH slightly inhibited the dehydrogenase. Rendzina alvar and Brown pseudopodzolic soils differ only in their organic matter and amorphous mineral-phase contents; their clay contents are the same. The dehydrogenase was more inhibited at lower organic matter and higher amorphous mineral-phase contents (i.e., in Brown pseudopodzolic soil). Organic matter and the amorphous mineral phase may therefore mask dehydrogenase inhibition by heavy metals.

Combined Effects of pH, OM, Clay, Cation Exchange Capacity (CEC), and Chemical Form of Metal

Reyes-Betancort et al., (2008) estimated the effects of different metals in different chemical forms (chloride, sulfate, and acetate salt) on soil phenoloxidase activity. This study investigated the soil enzyme inhibition by heavy metals while the researchers considered many factors and examined many heavy metals. The study results led to the conclusion that soil enzyme inhibition by heavy metals depends on: (1) the heavy metal and its concentration; (2) soil texture (clay content); and (3) the chemical form of the heavy metal (Karaca et al., 2000).

Combined Effects of Metal, Metal Oxidation State, and Organic Matter

Senwo and Tabatabai (1999) conducted a study on the effects of heavy metals on aspartase activity in soils. They concluded that: (1) the most effective inhibitors of aspartase activity were Ag(I) and Hg(I); (2) aspartase activity was significantly correlated with organic carbon, total nitrogen, and clay content; (3) activity inhibition was higher in air-dried soils than in field-moist soils because the air-dried soils provided more exposure of the enzyme to heavy metals and can be summarized as follows. (1) Higher organic matter and clay contents along with a higher soil pH result in less inhibition of aspartase activity. (2) Higher oxidation states of heavy metals are less inhibitory than are lower oxidation states. (3) Ag and Hg are highly toxic elements.

5 Mitigation of Heavy Metal Uptake by Plants

Some technologies that have been used are high-temperature incineration and various types of chemical decompositions (e.g., base-catalyzed dechlorination, ultraviolet (UV) oxidation). They can be highly effective at reducing levels of a range of contaminants but have several weaknesses, principally their technological complication, the cost for small-scale application, and lack of public acceptance, especially for incineration, which may increase the exposure to contaminants for both workers at the site and nearby residents. Bioremediation is a natural process, which relies on bacteria, fungi, and plants to alter contaminants as these organisms carry out their normal life functions. Metabolic processes of these organisms are capable of using chemical contaminants as an energy source, rendering the contaminants harmless or less toxic products in most cases. Thus, bioremediation provides an alternative tool to destroy or render the harmful contaminants through biological activity, and this method is also cost-effective (Kamaluddin and Zwiazek 2003).

Bioremediation/Phytoremediation and Rhizoremediation, Microflora associated with plants, endophytic bacteria, rhizosphere bacteria, and mycorrhizae have the potential to degrade heavy metals into their associated ions in plants, and this process is termed "rhizoremediation". Thus, bioremediation, phytoremediation, and rhizoremediation significantly contribute to the fate of hazardous wastes (heavy metals) and can be used to remove these unwanted compounds from the biosphere.

Bioremediation processes can also be assessed through multifaceted approaches such as natural attenuation, sensing environmental pollution, metabolic pathway engineering, applying phyto and microbial diversity to problematic sites, plant–endophyte partnerships, and systems biology (Asha and Anju 2013). Enhancement of these polluted soil residues with different organic amendments like manure composts, biosolids, and municipal solid wastes (MSWs) will lead to increased bioavailability, which, in turn, will act as nutrients for microorganisms and also a conditioner to improve the physical properties and fertility of the soils (Jin et al., 2011).

Phytoremediation, a fast-emerging technology, is an eco-friendly, low-tech, cost-effective, green alternative to the problem (Meagher 2000). The specific plant and wild species that are used in this technique accumulate increasing amounts of toxic heavy metals by their roots and transport/translocate them through various plant tissues where they can be metabolized, sequestered, and volatilized (Gurol et al., 2006, Doty et al., 2000). These plants are known as hyperaccumulators. Phytoremediation can be carried out in different ways such as rhizofiltration, phytostabilization, phytovolatilization, phytodegradation (Long and Liu 2002), and phytoextraction.

Phytoremediation is the use of plants to remove pollutants from the environment, whereas microbial remediation refers to the use of microbes. These two approaches are preferred over chemical/physical remediation because of their cost-effectiveness, environmental friendliness, and less side effects. Phytoremediation cannot be performed alone by the plant, as there is always a close interaction between the microorganisms in the rhizosphere and the plant, which leads to an increased activity related to soil remediation (Compant et al., 2010). Hence, search for hyperaccumulating plants in combination with a beneficial rhizo- and/or endospheric microbial community holds great promise for low-cost cleaning of contaminated sites. Arbuscular mycorrhizal fungi (AMF) are one of the important endophytic fungi living in the roots of most terrestrial plants. This symbiosis directly confers benefits on the host plant's growth and development through the acquisition of phosphorus and other mineral nutrients from the soil by the fungus. In addition, they may also enhance the plant's resistance to biotic and abiotic stresses (Harrier and Sawczak, 2000). The potential roles of AMF associations have constantly been verified to alleviate metal stress of plants (Hildebrandt et al., 2007; Awotoye et al., 2009, 2011).

5.1 *Degradation by Genetically Engineered Microorganisms*

In the late 1970s and early 1980s, bacterial genes encoding catabolic enzymes for recalcitrant compounds started to be cloned and characterized. Presently, many microbiologists and molecular biologists have realized the potential of genetic engineering for addressing biodegradation (Kasper et al., 2005). A genetically engineered microorganism (GEM) or a genetically modified microorganism (GMM) is a microorganism whose genetic material has been altered using genetic engineering techniques encouraged by the natural genetic exchange between microorganisms. These techniques are generally known as recombinant DNA technology. Genetically engineered microorganisms (GEMs) have shown potential for bioremediation of soil, groundwater, and activated sludge, exhibiting the enhanced degrading capabilities of a wide range of chemical contaminants (Sayler and Ripp 2000). As soon as the prospect of releasing genetically modified microorganisms for bioremediation became a reality, much of the research efforts in the field were aimed at biosafety.

5.2 *Microbial Remediation of Heavy Metals*

The term "biodegradation" is frequently used in relation to ecology and waste management and is mostly associated with environmental remediation (bioremediation). The bioremediation process can be divided into three phases or levels. First, through natural attenuation, contaminants are reduced by native microorganisms without any human augmentation. Second, biostimulation is employed where nutrients and oxygen are applied to the systems to improve their effectiveness and to accelerate biodegradation. Finally, during bioaugmentation, microorganisms are added to the systems. These supplementary organisms should be more effective at decomposing the target pollutant than native flora (Marinescu et al., 2009). A feasible remedial technology requires that microorganisms be capable of quick adaptation and that pollutants of interest be efficiently used in a particular case in a reasonable period of time. In recent years, considerable interest has been paid to rhizobacteria, which are aggressive root colonizers and produce siderophores. Siderophores provide an advantage for the survival of both plants and bacteria (Narendra et al., 2015). Many factors influence microorganisms to use pollutants as substrates or to metabolize them; then, the genetic potential and certain environmental factors, such as temperature, pH, and available nitrogen and phosphorus sources, seem to determine the rate and the extent of degradation (Fritsche et al., 2008). Therefore, applications of genetically engineered microorganisms (GEMs) in bioremediation have received a great deal of attention. These GEMs have a higher degradative capacity and have been successfully demonstrated for the degradation of various pollutants under defined conditions. However, ecological and environmental concerns and regulatory constraints are major obstacles for testing GEMs in the field (Menn et al., 2008). In microbial remediation or bioremediation, microbial communities are of primary importance. The process is a cost-effective one, with nonhazardous end products (Ahmedna et al., 2004). During pollutant removal, the microbe(s) alter the metal chemistry and mobility through reduction, accumulation, mobilization, or immobilization (Faryal and Hameed, 2005). Based on the high level of heavy metal resistances, study have revealed/identified five bacterial isolates. Based on the high level of heavy metal resistances, study have revealed/identified five bacterial isolates. *Proteus vulgaris* (MR1), *Bacillus cereus* (MR2), *Bacillus decolorationis* (MR3), *Pseudomonas fluorescens* (SS4), and *Pseudomonas fluorescens* (SS5) were identified based on their appearance (SS5). The soil isolates showed optimum growth at pH 7.0 and 30 °C. The identified isolates were resistant to cadmium (Cd), nickel (Ni), lead (Pb), arsenic (As), and chromium (Cr). The minimal inhibitory concentration (MIC) of soil isolates against Cd, Cr, Ni, Pb, and As was determined in solid media (Narendra et al., 2016). The identified heavy metal-resistant bacteria could be effective and useful in the bioremediation of heavy metal-contaminated soil. The major groups of microorganisms that have been implicated in heavy metal remediation are bacteria (such as *Arthrobacter*, *Bacillus* sp., *Citrobacter*, *Cupriavidus metallidurans*, *Cyanobacteria*, *Enterobacter cloacae*, *Pseudomonas aeruginosa*, *Streptomyces* sp., *Zoogloea ramigera*, *Alcaligenes*, *Sphingomonas*, *Rhodococcus*, *Mycobacterium*, and *Arthrobacter*) and fungi (such as *Aspergillus tereus*, *Penicillium chrysogenum*, *Candida utilis*, *Hansenula anomala*, and *Rhodotorula mucilaginosa*).

Besides bacteria and fungi, certain protozoa, such as *Euplotes mutabilis*, and algae, such as *Oscillatoria* sp., *Chlorella vulgaris*, and *Chlamydomonas* sp., have been reported to possess metal-reducing capabilities (Ramasamy et al., 2006).

The microbial remediation of toxic metals is said to occur in two ways: direct and indirect reduction (Rastogi and Sinha et al., 2009). Microbial remediation can be in the form of bioaugmentation, biosorption, or sparging. Bioaugmentation entails the introduction of a microbial strain, which has a high degradation factor to assist the indigenous microbe in the active degradation process of the contaminated environment. It is mostly used in municipal wastewater to restart an activated sludge bioreactor (Vasanthavigar et al., 2010). Soil microorganisms vary widely in their tolerance to heavy metal contamination, and the proportion of culturable resistant microorganisms can range from 10% to nearly 100%. The activities of enzymes in soil may serve as indicators of heavy metal contamination, as there are generally high correlations between reduced enzyme activities (of, e.g., dehydrogenases, acid phosphatases, and ureases) and increased heavy metal contamination (Ahirwar and Shukla 2018). In our previous studies, we have reported a higher reduction of chromium for lower initial concentrations by *Bacillus cereus*, *Bacillus decolorationis*, and *P. fluorescens*. Seed germination and plant growth ability were analyzed in different experimental groups using *Pseudomonas fluorescens*, *Bacillus cereus*, and *Bacillus decolorationis*. *Pseudomonas fluorescens* (95%) and *Bacillus cereus* (90%) have shown the maximum noted seed germination and plant growth ability compared to *Bacillus decolorationis* (84%)-inoculated strain in Cr-contaminated soil (Ahirwar and Dehariya 2013).

In biosorption, there is the immobilization of metals by microbial cells. Its technique involves the sequestration of positively charged heavy metal ions to the negatively charged microbial cell membranes and polysaccharides, which are secreted (Rastogi and Sinha et al., 2009). The mechanisms of heavy metal removal from soil by microorganisms can be based on microbial precipitation, complexation, ion exchange, and intracellular accumulation. During biosparging, also known as air sparging, there is an injection of air by pressure to water to enhance the activation of oxygen concentration by the microorganism, which can increase biological degradation of the contaminant. Apart from the encouragement of aerobic bacterial growth, air sparging also leads to the volatilization of contaminants from the liquid to the vapor phase (Sharma et al., 2012).

A wide variety of synthetic organic compounds contaminate the environment from chemical and industrial processes. Microbial degradation is dependent upon physical and chemical environmental variables as well as on the toxicity of the chemical. Physical and chemical factors may render a given compound more or less susceptible to microbial degradation. For example, irradiation in the visible and ultraviolet ranges can aid in the degradation of polymerized plastics and dechlorination of halogenated substrates and, perhaps, in the cleavage of alkylated biphenyls and fused aromatic ring systems. Photodegradation has also been implicated in the potential formation of chlorinated dibenzofurans from chlorinated biphenyls, producing more toxic compounds of unknown biodegradative potential (Crosby and Wong 1973). Especially attractive is the potential for early warning of environmental change since microbiological responses are rapid and can be detected within hours or days. The microbial potential, perhaps measured as a community structure index, or other mathematical formulations, should be more fully investigated as an ecotoxicological yardstick of health. Noticeably, the microbial aspects of ecotoxicology should be explored since here lies, indeed, a fertile soil for discovery and application in environmental pollution and risk assessment (Cases et al., 2005).

6 Conclusions

This review, which was aimed at discussing soil microbial enzymes and mitigation of heavy metal uptake by plants, revealed that soil microbial enzymes are a useful tool for assessing the functional diversity of soil microbial communities or soil organic mass turnover. Moreover, the quantification of enzymes has become an effective soil bioindicator, since enzymatic activity can be used to assess the activity of the microbiota. Some studies reported that ecological dose 50 values may be more appropriate indicators of the sensitivity of an ecosystem to stress because a 50% decrease in the basic ecological process may be too extreme for its continued functioning. However, the two main sources of heavy metals in soil were revealed as natural and anthropogenic/human. The natural factors include soil erosion, volcanic activities, urban runoffs, and aerosol particulate, whereas the human factors include metal finishing and electroplating processes, mining extraction operations, textile industries, and nuclear power.

Inadequately treated heavy metal-contaminated soils create a variety of health and environmental impacts on humans, animals, and plants. Moreover, heavy metals have a negative consequence on the growth of plants. To maintain the safety of soil and for environmental sustainability, an array of biological treatment processes is employed for the extraction of heavy metals from soil, with the most common being microbial remediation and phytoremediation. Biological confiscation of heavy metals from soil is a selective technique that utilizes the operational flexibility of microorganisms and plants for the elimination of pollutants from soil.

Microbial remediation may require ex situ and in situ applications. In phytoremediation, plants play a great role in the biological process as they break down, reduce, degrade, and remove these contaminants using different parts, such as the root, leaves, stomata, cell wall, and the shoot. The microbial remediation of toxic metals is said to occur in two ways: direct and indirect reduction. Microbial remediation can be in the form of bioaugmentation, biosorption, or biosparging. In phytoremediation, green plants are employed in the in situ treatment of contaminants. Such plants have the advantage of accumulating and degrading mechanisms of such contaminants. The commonest phytoremediation processes are rhizofiltration, phytostabilization, phytoextraction, phytovolatilization, phytodegradation, and rhizodegradation.

References

Adriano, D. C. (2001). Bioavailability of trace metals. In *Trace elements in terrestrial environments* (pp. 61–89). Springer.

Ahirwar, S., & Dehariya, K. (2013). Isolation and characterization of hydrocarbon degrading microorganisms from petroleum oil contaminated soil sites. *Bulletin of Environmental and Scientific Research, 2*(4), 5–10.

Ahirwar, S., & Shukla, J. P. (2018). Assessment of groundwater vulnerability in upper Betwa river watershed using GIS based DRASTIC model. *Journal of the Geological Society of India, 91*(3), 334–340.

Ahmedna, M., Marshall, W. E., Husseiny, A. A., Rao, R. M., & Goktepe, I. (2004). The use of nutshell carbons in drinking water filters for removal of trace metals. *Water Research, 38*(4), 1062–1068.

Amarlal, D., Rayen, R., & Muthu, M. S. (2006). Macroabrasion in pediatric dentistry. *Journal of Clinical Pediatric Dentistry, 31*(1), 9.

Asha, S., & Anju, D. (2013). Nickel and cadmium toxicity in plants. *Journal of Pharmaceutical and Scientific Innovation (JPSI), 2*(2), 20–24.

Awotoye, O. O., Ekanade, O., & Airouhudion, O. O. (2009). Degradation of the soil physicochemical properties resulting from continuous logging of Gmelina arborea and Tectona grandis plantations. *African Journal of Agricultural Research, 4*(11), 1317–1324.

Awotoye, O. O., Ogunkunle, C. O., & Adeniyi, S. A. (2011). Assessment of soil quality under various land use practices in a humid agro-ecological zone of Nigeria. *African Journal of Plant Science, 5*(10), 565–569.

Bååth, E. (1989). Effects of heavy metals in soil on microbial processes and populations (a review). *Water, Air, and Soil Pollution, 47*(3), 335–379.

Babich, H., Bewley, R. J. F., & Stotzky, G. (1983). Application of the "ecological dose" concept to the impact of heavy metals on some microbe-mediated ecologic processes in soil. *Archives of Environmental Contamination and Toxicology, 12*(4), 421–426.

Baihong, L., Juan, Z., Zhen, F., Lei, G., Xiangru, L., Guocheng, D., & Jian, C. (2013). Enhanced thermostability of keratinase by computational design and empirical mutation. *Journal of Industrial Microbiology & Biotechnology, 40*, 697–704.

Bandick, A. K., & Dick, R. P. (1999). Field management effects on soil enzyme activities. *Soil Biology and Biochemistry, 31*(11), 1471–1479.

Bornscheuer, U. T., Huisman, G. W., Kazlausaks, R. J., Lutz, S., Moore, J. C., & Robins, K. (2012). Engineering the third wave of biocatalysis. *Nature, 485*, 185–194.

Boyd, S. A., & Mortland, M. M. (1990). Enzyme interactions with clays and clay-organic matter complexes. *Soil Biochemistry, 1*.

Brandelli, A., Daroit, D. J., & Riffel, A. (2010). Biochemical features of microbial keratinases and their production and applications. *Applied Microbiology and Biotechnology, 85*, 1735–1750.

Burns, R. G. (1978). Enzymes activity in soil: Some theoretical and practical considerations.

CACAR. (2003). *Canadian contaminants assessment report II, sources, occurrence, trends and pathways in the physical environment*. Northern Contaminants Program, Minister of Indian Affairs and Northern Development, Minister of Public Works and Government Services Canada, 332.

Campbell, I. M. (1988). *Catalysis at surfaces*. Springer.

Chirumamilla, R. R., Muralidhar, R., Marchant, R., & Nigam, P. (2001). Improving the quality of industrially important enzymes by directed evolution. *Molecular and Cellular Biochemistry, 224*, 159–168.

Chudasama, C. J., Jani, S. A., Jajda, H. M., & Pate, H. N. (2010). Optimization and production of alkaline protease from Bacillus thuringiensis CC7. *Cell and Tissue Research, 10*, 2257–2262.

Collins, T., Gerday, C., & Feller, G. (2005). Xylanases, xylanase families and extremophilic xylanases. *FEMS Microbiology Reviews, 29*, 3–23.

Compant, S., Clément, C., & Sessitsch, A. (2010). Plant growth-promoting bacteria in the rhizo-and endosphere of plants: Their role, colonization, mechanisms involved and prospects for utilization. *Soil Biology and Biochemistry, 42*(5), 669–678.

Crosby, D. G., & Wong, A. S. (1973). Photodecomposition of 2, 4, 5-trichlorophenoxyacetic acid (2, 4, 5-T) in water. *Journal of Agricultural and Food Chemistry, 21*(6), 1052–1054.

Cushnie Jr, G. C. (1985). Electroplating wastewater pollution control technology.

Dahiya, J. S., Singh, D., & Nigam, P. (1998). Characterisation of laccase produced by Coniotherium minitans. *Journal of Basic Microbiology, 38*, 349–359.

D'Ascoli, R., Rao, M. A., Adamo, P., Renella, G., Landi, L., Rutigliano, F. A., ... Gianfreda, L. (2006). Impact of river overflowing on trace element contamination of volcanic soils in south Italy: Part II. Soil biological and biochemical properties in relation to trace element speciation. *Environmental Pollution, 144*(1), 317–326.

Dick, M. W. (1997). Fungi, flagella and phylogeny. *Mycological Research, 101*(4), 385–394.

Dick, W. A., & Tabatabai, M. A. (1977). An alkaline oxidation method for determination of total phosphorus in soils. *Soil Science Society of America Journal, 41*(3), 511–514.

de Araújo, A. S. F., & Monteiro, R. T. R. (2007). Indicadores biológicos de qualidade do solo. *Bioscience Journal, 23*(3).

Diels, L., Van der Lelie, N., & Bastiaens, L. (2002). New developments in treatment of heavy metal contaminated soils. *Reviews in Environmental Science and Biotechnology, 1*(1), 75–82.

Doty, S. L., Shang, T. Q., Wilson, A. M., Tangen, J., Westergreen, A. D., Newman, L. A., ... Gordon, M. P. (2000). Enhanced metabolism of halogenated hydrocarbons in transgenic plants containing mammalian cytochrome P450 2E1. *Proceedings of the National Academy of Sciences, 97*(12), 6287–6291.

Dumon, C., Songa, L., Bozonneta, S., Fauréa, R., & O'Donohue, M. J. (2012). Progress and future prospects for pentose-specific biocatalysts in biorefining. *Process Biochemistry, 47*, 346–357.

Dwivedi, P., Vivikanand, V., Pareek, N., Sharma, A., & Singh, R. P. (2010). Bleach enhancement of mixed wood pulp by xylanase-laccase concoction derived through co-culture strategy. *Applied Biochemistry and Biotechnology, 160*, 255–268.

Effron, D., De la Horra, A. M., Defrieri, R. L., Fontanive, V., & Palma, R. M. (2004). Effect of cadmium, copper, and lead on different enzyme activities in a native forest soil. *Communications in Soil Science and Plant Analysis, 35*(9–10), 1309–1321.

Eivazi, F., & Tabatabai, M. A. (1988). Glucosidases and galactosidases in soils. *Soil Biology and Biochemistry, 20*(5), 601–606.

Esterbauer, H., Schaur, R. J., & Zollner, H. (1991). Chemistry and biochemistry of 4-hydroxynonenal, malonaldehyde and related aldehydes. *Free Radical Biology and Medicine, 11*(1), 81–128.

Faryal, R. A. N. I., & Hameed, A. (2005). Isolation and characterization of various fungal strains from textile effluent for their use in bioremediation. *Pakistan Journal of Botany, 37*(4), 1003.

Frankenberger, W. T., Jr., & Tabatabai, M. A. (1980). Amidase activity in soils: II. Kinetic parameters. *Soil Science Society of America Journal, 44*(3), 532–536.

Fritsche, G., Nairz, M., Werner, E. R., Barton, H. C., & Weiss, G. (2008). Nramp1-functionality increases iNOS expression via repression of IL-10 formation. *European Journal of Immunology, 38*(11), 3060–3067.

Fusi, P., Ristori, G. G., Calamai, L., & Stotzky, G. (1989). Adsorption and binding of protein on "clean"(homoionic) and "dirty"(coated with Fe oxyhydroxides) montmorillonite, illite and kaolinite. *Soil Biology and Biochemistry, 21*(7), 911–920.

Gali, N. K., & Kotteazeth, S. (2013). Biophysical characterization of thermophilic laccase from the xerophytes: Cereus pterogonus and Opuntia vulgaris. *Cellulose, 20*, 115–125.

Gama-Rodrigues, E. F., Ramachandran Nair, P. K., Nair, V. D., Gama-Rodrigues, A. C., Baligar, V. C., & Machado, R. C. (2010). Carbon storage in soil size fractions under two cacao agroforestry systems in Bahia, Brazil. *Environmental Management, 45*(2), 274–283.

Garg, A. P., Roberts, J. C., & McCarthy, A. (1998). Bleach boosting effect of cellulase free xylanase of Streptomyces thermoviolaceus and its comparison with two commercial enzyme preparations on birchwood Kraft pulp. *Enzyme Microbial Biotechnology, 22*, 594–598.

Geiger, M., Walch-Liu, P., Engels, C., Harnecker, J., Schulze, E. D., Ludewig, F., ... Stitt, M. (1998a). Enhanced carbon dioxide leads to a modified diurnal rhythm of nitrate reductase activity in older plants, and a large stimulation of nitrate reductase activity and higher levels of amino acids in young tobacco plants. *Plant, Cell & Environment, 21*(3), 253–268.

Geiger, G., Brandl, H., Furrer, G., & Schulin, R. (1998b). The effect of copper on the activity of cellulase and β-glucosidase in the presence of montmorillonite or Almontmorillonite. *Soil Biology and Biochemistry, 30*(12), 1537–1544.

Gianfreda, L., Rao, M. A., & Violante, A. (1991). Invertase β- fructosidase): Effects of montmorillonite, AL-hydroxide and AL (OH) x-montmorillonite complex on activity and kinetic properties. *Soil Biology and Biochemistry, 23*(6), 581–587.

Gianfreda, L., Xu, F., & Bollag, J. M. (1999). Laccases: A useful group of oxidoreductive enzymes. *Bioremediation Journal, 3*(1), 1–26.

Gupta, A., Joseph, B., Mani, A., & Thomas, G. (2008). Biosynthesis and properties of an extracellular thermostable serine alkaline protease from Virgibacillus pantothenticus. *World Journal of Microbiology and Biotechnology, 24*, 237–243.

Gupta, R., Beg, Q. K., & Lorenz, P. (2002). Bacterial alkaline proteases: Molecular approaches and industrial applications. *Applied Microbiology and Biotechnology, 59*, 15–32.

Gupta, R., Sharma, R., & Beg, Q. K. (2013). Revisiting microbial keratinases: Next generation proteases for sustainable biotechnology. *Critical Reviews in Biotechnology, 33*, 216–228.

Gurol, M. E., Irizarry, M. C., Smith, E. E., Raju, S., Diaz-Arrastia, R., Bottiglieri, T., . . . Greenberg, S. M. (2006). Plasma β-amyloid and white matter lesions in AD, MCI, and cerebral amyloid angiopathy. *Neurology, 66*(1), 23–29.

Gurumurthy, G. P., Balakrishna, K., Riotte, J., Braun, J. J., Audry, S., Shankar, H. U., & Manjunatha, B. R. (2012). Controls on intense silicate weathering in a tropical river, southwestern India. *Chemical Geology, 300*, 61–69.

Gushterova, A., Vasileva-Tonkova, E., Dimova, E., Nedkov, P., & Haertle, T. (2005). Keratinase production by newly isolated Antarctic actinomycete strains. *World Journal of Microbiology and Biotechnology, 21*, 831–834.

Hadj-Ali, N. E., Rym, A., Basma, G. F., Alya, S. K., Safia, K., & Moncef, N. (2007). Biochemical and molecular characterization of a detergent stable alkaline serineprotease from a newly isolated Bacillus licheniformis NH1. *Enzyme and Microbial Technology, 40*, 515–523.

Hagberg, M., Vilhemsson, R., Tornqvist, E. W., & Toomingas, A. (2007). Incidence of self-reported reduced productivity owing to musculoskeletal symptoms: Association with workplace and individual factors among computer users. *Ergonomics, 50*(11), 1820–1834.

Hamouda, A. E., & Adia, A. R. (2009). Three sixty appraisal system for software houses. *44th Annual Conference on Statistic Sciences and Operation Research, 44*(30), 7–9.

Hardiman, E., Gibbs, M., Reeves, R., & Bergquist, P. (2010). Directed Evolution of a thermophilic beta-glucosidase for cellulosic bioethanol production. *Applied Biochemistry and Biotechnology, 2010*(161), 301–312.

Harrier, L., & Sawczak, J. (2000). Detection of the 3-phosphoglycerate kinase protein of Glomus mosseae. *Mycorrhiza, 10*(2), 81–86.

Hemida, S. K., Omar, S. A., & Abdel-Mallek, A. Y. (1997). Microbial populations and enzyme activity in soil treated with heavy metals. *Water, Air, and Soil Pollution, 95*(1), 13–22.

Hildebrandt, U., Regvar, M., & Bothe, H. (2007). Arbuscular mycorrhiza and heavy metal tolerance. *Phytochemistry, 68*(1), 139–146.

Indhuja, S., Shiburaj, S., Pradeep, N. S., Thankamani, V., & Abraham, T. K. (2012). Extracellular keratinolytic proteases from an alkalophilic Streptomyces albidoflavus TBG-S13A5: Enhanced production and characterization. *Journal of Pure and Applied Microbiology, 6*, 1599–1607.

Irha, N., Slet, J., & Petersell, V. (2003). Effect of heavy metals and PAH on soil assessed via dehydrogenase assay. *Environment International, 28*(8), 779–782.

Jhamb, K., & Sahoo, D. K. (2012). Production of soluble recombinant proteins in Escherichia coli: Effects of process conditions and chaperone co-expression on cell growth and production of xylanase. *Bioresource Technology, 123*, 135–143.

Jin, L. J., Armitage, G. C., Klinge, B., Lang, N. P., Tonetti, M., & Williams, R. C. (2011). Global oral health inequalities: Task group-periodontal disease. *Advances in Dental Research, 23*, 221–226.

Juma, N. G., & Tabatabai, M. A. (1977). Effects of trace elements on phosphatase activity in soils. *Soil Science Society of America Journal, 41*(2), 343–346.

Kabata-Pendias, A., & Pendias, H. (2001). *Trace elements in soils and plants*, 3rd edn CRC Press. Boca Raton, FL, USA.

Kamaluddin, M., & Zwiazek, J. J. (2003). Fluoride inhibits root water transport and affects leaf expansion and gas exchange in aspen (Populus tremuloides) seedlings. *Physiologia Plantarum, 117*(3), 368–375.

Kandeler, E., Tscherko, D., & Spiegel, H. (1999). Long-term monitoring of microbial biomass, N mineralisation and enzyme activities of a Chernozem under different tillage management. *Biology and Fertility of Soils, 28*, 343–351.

Karaca, M., Deniz, A., & Tayanç, M. (2000). Cyclone track variability over Turkey in association with regional climate. *International Journal of Climatology: A Journal of the Royal Meteorological Society, 20*(10), 1225–1236.

Karaca, H. E., Karaman, I., Basaran, B., Chumlyakov, Y. I., & Maier, H. J. (2006). Magnetic field and stress induced martensite reorientation in NiMnGa ferromagnetic shape memory alloy single crystals. *Acta Materialia, 54*(1), 233–245.

Karaca, A., Cetin, S. C., Turgay, O. C., & Kizilkaya, R. (2010). Effects of heavy metals on soil enzyme activities. In *Soil heavy metals* (pp. 237–262). Springer.

Kasper, D., Planells-Cases, R., Fuhrmann, J. C., Scheel, O., Zeitz, O., Ruether, K., . . . Jentsch, T. J. (2005). Loss of the chloride channel ClC-7 leads to lysosomal storage disease and neurodegeneration. *The EMBO Journal, 24*(5), 1079–1091.

Khan, A., Fux, B., Su, C., Dubey, J. P., Dardé, M. L., Ajioka, J. W., . . . Sibley, L. D. (2007). Recent transcontinental sweep of toxoplasma gondii driven by a single monomorphic chromosome. *Proceedings of the National Academy of Sciences, 104*(37), 14872–14877.

Kohli, U., Nigam, P., Singh, D., & Chaudhary, K. (2001). Thermostable, alkalophilic and cellulase free xylanase production by Thermonoactinomyces thalophilus subgroup C. *Enzyme and Microbial Technology, 28*, 606–610.

Kundu, S., Bhattacharyya, R., Prakash, V., Ghosh, B. N., & Gupta, H. S. (2007). Carbon sequestration and relationship between carbon addition and storage under rainfed soybean–wheat rotation in a sandy loam soil of the Indian Himalayas. *Soil and Tillage Research, 92*(1–2), 87–95.

Kunito, T., Saeki, K., Goto, S., Hayashi, H., Oyaizu, H., & Matsumoto, S. (2001). Copper and zinc fractions affecting microorganisms in long-term sludge-amended soils. *Bioresource Technology, 79*(2), 135–146.

Kuperman, R. G., & Carreiro, M. M. (1997). Soil heavy metal concentrations, microbial biomass and enzyme activities in a contaminated grassland ecosystem. *Soil Biology and Biochemistry, 29* (2), 179–190.

Li, Y., Niu, D., Zhang, L., Wang, Z., & Shi, G. (2013). Purification, characterization and cloning of a thermotolerant isoamylase produced from Bacillus sp. CICIM 304. *Journal of Industrial Microbiology & Biotechnology, 40*, 437–446.

Lisboa, I. P., Cherubin, M. R., Satiro, L. S., Siqueira-Neto, M., Lima, R. P., Gmach, M. R., . . . Cerri, C. E. (2019). Applying soil management assessment framework (SMAF) on short-term sugarcane straw removal in Brazil. *Industrial Crops and Products, 129*, 175–184.

Liu, B., Zhang, J., Li, B., Liao, X., Du, G., & Chen, J. (2013). Expression and characterization of extreme alkaline, oxidation-resistant keratinase from Bacillus licheniformis in recombinant Bacillus subtilis WB600 expression system and its application in wool fiber processing. *World Journal of Microbiology and Biotechnology, 2013*(29), 825–832.

Long, G. L., & Liu, X. S. (2002). Theoretically efficient high-capacity quantum-key-distribution scheme. *Physical Review A, 65*(3), 032302.

Lorenz, R. D., Wall, S., Radebaugh, J., Boubin, G., Reffet, E., Janssen, M., . . . West, R. (2006). The sand seas of Titan: Cassini RADAR observations of longitudinal dunes. *Science, 312*(5774), 724–727.

Lothenbach, B., Krebs, R., Furrer, G., Gupta, S. K., & Schulin, R. (1998). Immobilization of cadmium and zinc in soil by Al-montmorillonite and gravel sludge. *European Journal of Soil Science, 49*(1), 141–148.

Luo, H., Li, J., Yang, J., Wang, H., Yang, Y., Huang, H., Shi, P., Yuan, T., Fan, Y., & Yao, B. A. (2009). thermophilic and acid stable family-10 xylanase from the acidophilic fungus Bispora sp. MEY-1. *Extremophiles, 13*, 849–857.

Luo, H., Wang, K., Huang, H., Shi, P., Yang, P., & Yao, B. (2012). Gene cloning, expression and biochemical characterization of an alkali-tolerant b-mannanase from Humicola insolens Y1. *Journal of Industrial Microbiology & Biotechnology, 39*, 547–555.

Maliszewska-Kordybach, B., & Smreczak, B. (2003). Habitat function of agricultural soils as affected by heavy metals and polycyclic aromatic hydrocarbons contamination. *Environment International, 28*(8), 719–728.

Mamo, G., Thunnissen, M., Hatti-Kaul, R., & Mattiasson, B. (2009). An alkaline active xylanase: Insights into mechanisms of high pH catalytic adaptation. *Biochimie, 91*, 1187–1196.

Marinescu, M., Dumitru, M., & Lăcătuşu, A. R. (2009). Biodegradation of petroleum hydrocarbons in an artificial polluted soil. *Research Journal of Agricultural Science, 41*(2), 157–162.

Marques, S., Alves, L., Ribeiro, S., Girio, F. M., & Amaralcollaco, M. T. (1998). Characterisation of a thermotolerant and alkalotolerant xylanase from a Bacillus sp. *Applied Biochemistry and Biotechnology, 73*, 159–172.

Mathieu, J., Rossi, J. P., Mora, P., Lavelle, P., Martins, P. D. S., Rouland, C., & Grimaldi, M. (2005). Recovery of soil macrofauna communities after forest clearance in Eastern Amazonia, Brazil. *Conservation Biology, 19*(5), 1598–1605.

McKnight, D. M., Boyer, E. W., Westerhoff, P. K., Doran, P. T., Kulbe, T., & Andersen, D. T. (2001). Spectrofluorometric characterization of dissolved organic matter for indication of precursor organic material and aromaticity. *Limnology and Oceanography, 46*(1), 38–48.

Meagher, R. B. (2000). Phytoremediation of toxic elemental and organic pollutants. *Current Opinion in Plant Biology, 3*(2), 153–162.

Megharaj, M., Avudainayagam, S., & Naidu, R. (2003). Toxicity of hexavalent chromium and its reduction by bacteria isolated from soil contaminated with tannery waste. *Current Microbiology, 47*(1), 0051–0054.

Mengel, K., & Kirkby, E. A. (1982). *Principles of plant nutrition*. International Potash Institute.

Menn, B., Massberg, D., & Lorentz, S. (2008). Sandflies and leishmaniasis in Germany. *Parasitology Research, 103*.

Mishra, A., & Kumar, S. (2009). Kinetic studies of laccase enzyme of Coriolus versicolor MTCC 138 in an inexpensive culture medium. *Biochemical Engineering Journal, 46*(3), 252–256.

Modaihsh, A. S., Ai-Swailem, M. S., & Mahjoub, M. O. (2004). Heavy metals content of commercial inorganic fertilizers used in the Kingdom of Saudi Arabia. *Journal of Agricultural and Marine Sciences [JAMS], 9*(1), 21–25.

Nannipieri, P. (1994). *The potential use of soil enzymes as indicators of productivity, sustainability and pollution. Soil biota: Management in sustainable farming systems* (pp. 238–244).

Narendra, V., Rocha, P. P., An, D., Raviram, R., Skok, J. A., Mazzoni, E. O., & Reinberg, D. (2015). CTCF establishes discrete functional chromatin domains at the Hox clusters during differentiation. *Science, 347*(6225), 1017–1021.

Narendra, V., Bulajić, M., Dekker, J., Mazzoni, E. O., & Reinberg, D. (2016). CTCF-mediated topological boundaries during development foster appropriate gene regulation. *Genes & Development, 30*(24), 2657–2662.

Nigam, P., & Pandey, A. (Eds.). (2009). *Biotechnology for agro-industrial residues utilisation* (pp. 1–466). Springer Science Business Media.

Nigam, P., & Prabhu, K. A. (1991). Effect of cultural factors on cellulase biosynthesis in submerged bagasse fermentation by basidiomycetes cultures. *Journal of Basic Microbiology, 31*, 285–292.

Nigam, P. S. N., & Prabhu, K. A. (1988). Isolation and recovery of cellulase and ligninase from crude enzymes produced by two basidiomycetes cultures in submerged bagasse fermentation. *Sharkara, 27*(14)), 40–46.

Nriagu, J. O. (1996). A history of global metal pollution. *Science, 272*(5259), 223–223.
Pandey, A., Selvakumar, P., Soccol, C. R., & Nigam, P. (1999). Solid-state fermentation for the production of industrial enzymes. *Current Science, 77*, 149–162.
Pandey, A., Nigam, P., Soccol, C. R., Soccol, V. T., Singh, D., & Mohan, R. (2000). Advances in microbial amylases. *Biotechnology and Applied Biochemistry, 31*(2), 135–152.
Pascual, J. A., Hernandez, T., Garcia, C., & Ayuso, M. (1998). Enzymatic activities in an arid soil amended with urban organic wastes: Laboratory experiment. *Bioresource Technology, 64*(2), 131–138.
Pflug, W. (1982). Effect of clay minerals on the activity of polysaccharide cleaving soil enzymes. *Zeitschrift für Pflanzenernährung und Bodenkunde, 145*(5), 493–502.
Polizeli, M. L., Rizzatti, A. C., Monti, R., Terenzi, H. F., Jorge, J. A., & Amorim, D. S. (2005). Xylanases from fungi: Properties and industrial applications. *Applied Microbiology and Biotechnology, 67*, 577–591.
Pregitzer, K. S., Zak, D. R., Curtis, P. S., Kubiske, M. E., Teeri, J. A., & Vogel, C. S. (1995). Atmospheric CO2, soil nitrogen and turnover of fine roots. *New Phytologist, 129*(4), 579–585.
Ramasamy, S., Singh, S., Taniere, P., Langman, M. J. S., & Eggo, M. C. (2006). Sulfidedetoxifying enzymes in the human colon are decreased in cancer and upregulated in differentiation. *American Journal of Physiology-Gastrointestinal and Liver Physiology, 291*(2), G288–G296.
Rastogi, R. P., & Sinha, R. P. (2009). Biotechnological and industrial significance of cyanobacterial secondary metabolites. *Biotechnology Advances, 27*(4), 521–539.
Reddivari, M., Chirumamilla, R., & Nigam, P. (2002). Understanding lipase stereoselectivity. *World Journal of Microbiology and Biotechnology, 18*, 81–97.
Renella, G., Ortigoza, A. R., Landi, L., & Nannipieri, P. (2003). Additive effects of copper and zinc on cadmium toxicity on phosphatase activities and ATP content of soil as estimated by the ecological dose (ED50). *Soil Biology and Biochemistry, 35*(9), 1203–1210.
Reyes-Betancort, J. A., Guerra, A. S., Guma, I. R., Humphries, C. J., & Carine, M. A. (2008). Diversity, rarity and the evolution and conservation of the Canary Islands endemic flora. *In Anales del Jardín Botánico de Madrid, 65*(1), 25–45.
Riva, S. (2013). The long wave of biocatalysis. *Trends in Biotechnology, 31*, 120–121.
Robinson, T., & Nigam, P. (2008). Remediation of textile dye-waste water using a white rot fungus Bjerkandera adusta through solid state fermentation (SSF). *Applied Biochemistry and Biotechnology, 151*, 618–628.
Robinson, T., Chandran, B., & Nigam, P. (2001). Studies on the production of enzymes by white-rot fungi for the decolourisation of textile dyes. *Enzyme and Microbial Technology, 29*(8–9), 575–579.
Sabiha-Javied, M. T., Chaudhry, M. M., Tufail, M., & Irfan, N. (2009). *Microchemical Journal, 91*, 94–99.
Sanadi, D. R. (1982). Mitochondrial coupling factor B: Properties and role in ATP synthesis. *Biochimica et Biophysica Acta (BBA)-Reviews on Bioenergetics, 683*(1), 39–56.
Sardar, K., Ali, S., Hameed, S., Afzal, S., Fatima, S., Shakoor, M. B., . . . Tauqeer, H. M. (2013). Heavy metals contamination and what are the impacts on living organisms. *Greener Journal of Environmental management and public safety, 2*(4), 172–179.
Sayler, G. S., & Ripp, S. (2000). Field applications of genetically engineered microorganisms for bioremediation processes. *Current Opinion in Biotechnology, 11*(3), 286–289.
Senwo, Z. N., & Tabatabai, M. A. (1999). Aspartase activity in soils: Effects of trace elements and relationships to other amidohydrolases. *Soil Biology and Biochemistry, 31*(2), 213–219.
Sharma, P., Jha, A. B., Dubey, R. S., & Pessarakli, M. (2012). Reactive oxygen species, oxidative damage, and antioxidative defense mechanism in plants under stressful conditions. *Journal of Botany, 2012*.
Shen, M., Yang, X., Cox-Foster, D., & Cui, L. (2005). The role of varroa mites in infections of Kashmir bee virus (KBV) and deformed wing virus (DWV) in honey bees. *Virology, 342*(1), 141–149.

Shuyun, W., Yimin, L., Xu, L., Guojun, H., Yanlong, G., Qiang, W., & Fangtao, T. (2013). Preparation and properties of boron carbide film using pulsed laser deposition. *强激光与粒子束, 25*(04), 895–897.

Silva, R. R., Corso, C. R., & Matheus, D. R. (2010). Effect of culture conditions on the biomass determination by ergosterol of Lentinus crinitus and Psilocybe castanella. *World Journal of Microbiology and Biotechnology, 26*, 841–846.

Silveira, M. L., Miyittah, M. K., & O'Connor, G. A. (2006). Phosphorus release from a manure impacted Spodosol: Effects of a water treatment residual. *Journal of Environmental Quality, 35* (2), 529–541. https://doi.org/10.2134/jeq2005.0139

Sivaramakrishnan, S., Gangadharan, D., Nampoothiri, K. M., Soccol, C. R., & Pandey, A. (2006). α-amylases from microbial sources – An overview on recent developments. *Food Technology and Biotechnology, 44*, 173–184.

Song, J. J., Smith, S. K., Hannon, G. J., & Joshua-Tor, L. (2004). Crystal structure of Argonaute and its implications for RISC slicer activity. *Science, 305*(5689), 1434–1437.

Speir, T. W., Kettles, H. A., Percival, H. J., & Parshotam, A. (1999). Is soil acidification the cause of biochemical responses when soils are amended with heavy metal salts? *Soil Biology and Biochemistry, 31*(14), 1953–1961.

Stevenson, F. J. (1994). *Humus chemistry: genesis, composition, reactions*. John Wiley & Sons.

Sunnotel, O., & Nigam, P. (2002). Pectinolytic activity of bacteria isolated from soil and two fungal strains during submerged fermentation. *World Journal of Microbiology and Biotechnology, 18*, 835–839.

Tang, C., & Yu, Q. (1999). Impact of chemical composition of legume residues and initial soil pH on pH change of a soil after residue incorporation. *Plant and Soil, 215*(1), 29–38.

Taylor, H. F. (1997). *Cement chemistry* (Vol. 2, p. 459). Thomas Telford.

Tejada, M., Gonzalez, J. L., García-Martínez, A. M., & Parrado, J. (2008a). Effects of different green manures on soil biological properties and maize yield. *Bioresource Technology, 99*(6), 1758–1767.

Tejada, M., Gonzalez, J. L., García-Martínez, A. M., & Parrado, J. (2008b). Application of a green manure and green manure composted with beet vinasse on soil restoration: Effects on soil properties. *Bioresource Technology, 99*(11), 4949–4957.

Tietjen, T. E., & Wetzel, R. G. (2003). Seasonal and spatial distribution of bacterial biomass and the percentage of viable cells in a reservoir of Alabama. *Journal of Plankton Research, 25*(12), 1521–1534.

Trincone, A. (2010). Potential biocatalysts originating from sea environments. *Journal of Molecular Catalysis B: Enzymatic, 66*, 241–256.

Trincone, A. (2013). Angling for uniqueness in enzymatic preparation of glycosides. *Biomolecules, 3*, 334–350.

Uthandi, S., Saad, B., Humbard, M. A., & Maupin-Furlow, J. A. (2010). LccA, an archaeal laccase secreted as a highly stable glycoprotein into the extracellular medium by Haloferax volcanii. *Applied and Environmental Microbiology, 76*, 733–743.

Vasanthavigar, M., Srinivasamoorthy, K., Vijayaragavan, K., Rajiv Ganthi, R., Chidambaram, S., Anandhan, P., ... Vasudevan, S. (2010). Application of water quality index for groundwater quality assessment: Thirumanimuttar sub-basin, Tamilnadu, India. *Environmental Monitoring and Assessment, 171*(1), 595–609.

Vig, K., Megharaj, M., Sethunathan, N., & Naidu, R. (2003). Bioavailability and toxicity of cadmium to microorganisms and their activities in soil: A review. *Advances in Environmental Research, 8*(1), 121–135.

Vijayalakshmi, S., Venkat Kumar, S., & Thankamani, V. (2011). Optimization and cultural characterization of Bacillus RV. B2.90 producing alkalophilic thermophilic protease. *Research Journal of Biotechnology, 6*, 26–32.

Vineis, P., & Xun, W. (2009). The emerging epidemic of environmental cancers in developing countries. *Annals of Oncology, 20*(2), 205–212.

Wang, L., & Qu, J. J. (2007). NMDI: A normalized multi-band drought index for monitoring soil and vegetation moisture with satellite remote sensing. *Geophysical Research Letters, 34*(20).

Wang, P. S., Aguilar-Gaxiola, S., Alonso, J., Angermeyer, M. C., Borges, G., Bromet, E. J., … Wells, J. E. (2007). Use of mental health services for anxiety, mood, and substance disorders in 17 countries in the WHO world mental health surveys. *The Lancet, 370*(9590), 841–850.

Wang, J., Liu, Y., Jiao, F., Lao, F., Li, W., Gu, Y., … Chen, C. (2008). Time-dependent translocation and potential impairment on central nervous system by intranasally instilled TiO2 nanoparticles. *Toxicology, 254*(1–2), 82–90.

Wei, H., Li, B., Li, J., Dong, S., & Wang, E. (2008). DNAzyme-based colorimetric sensing of lead (Pb2+) using unmodified gold nanoparticle probes. *Nanotechnology, 19*(9), 095501.

Weinstein, R. A., Choukas, N. C., & Wood, W. S. (1974). Cancrum oris-like lesion associated with acute myelogenous leukemia. *Oral Surgery, Oral Medicine, Oral Pathology, 38*(1), 10–14.

Wongwilaiwalin, S., Rattanachomsri, U., Laothanachareon, T., Eurwilaichitr, L., Igarashi, Y., & Champreda, V. (2010). Analysis of a thermophilic lignocellulose degrading microbial consortium and multi-species lignocellulolytic enzyme system. *Enzyme and Microbial Technology, 47*, 283–290.

Wood, T. M., & Garcia-Campayo, V. (1990). Enzymology of cellulose degradation. *Biodegradation, 1*(2), 147–161.

Wuana, R. A., & Okieimen, F. E. (2011). *Heavy metals in contaminated soils: A review of sources, chemistry, risks and best available strategies for remediation*. International Scholarly Research Notices.

Yang, L., & Han, J. (2000). Electronic structure of deformed carbon nanotubes. *Physical Review Letters, 85*(1), 154.

Yang, R. Y., Chang, L. C., Hsu, J. C., Weng, B. B., Palada, M. C., Chadha, M. L., & Levasseur, V. (2006). Nutritional and functional properties of Moringa leaves–from germplasm, to plant, to food, to health. In C. D. E. Moringanews (Ed.), *Moringa leaves: Strategies, standards and markets for a better impact on nutrition in Africa* (pp. 1–9). CTA, GFU.

Zeng, F., Hou, C., Wu, S., Liu, X., Tong, Z., & Yu, S. (2007). Silver nanoparticles directly formed on natural macroporous matrix and their antimicrobial activities. *Nanotechnology, 18*(5), 055605.

Zhang, Y. H., Hu, M., Zhong, L. J., Wiedensohler, A., Liu, S. C., Andreae, M. O., … Fan, S. J. (2008). Regional integrated experiments on air quality over Pearl River Delta 2004 (PRIDE-PRD2004): Overview. *Atmospheric Environment, 42*(25), 6157–6173.

Zhou, D. M., Nigam, P., Marchant, R., & Jones, J. (1995). Production of salicylate hydroxylase from Pseudomonas putida UUC-1 and its application in the construction of biosensor. *Journal of Chemical Technology and Biotechnology, 64*, 331–338.

Żyszkowska, M., & Paszkiewicz-Jasińska, A. (2011). An assessment of the natural value of selected meadow-pasture communities in the middle Sudetes region. *Journal of Water and Land development*, 53–64.

Communities of Microbial Enzymes and Biodegradation of Persistent Environmental Pollutants

Oluwafemi Adebayo Oyewole, Muhammed Muhammed Saidu, Abdullahi Dabban Idris, Japhet Gaius Yakubu, and Aisha Bisola Bello

Abstract Enzymes are biocatalysts that potentiate the rate of substrate conversion into products. They are composed of amino acids with one or more polypeptide moieties. Microbial enzymes are the various enzymes of microorganisms' source, which have wide scope of applications in medicine and industries, including the degradation of persistent environmental wastes. Persistent environmental pollutants have become a global environmental and health concern. Owing to the rapid technological advancement and development in industries, large quantities of persistent environmental pollutants are being let out into the ecosystem posing serious threats to living organisms, thereby deteriorating the environment. Several microbial enzymes are widely used in the decomposition of recalcitrant organic and inorganic wastes. Oxidoreductases and hydrolases constitute the major class of microbial enzymes utilized in biodegradation of environmental pollutants; oxygenases, laccases, and peroxidases are the superfamilies of the oxidoreductase class, whereas lipases, cellulases, and proteases constitute the superfamilies of hydrolytic enzymes widely employed for bioremediation. Bioremediation involves the use of enzymes of microbial origin or the whole cell in the breakdown or transformation of environmental pollutants into less toxic or nontoxic products. Polymeric compounds such as polyethylene, polypropylene, polystyrene, polyvinyl chloride (PVC), polyurethane (PUR), and polyethylene terephthalate (PET) have been degraded using microbial enzymes. The biodegradation process is, however, often impeded due to the incapability of microbial enzymes to hydrolyze the functional groups present.

O. A. Oyewole (✉) · M. M. Saidu · J. G. Yakubu
Department of Microbiology, Federal University of Technology Minna, Minna, Nigeria
e-mail: oa.oyewole@futminna.edu.ng; muhammad.m1603943@st.futminna.edu.ng

A. D. Idris · A. B. Bello
Department of Biological Sciences, The Federal Polytechnic Bida Niger State, Bida, Nigeria

N. R. Maddela et al. (eds.), *Ecological Interplays in Microbial Enzymology*,
Environmental and Microbial Biotechnology,
https://doi.org/10.1007/978-981-19-0155-3_12

1 Introduction

The conjoined effect of the upsurge in human population and the growing industrial sector has mounted pressure on the limited natural resources (land, air, and water). The environment has not been adequately and judiciously managed and maintained since the industrial and technological development. Therefore, there have been indiscriminate increases in environmental pollution globally as a result of the increasing population, rapid industrialization, and other anthropogenic factors including agricultural practices that are potentially hazardous. These have led to the accumulation of a variety of persistent environmental pollutants such as polymeric compounds, organic wastes, industrial effluents, heavy metal-containing wastes, and other inorganic wastes. These pollutants often contain traces of heavy metals (such as chromium, lead, cadmium, nickel, iron, zinc, copper), petroleum products, pesticides, and various organic compounds such as hydrocarbons, organophosphorus, and phenolic compounds (Megharaj et al., 2011).

Persistent environmental pollutants are highly lipophilic compounds, which are potentially harmful to humans and affect the stability of the ecosystem when they biomagnify through the food chain. The level of environmental quality greatly influences the safety and well-being of every living organism in that environment. Persistent environmental pollutants have become a global environmental and health concern (Guo et al., 2019). Science and technological advancements coupled with the industrial revolution have, however, led to large quantities of recalcitrant pollutants with varying degrees of toxicity including organic and radioactive waste being effluxed into the environment, thereby endangering several life forms and affecting the quality of the environment. An effective means of getting rid of these persistent pollutants is of immense significance (Aislabie et al., 2005; Guo et al., 2019). In ancient times, wastes were usually eradicated from the environment by burying them in soil. Owing to the inadequacy of new places to dump and bury wastes, this technique of waste treatment became obsolete. With the advent of new techniques, the method of waste burying was gradually replaced with techniques like use of elevated temperatures to incinerate waste and chemical decomposition of waste, examples of which include ultraviolet (UV) oxidation and base-catalyzed dechlorination. These technologies, despite their effectiveness at reducing pollutants, have their own drawbacks. They are not environmentally friendly, are uneconomical and complex, and are not generally acceptable. The use of microbial enzymes to bioremediate pollutants is now seen as a suitable alternative to elevated temperature and use of chemicals for waste eradication.

Degradation of environmental pollutants by making use of microorganisms or microbial enzymes has gained increasing recognition over the years as a promising technique for waste removal. Generally, bioremediation approaches of waste removal are environmental transformations, in the form of supply of nutrients (biostimulation) and oxygen (bioventing or biosparging) and introduction of effective degraders (bioaugmentation). The enzymatic degradation of recalcitrant environmental pollutants offers numerous merits and has high preference over the physical and chemical waste remediation approach, particularly for less toxic and large quantities of wastes. One of the most attractive advantages of the bioremediation strategy is the in situ treatment, which entails treating the contaminated soil at the contamination site without transporting the contaminated material. Ryan et al. (1991) recounted that several reports have been documented on the successful treatment of petroleum-contaminated sites using microbial enzymes. Microbial enzymes have been extensively utilized in the management and treatment of hazardous wastes such as polychlorinated biphenils, trichloroethylene, and benzene, toluene, ethylene, and xylene (BTEX).

2 Biodegradation of Persistent Environmental Pollutants

Biodegradation is the breakdown, disintegration, or transformation of pollutants into nontoxic or less toxic substances by the metabolic or enzymatic action of microorganisms. The use of various microorganisms in the degradation of a wide variety of environmental wastes has been documented (Vidali, 2001; Leung, 2004). The biodegradation process is solely dependent on the enzymatic action of microorganisms to biotransform pollutants into nontoxic or innocuous substances. The effectiveness of bioremediation is a measure of the suitability of environmental and nutritional conditions for the proliferation and activity of microorganisms, and as such oftentimes requires the conditions to be manipulated to achieve a faster rate of pollutant degradation.

Various microorganisms having numerous hereditary determinants such as extrachromosomal DNA and catabolic genes adapt to adverse environmental conditions by genetic recombination, duplication, mutation, and hypermutation. Chakraborty and Das (2016) reported that the microbial metagenome constitutes the largest genetic reservoir with miscellaneous enzymatic activities implicated in biodegradation. Some major persistent environmental pollutants catabolic genes involve in biodegradation include biphenyl dioxygenase for degradation of biphenyl, 2,3-dioxygenase for breaking down of organochlorine pesticides and angular dioxygenase for degradation of dioxins or furans (Chakraborty & Das, 2016). However, bioremediation is a gradual process, with only few strains of bacteria and fungi proven to be potent pollutant degraders, while others could only be effective under laboratory conditions.

2.1 *Principles of Biodegradation*

Biodegradation involves the disintegration and detoxification of toxic environmental pollutants by the action of living organisms such as plants, animals, and microorganisms (including bacteria, fungi, and algae), which biodegrade or detoxify and transform the toxic environmental pollutants into products such as CO_2, H_2O, and metabolites, which are mostly innocuous or less toxic than the original compound. During bioremediation, the degrading species can be from the site of contamination or may be introduced from elsewhere to the site for bioremediation through the process of bioaugmentation. Microorganisms obtain energy for their growth and development by degrading or transforming these pollutants through enzymatic action.

The rate of biodegradation is influenced by a complex of interwoven factors, which are population density of the degrading species, environmental conditions, the complexity of the pollutants, and, particularly, the bioavailability of the pollutants to the organisms; therefore, the manipulation of environmental conditions is immensely important to bring about an increase in microbial growth rate and thus fast and effective biodegradation.

2.2 Limiting Factors of Bioremediation

Bacterial growth, which ultimately determines the rate of bioremediation, is often limited by various factors such as:

- pH
- Oxygen
- Temperature
- Poor bioavailability of pollutants
- Moisture
- Soil structure
- Inadequate nutrients
- Other bactericidal or bacteriostatic compounds may be present

Microorganisms can thrive in extreme environments; however, a vast majority of bacterial and fungal species prefer optimal conditions of growth. Such a situation is not easily set up and maintained in the field or site (Dua et al., 2002; Dana & Bauder, 2011). The bioremediation process usually takes place under aerobic conditions, although microbial degradation of recalcitrant pollutants can also occur in anaerobic environments. The involvement of various microbial enzymes is extremely crucial for degradation of persistent lignin and organic pollutants by bacteria and fungi, respectively (Vidali, 2001; Lehninger et al., 2004).

2.3 Microbial Enzymes in Bioremediation

Enzymes are biochemical substances that catalyze the rate of substrate transformation into products by lowering the energy of activation of the reaction. Enzymes are usually composed of several amino acids; hence, they are proteinous in nature with one or more polypeptide moieties. The catalytic site of an enzyme in which a reaction takes place is called the active site. A holoenzyme is composed of an apoenzyme (glycoprotein moiety of an enzyme) and a prosthetic group (nonprotein moiety of an enzyme).

Microbial enzymes are enzymes derived from microorganisms, bacterial, algal, or fungal species, with numerous uses in the field of medicine and in several industries. Microbial enzymes are widely used in the degradation of recalcitrant pollutants.

2.4 Sources of Microbial Enzymes

The application of microbial enzymes has gained attention over the years due to their versatility, specificity, selectivity (chemo–regio–enantio), and catalysis of diverse reactions. A number of enzymes are produced by various microorganisms that catalyze the degradation of complex natural polymeric compounds into simple ones. Microorganisms producing extracellular enzymes are ubiquitous in nature, and studies have reported their isolation from various environments. Pollutant-degrading strains are widespread in the environment including air, water, soil, sludge, industrial effluents, and also as normal flora in the human body. Several microbial enzymes known to degrade pesticides and hydrocarbons are produced by strains of *Mycobacterium*, *Alcaligenes*, *Sphingomonas*, and *Pseudomonas* amongst other aerobic bacterial species. Lipases, a group of microbial enzymes capable of degrading polyurethane and other recalcitrant pollutants, are produced by various microbes such as *Comamonas acidovorans* TB-35, *Curvularia senegalensis*, *Aureobasidium pullulans*, *Fusarium solani*, *Cladosporium* sp., *Pseudomonas chlororaphis*, *Pseudomonas stutzeri*, and *Pestalotiopsis microspora* (Roohi Kulsoom et al., 2017). Different bacterial and fungal species such as *Bacillus* sp., *Pseudomonas* sp., *Aspergillus nidulans*, *Aspergillus niger*, *Penicillium simplicissimum* YK, and *Enterobacter* sp. produce esterases, a class of hydrolases involved in the breakdown of polyvinyl chloride (PVC) and polyethylene amongst other plastics. The microbial flora, which secrete extracellular enzymes for the degradation of plastics, include *Streptococcus*, *Pseudomonas*, *Staphylococcus*, *Micrococcus*, and *Moraxella* among the bacterial species, and the fungal species are *Aspergillus niger* and *Aspergillus glaucus*, whereas *Actinomycetes* sp. and *Saccharomonos poragenus* constitute the yeast species. Chlorinated aliphatics including di- and trichloroethylene have also been degraded by certain aerobic methylotrophs. Complete degradation of chloroform, PCBs (polychlorinated biphenyls), and chlorinated solvents has been achieved by various extracellular enzymes produced by certain anaerobes (Roohi Kulsoom et al., 2017).

3 Classes of Microbial Enzymes in Bioremediation

The various classes of microbial enzymes include:

3.1 Microbial Oxidoreductases

Oxidoreductases in bacteria and fungi mediate the detoxification of toxic pollutants by oxidizing substrates (Gianfreda et al., 1999). Microorganisms derive carbon and energy by oxidizing the contaminants into completely harmless compounds; this is achieved using oxidoreductases to hydrolyze the linkages so as to aid electron transfer from the donor (usually the reduced substrate) to the recipient (another chemical compound).

The humification of various phenolic compounds obtained from degradation of lignin is catalyzed by oxidoreductases. In addition, various toxic xenobiotic compounds can be detoxified by oxidoreductases; detoxification of phenol or aniline-containing compounds is carried out by polymerizing the compound with other substrates or by linking the compounds to humic substances (Park et al., 2006). Degradation and decolorization of azo dyes by microbial enzymes have also been reported (Vidali, 2001; Husain, 2006). Chlorinated phenolic compounds are recalcitrant pollutants and the major constituents of wastes from the pulp and paper industry. During pulp bleaching process, incomplete breakdown of lignin results in the formation of chlorinated phenolic compounds, and they are removed from the contaminated environment by many fungal species using various extracellular oxidoreductases produced in the mycelium such as laccases, manganese peroxidases, and lignin peroxidases. Rubilar et al. (2008) observed that fungi degrade soil pollutants more readily than do bacteria due to their filamentous nature.

Oxidoreductases have the ability to catalyze the coupling of carbon to carbon and carbon to other elements like oxygen and nitrogen, and, sometimes, coupled with polymerization of carbon units and removal of methyl or halogen group, they have the ability to form reactive radicals that hydrolyze various chemical bonds. These reactions facilitate the purification and transformation of recalcitrant environmental pollutants by enzymes into nontoxic or less toxic substances that can be easily evicted from the environment in subsequent treatment procedures. Various studies have shown the potential applications of oxidoreductases mainly in bioremediation of soil pollutants and industrial wastewater remediation. However, intracellular enzymes commonly produced by most fungi, such as cytochrome P450 monooxygenase, may catalyze degradation of various organic pollutants (Bezalel et al., 1996); white rot fungi produce several ligninases, which are extracellularly active and so are better agents of bioremediation of highly polar pollutants (Field et al., 1993). Therefore, laccases and peroxidases are the main enzymes that have been used, whereas tyrosinases and catechol oxidases have been found to have limited applications in bioremediation.

Microbial Oxygenases

Oxygenases are microbial enzymes belonging to the oxidoreductases class of enzymes. They oxidize the reduced substrate by incorporating an oxygen molecule using flavin adenine dinucleotide (FAD)/nicotinamide adenine dinucleotide (NADH)/nicotinamide adenine dinucleotide phosphate (NADPH) as the cofactors. They are classified into two based on the number of oxygen atoms used to oxidize the substrate, namely, monooxygenases and dioxygenases. The most studied groups of microbial enzymes used in bioremediation are bacterial mono- and dioxygenases. They play important roles in the breakdown of organic compounds by their increased reactivity, high catalytic efficiency, and water solubility, to bring about the cleavage of the aromatic ring by introducing oxygen atoms into the organic compound. Oxygenases act on diverse group of substrates including chlorinated aliphatics. Oxygenases catalyze the degradation of numerous halogenated compounds, which constitute the largest group of persistent environmental pollutants owing to their widespread use as pesticides, hydraulic and heat transfer fluids, plasticizers, and intermediates for chemical synthesis. Dehalogenation of alkyl halides such as methyl halide, ethyl halide, and halogenated ethylene is catalyzed by oxygenases with conjoined action of multifunctional enzymes (Fetzner & Lingens, 1994).

Microbial Monooxygenases

Monooxygenases oxygenate substrates by adding an oxygen molecule to them. Monooxygenases, owing to their high regioselectivity and stereoselectivity on a broad range of substrates, function as biological catalysts in the bioremediation process. Monooxygenases catalyze various reactions such as removal of sulfur, halogen, and nitrate and incorporation of ammonia and hydroxyl group as well as transformation of degradation of aliphatic and aromatic compounds. Several researches have been carried out on monooxygenases for their high catalytic efficiency in biodegradation of aromatic compounds; the degradation of phenol by monooxygenase is shown in Fig. 1. Of all the monooxygenases, methane monooxygenase is the best and most used monooxygenase. The enzyme catalyzes the transformation and degradation of hydrocarbons such as tetrachloromethanes, alkanes, cycloalkanes, alkenes, haloalkenes, ethers, and aromatic and heterocyclic hydrocarbons (Fox et al., 1990; Grosse et al., 1999). Monooxygenases catalyze the oxidative dehalogenation reactions under aerobic conditions and reductive dechlorination at low oxygen levels.

Monooxygenases are subdivided into flavin-dependent monooxygenases and P450 monooxygenases on the basis of the cofactor present. Flavin is present in flavin-dependent monooxygenases as a prosthetic group, and, so, NADP or NADPH is required as a coenzyme. P450 monooxygenases are found in eukaryotes and prokaryotes and are known to contain heme. Most of the monooxygenases that

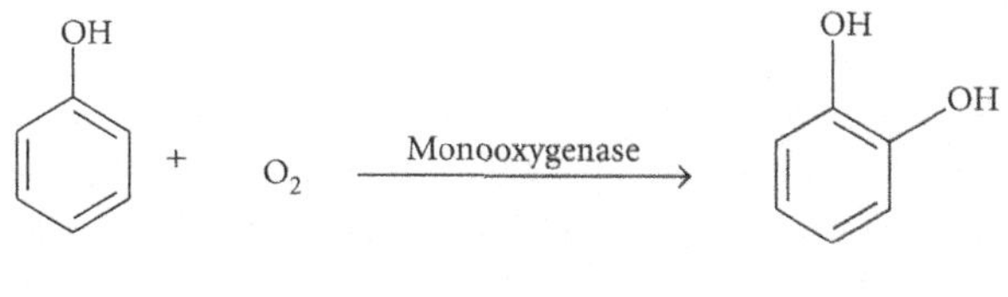

Fig. 1 Degradation of phenol by monooxygenase (Arora et al., 2010)

have been previously identified have a cofactor; however, certain monooxygenases have the unique ability of functioning independently of a cofactor. These monooxygenases only require molecular oxygen for their activities, utilizing the substrate as a reducing agent (Arora et al., 2010). Monooxygenases include a versatile superfamily of enzymes that are involved in the oxidation of a diverse group of substrates ranging from simple hydrocarbons as alkanes to steroids, fatty acids, and other complex molecules.

Microbial Dioxygenases

Dioxygenases oxidize substrates by incorporating oxygen molecule into the substrate. Dioxygenases are a group of Rieske nonheme iron oxygenases. They are involved in the oxidation of a broad range of substrates, mainly aromatic compounds, and, so, they have important application in bioremediation of environmental pollutants. Dioxygenases contain one or two electron transport proteins, which precede their oxygenase components. The presence of the Rieske (2Fe–2S) cluster and mononuclear iron in the crystal structure of naphthalene dioxygenase has been confirmed in each alpha subunit (Dua et al., 2002).

Catechol dioxygenases form part of nature's strategy for degrading aromatic molecules in the environment. These enzymes are present in various soil bacterial species, and they transform aromatic precursors into aliphatic products. The enzymes that cleave intradiol use Fe(III), whereas the enzymes that cleave extradiol make use of Fe(II) and Mn(II), as illustrated in Fig. 2 (Que & Ho, 1996).

Microbial Laccases

Laccases are a group of multicopper oxidase enzymes widely distributed in higher plants, insects, and especially found in various bacterial and fungal species. They catalyze the reduction of oxygen molecules to water and oxidation of reduced phenolic compounds (Gianfreda et al., 1999; Mai et al., 2000). Microbial laccases

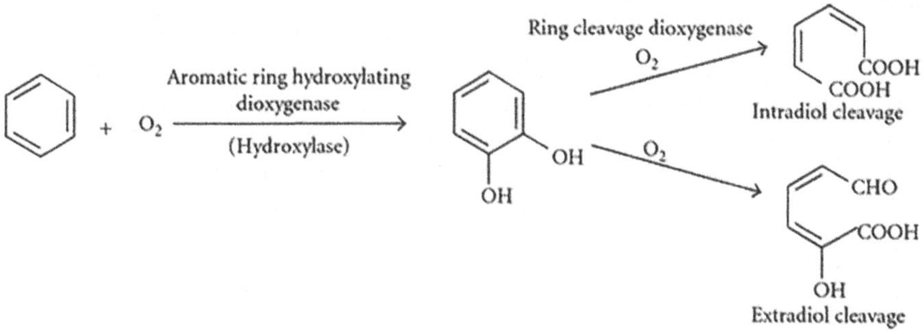

Fig. 2 Degradation of aromatic compounds by dioxygenase (Que & Ho, 1996; Arora et al., 2009)

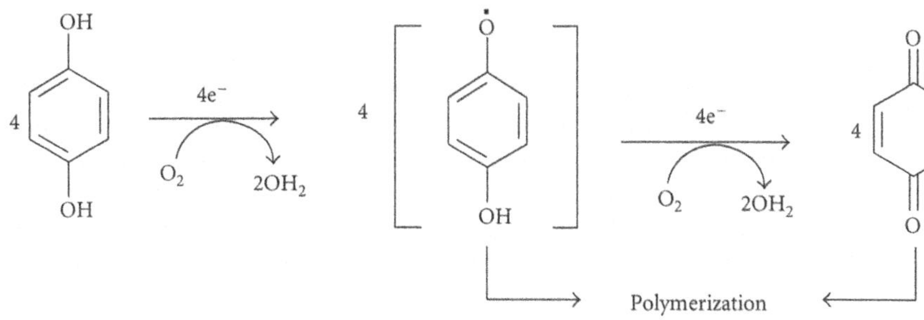

Fig. 3 General reaction mechanism for phenol oxidation by laccase (Dedeyan et al., 2000)

are a glycosylated polyphenol group of oxidoreductases containing four Cu ions in each molecule, which catalyze the oxidation of phenols and other aromatic compounds. Each polymeric laccase molecule has type 1, type 2, and type 3 Cu subunits, of which type 2 and type 3 form a trinuclear Cu cluster. Certain bacterial and fungal species produce numerous intracellular and extracellular laccases, which have the potential to oxidize O- and P-diphenol, aminophenols, polyphenols, polyamines, lignins, and aryl diamines as well as some inorganic ions as depicted in Fig. 3 (Ullah et al., 2000; Couto & Toca Herrera, 2006). Isolation of laccases has been reported from various species of Ascomycetes, Deuteromycetes, and Basidiomycetes fungi having more than 60 fungal strains. Laccases play a vital role in biodegradation of phenolic pollutants and removal of endocrine disruptors (Couto & Toca Herrera, 2006). Laccases are widely used for breaking down lignin into pulp, degradation of insecticides and pesticides, organic synthesis, detoxification of wastes, and transformation of textile dyes. When a laccase oxidizes a compound, a single electron is lost often in the form of a free radical, which may undergo further oxidization or other nonenzymatic reactions such as hydration, disproportionation, and polymerization (Faccelo & Cruz, 2008). Oxidation, decarboxylation, and demethylation of phenolic and methoxyphenolic acids using microbial laccases have been reported. Lignins are also depolymerized by microbial laccases to yield a variety of phenols. Laccases express the unique ability to degrade a broad range of environmental pollutants and offer great potential for bioremediation applications (Gianfreda et al., 1999). The specificity and affinity of laccases to substrates vary with pH. Various reagents are capable of inhibiting laccase enzymes, such as halides (with the exception of iodide), azides, cyanides, and hydroxides (Xu, 1996).

Properties of Microbial Laccases

Laccases are glycoproteins composed of one, two, or numerous monomers. Glycosylation enhances the ability of laccases to retain copper, remain unchanged with temperature variation, and be susceptible to degradation and secretion. Laccases show considerable heterogeneity upon purification. The growth medium composition determines the glycoprotein composition and glycosylation content.

Sources of Microbial Laccases

The major sources of laccases are higher plants and fungi; however, various laccases have been recently isolated from different bacterial species, particularly *Streptomyces* sp. and *Marinomonas mediterranea* (Arias et al., 2003; Jimenez-Juarez et al., 2005). Fungi, however, have been found to have more laccases than higher plants. Various Basidiomycetes like *Phanerochaete chrysosporium*, *Theiophora terrestris*, and *Lenzites betulina* (Viswanath et al., 2008) and other species of white rot fungi (Kiiskinen et al., 2004a, b) such as *Phlebia radiata* (Niku-Paavola et al., 1998), *Pleurotus ostreatus* (Palmieri et al., 2000), and *Trametes versicolour* (Bourbonnais et al., 1995) have shown laccase activity. Laccases have also been isolated from many species of *Trichoderma* including *Trichoderma atroviride*, *Trichoderma harzianum* (Holker et al., 2002), and *Trichoderma longibrachiatum* (Velazquez-Cedéno et al., 2004). Laccase isolated from a species of Ascomycetes (*Monocillium indicum*) was the first laccase enzyme to be studied and characterized, which showed peroxidase activity (Thakker et al., 1992). Laccase produced by *Pycnoporus cinnabarinus* has lignin-degrading potential, whereas laccase isolated from *Pycnoporus sanguineus* has the ability to oxidize phenolic compounds (Pointing & Vrijmoed, 2000). Contrary to the role of laccase in plants, which catalyzes lignification, laccase in fungi plays various roles in lignin breakdown, formation of spores, pigment and fruiting body, and causing of various plant diseases (Yaver et al., 2001).

Mechanism of Microbial Laccase Activity

Laccases catalyze reactions by reducing an oxygen molecule to water and oxidizing an electron with a broad range of aromatic compounds including polyphenols, methoxyphenols, and aromatic amines. Laccases have copper atoms depicted as Cu T1 (to which the substrate binds) and a trinuclear copper cluster T2/T3 (shuttling of electron between the three Cu ions that results in oxygen reduction to water) (Gianfreda et al., 1999). The Cu ions are grouped into three types: Type 1 (T1), Type 2 (T2), and Type 3 (T3). Electron paramagnetic resonance (EPR) spectroscopy and UV light are commonly used to differentiate between the three types. A trinuclear center is formed by Type 2 and Type 3, which catalyzes various reaction mechanisms. Asymmetric activation of the trinuclear center is brought about by the binding of the oxygen molecule to prevent the oxidizing agents from binding. The catalytic activity of laccase involves the reduction of oxygen in a steady state. Laccase serves as a power house that stores electrons for oxidation reactions to bring about reduction of the oxygen molecule. Therefore, for the complete reduction of molecular oxygen to water to occur, four reducing substrate molecules must be oxidized by laccase enzymes. Free radicals are generated when laccase oxidizes the substrate.

Catalysis of the substrate mediated by laccase is applied to non-phenolic compounds with the addition of a mediator. A mediator is an organic compound with low molecular weight oxidized by laccase. The highly active cation radicals oxidize the non-phenolic compounds that laccase alone cannot oxidize. 1-Hydroxy benzotriazole (HOBT), *N*-hydroxyphthalimide (NHPI), 2,2-azinobis-(3-ethylbenzothiazoline-6-sulfonate) (ABTS), and 3-hydroxyanthranilic acid are the most common synthetic mediators (Gochev & Krastanov, 2007).

Production of Microbial Laccases

Laccases are produced extracellularly by several species of fungi during secondary metabolism, with the exception of Zygosaccharomyces and Chytridiomycetes (Morozova et al., 2007). Some species of soil and freshwater Ascomycetes have been reported to produce laccase (Junghanns et al., 2005). Moreover, laccase has been isolated from various organisms including *Gaeumannomyces graminis*, *Magnaporthe grisea*, *Melanocarpus albomyces*, *Monocillium indicum*, *Neurospora crassa*, *Ophiostoma novo-ulmi*, and *Podospora anserina* (Iyer & Chattoo, 2003; Palonen et al., 2003). A dimethoxyphenol oxidizing enzyme produced by *Botryosphaeria* is a true laccase. In plant biomass decay, syringaldazine is oxidized by the laccase-producing Ascomycetes species (Lyons et al., 2003). Phenols and aminophenols are oxidized by laccase produced by a Basidiomycetes yeast (*Cryptococcus neoformans*), although tyrosine is unaffected. *Saccharomyces cerevisiae* produces oxidase with a membrane-bound multicopper plasma, which is homologous to fungal laccase (Stoj & Kosman, 2003). Studies have shown that the fungal species known for producing appreciable amounts of laccases in varying quantities are Basidiomycetes and saprotrophic fungi (Hatakka, 2001). Laccase is also produced by *Pycnoporus cinnabarinus* as the only ligninase that breaks down lignin (Eggert et al., 1996). Although the laccase-producing capacity of brown rot fungi remains largely unknown and no laccase undergoes purification, research has shown that *Coniophora puteana* oxidizes the syringaldazine (Lee et al., 2004) and assists in oxidizing ABTS in *Laetiporus sulphureus* (Schlosser & Hofer, 2002). Gayazov and Rodakiewicz-Nowak (1996) concluded that laccase production is usually influenced by a number of factors, which include the cultivation type (either submerged or solid-state), limiting factor, and carbon and nitrogen source.

Microbial Peroxidases

Peroxidases belong to a subclass of oxidoreductases that catalyzes the reduction of peroxide compounds such as hydrogen peroxide (H_2O_2) and also oxidize several organic as well as inorganic compounds. Peroxidases are a widely available group of enzymes derived from numerous sources, which bring about degradation of lignin and several phenolic compounds by reducing the hydrogen peroxide in the presence of a mediator. Toxic compounds such as ferricyanide and ascorbate are also degraded by peroxidases to yield nontoxic compounds by donating electrons that bind to the substrate (Hamid & Rehman, 2009). Peroxidases are either heme or nonheme proteins. Peroxidases also regulate hormone and immune responses in mammals. Peroxidases are widely used to reduce pollutants, including remediation of industrial effluents heavily laden with phenols, cresols, and chlorinated phenols, and also in degrading and decolorizing synthetic dyes. Extracellular peroxidase secreted by white rot fungi catalyzes the degradation of lignin by the unspecific free radical, which brings about oxidation reactions (Lundell et al., 2010). Peroxidases oxidize various compounds including amines, dimethoxybenzene, lignin, phenols, and several other aromatic alcohols without the presence of a mediator (Mn(II)). Phenolic as well as non-phenolic compounds are also oxidized by peroxidases; a dye-decolorizing peroxidase from *Agaricus* was found to oxidize dyes and phenolic compounds (Hofrichter et al., 2010). Studies have shown great potential of *Phanerochaete chrysosporium* to degrade a broad range of pollutants (including dioxins, PCBs, hydrocarbon compounds, industrial effluents, pesticides, and trinitrotoluene commonly used in making munitions), which is attributed to various peroxidases, which are nonspecific in activity (Marco-Urrea & Reddy, 2012). Heme peroxidases are broadly classified into two categories; the first category is found only in animals, plants, fungi, and prokaryotes, whereas the second category has three subclasses based on sequence comparison. Class I includes intracellular enzymes like yeast cytochrome c peroxidase, ascorbate peroxidase from plants, and catalase peroxidase from bacterial duplicated genes. Class II consists of peroxidases secreted by fungi such as lignin peroxidase (LiP) and manganese peroxidase (MnP) by *Phanerochaete chrysosporium* and *Coprinus cinereus* peroxidase or *Arthromyces ramosus* peroxidase (ARP). Class II peroxidases are responsible for lignin degradation in wood. Class III includes peroxidases of plant origin including from horseradish (HRP), barley, or soybean. These peroxidases are biosynthetic enzymes that catalyze various processes in plants such as plant cell wall formation and lignin formation (Hiner et al., 2002; Koua et al., 2009).

There is no evolutionary link between nonheme peroxidases, so they constitute five broad independent families, which include thiol peroxidases, alkylhydroperoxidases, nonheme haloperoxidases, manganese catalases, and NADH peroxidase. The largest of the five families is thiol peroxidase, which is further subdivided into two subfamilies, glutathione peroxidases and peroxiredoxins (Koua et al., 2009).

Sources of Microbial Peroxidases

Peroxidases are ubiquitous and are obtained from numerous sources including a variety of plants, animals, and microorganisms. Microbial peroxidases are isolated from bacteria, cyanobacteria, fungi, actinomycetes, and yeasts. The predominant peroxidase-producing bacterial species include *Bacillus* sp., *Pseudomonas* sp., and *Citrobacter* sp., whereas *Phanerochaete chrysosporium*, *Candida krusei*, and *Coprinopsis cinerea* are the main peroxidase-producing fungi, *Streptomyces* sp. and *Thermobifida fusca* constitute the *Actinomycetes* species, and cyanobacteria include various species of *Anabaena*. Yeasts are used in the biomineralization of pollutants, decolorization of dyes, feed production, as bioindicators, and are also used extensively as raw materials in the food, chemical, paper, and pulp industries for degradation of lignin, decolorization of textile dyes, and remediation of sewage.

Properties of Microbial Peroxidases

Peroxidases are oxidoreductases involved in the catalysis of various reactions, particularly peroxide reduction and also the oxidation of several organic as well as inorganic compounds. They are heme proteins with a prosthetic group containing iron (III) protoporphyrin IX. Peroxidases include various specific enzymes such as NADH peroxidase, glutathione peroxidase, and iodine peroxidase and other nonspecific enzymes.

Subclasses of Microbial Peroxidases

Microbial peroxidases are subdivided into many types based of their source and activity. The most studied microbial peroxidases include lignin peroxidase (LiP), manganese-dependent peroxidase (MnP), and versatile peroxidase (VP), attributed to their high catalytic ability and ubiquity.

Microbial Lignin Peroxidases

Lignin peroxidases (LiPs) are heme peroxidases that are produced by *Phanerochaete chrysosporium* during secondary metabolism. LiPs act on lignin along with several phenolic substrates in the presence of hydrogen peroxide as a cosubstrate and veratryl alcohol as the mediator. During the process, reduction of hydrogen peroxide to water occurs when it accepts an electron from LiP while the LiP gets oxidized. The LiP thereafter accepts an electron from veratryl alcohol and gets reduced to its original form, whereas veratryl alcohol forms veratryl aldehyde. Finally, the veratryl aldehyde accepts an electron from the substrate and gets reduced to veratryl alcohol. The overall process yields oxidized halogenated phenolic compounds or polycyclic or other aromatic compounds accompanied by a series of nonenzymatic reactions (Yoshida, 1998; Ten Have & Teunissen, 2001).

LiPs play an important role in the breakdown of the lignin in the plant cell wall. Aromatic compounds are also oxidized by lignin peroxidases having redox potentials higher than 1.4 V normal hydrogen electrode (NHE) by single-electron abstraction, although the exact mechanism of the redox reaction is not well understood (Piontek et al., 2001).

Microbial Manganese Peroxidases

Manganese peroxidases (MnPs) are extracellular heme peroxidases secreted by the lignin-degrading Basidiomycetes species of fungi. They oxidize the manganate (II) ion [Mn^{2+}] to manganate (III) ion [Mn^{3+}] in a series of reactions. Manganese peroxidase production is triggered by the manganate (II) ion, which acts as the substrate for manganese peroxidase. The manganate (III) ion formed as a result of oxidation of the manganate (II) ion mediates the oxidization of a variety of phenolic compounds (Ten Have & Teunissen, 2001).

Microbial Versatile Peroxidases

Versatile peroxidases (VPs) are enzymes that can directly oxidize manganate (II) ion, methoxybenzenes, phenolic aromatic substrates similar to lignin peroxidase, manganese peroxidase, and horseradish peroxidase, but, unlike other peroxidases, versatile peroxidases are capable of oxidizing these substrates without manganese as the mediator. Versatile peroxidase has exceptionally broad substrate specificity and is found to catalyze the oxidation of phenolic as well as non-phenolic compounds (Ruiz-Duenas et al., 2007). As such, large-scale production of versatile peroxidases is required in industrial processes for various biotechnological applications and biodegradation of recalcitrant wastes (Tsukihara et al., 2006; Wong, 2009).

Applications of Peroxidases in Degradation of Environmental Pollutants

Biodegradation of Synthetic Dyes

Synthetic dyes constitute a problematic and recalcitrant class of environmental pollutants, which are not easily degraded (Ong et al., 2011). Dyes have important applications in the textile, paper and pulp, and petroleum industries and in color photography, amongst others. These compounds greatly contribute to environmental pollution when released into industrial effluents. The white rot fungus is considered a valuable alternative to biodegradation of environmentally hazardous compounds. Oxidative enzymes such as laccase, LiP, and MnP produced by *Phanerochaete chrysosporium* are found to oxidize the substrates. Peroxidases and oxidases are used as efficient oxidizing agents, which degrade dyes. Synthetic textile dyes have been completely decolorized using several bacterial peroxidases. *Brevibacterium* has been used to completely remove chromate Cr (VI) and azo dye Acid Orange 7 (AO7) under nutrient-limiting conditions. The reducing enzyme of *Brevibacterium casei* uses the AO7 as an electron donor to reduce Cr (VI). A purple color complex intermediate is formed by the oxidized AO7 and the reduced Cr (III) (Ng et al., 2010). Studies have shown the unique ability of *Phanerochaete chrysosporium* RP 78 to decolorize several azo dyes under optimal conditions (Ghasemi et al., 2010). Dawkar et al. (2008) observed that the *Bacillus* sp. VUS from soil contaminated with textile effluents expressed the potential to degrade various dyes. Several other peroxidases have been isolated from microorganisms in addition to LiPs that biodegrade synthetic dyes. The decolorization of Remazol Brilliant Blue was achieved by the extracellular peroxidase isolated from *Pleurotus ostreatus* along with other synthetic dyes including triarylmethane, heterocyclic azo, and polymeric dyes. The decolorization of bromophenol blue is best at 98%, and methylene blue as well as toluidine blue O both decolorize best at 10%.

Bioremediation of Wastewater

Industrial pollution is a major global concern being one of the principal sources of environmental pollutants, which deteriorate surface and groundwater and affect the general well-being of the environment. Pure water serves as a deterrent to infectious agents, which are the causative agents of numerous waterborne diseases. Microbial peroxidases have been effectively used to bioremediate wastewater containing various aromatic compounds (Hamid & Rehman, 2009; Ong et al., 2011). The major constituents of wastewater of several industries include phenols, aromatic amines, and other aromatic compounds (Kaušpediene et al., 2010).

Peroxidases are a class of oxidoreductase enzymes involved in detoxification of a wide range of phenolic compounds through oxidative coupling reactions (Mui et al., 2010). *Phanerochaete chrysosporium* produces a lignin peroxidase, *Bjerkandera adusta* produces lactoperoxidase, a versatile peroxidase, and *Caldariomyces fumago* produces chloroperoxidase that catalyzes the oxidative dehalogenation of pentachlorophenol to tetrachloro-1,4-benzoquinone in the presence of H_2O_2. *Pleurotus eryngii* and *P. ostreatus* produce another versatile peroxidase that oxidizes Mn^{2+} into Mn^{3+}, in a similar manner as MnP, and like the action of LiP, it oxidizes the high redox potential of aromatic compounds and has broad specificity and also oxidizes non-phenolic compounds (Ruiz-Duenas et al., 2009).

Mechanism of the Horseradish–H_2O_2–PhOH Reaction

The reaction of horseradish peroxidase with phenolic compounds follows a cyclic path. The summary of the reactions is given as:

$$E + H_2O_2 \rightarrow Ei + H_2O, \tag{1}$$

$$Ei + PhOH' \rightarrow Eii + PhO, \tag{2}$$

$$Eii + PhOH'' \rightarrow E + PhO + H_2O. \tag{3}$$

The reaction begins with the original form of the enzyme (E), and, thereafter, the enzyme gets oxidized by H_2O_2 to yield compound A (Ei), an active intermediate compound. Compound A then oxidizes one phenol molecule (PhOH) to form its free radical (PhO), and, then, compound A becomes compound B (Eii). The second phenol molecule is further oxidized by compound B to yield another free radical of phenol (PhO). The enzyme returns to its original state (E) to complete the cycle. The free radicals (PhO) are polymerized to yield an insoluble precipitate (Mossallam et al., 2009). The equation is given as:

$$\text{PhO} + \text{PhO} \rightarrow \text{Aromatic polymer.} \tag{4}$$

Biodegradation of Polycyclic Aromatic Hydrocarbon (PAH) Pesticides

Pesticides include all classes of synthetic chemical compounds employed in the control of pest organisms such as insects (insecticides), rodents (rodenticides), birds (avicides), weeds (herbicides), and fungi (fungicides). Several health malignancies are associated with exposure to different pesticides, which include, but are not limited to, memory disorders, dermatological conditions, respiratory disorders, neurological deficits, cancer, depression, miscarriages, and birth defects (McCauley et al., 2006). Biodegradation of pesticides using microorganisms or their enzymes is seen as the most promising approach to eliminate the toxic products of pesticides amassed in the environment. Microorganisms bring about the structural change and degradation of these compounds in the environment by physically and chemically interacting with the substrates. Fungal peroxidases have been used to effectively detoxify various pesticides into innoxious compounds. Studies have shown the great potential of *P. chrysosporium* to transform organophosphorus pesticides (Jauregui et al., 2003), and the enzyme chloroperoxidase from *Caldariomyces fumago* has also been found to biotransform organophosphorus pesticides. Polycyclic aromatic hydrocarbons are acted upon by phenol oxidases and peroxidases to reduce them to simpler and less toxic forms that can easily be degraded. Peroxidases such as LiPs and MnPs catalyze the oxidation of polycyclic aromatic hydrocarbons (Harford-Cross et al., 2000).

Biodegradation of Chlorinated Alkanes and Alkenes

Chlorinated alkenes such as trichloroethylene (TCE) and perchloroethylene (PCE) widely utilized as degreasing solvents have contributed significantly to soil and aquifer contamination, thereby posing severe threat to human health. The LiP of *P. chrysosporium* catalyzes the in vitro reductive dehalogenation of TCE to yield its corresponding chlorinated radicals in the presence of hydrogen peroxide, tertiary alcohol, and oxalate (or ethylenediaminetetraacetic acid (EDTA)) (Yadav et al., 2000). An imazethapyr (IMZT)-degrading strain of bacterium IM-4 was isolated from the soil heavily laden with IMZT. In addition to imazethapyr, several imidazolinone herbicides including imazapic, imazapyr, and imazamox are also degraded by the strain (Huang et al., 2009). *Tinea versicolor*, which produces an extracellular hydroxyl radical through quinone redox cycling, also has the ability to reduce PCE and TCE (Marco-Urrea et al., 2009). *P. chrysosporium* cultures grown under aerobic conditions are capable of mineralizing TCE.

Biodegradation of Phenoxyalkanoid Acid and Triazine Herbicides

Herbicides are generally used in agricultural settings around the world, and the most widely used herbicides include 2,4-D (2,4-dichlorophenoxyacetic acid) and 2,4,5-T (2,4,5-trichlorophenoxyaceticacid). Agent Orange used by the US forces as defoliants in the Vietnam War contains 2,4-D and 2,4,5-T. 2,4-D is readily eradicated from the environment since it are easily degraded by bacterial species. On the contrary, 2,4,5-T is difficult to biodegrade by microorganisms, and, so, it persists more in the environment. Serious illnesses were attributed to 2,4,5-T during the Vietnam War for their exposure to Agent Orange. They are extremely toxic to humans and are most often considered as mutagenic agents. *P. chrysosporium* and *Dichomitus squalens* produce ligninolytic peroxidases that are involved in detoxification of 2,4-D- and 2,4,5-T-chlorinated phenolic intermediates. Laccases and peroxidases produced by *P. chrysosporium* were reported to biodegrade atrazine, which is a commonly used triazine herbicide (Bending et al., 2002).

Biodegradation of Chlorinated Dioxins

Polychlorinated dibenzodioxins are highly toxic environmental pollutants that are carcinogenic in nature and are found bioaccumulating in humans and animals as a result of their lipophilic properties. Various species of white rot fungi have been used to degrade chlorinated compounds like polychlorinated dibenzodioxins and polychlorinated dibenzofurans, indicating the activity of LiP and MnP (Kasai et al., 2010). Dioxins have also been effectively degraded using MnP produced by *Phanerochaete sordida*.

Biodegradation of Chlorinated Insecticides

Lindane was extensively used as an insecticide in the past century with a global production of approximately 600,000 tons between the year 1950 and 2000. Lindane has been banned globally, owing to its high resistance to degradation leading to its persistence. Under ligninolytic conditions, lindane is partially mineralized by *P. chrysosporium* in a broth medium and in corncob-amended soils (Quintero et al., 2008). However, degradation of lindane has not been extensively studied in vitro using LiP and MnP from *P. chrysosporium*. Excessive use of dichlorodiphenyltrichloroethane (DDT) (the first of the chlorinated organic insecticides) was reported after World War II. Heavy contamination of agricultural soils with DDT poses severe threats to the safety of food and to human health. DDT has been found to be susceptible to attacks by *P. chrysosporium*, *P. ostreatus*, *T. versicolor*, and *Phellinus weirii*.

3.2 Microbial Hydrolytic Enzymes

Soil and water pollution by industrial effluents having ample amount of hydrocarbons and heavy metals is a weighty problem of the modern world. These compounds constitute the major contaminants of aquatic and terrestrial environments, as a result of their extensive use. The technique of eliminating them by the use of microorganisms or microbial enzymes is seen as a safe and cost-effective approach. The action of bacteria on the pollutants is the major process of breakdown of organic pollutants. Vasileva-Tonkova and Galabova (2003) reported that the activity of extracellular enzymes is the basic step in the biodegradation of these organic compounds.

Hydrolases constitute the Class III enzymes and are further subdivided based on the type of bond they hydrolyze. The chemical bonds in toxic pollutants are cleaved by the action of hydrolytic enzymes, which ultimately reduce their toxicity. Oil spills and insecticides (carbamate and organophosphate) are degraded by cleavage of bonds in the compounds. Hydrolases are also involved in condensations and alcoholysis amongst several other related reactions. Hydrolase enzymes have numerous advantages such as ubiquity, lack of cofactor stereoselectivity, and tolerance to water-miscible solvents. Hydrolytic enzymes such as lipases, cellulases, proteases, and amylases amongst other extracellular enzymes are widely used in various industries including the food and beverage industry, as feed additives, and in the pharmaceutical and chemical industry (Sanchez-Porro et al., 2003). Important enzymes such as hemicellulase, cellulase, and glycosidase are widely used in biomass degradation (Schmidt, 2006). The various microbial hydrolytic enzymes in bioremediation include microbial lipases, microbial cellulases, and microbial proteases.

Microbial Lipases

Lipase enzymes degrade lipids. Lipase is found in bacteria, actinomycetes, and in plant and animal cells, but, of all these, microbial lipase is the most versatile and widely used for industrial applications. Sharma et al. (2011) stressed the involvement of lipase in a series of reactions including hydrolysis, esterification, interesterification, aminolysis, and solvolysis with alcohol. Lipases are widespread and are obtained from numerous sources; they are involved in the breakdown of triacylglycerols to yield fatty acids and glycerol. Lipases are responsible for the depletion of hydrocarbon compounds in contaminated soils, and, as such, lipase is considered the most effective indicator parameter for measuring the rate of degradation of hydrocarbons in soil (Riffaldi et al., 2006). Lipase is not only used in assessing the rate of bioremediation but also has several potent industrial applications such as in the food processing industry, chemical and detergent manufacturing, cosmetic formulation, and the paper and pulp industry; however, its industrial uses are limited by the cost of its production (Joseph et al., 2006; Sharma et al., 2011).

Sources of Microbial Lipases

Lipases have widespread occurrence in a variety of plants and animals; however, they are more abundant in microbial flora comprising bacteria, fungi, and yeasts. Lipases of microbial origin have gained much attention, and they have the most biotechnological applications among various classes of microbial enzymes, as lipase is produced by numerous microbial strains. Prominent lipase-producing microorganisms include *Candida* sp., *Pseudomonas* sp., and *Rhizopus* sp. *Candida rugosa* is the most frequently used organism for lipase synthesis. The commonly adopted in situ techniques for lipase synthesis are submerged and solid-state fermentations. Various microorganisms including filamentous fungi, yeasts, and bacteria have been extensively studied for the production of lipases.

Filamentous Fungi

The production of lipase by filamentous fungi is influenced by various factors such as the strains, growth medium composition, and conditions including temperature, pH, oxygen content, and source of carbon and nitrogen (Cihangir & Sarikaya, 2004). Commercially important lipase-producing fungi include strains such as *Rhizopus*, *Aspergillus*, *Penicillium*, *Geotrichum*, *Mucor*, and *Rhizomucor*.

Yeasts

Vakhlu and Kour (2006) concluded that the main terrestrial lipase-producing yeasts include the species of *Candida*, *Rhodotorula*, *Pichia*, and *Yarrowia lipolytica*. Wang et al. (2007) reported the cloning and overexpression of the lipase-coding genes in the species of *Candida*, *Geotrichum*, *Trichosporon*, and *Y. lipolytica*. There is widespread use of lipase produced from different strains of *C. rugosa* and *Candida antarctica*; however, recent research has shown the potential of other yeasts to produce lipases. It has been recently discovered that the contamination of freshly produced olive oil with a rich microflora has resulted in the synthesis of enzymes that regulate the organoleptic and physicochemical attributes of the oil (Ciafardini et al., 2006). Strains including those of *Saccharomyces cerevisiae*, *Candida wickerhamii*, *Candida boidinii*, and *Williopsis californica*, in addition to other strains of yeasts, were identified among the microorganisms isolated from the contaminated oil, with *S. cerevisiae* and *W. californica* reported as effective lipase-producing strains. Lipase production in *S. cerevisiae* was intracellular, whereas extracellular lipase activity was found in *W. californica*.

Bacteria

Several bacterial species have been explored of which the *Bacillus* species expressed the extraordinary ability of lipase production, making them immensely recommended strains for biotechnological applications. Prominent lipase-producing Bacilli include *Bacillus subtilis*, *Bacillus pumilus*, *Bacillus stearothermophilus*, *Bacillus licheniformis*, *Bacillus coagulans*, and *Bacillus alcalophilus*. Other species of bacteria capable of producing lipase enzymes are *Pseudomonas* sp., *Burkholderia* sp., and *Staphylococcus* sp.

Ertugrul et al. (2007) studied 17 different strains of bacteria that can utilize tributyrin in a tributyrin medium and concluded that the best lipase-producing strain was a *Bacillus* strain.

Microbial Cellulases

Cellulose is a complex carbohydrate with numerous industrial applications such as in the manufacture of textiles, paper, explosives, and pharmaceutical products and therefore contributes to the economic development worldwide. Raw materials are largely composed of celluloses, which constitute more than 94% of cotton and more than 50% of wood. Cellulose is extensively used as a major raw material in numerous industries; however, abundant quantities of cellulosic materials are yet to be well explored or used more efficiently. Agricultural wastes, industrial effluents, and biosolids have been reported to contain substantial quantities of cellulose, which serve as a carbon source for anaerobic degradation of biosolids to yield methane. Celluloses in combination with other complex carbohydrate polymers such as hemicellulose and lignin along with traces of extractives constitute the cellulosic biomass.

The conversion of cellulose waste material into food by cellulases to meet the demands of the growing population has become a field of research interest (Bennet et al., 2002). Cellulose is hydrolyzed by a variety of microorganisms using a multienzyme system. Some microorganisms secrete cellulases that are cell-bound, associated with envelope or extracellular cellulases. Some bacterial and fungal species express low levels of extracellular cellulases, hemicellulases, and pectinases (Adriano-Anaya et al., 2005). Several enzymes usually make up the cellulase class of enzymes, with at least three distinct groups of cellulases catalyzing any hydrolysis reaction:

1. Endoglucanase (endo-1,4-D-glucanohydrolase) hydrolyzes cellulose from the region of low crystallinity, resulting in free chain ends.
2. Exoglucanase (1,4-β-D-glucan cellobiohydrolase) acts by extracting the cellobiose units from various free chain ends to further hydrolyze cellulose.
3. β-Glucosidase acts along with other enzymes to produce several glucose units by breaking down cellobiose.

Sun and Cheng (2002) reported that cellulose is degraded by cellulase enzyme to simple sugar, which is further acted upon by bacteria and/or yeasts to yield ethanol. Crystalline cellulose is also degraded by cellulase enzymes to glucose.

Sources of Microbial Cellulases

Fungi are the major source of microbial cellulase; however, cellulase has been isolated from few bacterial and actinomycete species. Cellulase-producing microorganisms generally degrade carbohydrates but lack the ability to break down proteins and lipids to obtain energy for growth and development (Lynd et al., 2002). The bacterial species *Cellulomonas* and *Cytophaga* and most other fungal species amongst the cellulolytic microbes have the ability to degrade other carbohydrates besides cellulose; however, cellulolytic anaerobes are found to utilize only cellulose and products of cellulose hydrolysis. Certain fungal strains produce a variety of extracellular cellulases attributed to the large amounts of extracellular proteins they secrete. *Trichoderma reesei* has been extensively studied and was found to hydrolyze both natural and synthetic celluloses to glucose. Prominent cellulose-degrading microorganisms include *Aspergillus* sp., *Humicola* sp., *Penicillium* sp., *Trichoderma* sp., *Phanerochaete chrysosporium*, *Fusarium solani*, and *Talaromyces emersonii* amongst the fungal species, whereas the bacterial species are *Cellulomonas* sp., *Cellvibrio* sp., *Microbispora bispora*, and *Thermomonospora* sp. amongst the aerobes and *Acetivibrio cellulolyticus*, *Bacteroides cellulosolvens*, *Bacteroides succinogenes*, *Clostridium thermocellum*, *Ruminococcus albus*, and *Ruminococcus flavefaciens* constitute the anaerobic cellulolytic bacteria. Cellulose is metabolized by numerous fungi for energy; however, the cellulase–enzyme complex required for hydrolyzing cellulose is produced only by few strains. Certain species of fungi including *Aspergillus* sp., *Penicillium* sp., *Humicola* sp., and *T. reesei* are capable of yielding substantial quantities of extracellular cellulases. Cellulolytic aerobic bacteria such as *Cellulomonas* and *Cytophaga* have the ability to degrade cellulose in pure cultures (Lynd et al., 2002). However, the microorganisms that are commercially used for the extraction of cellulases are mainly *T. reesei*, *Humicola insolens*, *A. niger*, *Thermomonospora fusca*, and the species of *Bacillus*. Saranraj et al. (2012) recounted that several researches has been conducted on bioremediation of organic wastes using bacteria, actinomycetes, and fungi (including *Trichoderma* sp., *Penicillium* sp., and *Aspergillus* sp.) to synthesize cellulase enzymes.

Mechanism of Cellulase Activity

Microorganisms bring about the degradation of cellulose by making use of several microbial enzymes, which constitute the multienzyme complex. These organisms include bacterial and fungal species consisting of aerobic and anaerobic and mesophilic and thermophilic organisms inhabiting different environments. A number of extracellular enzymes are secreted by aerobic bacteria, with unique binding sites to accommodate different conformations of cellulose. Cellulosome, an extracellular multienzyme complex composed of several enzymes, is produced by anaerobic bacteria. The activity of the single components of the multienzyme complex is stimulated toward the crystalline substrate following its binding to a noncatalytic structural protein (scaffolding). *Clostridium thermocellum*, which is a thermophilic bacterium, is the most complex and best investigated cellulosome.

Cellulase enzymes are found to degrade filter paper and other natural celluloses as well as synthesized celluloses like carboxymethyl or hydroxyethyl cellulose. Cellulases attack cellulose at 1,4-β linkages, whereas lignin and cereal are at β-D-glucans. The newly generated chain ends of cellulose are acted upon by exoglucanases to mainly produce cellobioses, which are repeating units of disaccharides with glucose units joined with a 1,4-β linkage, whereas β-glucosidase acts on cellulose from the ends by cleaving the terminal β-D-glucose residues. Cellulose usually occurs alongside other components like hemicellulose, lignin, and pectin, which are degraded by cellulase enzymes. Several D-glucose units present in polysaccharides linked by 1,4-α-D-glucosidic linkages are hydrolyzed by amylases, whereas the 1,4-α-D-galactosiduronic linkages in galacturans are randomly broken down by pectinases. Environmental pollutants containing lignocellulosic materials are degraded by a variety of cellulases (Saranraj et al., 2012).

Microbial Proteases

Microbial proteases constitute a group of microbial enzymes that cleave peptide linkages in aqueous solution and catalyze the synthesis of peptide linkages in a nonaqueous medium. Proteases catalyze the hydrolysis of peptide bonds in substances high in protein content, which are often released into the air by shedding and molting of feathers and other appendages, decay of dead animals, and as by-products of various industries including the textile and food industry. Singh (2003) and Beena and Geevarghese (2010) recounted that proteases are widely used in several industries including the food, textile, detergent, and pharmaceutical industry. Proteases are divided, on the basis of their catalytic action on peptide bonds, into endopeptidases and exopeptidases. Endopeptidases are further subdivided, on the basis of the position of the active site, into metallopeptidases and serine, cysteine, and aspartic endopeptidases.

Sources of Microbial Proteases

Proteases are widely available in a variety of plants, animals, and microorganisms; however, protease from a microbial source has more preference over other sources due to its easy accessibility, cost-effectiveness, and large-scale production as a result of fast reproducibility of microorganisms and easy manipulation for generation of recombinant enzymes with unique properties. Kumar and Takagi (1999) affirmed that the two-third share of global production of commercial proteases in the enzyme market is of microbial origin. Microorganisms degrade proteins and utilize the products for growth and development. Proteinases (endopeptidases) produced by various species of microorganisms initiate the degradation process, and peptidases (exopeptidases) catalyze further hydrolysis at various locations within and between cells. Different varieties of microbial proteases are produced by various species of an organism and also by different strains of the same species. The most common class of microbial proteases, produced by all groups of microorganisms (bacteria, fungi, yeast, and actinomycetes), are the alkaline serine proteases.

Fungal proteases: Proteases of fungal origin are of great interest to researchers due to their specificity to a wide range of substrates, high diversity, and thermostable nature. A vast array of fungal proteases produced by various species of fungi, such as *Aspergillus* sp., *Chrysosporium keratinophilum*, *Conidiobolus coronatus*, *Entomophthora coronata*, *Fusarium eumartii*, *Paecilomyces lilacinus*, *Scedosporium apiospermum*, *Rhizopus oligosporus*, *Cephalosporium* sp. KSM 388, and *Tritirachium album Limber*, has been studied (Veloorvalappil et al., 2013). Separation of the mycelium can be achieved by simple filtration, making it extremely advantageous.

Bacterial proteases: Bacterial proteases have extensive uses in various industries, such as the food, textile, and pharmaceutical industries, and are generally utilized in biodegradation of protein-containing wastes as a result of their large production scale and catalytic efficiency. The maximum catalytic efficiency of bacterial protease is achieved at a high pH level (8–12) with an optimum temperature of 50–70 °C. These unique features of bacterial proteases make them more suitable for biodegradation of pollutants. Prolific sources of microbial enzymes include several bacterial species including *Alteromonas* sp., *Mycobacterium* sp., *Pseudomonas* sp., *Streptomyces* sp., *Thermoactinomyces* sp., *Arthrobacter protophormiae*, *Lactobacillus helveticus*, *Xanthomonas maltophilia*, *Vibrio alginolyticus*, *Brevibacterium linens*, *Staphylothermus marinus*, and *Salinivibrio* sp. strain AF-2004.

4 Microbial Degradation of Plastics

Biodegradation of polymeric substances (polyethylene, polypropylene, polystyrene, polyvinyl chloride) is greatly impeded due to the absence of hydrolyzable functional groups in their backbone (Restrepo-Florez et al., 2014; Krueger et al., 2015). The decomposition and mass reduction of polymeric compounds is initiated by the conjoined action of microorganisms and climatic factors including temperature, humidity, rainfall, pressure, and other physical factors (Eubeler et al., 2010; Restrepo-Florez et al., 2014). Koutny et al. (2006a, b) and Fontanella et al. (2010) observed that the effects of UV irradiation and oxidizing agents result in the formation of carbonyl groups, which are easily accessible for further microbial attacks.

Numerous researches on biodegradation of the various types of polyethylene have been conducted in the past few decades (Restrepo-Florez et al., 2014; Sen &

Raut, 2015). Ligninases, which degrade lignin (a complex noncarbohydrate aromatic compound) in the cell wall of plants (Suhas et al., 2007), have been found to degrade polyethylene (Restrepo-Florez et al., 2014; Krueger et al., 2015); however, there is difficulty in attaining a complete and efficient degradation, since the degradation of lignin requires a lower redox potential compared to the homologous covalent linkages in the backbones of polyethylene (Krueger et al., 2015). Examples of such microbial enzymes are lignin peroxidases, manganese peroxidases, and laccases.

UV-irradiated polyethylene films in cell extracts as well as culture supernatants were degraded by a thermostable laccase produced by *Rhodococcus ruber* C208 in the presence of copper (Santo et al., 2013). Similarly, the molecular weight of a polyethylene membrane according to Fujisawa et al. (2001) has been effectively reduced by laccase produced from *Trametes versicolor* in the presence of 1-hydroxybenzotriazole, which acted as a cofactor in the reaction.

4.1 Microbial Degradation of Polyurethane

Polyurethane (PUR) is a polymeric substance containing urethane (carbamate) linkages between adjacent polyols and di- or polyisocyanate (Seymour & Kauffman, 1992). Polyurethanes are classed as either polyether polyurethanes or polyester polyurethanes depending on polyol (which is the amorphous part of the compound composed of a polyether or polyester) used for polycondensation reactions (Urgun-Demirtas et al., 2007). Loredo-Trevino et al. (2012) and Cregut et al. (2013) asserted that various microbial enzymes capable of hydrolyzing urethane linkages to depolymerize polyurethane include microbial ureases, esterases, and proteases. A variety of enzymes have been isolated from bacteria (Howard et al., 2012) and fungi (Russell et al., 2011) with the potential to degrade polyester polyurethane. Several studies have shown that carbamate and amide linkages are hydrolyzable by proteases and ureases that cleave the urea linkages (Matsumiya et al., 2010). The major enzymatic depolymerization of polyester polyurethane is the cleavage of the ester bonds by esterases and proteases (Howard, 2002). Christenson et al. (2006) reported that the urethane bonds present in polyether polyurethane may be hydrolyzed by hydrolases from bacteria and fungi; however, the class of polyurethane is much more resistant to enzymatic degradation than polyester polyurethane.

4.2 Microbial Degradation of Polyethylene Terephthalate

Polyethylene terephthalate (PET) is a polymeric substance made from polymerization of ester-linked terephthalic acid (an aromatic dicarboxylic acid) and ethylene glycol (Webb et al., 2013). The production of polyethylene terephthalate according to Research and Markets (2015) had surpassed 41.6 million tones worldwide as of

2014, and it is widely used in manufacturing beverage bottles, as packaging materials, and in the textile industry. Polyethylene terephthalate is highly durable and resistant to microbial degradation due to the repeating units of aromatic terephthalate in its backbone (Marten et al., 2003, 2005). The polyethylene terephthalate polymer is semicrystalline in nature (partially crystalline and partially amorphous); this also contributes to its resistance to degradation.

5 Conclusions

Environmental pollutants, which have become a serious global concern, are treated either by various physicochemical procedures or by enzymatic degradation. Application of microbial enzymes in degradation of persistent pollutants is more effective, efficient, economical, and eco-friendly and thus is the more acceptable and preferable approach to elimination of environmental pollutants. Microbial enzymes used to efficiently biodegrade persistent environmental pollutants include oxygenases, laccases, and peroxidases amongst the oxidoreductases, whereas the hydrolytic enzymes are lipases, cellulases, and proteases. Although enzymatic degradation is slow, it ensures complete depletion of conversion of pollutants into less harmful products. However, with the advent of recombinant DNA technology, microorganisms can easily by manipulated to produce enzymes with a broad spectrum of activity that will catalyze the depletion or conversion of persistent environmental pollutants into value-added products at less time.

References

Adriano-Anaya, M., Salvador-Figueroa, M., Ocampo, J. A., & García-Romera, I. (2005). Plant cell-wall degrading hydrolytic enzymes of *Gluconacetobacter diazotrophicus*. *Symbiosis, 40*(3), 151–156.

Aislabie, J., Bej, A. K., Ryburn, J., Lloyd, N., & Wilkins, A. (2005). Characterization of *Arthrobacter nicotinovorans* HIM, an atrazine-degrading bacterium, from agricultural soil New Zealand. *FEMS Microbiology Ecology, 52*(2), 279–286.

Arias, M. E., Arenas, M., Rodríguez, J., Soliveri, J., Ball, A. S., & Hernandez, M. (2003). Kraft pulp biobleaching and mediated oxidation of a nonphenolic substrate by laccase from *Streptomyces cyaneus* CECT 3335. *Applied and Environmental Microbiology, 69*(4), 1953–1958.

Arora, P. K., Kumar, M., Chauhan, A., Raghava, G. P., & Jain, R. K. (2009). OxDBase: A database of oxygenases involved in biodegradation. *BMC Research Notes, 2*, 67.

Arora, P. K., Srivastava, A., & Singh, V. P. (2010). Application of Monooxygenases in dehalogenation, desulphurization, denitrification and hydroxylation of aromatic compounds. *Journal of Bioremediation & Biodegradation, 1*, 1–8.

Beena, A. K., & Geevarghese, P. I. (2010). A solvent tolerant thermostable protease from a psychrotrophic isolate obtained from pasteurized milk. *Developmental Microbiology and Molecular Biology, 1*, 113–119.

Bending, G. D., Friloux, M., & Walker, A. (2002). Degradation of contrasting pesticides by white rot fungi and its relationship with ligninolytic potential. *FEMS Microbiology Letters, 212*(1), 59–63.

Bennet, J. W., Wunch, K. G., & Faison, B. D. (2002). *Use of fungi biodegradation*. ASM Press.

Bezalel, L., Hadar, Y., & Cerniglia, C. E. (1996). Mineralization of polycyclic aromatic hydrocarbons by the white rot fungus *Pleurotus ostreatus*. *Applied and Environmental Microbiology, 62*, 292–295.

Bourbonnais, R., Paice, M. G., Reid, I. D., Lanthier, P., & Yaguchi, M. (1995). Lignin oxidation by laccase isozymes from *Trametes versicolor* and role of the mediator 2,22-azinobis (3-ethylbenzthiazoline-6-sulfonate) in kraft lignin depolymerization. *Applied and Environmental Microbiology, 61*(5), 1876–1880.

Chakraborty, J., & Das, S. (2016). Molecular perspectives and recent advances in microbial remediation of persistent organic pollutants. *Environmental Science and Pollution Research, 23*, 16883–16903. https://doi.org/10.1007/s11356-016-6887-7

Christenson, E. M., Patel, S., Anderson, J. M., & Hiltner, A. (2006). Enzymatic degradation of poly (ether urethane) and poly(carbonate urethane) by cholesterol esterase. *Biomaterials, 27*, 3920–3926.

Ciafardini, G., Zullo, B. A., & Iride, A. (2006). Lipase production by yeasts from extra virgin olive oil. *Food Microbiology, 23*, 60–67. https://doi.org/10.1016/j.fm.2005.01.009

Cihangir, N., & Sarikaya, E. (2004). Investigation of lipase production by a new isolated of *Aspergillus* sp. *World Journal of Microbiology and Biotechnology, 20*, 193–197. https://doi.org/10.1023/B:WIBI.0000021781.61031.3a

Couto, S. R., & Toca Herrera, J. L. (2006). Industrial and biotechnological applications of laccases: A review. *Biotechnology Advances, 24*(5), 500–513.

Cregut, M., Bedas, M., Durand, M. J., & Thouand, G. (2013). New insights into polyurethane biodegradation and realistic prospects for the development of a sustainable waste recycling process. *Biotechnological Advancement, 31*, 1634–1647.

Dana, L. D., & Bauder, J. W. (2011). *A general essay on bioremediation of contaminated soil*. Montana State University.

Dawkar, V. V., Jadhav, U. U., Jadhav, S. U., & Govindwar, S. P. (2008). Biodegradation of disperse textile dye Brown 3REL by newly isolated *Bacillus* sp. VUS. *Journal of Applied Microbiology, 105*(1), 14–24.

Dedeyan, B., Klonowska, A., & Tagger, S. (2000). Biochemical and molecular characterization of a laccase from Marasmius quercophilus. *Applied and Environmental Microbiology, 66*(3), 925–929.

Dua, M., Singh, A., Sethunathan, N., & Johri, A. (2002). Biotechnology and bioremediation: Successes and limitations. *Applied Microbiology and Biotechnology, 59*(2–3), 143–152.

Eggert, C., Temp, U., & Eriksson, K. E. L. (1996). The ligninolytic system of the white rot fungus *Pycnoporus cinnabarinus*: Purification and characterization of the laccase. *Applied and Environmental Microbiology, 62*(4), 1151–1158.

Ertugrul, S., Donmez, G., & Takaç, S. (2007). Isolation of lipase producing *Bacillus* sp. from olive mill wastewater and improving its enzyme activity. *Journal of Hazardous Materials, 149*, 720–724. https://doi.org/10.1016/j.jhazmat.2007.04.034

Eubeler, J. P., Bernhard, M., & Knepper, T. P. (2010). Environmental biodegradation of synthetic polymers II. Biodegradation of different polymer groups. *TrAC Trends in Analytical Chemistry, 29*, 84–100.

Faccelo, J., & Cruz, O. (2008). *Banana skin: A novel material for a low-cost production of laccase*. M.S. Thesis, Universitat Roviral Virgili.

Fetzner, S., & Lingens, F. (1994). Bacterial dehalogenases: Biochemistry, genetics, and biotechnological applications. *Microbiological Reviews, 58*(4), 641–685.

Field, J. A., de Jong, E., Costa, G. F., & de Bont, J. A. M. (1993). Screening for ligninolytic fungi applicable to the biodegradation of xenobiotics. *Trends in Biotechnology, 11*, 44–49.

Fontanella, S., Bonhomme, S., Koutny, M., Husarova, L., Brusson, J. M., & Courdavault, J. P. (2010). Comparison of the biodegradability of various polyethylene films containing pro-oxidant additives. *Polymer Degradation and Stability, 95*, 1011–1021.

Fox, B. G., Borneman, J. G., Wackett, L. P., & Lipscomb, J. D. (1990). Haloalkene oxidation by the soluble methane monooxygenase from *Methylosinus trichosporium* OB3b: Mechanistic and environmental implications. *Biochemistry, 29*(27), 6419–6427.

Fujisawa, M., Hirai, H., & Nishida, T. (2001). Degradation of polyethylene and nylon-66 by the laccase-mediator system. *Journal of Polymer and the Environment, 9*, 103–108.

Gayazov, R., & Rodakiewicz-Nowak, J. (1996). Semi-continuous production of laccase by *Phlebia radiata* in different culture media. *Folia Microbiologica, 41*(6), 480–484.

Ghasemi, F., Tabandeh, F., Bambai, B., & Sambasiva Rao, K. R. S. (2010). Decolorization of different azo dyes by Phanerochaete chrysosporium RP78 under optimal condition. *International journal of Environmental Science and Technology, 7*(3), 457–464.

Gianfreda, L., Xu, F., & Bollag, J. M. (1999). Laccases: A useful group of oxidoreductive enzymes. *Bioremediation Journal, 3*(1), 1–25.

Gochev, V. K., & Krastanov, A. I. (2007). Fungal laccases. *Bulgarian Journal of Agricultural Science, 13*, 75–83.

Grosse, S., Laramee, L., Wendlandt, K. D., McDonald, I. R., Miguez, C. B., & Kleber, H. P. (1999). Purification and characterization of the soluble methane monooxygenase of the type II methanotrophic bacterium Methylocystis sp. strain WI 14. *Applied and Environmental Microbiology, 65*(9), 3929–3935.

Guo, X., Xie, C., Wang, L., Li, Q., & Wang, Y. (2019). Biodegradation of persistent environmental pollutants by *Arthrobacter* sp. *Environmental Science and Pollution Research, 26*, 8429–8443. https://doi.org/10.1007/s11356-019-04358-0

Hamid, H., & Rehman, K. U. (2009). Potential applications of peroxidases. *Food Chemistry, 115*(4), 1177–1186.

Harford-Cross, C. F., Carmichael, A. B., Allan, F. K., England, P. A., Rouch, D. A., & Wong, L. (2000). Protein engineering of cytochrome P458(cam) (CYP101) for the oxidation of polycyclic aromatic hydrocarbons. *Protein Engineering, 13*(2), 121–128.

Hatakka, A. (2001). Biodegradation of lignin. In M. Hofrichter & A. Steinbuchel (Eds.), *Lignin, humic substances and coal* (pp. 129–179). Wiley-VCH.

Hiner, A. N. P., Ruiz, J. H., & Rodri, J. N. (2002). Reactions of the class II peroxidases, lignin peroxidase and *Arthromyces ramosus* peroxidase, with hydrogen peroxide: Catalase-like activity, compound III formation, and enzyme inactivation. *The Journal of Biological Chemistry, 277*(30), 26879–26885.

Hofrichter, M., Ullrich, R., Pecyna, M. J., Liers, C., & Lundell, T. (2010). New and classic families of secreted fungal heme peroxidases. *Applied Microbiology and Biotechnology, 87*(3), 871–897.

Holker, U., Dohse, J., & Hofer, M. (2002). Extracellular laccases in ascomycetes *Trichoderma atroviride* and *Trichoderma harzianum*. *Folia Microbiologica, 47*(4), 423–427.

Howard, G. T. (2002). Biodegradation of polyurethane: A review. *International Journal of Biodeterioration and Biodegradation, 49*, 245–252.

Howard, G. T., Norton, W. N., & Burks, T. (2012). Growth of *Acinetobacter gerneri* P7 on polyurethane and the purification and characterization of a polyurethanase enzyme. *Biodegradation, 23*, 561–573.

Huang, X., Pan, J., Liang, B., Sun, J., Zhao, Y., & Li, S. (2009). Isolation, characterization of a strain capable of degrading imazethapyr and its use in degradation of the herbicide in soil. *Current Microbiology, 59*(4), 363–367.

Husain, Q. (2006). Potential applications of the oxidoreductive enzymes in the decolorization and detoxification of textile and other synthetic dyes from polluted water: A review. *Critical Reviews in Biotechnology, 26*(4), 201–221.

Iyer, G., & Chattoo, B. B. (2003). Purification and characterization of laccase from the rice blast fungus, *Magnaporthe grisea*. *FEMS Microbiology Letters, 227*(1), 121–126.

Jauregui, J., Valderrama, B., Albores, A., & Vazquez-Duhalt, R. (2003). Microsomal transformation of organophosphorus pesticides by white rot fungi. *Biodegradation, 14*(6), 397–406.
Jimenez-Juarez, N., Roman-Miranda, R., Baeza, A., Sanchez-Amat, A., Vazquez-Duhalt, R., & Valderrama, B. (2005). Alkali and halide-resistant catalysis by the multipotent oxidase from *Marinomonas mediterranea*. *Journal of Biotechnology, 117*(1), 73–82.
Joseph, B., Ramteke, P. W., & Kumar, P. A. (2006). Studies on the enhanced production of extracellular lipase by *Staphylococcus epidermidis*. *Journal of General and Applied Microbiology, 52*(6), 315–320.
Junghanns, C., Moeder, M., Krauss, G., Martin, C., & Schlosser, D. (2005). Degradation of the xenoestrogen nonylphenol by aquatic fungi and their laccases. *Microbiology, 151*(1), 45–57.
Kasai, N., Ikushiro, S., & Shinkyo, R. (2010). Metabolism of mono and dichloro-dibenzo-p-dioxins by *Phanerochaete chrysosporium* cytochromes P450. *Applied Microbiology and Biotechnology, 86*(2), 773–780.
Kaušpediene, D., Kazlauskiene, E., Gefeniene, A., & Binkiene, R. (2010). Comparison of the efficiency of activated carbon and neutral polymeric adsorbent in removal of chromium complex dye from aqueous solutions. *Journal of Hazardous Materials, 179*(1–3), 933–939.
Kiiskinen, L. L., Ratto, M., & Kruus, K. (2004a). Screening for novel laccase-producing microbes. *Journal of Applied Microbiology, 97*(3), 640–646.
Kiiskinen, L. L., Kruus, K., Bailey, M., Ylosmaki, E., Siika-aho, M., & Saloheimo, M. (2004b). Expression of *Melanocarpus albomyces* laccase in *Trichoderma reesei* and characterization of the purified enzyme. *Microbiology, 150*(9), 3065–3074.
Koua, D., Cerutti, L., & Falquet, L. (2009). PeroxiBase: A database with new tools for peroxidase family classification. *Nucleic Acids Research, 37*(Suppl 1), D261–D266.
Koutny, M., Sancelme, M., Dabin, C., Pichon, N., Delort, A. M., & Lemaire, J. (2006a). Acquired biodegradability of polyethylenes containing pro-oxidant additives. *Polymers Degradation and Stabilization, 91*, 1495–1503.
Koutny, M., Lemaire, J., & Delort, A. M. (2006b). Biodegradation of polyethylene films with prooxidant additives. *Chemosphere, 64*, 1243–1252.
Krueger, M. C., Harms, H., & Schlosser, D. (2015). Prospects for microbiological solutions to environmental pollution with plastics. *Applied Microbiology and Biotechnology, 99*, 8857–8874.
Kumar, C. G., & Takagi, H. (1999). Microbial alkaline proteases: From a bioindustrial viewpoint. *Biotechnology Advances, 17*, 561–594. https://doi.org/10.1016/S0734-9750(99)00027-0
Lee, K. H., Wi, S. G., Singh, A. P., & Kim, Y. S. (2004). Micromorphological characteristics of decayed wood and laccase produced by the brown-rot fungus *Coniophora puteana*. *Journal of Wood Science, 50*(3), 281–284.
Lehninger, A. L., Nelson, D. L., & Cox, M. M. (2004). *Lehninger's principles of biochemistry* (4th ed.). W. H. Freeman.
Leung, M. (2004). Bioremediation: Techniques for cleaning up a mess. *Journal of Biotechnology, 2*, 18–22.
Loredo-Trevino, A., Gutierrez-Sanchez, G., Rodruguez-Herrera, R., & Aguilar, C. N. (2012). Microbial enzymes involved in polyurethane biodegradation: A review. *Journal of Polymers and the Environment, 20*, 258–265.
Lundell, T. K., Makela, M. R., & Hilden, K. (2010). Lignin-modifying enzymes in filamentous basidiomycetes-ecological, functional and phylogenetic review. *Journal of Basic Microbiology, 50*(1), 5–20.
Lynd, L. R., Weimer, P. J., van Zyl, W. H., & Pretorious, I. S. (2002). Microbial cellulase utilization: Fundamentals and biotechnology. *Microbiology and Molecular Biology Reviews, 66*, 506–577.
Lyons, J. I., Newell, S. Y., Buchan, A., & Moran, M. A. (2003). Diversity of ascomycete laccase gene sequences in a southeastern US salt marsh. *Microbial Ecology, 45*(3), 270–281.

Mai, C., Schormann, W., Milstein, O., & Huttermann, A. (2000). Enhanced stability of laccase in the presence of phenolic compounds. *Applied Microbiology and Biotechnology, 54*(4), 510–514.

Marco-Urrea, C., & Reddy, C. A. (2012). Degradation of chloroorganic pollutants by white rot fungi microbial degradation of xenobiotics. *Environmental Science and Engineering, 2*, 31–66.

Marco-Urrea, E., Aranda, E., Caminal, G., & Guillen, F. (2009). Induction of hydroxyl radical production in *Trametes versicolor* to degrade recalcitrant chlorinated hydrocarbons. *Bioresource Technology, 100*(23), 5757–5762.

Marten, E., Muller, R. J., & Deckwer, W. D. (2003). Studies on the enzymatic hydrolysis of polyesters i. Low molecular mass model esters and aliphatic polyesters. *Polymer Degradation and Stability, 80*, 485–501.

Marten, E., Muller, R. J., & Deckwer, W. D. (2005). Studies on the enzymatic hydrolysis of polyesters. II. Aliphatic-aromatic copolyesters. *Polymer Degradation and Stability, 88*, 371–381.

Matsumiya, Y., Murata, N., Tanabe, E., Kubota, K., & Kubo, M. (2010). Isolation and characterization of an ether-type polyurethane-degrading micro-organism and analysis of degradation mechanism by *Alternaria* sp. *Journal of Applied Microbiology, 108*, 1946–1953.

McCauley, L. A., Anger, W. K., Keifer, M., Langley, R., Robson, M. G., & Rohlman, D. (2006). Studying health outcomes in farmworker populations exposed to pesticides. *Environmental Health Perspectives, 114*(6), 953–960.

Megharaj, M., Ramakrishnan, B., Venkateswarlu, K., Sethunathan, N., & Naidu, R. (2011). Bioremediation approaches for organic pollutants: A critical perspective. *Environment International, 37*(8), 1362–1375.

Morozova, O. V., Shumakovich, G. P., Gorbacheva, M. A., Shleev, S. V., & Yaropolov, A. I. (2007). Blue laccases. *Biochemistry (Moscow), 72*(10), 1136–1150.

Mossallam, K. F., Sultanova, F. M., & Salemova, N. A. (2009). Peroxidase catalysed the removal of phenol from synthetic waste water. In *Proceedings of the 13th International Water Technology Conference (IWTC 13), Hurghada, Egypt* (pp. 1009–1020).

Mui, E. L. K., Cheung, W. H., Valix, M., & McKay, G. (2010). Dye adsorption onto activated carbons from tyre rubber waste using surface coverage analysis. *Journal of Colloid and Interface Science, 347*(2), 290–300.

Ng, T. W., Cai, Q., Wong, C., Chow, A. T., & Wong, P. (2010). Simultaneous chromate reduction and azo dye decolourization by *Brevibacterium casei*: Azo dye as electron donor for chromate reduction. *Journal of Hazardous Materials, 182*(1–3), 792–800.

Niku-Paavola, M. L., Karhunen, E., Salola, P., & Raunio, V. (1998). Ligninolytic enzymes of the white-rot fungus *Phlebia radiata*. *Biochemical Journal, 254*(3), 877–884.

Ong, S. T., Keng, P. S., Lee, W. N., Ha, S. T., & Hung, W. T. (2011). Dye waste treatment. *Water, 3*(1), 157–176.

Palmieri, G., Giardina, P., Bianco, C., Fontallella, B., & Sannina, G. (2000). Copper induction of laccase isoenzyme in the lignolytic fungus *Pleurotus ostreatus*. *Applied Microbiology and Biotechnology, 66*, 920–924.

Palonen, H., Saloheimo, M., Viikari, L., & Kruus, K. (2003). Purification, characterization and sequence analysis of a laccase from the ascomycete *Mauginiella* sp. *Enzyme and Microbial Technology, 33*(6), 854–862.

Park, J. W., Park, B. K., & Kim, J. E. (2006). Remediation of soil contaminated with 2,4-dichlorophenol by treatment of minced shepherd's purse roots. *Archives of Environmental Contamination and Toxicology, 50*(2), 191–195.

Piontek, K., Smith, A. T., & Blodig, W. (2001). Lignin peroxidase structure and function. *Biochemical Society Transactions, 29*(2), 111–116.

Pointing, S. B., & Vrijmoed, L. L. P. (2000). Decolorization of azo and triphenylmethane dyes by *Pycnoporus sanguineus* producing laccase as the sole phenoloxidase. *World Journal of Microbiology and Biotechnology, 16*(3), 317–318.

Que, L., & Ho, R. Y. N. (1996). Dioxygen activation by enzymes with mononuclear non-heme iron active sites. *Chemical Reviews, 96*(7), 2607–2624.

Quintero, J. C., Moreira, M. T., Feijoo, G., & Lema, J. M. (2008). Screening of white rot fungal species for their capacity to degrade lindane and other isomers of hexachlorocyclohexane (HCH). *Ciencia e Investigacion Agraria, 35*(2), 123–132.

Research and Markets. (2015). *Global polyethylene terephthalate Market (PET resin) - By end-Use industries, products, and regions - Market size, demand forecasts, industry trends and updates (2014-2020).* Retrieved from http://www.researchandmarkets.com/reports/3505772/global-polyethylene-terephtalate-market-pet

Restrepo-Florez, J. M., Bassi, A., & Thompson, M. R. (2014). Microbial degradation and deterioration of polyethylene – A review. *International Biodeterioration and Biodegradation, 88*, 83–90.

Riffaldi, R., Levi-Minzi, R., Cardelli, R., Palumbo, S., & Saviozzi, A. (2006). Soil biological activities in monitoring the bioremediation of diesel oil-contaminated soil. *Water, Air, and Soil Pollution, 170*(1–4), 3–15.

Roohi Kulsoom, B., Mohammed, K., Mohd, R. Z., Qamar, Z., Mohd, F. K., Ghulam, M. A., Anamika, G., & Gjumrakch, A. (2017). Microbial enzymatic degradation of biodegradable plastics. *Current Pharmaceutical Biotechnology, 18*(5), 429. https://doi.org/10.2174/1389201018666170523165742

Rubilar, O., Diez, M. C., & Gianfreda, L. (2008). Transformation of chlorinated phenolic compounds by white rot fungi. *Critical Reviews in Environmental Science and Technology, 38*(4), 227–268.

Ruiz-Duenas, F. J., Morales, M., & Perez-Boada, M. (2007). Manganese oxidation site in Pleurotus eryngii versatile peroxidase: A site-directed mutagenesis, kinetic, and crystallographic study. *Biochemistry, 46*(1), 66–77.

Ruiz-Duenas, F. J., Morales, M., García, E., Miki, Y., Martínez, M. J., & Martínez, A. T. (2009). Substrate oxidation sites in versatile peroxidase and other basidiomycete peroxidases. *Journal of Experimental Botany, 60*(2), 441–452.

Russell, J. R., Huang, J., Anand, P., Kucera, K., Sandoval, A. G., & Dantzler, K. W. (2011). Biodegradation of polyester polyurethane by endophytic fungi. *Applied and Environmental Microbiology, 77*, 6076–6084.

Ryan, J. R., Loehr, R. C., & Rucker, E. (1991). Bioremediation of organic contaminated soils. *Journal of Hazardous Materials, 28*, 159–169.

Sanchez-Porro, C., Martın, S., Mellado, E., & Ventosa, A. (2003). Diversity of moderately halophilic bacteria producing extracellular hydrolytic enzymes. *Journal of Applied Microbiology, 94*(2), 295–300.

Santo, M., Weitsman, R., & Sivan, A. (2013). The role of the copper-binding enzyme - Laccase - In the biodegradation of polyethylene by the actinomycete *Rhodococcus ruber*. *International Journal of Biodeterioration and Biodegradation, 84*, 204–210.

Saranraj, P., Stella, D., & Reetha, D. (2012). Microbial cellulases and its applications: A review. *International Journal of Biochemistry & Biotech Science, 1*, 1–12.

Schlosser, D., & Hofer, C. (2002). Laccase-catalyzed oxidation of Mn2+ in the presence of natural Mn3+ chelators as a novel source of extracellular H2O2 production and its impact on manganese peroxidase. *Applied and Environmental Microbiology, 68*(7), 3514–3521.

Schmidt, O. (2006). *Wood and tree fungi*. Springer.

Sen, S. K., & Raut, S. (2015). Microbial degradation of low density polyethylene (LDPE): A review. *Journal of Environmental Chemical Engineering, 3*, 462–473.

Seymour, R. B., & Kauffman, G. B. (1992). Polyurethanes: A class of modern versatile materials. *Journal of Chemistry Education, 69*, 909.

Sharma, D., Sharma, B., & Shukla, A. K. (2011). Biotechnological approach of microbial lipase: A review. *Biotechnology, 10*(1), 23–40.

Singh, C. J. (2003). Optimization of an extracellular protease of *Chrysosporium keratinophilum* and its potential in bioremediation of keratinic wastes. *Mycopathologia, 156*(3), 151–156.

Stoj, C., & Kosman, D. J. (2003). Cuprous oxidase activity of yeast Fet3p and human ceruloplasmin: Implication for function. *FEBS Letters, 554*(3), 422–426.

Suhas, Carrott, P. J. M., & Ribeiro Carrott, M. M. L. (2007). Lignin – From natural adsorbent to activated carbon: A review. *Bioresource Technology, 98*, 2301–2312.

Sun, Y., & Cheng, J. (2002). Hydrolysis of lignocellulosic materials for ethanol production: A review. *Bioresource Technology, 83*(1), 1–11.

Ten Have, R., & Teunissen, P. J. M. (2001). Oxidative mechanisms involved in lignin degradation by white-rot fungi. *Chemical Reviews, 101*(11), 3397–3413.

Thakker, G. D., Evans, C. S., & Koteswara Rao, K. (1992). Purification and characterization of laccase from *Monocillium indicum* Saxena. *Applied Microbiology and Biotechnology, 37*(3), 321–323.

Tsukihara, T., Honda, Y., Sakai, R., Watanabe, T., & Watanabe, T. (2006). Exclusive overproduction of recombinant versatile peroxidase MnP2 by genetically modified white rot fungus, Pleurotus ostreatus. *Journal of Biotechnology, 126*(4), 431–439.

Ullah, M. A., Bedford, C. T., & Evans, C. S. (2000). Reactions of pentachlorophenol with laccase from Coriolus versicolor. *Applied Microbiology and Biotechnology, 53*(2), 230–234.

Urgun-Demirtas, M., Singh, D., & Pagilla, K. (2007). Laboratory investigation of biodegradability of a polyurethane foam under anaerobic conditions. *Polymer Degradation and Stability, 92*, 1599–1610.

Vakhlu, J., & Kour, A. (2006). Yeast lipases: Enzyme purification, biochemical properties and gene cloning. *Electronic Journal of Biotechnology, 9*, 1–17. https://doi.org/10.2225/vol9-issue1-fulltext-4

Vasileva-Tonkova, E., & Galabova, D. (2003). Hydrolytic enzymes and surfactants of bacterial isolates from lubricant-contaminated wastewater. *Zeitschrift für Naturforschung, 58*(1–2), 87–92.

Velazquez-Cedéno, M. A., Farnet, A. M., Ferre, E., & Savoie, J. M. (2004). Variations of lignocellulosic activities in dual cultures of *Pleurotus ostreatus* and *Trichoderma longibrachiatum* on unsterilized wheat straw. *Mycologia, 96*(4), 712–719.

Veloorvalappil, N. J., Robinson, B. S., Selvanesan, P., Sasidharan, S., Kizhakkepawothail, N. U., Sreedharan, S., Prakasan, P., & Moolakkariyil, S. J. (2013). Versatility of microbial proteases. *Advances in Enzyme Research, 1*(3), 39–51. https://doi.org/10.4236/aer.2013.13005

Vidali, M. (2001). Bioremediation. An overview. *Pure and Applied Chemistry, 73*(7), 1163–1172.

Viswanath, B., Subhosh Chandra, M., Pallavi, H., & Rajasekhar Reddy, B. (2008). Screening and assessment of laccase producing fungi isolated from different environmental samples. *African Journal of Biotechnology, 7*(8), 1129–1133.

Wang, L., Chi, Z. M., Wang, X. H., Liu, Z. Q., & Li, J. (2007). Diversity of lipase-producing yeasts from marine environments and oil hydrolysis by their crude enzymes. *Annals of Microbiology, 4*, 2–7.

Webb, H., Arnott, J., Crawford, R., & Ivanova, E. (2013). Plastic degradation and its environmental implications with special reference to poly (ethylene terephthalate). *Polymers, 5*, 1.

Wong, D. W. S. (2009). Structure and action mechanism of ligninolytic enzymes. *Applied Biochemistry and Biotechnology, 157*(2), 174–209.

Xu, F. (1996). Catalysis of novel enzymatic iodide oxidation by fungal laccase. *Applied Biochemistry and Biotechnology, 59*(3), 221–230.

Yadav, J. S., Bethea, C., & Reddy, C. A. (2000). Mineralization of trichloroethylene (TCE) by the white rot fungus Phanerochaete chrysosporium. *Bulletin of Environmental Contamination and Toxicology, 65*(1), 28–34.

Yaver, D. S., Berka, R. M., Brown, S. H., & Xu, F. (2001). *The Presymposium on recent advances in lignin biodegradation and biosynthesis* (Vol. 3–4). Vikki Biocentre, University of Helsinki.

Yoshida, S. (1998). Reaction of manganese peroxidase of Bjerkandera adusta with synthetic lignin in acetone solution. *Journal of Wood Science, 44*(6), 486–490.

Implication of Enzymes in the Adaptation of Extremophilic Microbes

Mounia Arab, Hafida Baoune, and Idris Hannous

Abstract Extremophiles belong to prokaryotes and eukaryotes, which can survive and thrive in hostile environments where conditions are supposed to be lethal. These organisms have developed different strategies to counteract the stress in their environment and to maintain their physiological properties, for instance, enzymes known as extremozymes are adapted to function even under unusual conditions. This chapter summarizes the enzyme adaptation mechanisms of extremophilic microbes, providing insights into the key role of these enzymes in the microbial adaptation to unfavorable conditions imposed by harsh environments. Extremophiles have certain modifications in their enzymes in order to retain their functions in adverse conditions, which, on the contrary, would aggregate, precipitate, or denature an enzyme from a non-extremophile. To maintain their function at high temperatures, thermophilic enzymes often have a prominent hydrophobic core and enhanced electrostatic contacts. At the same time, to conserve their flexibility and function at low temperatures, psychrophilic enzymes have a reduced hydrophobic core and less charged protein surface. Furthermore, halophilic, acidophilic, and alkaliphilic enzymes are characterized by increased negative surface charge, thus enhancing their acidic amino acid content and peptide insertions. This would compensate for the extreme ionic conditions. Enzymes from piezophilic microbes are generally characterized by low stability and high compressibility. Understanding extremophilic enzyme functioning and adaption mechanisms enables not only the understanding of the origins of life on Earth but also opens new prospects for

M. Arab
Laboratory of Applied Microbiology in Food, Biomedical and Environment, Department of Biology, Faculty of Nature Sciences and Life, University of Tlemcen, Tlemcen, Algeria

Faculty of Biological Sciences, University of Sciences and Technology Houari Boumediene, Algiers, Algeria

H. Baoune (✉)
Faculty of Natural and Life Sciences, Université Kasdi Merbah Ouargla, Ouargla, Algeria

I. Hannous
Service de diabétologie, Hôpital Central de l'Armée, Alger, Algeria

N. R. Maddela et al. (eds.), *Ecological Interplays in Microbial Enzymology*, Environmental and Microbial Biotechnology,
https://doi.org/10.1007/978-981-19-0155-3_13

developing and employing a new generation of enzymes required in biotechnological processes.

1 Introduction

The conditions of human development define the conventional normal environment; therefore, any variation is described as extreme. Organisms are divided into three domains: Bacteria, Archaea, and Eukarya (López-García, 2011). Microbes are well known for their challenges in relation to hostile biotopes, and, as a result, these microbes are known as extremophiles (Kristjansson & Hreggvidsson, 1995).

According to the literature, extremophilic microorganisms colonize terrestrial or subterrestrial marine geothermal springs, salt marshes, and acidic or basic springs or lakes. In fact, based on the characteristics of their environment, microbes can be classified into the following groups (the major groups and their survival conditions are provided in more detail in Fig. 1):

Thermophiles and hyperthermophiles are microorganisms that grow at temperatures of 45–80 °C and up to 80 °C, respectively (Madigan et al., 2000; Berenguer, 2011). Psychrophiles are able to grow at temperatures below 10 °C (Siddiqui et al., 2013). Acidophiles and alkaliphiles grow at pH values <5 and >9, respectively. Halophiles are able to grow at a concentration of 200–5900 mM of NaCl (Edbeib et al., 2016). Piezophiles grow at pressures greater than atmospheric pressure (0.1 MPa) (Yayanos, 2002). Xerophiles are capable of surviving in arid climates (water activity <0.75) (Connon et al., 2007). Radiotolerant (ultraviolet (UV)-resistant) extremophiles are resistant to long-term exposure to potentially harmful ultraviolet light (Gabani et al., 2014). Besides, in the absence of oxygen, anaerobic microbes belonging to bacteria and archaea are classified as extremophiles, since anaerobes were the earliest living forms on Earth (Pikuta et al., 2007).

Overall, extremophiles are often described by a single extreme state; several natural habitats have two or more severe situations. Thus, the microbiota that thrive in those environments are known as polyextremophiles (Urbieta et al., 2015b; Arab et al., 2018).

Unlike resistance to extreme conditions, the concept of extremophiles implies cellular machinery that can adapt to extreme conditions, and cells function optimally under these conditions, suggesting that extremophiles implement original strategies, such as specific enzymes called "extremozymes" (Li et al., 2005), which have extraordinary properties, catalyzing substrates under extreme conditions that are supposed to denature or inhibit them. Indeed, extremozymes are highly resistant molecules allowing the adaptation of extremophilic microorganisms to the physical and chemical stresses they face. Many of them are currently being applied in industrial biotechnology (Maddela & García, 2021; Maddela et al., 2021). Actually, such applications are widely described in the literature (Antranikian et al., 2005; Champdoré et al., 2006; Raddadi et al., 2015; Kour et al., 2019).

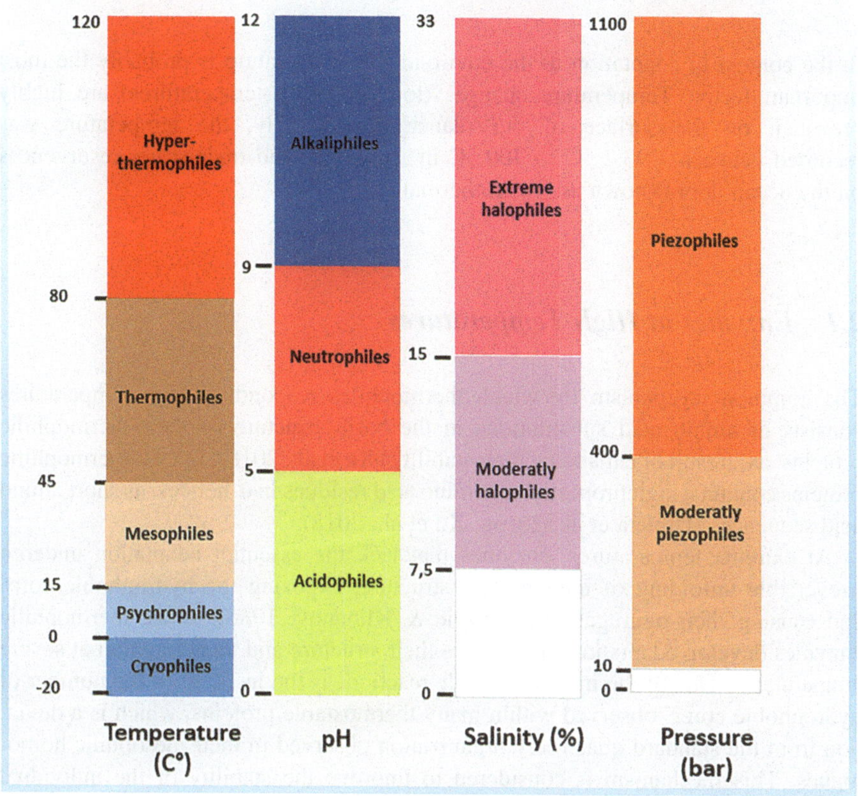

Fig. 1 Wide growth intervals of extremophilic bacteria in terms of temperature, pH, salinity, and pressure

Nevertheless, the fundamental aspects related to the functioning of extremozymes not only provide the key to understanding life processing under stress conditions and unravel the mechanisms developed by biological systems to overcome harsh conditions but also can elucidate mysteries about the emergence of life on Earth.

This chapter will therefore highlight the role of these extremozymes in the adaptation of microorganisms to extreme environmental conditions, focusing on their strategies to function under such conditions including the different changes that could occur in their enzymes, which lead to increase in their resistance.

2 Enzymes Against Extreme Temperatures

In the context of adaptation to the environment, temperature is probably the most important factor. Temperature changes (low and high temperatures) are highly prevalent on the surface of our planet. Respectively, the temperature was recorded between −93.2 °C to 400 °C in Antarctica and underwater resurgences on the ocean floor (known as "hydrothermal vents").

2.1 *Enzymes at High Temperatures*

The common mechanism by which thermophiles respond to high temperatures consists of amino acid substitutions in the basic structure of their thermophilic proteins, as a result of enhancing their stability (Xu et al., 2018). In fact, thermophilic proteins contain a high proportion of amino acid residues in α-helices, as short amino acid sequences (Urbieta et al., 2015a; Xu et al., 2018).

At extreme temperatures, enzymes that lack the essential adaptation undergo irreversible unfolding of their protein structure, exposing the hydrophobic cores and causing their aggregation (Tomazic & Klibanov, 1988). Thus, thermophilic enzymes develop adaptation to conserve their structure and their function at severe temperatures. One of the most noticeable reactions is the increase of the number of hydrophobic cores, observed within many thermostable proteins, which is a deviation from the standard quaternary organization observed in their mesophilic homologues. This mechanism is considered to improve the stability of the individual subunits, promoting tighter packing of the hydrophobic cores and reducing the exposure of hydrophobic residues to solvents (Vieille & Zeikus, 2001).

Moreover, the increase of the number of disulfide bonds between cysteine residues in the tertiary structure is of paramount importance in the determination of the overall structure of proteins. Among thermostable enzymes, Boutz et al. (2007) and Cacciapuoti et al. (2012) demonstrated the importance of these structural bonds, showing that their properties increase stability within thermophilic proteins and prevent the alteration of the quaternary structure.

Salt bridging is a common feature of most thermophilic enzymes in comparison to their mesophilic counterparts (Karshikoff & Ladenstein, 2001). This contradicts that salt bridging may destabilize mesophilic proteins and is disadvantageous compared to hydrophobic interactions (Hendsch & Tidor, 1994). In fact, at higher temperatures, the entropic cost and desolvation penalties associated with the ion pairing present in salt bridges are more readily overcome. If these thermodynamic aspects are negated, then salt bridges become a structure stabilizer, and the thermal capacity of proteins is increased by favorable charge–charge interactions (Chan et al., 2011; Reed et al., 2013).

The increase of surface charge residues is also often observed within thermostable enzymes (Fukuchi & Nishikawa, 2001). Based on several factors, replacement of polar uncharged surface residues with polar charged residues may lead to increased protein stability overall. At higher temperatures, polar residues such as glutamine and asparagine can undergo deamination, reducing their stability (Fukuchi & Nishikawa, 2001). The substitution of these and other thermolabile residues increases both short- and long-range charge interactions, avoiding thermal denaturation (Lee et al., 2005).

Some bacteria, such as *Bacillus pseudalcaliphilus*, an alkaliphilic halotolerant bacterium, exhibit cyclodextrin glucanotransferase enzyme activity that improves adaptation in high temperatures. It was demonstrated that preheating at 40–60 °C for 2 h induces the enzymatic activity of cyclodextrin glucanotransferase, which transforms raw corn starch to cyclodextrins. In addition, at pH 5–11, this enzyme retains 80% of its activity and shows a half-life of 2 h at 70 °C in the presence of 5 mM Ca^{2+} (Kitayska et al., 2011).

With regard to high temperature adaptability, hyperthermophiles are an ancestral type of archaea that diverged throughout archaeal evolution into many different lineages, including haloarchaea and methanogenic archaea (Di Giulio, 2005). Horizontal gene transfer (HGT) is required for these organisms to acquire numerous genes involved in the metabolism and cell envelope biogenesis from mesophilic microbes (López-García et al., 2015). Consequently, the protein structure has been modified in order to maintain functional activity in a high-temperature and salt-rich environment (Di Giulio, 2005; Reed et al., 2013).

2.2 *Enzymes at Low Temperatures*

In general, low temperatures significantly inhibit almost all enzyme-catalyzed processes and also reduce molecular movements related to protein activity (Feller, 2013). However, psychrophilic enzymes optimize high activity at low temperatures due to their ability to migrate and change their conformation more easily (Smalas et al., 2000).

At low temperatures, cold denaturation and reduction in the reaction speed of enzymes are considered major problems that must be overcome. In cold denaturation, water molecules surround proteins' surface, reducing their connection and changing them to unfolded form (De Maayer et al., 2014). For reaction speed, the Boltzmann equation states that reaction rates increase with the increase of temperature and drop two to threefold for every 10 °C reduction. Thus, the decrease of temperature will lead to an exponential decrease in the reaction rate (Gerday, 2013; Santiago et al., 2016).

Psychrophilic enzymes have great flexibility due to their weak protein interactions (Feller, 2010); variations in the amino acid composition between mesophilic and thermophilic proteins are shown in Fig. 2. Indeed, in cold active proteins, the stabilizing connections present inside a protein are diminished or absent. Feller (2010) summarized the different adaptations occurring in psychrophilic active proteins in four stages: (1) increasing glycine residues, (2) reducing proline residues,

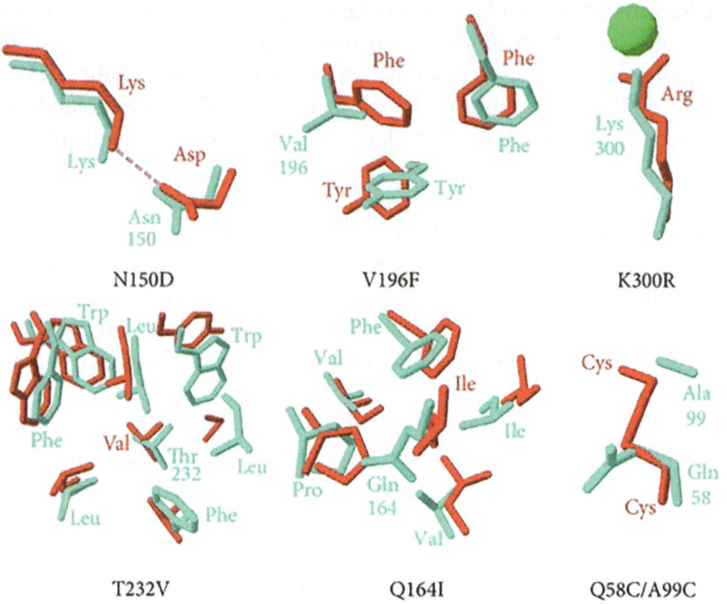

Fig. 2 Amino acid substitutions in the structure of a psychrophilic α-amylase (blue) compared with its mesophilic homologue (red) (Feller, 2013)

thereby providing conformational rigidity in loop regions, (3) decreasing the number of salt bridges and hydrogen bond-producing arginine residues, and (4) reducing the size of nonpolar residues in the protein core to generate weaker hydrophobic contacts (Feller, 2010).

It is common knowledge that weaker interactions between amino acid residues in psychrophilic enzymes prevent them from being "frozen" in a particular conformation, thus making the required molecular movement for catalysis. Thus, enzyme stability results from these weaker connections where cold-adapted proteins unfold at lower temperatures compared to mesophilic proteins (D'Amico et al., 2001; Georlette et al., 2003; Feller, 2010).

Despite the reduction of reaction rates at low temperatures, the specific activity of psychrophilic enzymes (*k*cat) is usually 10 times greater than that of mesophilic enzymes. In fact, the flexibility of their structure substantially increases their catalytic activity (Georlette et al., 2003; Feller, 2010). The highest *k*cat is explained by the increase of the binding site size in psychrophilic proteins (Smalas et al., 2000). Psychrophilic enzymes have a large substrate-binding area based on reaction type, whereas catalytic residues remain unaltered. Some of these reactions include glycine residues at specific locations near functional sites (Feller, 2010), the elimination of loops near the binding site (Russell et al., 1998), and increasing substrate accessibility by pushing out the protein backbone (Aghajari et al., 2003). Consequently, substrates do not effectively bind to cold enzymes, resulting in a high Michaelis–Menten constant (*Km*) of psychrophilic enzymes. Low substrate affinity boosts enzyme activity at low temperatures by decreasing the enzyme activation energy (D'Amico et al., 2006; Feller, 2010).

High-resolution models of psychrophilic proteins have revealed that cold-adapted proteins have a higher number and size of cavities than do their mesophilic homologues (Paredes et al., 2011). Cavities can retain high concentration of hydrophilic groups, which bind to a high number of water molecules, thus increasing internal solvation and enzyme flexibility (Paredes et al., 2011).

Jung et al. (2008) demonstrated that the destabilizing surface cavity of the cold-adapted M37 lipase of *Photobacterium lipolyticum* provides flexibility to the helical lid, leading to higher lateral movement followed by substrate attachment. Additionally, comparing M37 with its orthologs from the mesophile *Rhizomucor miehei* showed that a large oxyanion hole presents in its structure, enabling the binding of extra water molecules, thus contributing to the reduction of the energy required to achieve the transitory tetrahedral intermediate. In another study, ornithine carbamoyltransferase (OTCase) of *Moritella abyssi*, a psychrophilic and piezophilic bacterium, showed lower thermoresistance than did its homologues in thermophilic prokaryotes found as stable trimers or dodecamers. Moreover, the OTCase homologues in *Pyrococcus furiosus* exhibited a low melting point and variations in denaturation enthalpy. Indeed, changes in its most conserved motifs produce a high Km value for ornithine carbamoyl phosphate transferase than for their mesophilic and thermophilic homologues (Xu et al., 2003).

3 Enzymes Against High Salt Concentrations

Salt has a substantial impact on a protein's solubility, stability, and shape and thus on its functions. Excessive salt concentrations increase hydrophobic interactions. Moreover, the electrostatic connections between charged amino acids are disrupted. Consequently, non-halophilic proteins become unstable, losing their function, resulting in global unfolding, aggregation, and precipitation (Karan et al., 2012).

The huge increase in acidic residues on the protein's surface, such as glutamic and aspartic acid, is one of the most striking differences between halophilic and non-halophilic proteins. The acidic residues are highly prevalent, showing the differences between halophilic and non-halophilic protein sequences (Zhang et al., 2013). These acidic residues may be involved in a variety of functions. In addition, the increase of negative charge on the protein's surface leads ions to compete for water molecules, keeping the protein in solution (Britton et al., 2006; Karan et al., 2012).

Bioinformatics studies of enzyme sequences revealed that a distinct hydrophobic residue and a high amount of acidic residues are found in halophilic enzymes compared to their homologues in mesophilic microbes (Siglioccolo et al., 2011). Smaller hydrophobic residues provide weaker hydrophobic interactions, increasing protein flexibility under high salt concentrations and preventing the hydrophobic core from becoming excessively stiff (Mevarech et al., 2000).

Nowadays, one of the main knowledge is halophilic protein adaptation based on salt concentrations. This discovery showed that the adaptation of halophilic proteins to high salt concentrations is not only limited by retaining the protein structure but also its function (Muller-Santos et al., 2009).

Besides, it was demonstrated that halophilic enzymes have a high number of acidic residues as their halophilic adaptation is followed by an alkaliphilic adaptation to pH variation (Gimenez et al., 2000; Falb et al., 2005). In addition, several unusual peptides are found in halophilic proteins; possibly, these peptides might keep the stability of proteins and enhance their efficiency in extremely halophilic environments. A usual peptide located near the active site of the catalytic domain of halophilic archaeal cysteinyl-tRNA synthetases increases the binding efficiency of cysteine to tRNA-Cys (Evilia & Hou, 2006). A number of studies have found that halophilic bacteria possess an unusual P45 enzyme that protects cellular proteins against denaturation when cells are exposed to hyposaline environments. Otherwise, P45 is resistant to deactivation in high salt concentrations required for the enzyme halophilic malate dehydrogenase (hMDH), which binds to P45 (Franzetti et al., 2001). In the case of halophilic malate dehydrogenase (hMDH) from *Haloarcula marismortui*, high NaCl or KCl concentrations required for its stability might be explained by a particular low affinity binding of salt ions to the folded protein. Thus, a suitable salt concentration is required to completely saturate these binding sites (Pikuta et al., 2007).

4 Enzymes Against pH Variations

In acidic conditions, acidophilic enzymes are catalytically active at low pH levels ($pH < 1$) due to their stable structure. Most of the known acidophile enzymes are also known as thermophiles, having thermophilic characteristics as well. However, pH adaptation of acidophilic proteins is still unknown and uneven (Reed et al., 2013).

A variety of acidophilic enzymes showed optimum activity at low pH values compared to intracellular pH. For instance, the acidophilic and thermostable endoglucanase of *Sulfolobus solfataricus* exhibited an optimal stability at pH 1.8. In fact, the abundance of glutamic and aspartic acid residues on the surface of model enzymes resulted in negative surface charges at pH 7. The repulsion of these extra negative charges has been linked to the instability of numerous acidic surface residues at high pH. On the other hand, endo-β-glucanase does not show any extra negative charges at pH higher than 2, which might improve its stability in acidic environments (Huang et al., 2005).

As an example of acidophile-specific protein adaptation, the outer membrane protein porin in *Thiobacillus ferrooxidans* has an abnormally wide external loop, which decreases the pore size and ion selectivity of the bacteria (Guiliani & Jerez, 2000). Moreover, the adenosine triphosphate (ATP) synthase of *Bacillus pseudofirmus*, an alkaliphilic soil bacterium that thrives in high pH environments, adapts structurally by changing the motif components from repeated glycine (GxGxGxG) to alanine (AxAxAxA) residues. ATP synthase is a critical enzyme in the mitochondrial oxidative phosphorylation reactions in the electron transport chain. This enzyme is important for life since it is a downstream component of the tricarboxylic acid (TCA) cycle that generates the cellular energy currency. Moreover, the ATP synthase c-ring stoichiometry is characterized by a pattern of repeated glycine residues (GxGxGxG), which are responsible for the enzyme effectiveness in ATP production (Preiss et al., 2013). Furthermore, the alkaliphilic phosphoserine aminotransferase is a homodimer-forming vitamin B6-dependent enzyme. Its structure is similar to that of its mesophilic counterpart. This enzyme showed subtle differences in terms of the increased hydrogen bonds in the hydrophobic interactions at the dimer interface, as negatively charged amino acid residues that contribute to its stability in an alkaline environment (Dubnovitsky et al., 2009).

5 Enzymes Against High Pressure

Studies on piezophilic enzymes and their structural origins are just beginning to appear. However, many hypotheses have been suggested: (1) a required protection against a negative pressure effect when the effect occurs during the organism's growth, as indicated by the activity–stability–flexibility hypothesis, (2) some adaptations to cold and high pressure can be mutually exclusive, making them difficult to disentangle, (3) not all modifications that improve pressure tolerance are necessarily harmful under atmospheric pressure, and (4) not all adaptations are universally applicable to homologous enzymes from other piezophiles (Ichiye, 2018). Apart from their temperature adaptation, low stability, and high compressibility of bacterial and archaeal piezophiles, enzymes could maintain flexibility at high pressure and increase the catalytic activity, which can contribute to improving cold and high-pressure adaptations as well as variations in cavity size at key places for specific proteins. Furthermore, the requirement for a large overall cavity volume coupled with high compressibility must be balanced against the requirement that no individual cavity is large enough to allow disruptive water penetration. Thus, the general properties of the protein, such as high compressibility, may protect against the effects of compaction and specific sequence determinants. Therefore, water penetration at critical points in a protein may be regulated. Otherwise, the enhanced enzyme activity may contribute to the adaptation to high pressure (Ichiye, 2018). Moreover, the prevalence of smaller hydrogen-bonding amino acids enhanced multimerization (Reed et al., 2013). Furthermore, the number of hydrophobic residues in the core of their proteins, such as tryptophan and tyrosine, decreases as the amino acid "size" (molecular weight) decreases. This phenomenon is in contrast to most thermophilic proteins, which have a large ratio of high-molecular-weight amino acids in their hydrophobic core. Nonetheless, this change is beneficial, since it allows for tight packing, and results in most pressure-resistant proteins (Di Giulio, 2005).

Using mutagenesis and structural analyses with nuclear magnetic resonance demonstrated that any modification may form a cavity or damage the hydrophobic structure of the protein core, decreasing protein pressure and thermostability (Fusi et al., 1997; Consonni et al., 1999).

Multimeric proteins are another way for proteins to deal with pressure. A piezophilic protein isolated from *Pyrococcus horikoshii* known as the TET3 peptidase (TET3) forms a discrete dodecamer rather than a barrel-shaped multimer, as it is more stable at high pressure (Rosenbaum et al., 2012). The development of a dodecamer was critical for this protein since this structure allows the individual monomers to be more compact. When high pressures are applied, the protein monomers become more compact, preventing the penetration of water molecules into the protein core. The trapped water molecules would subsequently cause protein structure disturbance (Rosenbaum et al., 2012).

Multimerization also protects hydrogen bonds between protein subunits. These bonds are less vulnerable to pressure than are ionic interactions (Rosenbaum et al., 2012). At higher pressures, ionic interactions, particularly electrostatic interactions, are more sensitive to solvation, which disturbs these intraprotein connections (Boonyaratanakornkit et al., 2002). Salt bridge instability is mitigated by the strength of hydrogen connections between protein subunits (Rosenbaum et al., 2012). Under both extreme pressure and temperature conditions, some thermophilic modifications, such as the increase of basic amino acids, particularly arginine, were found to be helpful to proteins. This has also been observed in *Pyrococcus abyssi* proteins (Di Giulio, 2005).

6 Conclusions

In an extreme environment, enzymes produced by microbes must functionally adapt to harsh conditions. As a result, microbes have developed various strategies to meet this challenge. Based on the environmental stress type, different changes occur on a protein scale, producing resistant enzymes. For thermophilic enzymes, the most important changes observed are the increasing number of hydrophobic cores, disulfide bonds, and surface charge residues. Indeed, thermophilic enzymes exhibit polyextremophily adaptations, allowing bacteria to survive in different environmental conditions at the same time. In psychrophilic enzymes, protein adaptation involves an increase of glycine residues and a reduction of proline residues, salt bridge number, and hydrogen bond size of nonpolar residues. At high salt concentrations and pH variations, extremozymes frequently present more acidic residues on their protein surfaces and small hydrophobic residues, whereas at high pressure, piezophilic enzymes are lowly stable and highly compressed and proteins are frequently found in multimeric form. The study of extreme enzymes has considerably contributed to deciphering the enigmas of the appearance of life on Earth, as they offer valuable molecules for many biotechnological applications.

References

Aghajari, N., Van Petegem, F., Villeret, V., Chessa, J. P., Gerday, C., Haser, R., & Van Beeumen, J. (2003). Crystal structures of a psychrophilic metalloprotease reveal new insights into catalysis by cold-adapted proteases. *Proteins, 50*(4), 636–647. https://doi.org/10.1002/prot.10264

Antranikian, G., Vorgias, C. E., & Bertoldo, C. (2005). Extreme environments as a resource for microorganisms and novel biocatalysts. *Marine Biotechnology, I*, 219–262. https://doi.org/10.1007/b135786

Arab, M., Bakour, S., Lalaoui, R., Aissaoui, N., Nas, F., Hoceini, A., & Klouche-Khelil, N. (2018). Diversity of aerobic bacilli analysis using molecular and culture-based approaches in debagh hot spring. *Geomicrobiology Journal, 36*, 137–147. https://doi.org/10.1080/01490451.2018.1520937

Berenguer, J. (2011). Thermophile. In M. Gargaud, R. Amils, & J. C. Quintanilla (Eds.), *Encyclopedia of astrobiology* (pp. 1666–1667). Springer. https://doi.org/10.1007/978-3-642-11274-4_1583

Boonyaratanakornkit, B. B., Park, C. B., & Clark, D. S. (2002). Pressure effects on intra- and intermolecular interactions within proteins. *Biochimica et Biophysica Acta, 1595*(1–2), 235–249. https://doi.org/10.1016/s0167-4838(01)00347-8

Boutz, D. R., Cascio, D., Whitelegge, J., Perry, L. J., & Yeates, T. O. (2007). Discovery of a thermophilic protein complex stabilized by topologically interlinked chains. *Journal of Molecular Biology, 368*(5), 1332–1344. https://doi.org/10.1016/j.jmb.2007.02.078

Britton, K. L., Baker, P. J., Fisher, M., Ruzheinikov, S., Gilmour, D. J., Bonete, M. J., Ferrer, J., Pire, C., & Esclapez Rice, D. W. (2006). Analysis of protein solvent interactions in glucose dehydrogenase from the extreme halophile *Haloferax mediterranei*. *Proceedings of the National Academy of Sciences of the United States of America, 103*(13), 4846–4851. https://doi.org/10.1073/pnas.0508854103

Cacciapuoti, G., Fuccio, F., Petraccone, L., Del Vecchio, P., & Porcelli, M. (2012). Role of disulfide bonds in conformational stability and folding of 5'-deoxy5'-methylthioadenosine phosphorylase II from the hyperthermophilic archaeon *Sulfolobus solfataricus*. *Biochimica et Biophysica Acta, 1824*(10), 1136–1143. https://doi.org/10.1016/j.bbapap.2012.06.014

Champdoré, M., de Staiano, M., Rossi, M., & D'Auria, S. (2006). Proteins from extremophiles as stable tools for advanced biotechnological applications of high social interest. *Journal of the Royal Society Interface, 4*(13), 183–191. https://doi.org/10.1098/rsif.2006.0174

Chan, C. H., Yu, T. H., & Wong, K. B. (2011). Stabilizing salt-bridge enhances protein thermostability by reducing the heat capacity change of unfolding. *PLoS One, 6*(6), e21624. https://doi.org/10.1371/journal.pone.0021624

Connon, S. A., Lester, E. D., Shafaat, H. S., Obenhuber, D. C., & Ponce, A. (2007). Bacterial diversity in hyperarid Atacama Desert soils. *Journal of Geophysical Research – Biogeosciences, 112*, G04S17. https://doi.org/10.1029/2006JG000311

Consonni, R., Santomo, L., Fusi, P., Tortora, P., & Zetta, L. (1999). A single-point mutation in the extreme heat- and pressure resistant Sso7d protein from *Sulfolobus solfataricus* leads to a major rearrangement of the hydrophobic core. *Biochemistry, 38*(39), 12709–12717. https://doi.org/10.1021/bi9911280

D'Amico, S., Gerday, C., & Feller, G. (2001). Structural determinants of cold adaptation and stability in a large protein. *The Journal of Biological Chemistry, 276*(28), 25791–25796. https://doi.org/10.1074/jbc.M102741200

D'Amico, S., Sohier, J. S., & Feller, G. (2006). Kinetics and energetics of ligand binding determined by microcalorimetry: Insights into active site mobility in a psychrophilic α-amylase. *Journal of Molecular Biology, 358*(5), 1296–1304. https://doi.org/10.1016/j.jmb.2006.03.004

De Maayer, P., Anderson, D., Cary, C., & Cowan, D. A. (2014). Some like it cold: Understanding the survival strategies of psychrophiles. *EMBO Reports, 15*(5), 508–517. https://doi.org/10.1002/embr.201338170

Di Giulio, M. (2005). A comparison of proteins from *Pyrococcus furiosus* and *Pyrococcus abyssi*: Barophily in the physicochemical properties of amino acids and in the genetic code. *Gene, 346*, 1–6. https://doi.org/10.1016/j.gene.2004.10.008

Dubnovitsky, A. P., Kapetaniou, E. G., & Papageorgiou, A. C. (2009). Enzyme adaptation to alkaline pH: Atomic resolution (1.08 Å) structure of phosphoserine aminotransferase from *Bacillus alcalophilus*. *Protein Science, 14*, 97–110. https://doi.org/10.1110/ps.041029805

Edbeib, M. F., Wahab, R. A., & Huyop, F. (2016). Halophiles: Biology, adaptation, and their role in decontamination of hypersaline environments. *World Journal of Microbiology and Biotechnology, 32*, 135. https://doi.org/10.1007/s11274-016-2081-9

Evilia, C., & Hou, Y. M. (2006). Acquisition of an insertion peptide for efficient aminoacylation by a halophile tRNA synthetase. *Biochemistry, 45*, 6835–6845. https://doi.org/10.1021/bi0521386

Falb, M., Pfeiffer, F., Palm, P., Rodewald, K., Hikkmann, V., Tittor, J., & Oesterhelt, D. (2005). Living with two extremes: Conclusions from the genome sequence of *Natronomonas pharaonic*. *Genome Research, 15*(10), 1336–1343. https://doi.org/10.1101/gr.3952905

Feller, G. (2010). Protein stability and enzyme activity at extreme biological temperatures. *New Journal of Physics, 22*(32), 323101. https://doi.org/10.1088/0953-8984/22/32/323101

Feller, G. (2013). Psychrophilic enzymes: From folding to function and biotechnology. *Scientifica, 2013*, 512840. https://doi.org/10.1155/2013/512840

Franzetti, B., Schoehn, G., Ebel, C., Gagnon, J., Ruigrok, R. W. H., & Zaccai, G. (2001). Characterization of a novel complex from Halophilic archaebacteria, which displays chaperone-like activities *in vitro*. *The Journal of Biological Chemistry, 276*, 29906–29914. https://doi.org/10.1074/jbc.M102098200

Fukuchi, S., & Nishikawa, K. (2001). Protein surface amino acid compositions distinctively differ between thermophilic and mesophilic bacteria. *Journal of Molecular Biology, 309*(4), 835–843. https://doi.org/10.1006/jmbi.2001.4718

Fusi, P., Goossens, K., Consonni, R., Grisa, M., Puricelli, P., Vecchio, G., Vanoni, M., Zetta, L., Heremans, K., & Tortora, P. (1997). Extreme heat and pressure-resistant 7-kDa protein P2 from the archaeon *Sulfolobus solfataricus* is dramatically destabilized by a single point amino acid substitution. *Proteins, 29*(3), 381–390. https://doi.org/10.1002/(sici)1097-0134(199711)29:3<381::aid-prot11>3.0.co;2-j

Gabani, P., Prakash, D., & Singh, O. V. (2014). Bio-signature of ultraviolet radiation-resistant extremophiles from elevated land. *American Journal of Microbiological Research, 2*, 94–104. https://doi.org/10.12691/ajmr-2-3-3

Georlette, D., Damien, B., Blaise, V., Dpiereux, E., Uversky, V. N., Gerday, C., & Feller, G. (2003). Structural and functional adaptations to extreme temperatures in psychrophilic, mesophilic and thermophilic DNA ligases. *The Journal of Biological Chemistry, 278*(39), 37015–37023. https://doi.org/10.1074/jbc.M305142200

Gerday, C. (2013). Psychrophily and catalysis. *Biology, 2*(2), 719–741. https://doi.org/10.3390/biology2020719

Gimenez, M. I., Studdert, C. A., Sanchez, J. J., & De Castro, R. E. (2000). Extracellular protease of Natrialba magadii: Purification and biochemical characterization. *Extremophiles, 4*(3), 181–188. https://doi.org/10.1007/s007920070033

Guiliani, N., & Jerez, C. A. (2000). Molecular cloning, sequencing, and expression of omp-40, the gene coding for the major outer membrane protein from the acidophilic bacterium *Thiobacillus ferrooxidans*. *Applied and Environmental Microbiology, 66*, 2318–2324. https://doi.org/10.1128/AEM.66.6.2318-2324.2000

Hendsch, Z. S., & Tidor, B. (1994). Do salt bridges stabilize proteins? A continuum electrostatic analysis. *Protein Science, 3*(2), 211–226. https://doi.org/10.1002/pro.5560030206

Huang, Y., Krauss, G., Cottaz, S., Driguez, H., & Lipps, G. (2005). A highly acid-stable and thermostable endo-β-glucanase from the thermoacidophilic archaeon *Sulfolobus solfataricus*. *The Biochemical Journal, 385*(2), 581–588. https://doi.org/10.1042/BJ20041388

Ichiye, T. (2018). Enzymes from piezophiles. *Seminars in Cell & Developmental Biology, 84*, 138. https://doi.org/10.1016/j.semcdb.2018.01.004

Jung, S. K., Jeong, D. G., Lee, M. S., Lee, J. K., Kim, H. K., Ryu, S. E., Park, B. C., Kim, J. H., & Kim, S. J. (2008). Structural basis for the cold adaptation of psychrophilic M37 lipase from *Photobacterium lipolyticum. Proteins, 71*, 476–484. https://doi.org/10.1002/prot.21884

Karan, R., Capes, M. D., & Dassarma, S. (2012). Function and biotechnology of extremophilic enzymes in low water activity. *Aquatic Biosystems, 8*(1), 4. https://doi.org/10.1186/2046-9063-8-4

Karshikoff, A., & Ladenstein, R. (2001). Ion pairs and the thermotolerance of proteins from hyperthermophiles: A traffic rule for hot roads. *Trends in Biochemical Sciences, 26*(9), 550–556. https://doi.org/10.1016/s0968-0004(01)01918-1

Kitayska, T., Petrova, P., Ivanova, V., & Tonkova, A. I. (2011). Purification and properties of a new thermostable cyclodextrin glucanotransferase from *Bacillus pseudalcaliphilus* 8SB. *Applied Biochemistry and Biotechnology, 165*, 1285–1295. https://doi.org/10.1007/s12010-011-9346-4

Kour, D., Rana, K. L., Kaur, T., Singh, B., Chauhan, V. S., Kumar, A., & Gupta, V. K. (2019). Extremophiles for hydrolytic enzymes productions: Biodiversity and potential biotechnological applications. In *Bioprocessing for biomolecules production* (pp. 321–372). Wiley. https://doi.org/10.1002/9781119434436.ch16

Kristjansson, J. K., & Hreggvidsson, G. O. (1995). Ecology and habitats of extremophiles. *World Journal of Microbiology and Biotechnology, 11*, 17–25. https://doi.org/10.1007/BF00339134

Lee, C. F., Makhatadze, G. I., & Wong, K. B. (2005). Effects of chargeto-alanine substitutions on the stability of ribosomal protein L30e from *Thermococcus celer. Biochemistry, 44*(51), 16817–16825. https://doi.org/10.1021/bi0519654

Li, W. F., Zhou, X. X., & Lu, P. (2005). Structural features of thermozymes. *Biotechnology Advances, 23*, 271–281. https://doi.org/10.1016/j.biotechadv.2005.01.002

López-García, P. (2011). Domain (taxonomy). In M. Gargaud et al. (Eds.), *Encyclopedia of astrobiology*. Springer. https://doi.org/10.1007/978-3-642-11274-4_454

López-García, P., Zivanovic, Y., Deschamps, P., & Moreira, D. (2015). Bacterial gene import and mesophilic adaptation in archaea. *Nature Reviews. Microbiology, 13*, 447–456. https://doi.org/10.1038/nrmicro3485

Maddela, N. R., & García, L. C. (2021). *Innovations in biotechnology for a sustainable future* (p. VIII). Springer Nature. https://doi.org/10.1007/978-3-030-80108-3. 502, ISBN: 978-3-030-80107-6. Retrieved from https://www.springer.com/gp/book/9783030801076

Maddela, N. R., Garcia, L. C., & Chakraborty, S. (2021). *Advances in the domain of environmental biotechnology* (p. XVIII). Springer Nature Singapore Pte Ltd. https://doi.org/10.1007/978-981-15-8999-7. 717, ISBN: 978-981-15-8999-7. Retrieved from https://www.springer.com/gp/book/9789811589980

Madigan, M. T., Martinko, J. M., & Parker, J. (2000). *Brock biology of microorganisms*. Prentice Hall.

Mevarech, M., Frolow, F., & Gloss, L. M. (2000). Halophilic enzymes: Proteins with a grain of salt. *Biophysical Chemistry, 86*(2–3), 155–164. https://doi.org/10.1016/s0301-4622(00)00126-5

Muller-Santos, M., de Souza, E. M., Pedrosa, F. D. O., Mitchel, D. A., Longhi, S., Carrière, F., Canaan, S., & Krieger, N. (2009). First evidence for the salt-dependent folding and activity of an esterase from the halophilic archaea *Haloarcula marismortui. Biochimica et Biophysica Acta, 1791*(8), 719–729. https://doi.org/10.1016/j.bbalip.2009.03.006

Paredes, D. I., Watters, K., Pitman, D. J., Bystroff, C., & Dordick, J. S. (2011). Comparative void-volume analysis of psychrophilic and mesophilic enzymes: Structural bioinformatics of psychrophilic enzymes reveals sources of core flexibility. *BMC Structural Biology, 11*, 42. https://doi.org/10.1186/1472-6807-11-42

Pikuta, E. V., Hoover, R. B., & Tang, J. (2007). Microbial extremophiles at the limits of life. *Critical Reviews in Microbiology, 33*(3), 183–209. https://doi.org/10.1080/10408410701451948

Preiss, L., Klyszejko, A. L., Hicks, D. B., Liu, J., Fackelmayer, O. J., Yildiz, O., Krulwich, T. A., & Meier, T. (2013). The c-ring stoichiometry of ATP synthase is adapted to cell physiological requirements of alkaliphilic *Bacillus pseudofirmus* OF4. *Proceedings of the National Academy of Sciences, 110*, 7874–7879. https://doi.org/10.1073/pnas.1303333110

Raddadi, N., Cherif, A., Daffonchio, D., Neifar, M., & Fava, F. (2015). Biotechnological applications of extremophiles, extremozymes and extremolytes. *Applied Microbiology and Biotechnology, 99*(19), 7907–7913. https://doi.org/10.1007/s00253-015-6874-9

Reed, C. J., Lewis, H., Trejo, E., Winston, V., & Evilia, C. (2013). Protein adaptations in archaeal extremophiles. *Archaea, 2013*, 373275. https://doi.org/10.1155/2013/373275

Rosenbaum, E., Gabel, F., Dura, M. A., Finet, S., Baraud, C. C., Masson, P., & Franzett, B. (2012). Effects of hydrostatic pressure on the quaternary structure and enzymatic activity of a large peptidase complex from *Pyrococcus horikoshii*. *Archives of Biochemistry and Biophysics, 517*(2), 104–110. https://doi.org/10.1016/j.abb.2011.07.017

Russell, R. J. M., Gerike, U., Danson, M. J., Hough, D. W., & Taylor, G. L. (1998). Structural adaptations of the cold-active citrate synthase from an Antarctic bacterium. *Structure, 6*(3), 351–362. https://doi.org/10.1016/s0969-2126(98)00037-9

Santiago, M., Ramírez-Sarmiento, C. A., Zamora, R. A., & Parra, L. P. (2016). Discovery, molecular mechanisms, and industrial applications of cold-active enzymes. *Frontiers in Microbiology, 7*, 1408. https://doi.org/10.3389/fmicb.2016.01408

Siddiqui, K. S., Williams, T. J., Wilkins, D., Yau, S., Allen, M., Brown, M., Lauro, F. M., & Cavicchioli, R. (2013). Psychrophiles. *Annual Review of Earth and Planetary Sciences, 41*, 87–115. https://doi.org/10.1146/annurev-earth-040610-133514

Siglioccolo, A., Paiardini, A., Piscitelli, M., & Pascarella, S. (2011). Structural adaptation of extreme halophilic proteins through decrease of conserved hydrophobic contact surface. *BMC Structural Biology, 11*, 50. https://doi.org/10.1186/1472-6807-11-50

Smalas, A. O., Leiros, H. K., Os, V., & Willassen, N. P. (2000). Cold adapted enzymes. *Biotechnology Annual Review, 6*, 1–57. https://doi.org/10.1016/s1387-2656(00)06018-x

Tomazic, S. J., & Klibanov, A. M. (1988). Mechanisms of irreversible thermal inactivation of *Bacillus* α-amylases. *The Journal of Biological Chemistry, 263*(7), 3086–3091. https://doi.org/10.1016/S0021-9258(18)69038-6

Urbieta, M. S., Donati, E. R., Chan, K. G., Sin, L. L., & Goh, K. M. (2015a). Thermophiles in the genomic era: Biodiversity, science, and applications. *Biotechnology Advances, 33*, 633–647. https://doi.org/10.1016/j.biotechadv.2015.04.007

Urbieta, M. S., Gonzalez-Toril, E., Bazán, A. A., Giaveno, M. A., & Donati, E. (2015b). Comparison of the microbial communities of hot springs waters and the microbial biofilms in the acidic geothermal area of Copahue (Neuquen, Argentina). *Extremophiles, 19*, 437–450. https://doi.org/10.1007/s00792-015-0729-2

Vieille, C., & Zeikus, G. J. (2001). Hyperthermophilic enzymes: Sources, uses, and molecular mechanisms for thermostability. *Microbiology and Molecular Biology Reviews, 65*(1), 1–43. https://doi.org/10.1128/MMBR.65.1.1-43.2001

Xu, Y., Nogi, Y., Kato, C., Liang, Z., Rüger, H. J., De Kegel, D., & Glansdorff, N. (2003). *Moritella profunda* sp. nov. and *Moritella abyssi* sp. nov., two psychropiezophilic organisms isolated from deep Atlantic sediments. *International Journal of Systematic and Evolutionary Microbiology, 53*, 533–538. https://doi.org/10.1099/ijs.0.02228-0

Xu, L., Wu, Y. H., Zhou, P., Cheng, H., Liu, Q., & Xu, X. W. (2018). Investigation of the thermophilic mechanism in the genus *Porphyrobacter* by comparative genomic analysis. *BMC Genomics, 19*, 385. https://doi.org/10.1186/s12864-018-4789-4

Yayanos, A. A. (2002). Barophiles and Piezophiles. In *Nature. Encyclopedia of life sciences*. Nature Publishing. https://doi.org/10.1038/npg.els.0000341

Zhang, G., Huihua, G., & Yi, L. (2013). Stability of halophilic proteins: From dipeptide attributes to discrimination classifier. *International Journal of Biological Macromolecules, 53*, 1–6. https://doi.org/10.1016/j.ijbiomac.2012.10.031

Applications of Microbial Enzymes in Industries and Medicine

Oluwafemi Adebayo Oyewole, Japhet Gaius Yakubu, Mordecai Gana, Binta Buba Adamu, and Abdullahi Dabban Idris

Abstract Biomolecules that accelerate or enhance the rate of chemical reactions are known as enzymes. These enzymes are required for breakdown as well as synthesis reactions by living organisms. Enzymes, just like chemical catalysts, increase the rate of reaction by lowering the activation energy (Ea); thus, products are formed at a faster rate with the reaction reaching its equilibrium more rapidly. The importance of enzymes cannot be overemphasized. There has been a drastic increase in the global production of enzymes with more than 53,000 tons produced per year, most of which (75%) are hydrolytic, causing the breakdown of linkages in polypeptides, lipids, and polysaccharides. These enzymes (i.e., lipases, peptidases, amylases, etc.) have profound applications in various industries, such as food, chemical, agricultural, leather, textile, cosmetics, and medicine among others, which function differently from biotransformation of raw materials to use as therapeutic agents. Enzymes can be obtained from varying sources (i.e., plants, animals, and microorganisms); however, microorganisms are the most preferred choice for enzyme production. Over the past few decades, microbial production of enzymes for industrial purposes and use in medicine has proven to be of high industrial value, since these enzymes are stable, efficient, cost-effective, and can be genetically upregulated for high-yield production. Microorganisms from extreme environments (i.e., the Polar Regions, volcanoes) are, however, yet to be fully explored for enzyme production. As such, advanced techniques involving genomics and metagenomics together with some classical techniques can be used to isolate and screen microorganisms with the potential to produce industrially important enzymes from such environments.

O. A. Oyewole (✉) · J. G. Yakubu · M. Gana
Department of Microbiology, Federal University of Technology Minna, Minna, Nigeria
e-mail: oa.oyewole@futminna.edu.ng

B. B. Adamu
Department of Bioentrepreneurship and Extension Services, National Biotechnology Development Agency, Abuja, Nigeria

A. D. Idris
Department of Biological Sciences, The Federal Polytechnic Bida Niger State, Bida, Nigeria

N. R. Maddela et al. (eds.), *Ecological Interplays in Microbial Enzymology*, Environmental and Microbial Biotechnology,
https://doi.org/10.1007/978-981-19-0155-3_14

1 Introduction

Biomolecules that accelerate or enhance the rate of chemical reactions are known as enzymes (Das & Goyal, 2014; Al-Manhel, 2018). These enzymes are required for breakdown as well as synthesis reactions by living organisms (Patel et al., 2017). Most processes that occur in a biological system at certain locations require the action of one or more enzymes (Das & Goyal, 2014). Enzymes are highly specific to a particular substrate; factors such as temperature, pH, and substrate concentration regulate the activities of enzymes (Vermelho et al., 2016). All enzymes are protein in nature, with the exception of ribozymes (i.e., RNA catalytic molecules) (Singh et al., 2016; Vandenberghe et al., 2016; Patel et al., 2017), and their activity is ascertained by the amount of product formed or substrate transformed per unit of time (Das & Goyal, 2014). Metabolic activities in all living things involve one or more enzymes; these enzymes help in maintaining the proper function of all living things (Vermelho et al., 2016). As a result, enzymes are termed "biological catalysts" since they possess the ability to convert specific substrates into desired products at a high reaction rate (Singh et al., 2016; Patel et al., 2017). Enzymes, just like chemical catalysts, increase the rate of reaction by lowering the activation energy (Ea) (Singh et al., 2016); thus, products are formed at a faster rate with the reaction reaching its equilibrium more rapidly. Reactions catalyzed by enzymes occur a million times faster than do those without enzymes as catalysts. Reactions that were supposed to take weeks, months, or even hundreds of years to occur can take minutes or even seconds when catalyzed by enzymes (Patel et al., 2017).

The importance of enzymes cannot be overemphasized. There has been a drastic increase in the global production of enzymes with more than 53,000 tons produced per year, most of which (75%) are hydrolytic with profound uses in various industries (Al-Manhel, 2018). So far, Denmark is leading the global production of enzymes with Danisco and Novozymes leading the way with a massive production of 70% of the total global enzyme production with an estimated sale of $625–-700 million from 1989 to 1990 (Al-Manhel, 2018). Currently, there are more than 4000 known different enzymes (Liu & Kokare, 2017). However, only 5% of those are used industrially for the production of more than 500 commercial products (Sanchez & Demain, 2017). The total global market value of enzymes is dependent on the time and sources consulted. In one case, the market reached $5.1 billion in 2009 (Sanchez & Demain, 2017), $4.2 billion in 2014 (Singh et al., 2016), and $5–5.5 billion in 2016 (Al-Manhel, 2018) and is predicted to rise to $6.45 billion per annum (Sanchez & Demain, 2017) with expectations to reach or exceed $7.6 billion by 2022 at a compound annual growth rate (CAGR) of 7% (Singh et al., 2016; Al-Manhel, 2018). The market reports on different sectors or industries through which enzymes are applicable including food and animal feed (34%), followed by cleaners and detergents (29%), and leather and textile industries (17%) while the paper and pulp industry shares 11% of the market value (Liu & Kokare, 2017).

Almost all living things (i.e., animals, plants, and microorganisms) contain one form of an enzyme or another, which is mostly used for metabolic activities (Singh et al., 2019). In animals and plants, enzymes are found to be present in minute quantities and, as such, cannot be exploited if one wants to produce them in large quantities for industrial usage. In contrast, microorganisms are the best sources for enzyme production; prokaryotes (archaea and bacteria), yeasts, and fungi have all been demonstrated to be important sources for industrial enzyme production (Vermelho et al., 2016). Microbial production of enzymes is of great importance to the industrial world since it offers numerous advantages over other sources such as being easy to handle, cost-effective, utilizes cheap substrates, can be produced in large quantities, production conditions can be optimized for high-quality produce, and can be genetically upregulated to meet the high demands placed on a particular enzyme of focus, among many other benefits (Vermelho et al., 2016; Singh et al., 2019; Mishra et al., 2020). Moreover, enzymes obtained from microbial sources for industrial purposes have high specificity, catalytic potential, stability, and are nontoxic and friendly to the environment (Singh et al., 2016, 2019).

Although there are more than 4000 known enzymes, approximately 200 of the microbial enzymes are currently being used for commercial purposes with approximately 20 of them being truly produced on an industrial scale (Liu & Kokare, 2017). In recent years, industrial production and use of enzymes has grown immensely. This can be seen by the estimated market value of industrial microbial enzymes being \$1 billion, \$3 billion, and \$3.74 billion for the years 2012, 2013, and 2015, respectively (Liu & Kokare, 2017). The major bulk of these technical enzymes (i.e., lipases, amylases, cellulases, and many more) is used for the manufacture of biofuels, paper, pulp, leather, textiles, and detergents and reached a revenue of \$1.2 billion in 2011, with an expected rise above that as time goes on (Sanchez & Demain, 2017). Other enzyme applications include fine chemicals, animal feeds, food products, household care, pharmaceuticals, and medicine. Microbial enzymes exhibit unique properties such as the ability to operate under extreme conditions, high yields, and low generation of waste materials. These, among other properties, ensure flexibility in terms of the operating conditions employed in the reactor (Sanchez & Demain, 2017).

The next section elucidates the microbial production of enzymes, the types of fermentation techniques used, the groups of microorganisms involved, and the types of enzymes produced.

2 Microbial Production of Enzymes

Over the past few decades, microbial production of enzymes for industrial purposes has proven to be of high industrial value; these enzymes are stable, efficient, cost-effective, and can be genetically upregulated for high-yield production (Niyonzima et al., 2020; Naureen et al., 2021). Microorganisms produce enzymes either intracellularly (i.e., yeasts and bacteria) or extracellularly (i.e., molds) (Al-Manhel, 2018), through fermentation technology with the latter reducing production cost (Liu & Kokare, 2017). Fermentation technologies simply involve the utilization of microorganisms in the bioconversion of complex substrates into enzymes and other simple valuable substances under controlled conditions (Liu & Kokare, 2017; Patel et al., 2017). Due to their ease of handling and robust yield, microorganisms used for this production are generally recognized as safe (GRAS) since some of them are directly involved in the production of both human and animal consumables (Patel et al., 2017). The processes involved in fermentation cut across various disciplines such as microbial physiology and chemical engineering among others for a successful scale-up (Patel et al., 2017). Two major methods exist in fermentation technology in terms of enzyme production. They include solid-state fermentation (SSF) and submerged fermentation (SmF) (Das & Goyal, 2014; Liu & Kokare, 2017; Al-Manhel, 2018).

2.1 Submerged Fermentation

As the name implies, submerged fermentation (SmF) is a type of fermentation technique in which substrates and microbial strains are submerged in excess of water (Patel et al., 2017; Naureen et al., 2021). SmF has been used in the industrial production of varying enzymes through careful selection of microorganisms (fungi and/or bacteria) in an enclosed vessel containing broth rich in nutrients and high oxygen concentration required for the fermentation period (Thakur et al., 2020; Naureen et al., 2021). As the microorganisms grow, they break down nutrients present in the fermentation medium and, in the process, release the desired enzyme into the solution (Naureen et al., 2021). Bacteria that demand high water activity (a_w) are best suited for SmF (Al-Manhel, 2018); however, fungi such as *Kluyveromyces marxianus* and *Aspergillus niger* have been reported to be used in SmF of industrial enzymes (Al-Manhel, 2018). Industrial production of enzymes often utilizes SmF technology due to its ease of handling on a large-scale basis than solid-state fermentation (SSF) (Patel et al., 2017). Fermenters used for large-scale SmF processes are usually large, with different carrying capacities ranging from thousands up to hundreds of thousands of liters. These fermenters are well developed with an online control system, which is used to monitor and control numerous operational parameters such as foam formation, dissolved oxygen (DO), temperature, and pH. Furthermore, there is little or no problem when it comes to mass transfer and the removal of heat from the fermentation system. Due to some of these benefits, SmF technology is highly regarded and generally accepted for industrial production of metabolites such as enzymes (Patel et al., 2017).

Due to the high cost of obtaining chemically defined media for large-scale production of enzymes via SmF, crude media are often utilized since they are cheap, easily obtainable, and provide a robust nutrient source for microbial activities. Examples of such crude media include whey and corn steep liquor to mention a few (Thakur et al., 2020). Since the fermentation medium in SmF is always in a liquid state, the microorganisms are always in contact with nutrients; oxygen supply, which is essential in SmF, is provided by the actions of the sparger. The continuous mixing of nutrients, gas, suspended particles, and biomass is ensured by the impellers and stirrers (Patel et al., 2017). Although bacteria work perfectly in SmF, molds can also be utilized in the production of enzymes (Thakur et al., 2020). The online provision to monitor and control the fermentation conditions allows for SmF to be adopted in various enzyme-producing industries. Enzymes such as lipases (from *Bacillus licheniformis*), proteases (from *Aspergillus niger*), and β-galactosidases (from *Kluyveromyces marxianus* and *A. niger*) among others have been reported to be produced through the SmF method (Al-Manhel, 2018). After downstream processing, what follows is enzyme purification. This can be achieved using varying chromatographic techniques such as ultrafast preparative and purification liquid chromatography (UFPLC), high-performance liquid chromatography (HPLC), and column chromatography (i.e., affinity, ion exchange, gel exclusion). For analyzing enzyme homogeneity and purity, gel electrophoresis is often carried out (Thakur et al., 2020). The two major running conditions involved in SmF are batch and continuous fermentation (Liu & Kokare, 2017; Al-Manhel, 2018; Thakur et al., 2020).

Batch Fermentation

In this type of SmF, all the nutrients required for the fermentation period are supplied at the beginning of cultivation and, as such, no nutrient is added subsequently (Patel et al., 2017; Naureen et al., 2021). However, control elements like bases, acids, and gasses are added during the fermentation process since it is a closed system; the nutrients are exhausted at the end. This particular batch SmF technique is best suited for rapid experiments such as those involving characterization of strains or nutrient medium optimization. However, the disadvantage of this method is the limitation of product and biomass yield. Due to the fact that oxygen and/or carbon transfer is most times the limiting factor, there is no long exponential phase for the fermentation microorganisms (Naureen et al., 2021).

Continuous Fermentation

Unlike the batch fermentation technique, a continuous culture experiences the addition of fresh nutrients into the batch system when the microorganisms are in the exponential growth phase with a corresponding removal medium containing the yielded products (Patel et al., 2017; Naureen et al., 2021), although there is little fluctuation of metabolites, nutrients, biomass, and cell numbers. However, there is a near balanced cultivation conditions like those found in batch fermentation (Patel et al., 2017; Naureen et al., 2021).

2.2 *Solid-State Fermentation*

An advanced technology employed in the fermentation process is known as solid-state fermentation (SSF). SSF has an edge over SmF from an economic point of view and serves as an ideal alternative to SmF during enzyme production (Al-Manhel, 2018). Unlike SmF, which requires abundant availability of water, SSF does not. SSF simply refers to a process of microbial growth on a solid substrate with the presence of moisture in the solid matrix only in the adsorbed form. This solid matrix either serves as a nutrient source or as an inert material acting as a support for microbial growth. This inert material is usually impregnated with growth solution in the absence of the liquid phase (Fernandes & Carvalho, 2017; Thakur et al., 2020). Most microbial enzymes are produced via the SSF process owing to the fact that enzymes produced through this route are usually stable and concentrated with high-yield products, which are above (5.5 times more) SmF. Since SSF requires low water content, it often helps in reducing or minimizing the cost of extracting the enzymes in their pure forms with an added high enzyme activity. There is also easy aeration within the medium due to low substrate weight per unit of the medium's volume. Furthermore, the extracellular nature of the enzymes makes them more stable across a wide range of temperatures and pH during application (Al-Manhel, 2018).

Owing to the peculiar nature of the SSF technique (i.e., it requires little water content), yeast and fungal cells are best suited in such environments due to the large low water tolerance capacity that they possess (Thakur et al., 2020). The substrates best suited for SSF are mostly residues of an agro-industrial origin. A great number of such substrates have been employed in the past for microbial production of enzymes. These substrates include cassava and tea waste, banana peel, coconut coir pith, corncob, saw dust, grapevine trimming dust, sago hampas, soy hull, rice husk, rice straw, wheat straw, gram bran, maize bran, rice bran, wheat bran, sugar cane bagasse, starch, peanut meal, apple pomace, sweet sorghum pulp, sugar beet pulp, aspen pulp, oil palm mill waste, mustard oil cake, coconut oil cake, rapeseed cake, steam-pretreated willow, steamed rice, corn flour, and wheat flour among many others (Patel et al., 2017; Al-Manhel, 2018). However, wheat band holds the key and has been commonly used in various fermentation processes (Patel et al., 2017).

Enzymes produced by the SSF technique are mostly extracellular, which makes the downstream processing and extraction easy and simple with water or a suitable buffer, followed by centrifugation to separate the solid (i.e., removal of the fungal mycelium) (Fernandes & Carvalho, 2017). The last decade has experienced increased recognition of enzyme production using the SSF process partly as a result of genetically modified organisms (GMOs) being able to produce enzymes more effectively through SSF (Liu & Kokare, 2017).

There are three major advantages attached to SSF. They are (a) it requires simple fermentation equipment with less generation of effluents, (b) products are relatively higher in concentration, and (c) higher volumetric productivity (Liu & Kokare, 2017; Patel et al., 2017). Unlike SmF, which is greatly affected by end-product repression of the fermentation medium, SSF is not affected by end-product repression and, as such, is of economic importance. Liu and Kokare (2017) documented that SSF has amenability potential to fermentation substrates by up to 20–30% unlike the 5% amenability demonstrated by SmF. Enzymes, such as pectinase (*Aspergillus niger*), cellulase (*Trichoderma viride*), lactase (*Aspergillus oryzae*), protease (*Lactococcus lactis*), and α-amylase (*Bacillus* sp.), have all been reported to be produced via SSF on an industrial scale (Thakur et al., 2020).

2.3 Microbial Enzymes, Enzymatic Action, and Types of Producer Microorganisms

Enzymes produced by microorganisms are essential commodities, which have found important applications as metabolic catalysts in numerous industries. These enzymes have gradually taken over the use of chemical catalysts. Industries, nowadays, employ microbial enzymes in the production of good quality products. Microbial sources of enzymes can be categorized, on the basis on individual production, as 20%, 35%, and 50% by bacteria, yeasts, and fungi, respectively (Sonali & Arora, 2020). Some of the enzymes, their actions, and producer microorganisms are discussed in the following sections.

Proteases

Proteases, also termed as "proteinases," "peptidases," or "proteolytic enzymes" (Singh et al., 2019; Mishra et al., 2020), are members of the family of hydrolases who function catalytically in the hydrolysis of protein's peptide bonds (Liu & Kokare, 2017; Pandey et al., 2017; Sindhu et al., 2018; Bhandari et al., 2021). There are two types of proteases based on their evolutionary relationship, type of reaction they catalyze, and the chemical nature of their active sites. They are (1) endopeptidases (responsible for cleaving the internal amino acid bonds) and (2) exopeptidases (responsible for the removal of amino acids from either the carboxy-terminal or the amino-terminal) (Liu & Kokare, 2017; Vieira & Delerue-Matos, 2020). A further subdivision of exopeptidases exists such as (a) carboxypeptidases (i.e., acting on the carboxy-terminal) and (b) aminopeptidases (i.e., acting on the amino-terminal by removing amino acids) (Mishra et al., 2020).

Various investigations have been carried out on proteases in protein engineering and protein chemistry. Proteases have also been investigated and used in practical applications such as in dehairing of animal hair, as food additives, and as cleaning agents (Pandey et al., 2017). Processes of traditional proteolytic fermentation are dependent on microorganisms that exist naturally in raw materials. Some of the commonly isolated proteolytic producing microbes include *Bacillus pumilus*,

Bacillus amyloliquefaciens, Bacillus stearothermophilus, Bacillus cereus, Bacillus lincheniformis, Bacillus subtilis, Streptococcus sp., *Clostridium* sp., *Caldicoprobacter guelmensis, Yarrowia lipolytica, Streptomyces* sp., *Rhizopus oligosporus, Rhizopus oryzae, Mucor racemosus, Penicillium roqueforti, Aspergillus egyptiacus, A. flavus*, and *A. oryzae* (Vermelho et al., 2016; Sonali & Arora, 2020).

Xylanases

Xylanases (EC 3.2.1.8), also known as endo-β-1,4-xylan-xylanohydrolase, which belong to the family of glycoside hydrolases (GH10), are responsible for catalyzing xylan hydrolysis into D-xylose, xylobiose, and xylooligosaccharides (Sindhu et al., 2018; Singh et al., 2019). Microorganisms produce this group of enzymes to cleave xylan, which is a major component of hemicellulose (Singh et al., 2019). Among bacteria, *B. amyloliquefaciens, B. subtilis, Bacillus polymyxa, Bacillus circulans*, and *B. stearothermophilus* are predominantly utilized for producing xylanases. Other genera of bacteria such as *Rhodococcus, Pseudoxanthomonas, Microbacterium, Thermotoga, Arthrobacter, Staphylococcus, Paenibacillus, Micrococcus*, and *Cellulomonas* have all been reported to produce xylanases in considerable amounts (Sindhu et al., 2018, Singh et al., 2019). Among the fungal strains to be well-known producers of xylanases are *Trichoderma, Cephalosporium, Paecilomyces, Geotrichum, Penicillium, Fusarium*, and *Aspergillus. Thermoactinomyces, Thermomonospora*, and *Streptomyces* have all been reported to produce reasonable amounts of xylanases (Sindhu et al., 2018, Singh et al., 2019). Vyas and Yakubu (2020) reported xylanases produced by *Streptomyces* L2001 at a pH of 5.3 and a temperature of 70 °C to possess bioleaching capability.

Laccases

Laccases (EC 1.10.3.2) belong to the group of phenol oxidases. They are phenol-oxidizing enzymes that oxidize aromatic and phenolic compounds including some esters, ethers, and amines through the mechanism of one electron (Pandey et al., 2017; Sindhu et al., 2018; Singh et al., 2019). Electron transfers, coupled with this blue multicopper enzyme oxidase, cause water molecule oxidation (Sindhu et al., 2018; Singh et al., 2019; Preethi et al., 2020). Laccase specificity of a substrate is wide, and, as such, the final electron acceptor is oxygen and, subsequently, peroxide or a cofactor is not required for any catalytic activity. As a result, investigations have been carried out for their possible application in numerous biotechnological processes (Singh et al., 2019).

Several microorganisms produce extracellular and intracellular laccases. These laccases are capable of facilitating the oxidation of aryl diamonds, lignins, polyamines, polyphenols, aminophenols, para- and orthodiphenols, and, likewise, some inorganic ions (Pandey et al., 2017). Most laccase-producing microorganisms have been reported among filamentous fungi such as Basidiomycetes and Ascomycetes. Among these, the white rot fungi are mostly found to be efficient in the breakdown of lignin using their enzymatic system rich in peroxidases, manganese-dependent peroxidases, and laccases (Preethi et al., 2020; Vyas & Yakubu, 2020). White rot fungi with high capacity to produce laccases include *Pleurotus tailandia*, *Pleurotus pulmonarius*, *Pleurotus ostreatus*, and *Pleurotus florida*. Other laccase-producing strains of fungi include *Lentinula* sp., *Grifola* sp., *Coriolopsis* sp., and *Trametes* sp. among many others (Singh et al., 2019).

Lipases

Lipases (EC 3.1.1.3) catalyze numerous reactions including aminolysis, alcoholysis, and esterification (Pandey et al., 2017; Vishnoi et al., 2020). Lipases enhance the hydrolysis of the bonds of carboxyl esters in triglycerides into glycerol and fatty acids (Liu & Kokare, 2017; Vieira & Delerue-Matos, 2020). Lipases from microorganisms are one of the most widely utilized classes of enzymes in biotechnological processes and organic chemistry (Vermelho et al., 2016; Chandra et al., 2020). Yeasts, fungi, and bacteria are good microbial sources of lipases. Approximately, 90% of the global production of lipases comes from microbial sources (Sindhu et al., 2018).

Lipase-producing fungi that are important for the commercial production of lipases mostly belong to the genera of *Rhizomucor*, *Mucor*, *Geotrichum*, *Penicillium*, *Aspergillus*, and *Rhizopus* sp. Yeast species of terrestrial origin that have been frequently associated with lipase production include, but are not limited to, *Candida curvata*, *Candida deformans*, *Candida parapsilopsis*, *Candida cylindracea*, *Candida antarctica*, *Candida tropicalis*, and *Candida rugosa* among many others (Vieira & Delerue-Matos, 2020). *Bacillus alcalophilus*, *B. stearothermophilus*, *Bacillus coagulans*, *B. licheniformis*, *B. pumilus*, and *B. subtilis* are majorly used for lipase production. Other bacteria such as *Staphylococcus caseolyticus*, *Burkholderia cepacia*, *Burkholderia multivorans*, and *Pseudomonas* sp. have all been utilized for the production of lipases (Vermelho et al., 2016; Vieira & Delerue-Matos, 2020; Vyas & Yakubu, 2020).

Chitinases

Chitinases (EC 3.2.1.14) catalyze the hydrolyzation of the monomer of chitin, *N*-acetyl-D-glucosamine. In nature, cellulose is the most abundant polysaccharide, and next to it is chitin. Chitin is important for various physiological functions with potential applications such as in the degradation and treatment of biowastes, chitooligosaccharide production, and phytopathogen control (Mishra et al., 2020; Vieira & Delerue-Matos, 2020). On the basis of function, chitinases are divided into two major types, exochitinases and endochitinases. Endochitinases (EC 3.2.1.14) randomly catalyze the cleavage of internal points across the entire length to produce

N-acetyl glucosamine multimer (i.e., chitotetraose and chitotriose) and dimer diacetylchitobiose. There are two types of endochitinases: chitobiosidases (EC 3.2.1.29) are responsible for the cleavage of nonreducing ends and, as a result, produce diacetylchitobiose in a stepwise manner. The second type is β-1,4-glucosaminidases (EC 3.2.1.30). This group of enzymes is responsible for the cleavage of oligomers obtained through monomers of *N*-acetyl glucosamine (Mishra et al., 2020).

Microorganisms with the potential of chitinase production include *Pseudomonas cepacia*, *Serratia marcescens*, *Micromonospora carbonacea*, *Paenibacillus ehimensis*, *Streptomyces viridodiasticus*, *Pseudomonas fluorescens*, *Bacillus aerius*, *B. subtilis*, *Pseudomonas putida*, *Geobacillus thermodenitrificans*, *Trichoderma asperellum*, *Aeromonas caviae*, and *Enterobacter agglomerans* (Mishra et al., 2020).

Inulinases

Inulinases (EC 3.2.1.7) catalyze the hydrolysis of the β-2,1 linkage of inulin into fructooligosaccharides or fructose (Vieira & Delerue-Matos, 2020). Inulinases belong to the family of glycoside hydrolases (GHs) 32 and 91. The GH32 family mainly consists of endoinulinases (EC 3.2.1.7), exoinulinases (EC 3.2.1.80), 1-exohydrolase (EC 3.2.1.153), 1,2-β-fructan 1F-fructosyltransferase (EC 2.4.1.100), and sucrose 1F-fructosyltransferase (EC 2.4.1.99). Production of inulinases using microorganisms has become a major industrial choice due to the advantages these microorganisms provide such as genetic manipulation, easy handling, high yield and rapid production, and considerable variability in the biochemical and biophysical characteristics (Singh et al., 2019).

Microbial production of inulinases is prominent among *Streptomyces* sp., *Pseudomonas* sp., *Bacillus* sp., *Kluyveromyces* sp., *Penicillium* sp., and *Aspergillus* sp. However, fungal inulinases are preferred over other sources since they have the capacity to withstand high temperature and low pH conditions. Furthermore, fungal strains require low substrate concentration for optimal growth and high product yield (Singh et al., 2019). Among the Aspergilli, the prominent producers of inulinases are *Aspergillus tubingensis*, *Aspergillus terreus*, *Aspergillus tamari*, *Aspergillus ficuum*, *Aspergillus niveus*, *Aspergillus tritici*, and *A. niger*, whereas among the Penicilli, they are *Penicillium trzebinski*, *Penicillium rugulosum*, *Penicillium purpurogenum*, *Penicillium subrubescens*, *Penicillium ocalicum*, and *Penicillium expansum*. Bacterial strains such as *Xanthomonas* sp., *Streptomyces* sp., and *Clostridium* sp. among many others have been demonstrated to produce inulinases due to their capacity to tolerate high salinity, alkalinity, acidity, and temperature (Singh et al., 2019). Yeast strains such as *Zygosaccharomyces cerevisiae*, *Cryptococcus aureus*, and *Meyerozyma guilliermondii* have also been demonstrated to be efficient producers of inulinases (Singh et al., 2019).

Amylases

Amylases belong to a class of enzymes that facilitates the hydrolysis of starch into sugars like maltose and glucose (Bhatt et al., 2020). There are three subclasses of amylases. They are α-, β-, and γ-amylases on the basis of the type of link/bond they cleave (Liu & Kokare, 2017; Bhatt et al., 2020). α-Amylases (EC 3.2.1.1) catalyze the hydrolysis of the internal α-1,4-*O*-glycosidic bonds in a polysaccharide; β-amylases (EC 3.2.1.2), on the other hand, enhance the hydrolysis of the bonds of α-1,4-glucan to give rise to successive units of maltose; and γ-amylases (EC 3.2.1.3) are responsible for the cleavage of the α-(1–4) glycosidic bonds and the α-(1–6) glycosidic bonds of amylopectin and amylose nonreducing ends (Vieira & Delerue-Matos, 2020). Amylases are of great importance to present day biotechnology with varying applications ranging from fermentation in breweries to food, paper, and textile industries.

There are different sources through which amylases are obtainable; however, microbial sources generally satisfy industrial demands. They have been derived from actinomycetes, yeasts, bacteria, and fungi (Vermelho et al., 2016). Among the aforementioned microbial sources, bacterial and fungal sources are mostly employed in industries. Amylases have been reported in *Thermomonospora fusca*, *Streptomyces* sp., *Bacillus flavothermus*, *B. amyloliquefaciens*, *B. subtilis*, *Lactobacillus plantarum*, *Saccharomyces cerevisiae*, *Lipomyces kononenkoae*, *Fusarium oxysporum*, *Thermomyces lanuginosus*, *Clavatia gigantea*, *Aspergillus oryzae*, and *Aspergillus awamori* (Vermelho et al., 2016; Vyas & Yakubu, 2020).

Tannases

Tannase (EC 3.1.1.20), also known as tannin acyl hydrolase (Singh et al., 2019; Vieira & Delerue-Matos, 2020), is a known hydrolytic enzyme that enhances the hydrolysis of depside bonds and esters present in the esters of gallic acid, complex tannins, ellagitannins, and gallotannins, with a resultant release of alcohols/glucose, ellagic acid, and gallic acid (Vieira & Delerue-Matos, 2020).

The enzyme tannase is ubiquitous, that is, it can be derived from various living things including animals, plants, and microbes. Microbial tannases, however, are preferred over other sources due to ease in their genetic manipulation, enzyme stability, and higher yield of enzymes. Microbial strains from yeast bacteria and fungi can produce tannases efficiently. Filamentous fungi, particularly those belonging to Aspergilli, such as *A. awamori*, *Aspergillus gallonyces*, *Aspergillus japonicus*, *A. oryzae*, and *A. niger*, have all been used efficiently in tannase production, whereas bacteria such as those belonging to the *Bacillus* strains such as *B. polymyxa* and *B. pumilus* have all been reported to be potent in tannase production. Similarly, other microbial strains such as *S. cerevisiae*, *Selenomonas ruminantium*, *Streptococcus bovis*, *Klebsiella pneumonia*, and *Corynebacterium* sp. have all been reported in tannase production (Singh et al., 2019).

β-Galactosidases

β-Galactosidases (EC 3.2.1.23) are hydrolase enzymes responsible for the hydrolysis of lactose to galactose and glucose (Vermelho et al., 2016; Sindhu et al., 2018). These enzymes also catalyze transglycosylation reactions, permitting the transfer of groups of hydroxyl galactose to disaccharide lactose (Vermelho et al., 2016). As a result of these enzyme activities, β-galactosidases are of utmost importance in industries, particularly the agro-food industry, where they are added to food products to help individuals who are lactose-intolerant as a result of lactase deficiency (Vermelho et al., 2016).

Microbial production of β-galactosidase is always a preferred choice over other sources (i.e., plants) because of high yield with low cost of production. The choice of source of β-galactosidase is sometimes influenced by its use or application. For example, β-galactosidase obtained from yeasts with an optimum pH of 6.5–7.0 is mostly used for hydrolyzation of lactose in whey or milk (Sindhu et al., 2018). β-Galactosidases used for commercial purposes are produced from yeasts, such as *Kluyveromyces marxianus* and *Kluyveromyces lactis*, and molds including *A. oryzae*, *A. niger*, *B. licheniformis*, *Enterobacter agglomerans*, *Sulfolobus solfataricus*, *Lactobacillus acidophilus*, *Thermotoga maritima*, *Lactococcus lactis*, and *Arthrobacter psychrolactophilus* are all bacterial strains that have been demonstrated to produce β-galactosidases (Vermelho et al., 2016).

Cellulases

The liberation of glucose units through hydrolyzation of polymeric cellulose and β-1,4 linkages is catalyzed by cellulases (Vermelho et al., 2016). Cellulases are grouped into three major classes: exo-(1,4)-β-D-glucanase (EC 3.2.1.4), endo-(1,4)-β-D-glucanase (EC 3.2.1.4), and β-glucosidases (EC 3.3.1.21) (Liu & Kokare, 2017; Sindhu et al., 2018; Mishra et al., 2020; Vieira & Delerue-Matos, 2020). Cellulase catalytic modules belong to the GH family and have been further classified into varying groups on the basis of different sequences of amino acids as well as the 3D structural features (Sindhu et al., 2018). The family of GH enzymes mainly utilizes the mechanism of acid–base catalysis for the cleavage of glycoside linkages in cellulose. Two residues, a nucleophile and a proton donor, in the enzyme's active site are utilized for achieving cellulase catalysis. Depending on the spatial positions of the aforementioned catalytic residues, the cellulase-catalyzed reaction occurs through the inversion and retention mechanism (Sindhu et al., 2018).

Several microorganisms such as fungi, actinomycetes, and bacteria have been reported to produce cellulases. However, strains of fungi are preferred choices for cellulase production owing to the fact that they are produced extracellularly. The most prominent producer of exo- and endoglucanases is *Trichoderma reesei*; however, it does not produce enough quantity of β-glucosidases. However, strains of *Aspergillus* such as *A. awamori*, *A. terreus*, *A. oryzae*, and *Aspergillus nidulans* are excellent producers of β-glucosidases (Bansal et al., 2012). Other microorganisms that produce cellulases include *Streptomyces lividans*, *Streptomyces drozdowiczii*, *Streptomyces veridobrunneus*, *B. licheniformis*, *B. amyloliquefaciens*, *B. pumilus*, *B. subtilis*, *Pleurotus ostreatus*, *Phanerochaete chrysosporium*, *Paecilomyces inflatus*, *Thermoascus aurantiacus*, *Mucor circinelloides*, *Trichoderma atroviride*, *Trichoderma longibrachiatum*, *Trichoderma harzianum*, *Penicillium echinulatum*, *Melanocarpus albomyces*, and *Humicola grisea* (Vermelho et al., 2016).

Peroxidases

Peroxidases (EC 1.11.1.7) are a group of oxidoreductase proteins containing a prosthetic group known as iron (III) protoporphyrin IX (Sindhu et al., 2018). The oxidation of a wide range of organic and inorganic compounds as well as peroxide reduction is catalyzed by peroxidases (Pandey et al., 2017; Sindhu et al., 2018). Peroxidases have the capacity to degrade highly potential toxic substances that occur in nature. As a result, manganese-dependent peroxidase (MnP) and lignin peroxidase (LiP) have been widely studied (Pandey et al., 2017). Animals, plants, and microorganisms experience peroxide activities. In plants, peroxidases are involved in a number of lignification processes and also form a defense against infectious and damaged tissues (Sindhu et al., 2018).

Phanerochaete chrysosporium is the microorganism with the highest capacity to produce peroxidases. Fungal peroxidases sometimes encounter challenges when it comes to industrial applications such as those associated with protein post-translational modifications. On the contrary, peroxidases produced by bacteria face little challenges in production; they are stable and suitable for industrial use (Sindhu et al., 2018).

Pectinases

Pectinases (EC 3.2.1.15) contain a group of enzymes, which are responsible for the degradative catalysis of pectic substances through deesterification (esterases) and depolymerization (lyases and hydrolases) reactions (Liu & Kokare, 2017). In terms of functionality, pectinases are grouped as polygalacturonases (that hydrolyze the α-(1–4) glycosidic bonds), pectin esterases (responsible for the removal of methyl and acetyl groups from pectin), pectate, and pectin lyases (Sindhu et al., 2018). Pectin has the properties of a hydrocolloid, which is hydrophilic and forms a gel easily with water under certain conditions. As a result, the addition of pectinase improves the pressability of pectin gel as well as reduces the viscosity (Vieira & Delerue-Matos, 2020). Pectinases are commonly used in processes that involve degradation of materials of plants such as improving or enhancing the extraction rate of fruit juice from fruits such as sapota and apples. Since the 1960s, pectinases have been employed in wine production for juice clarification (Vermelho et al., 2016). Varying microorganisms have been reported to produce pectinases in the past; they include filamentous fungi, yeasts, actinobacteria, and bacteria. Examples of microbial strains that produce pectinases are *Lactobacillus lactis*, *Pseudomonas solanacearum*, *Erwinia chrysanthemi*, *Penicillium occitanis*, *A. niger*, *S. cerevisiae*, and *Rhodotorula* sp. In the commercial preparations of pectinases, more than one pectinolytic microorganism is used, although it depends on the final application of enzymes such as amylases, peptidases, hemicellulases, and cellulases (Vermelho et al., 2016).

Catalases

Catalases (EC 1.11.1.6) are enzymes that are components of aerobic microorganisms. Catalases are proteins that are tetrameric in nature (Sindhu et al., 2018); they catalyze a reaction to cause the conversion of hydrogen peroxide (H_2O_2) to a molecule of water and oxygen (H_2O + O) (Liu & Kokare, 2017; Vieira & Delerue-Matos, 2020). Catalases help in protecting cellular proteins against any form of reacting oxygen species (ROS) by protecting the oxidative inactivation of glucose-6-phosphate dehydrogenase (Vieira & Delerue-Matos, 2020).

Catalase production is carried out the most by *Micrococcus luteus* and *A. niger* (Sindhu et al., 2018). Others include *Comamonas terrigena*, *Comamonas testosteroni*, *Oceanobacillus oncorhynchi*, *Bacillus halodurans* LBK, *Bacillus marocanus*, *Pyrobaculum calidifontis*, *Enterococcus faecalis*, *Bacteroides fragilis*, *Rhizobium radiobacter*, *Psychrobacter piscatorii* T-3, and *Metarhizium anisopliae* (Sindhu et al., 2018).

Glucose Isomerases

Glucose isomerases (EC 5.3.1.9) are used to catalyze aldehyde sugar glucose conversion to their respective ketone sugar fructose that is twice as sweet. This particular feature is the major reason why glucose isomerases are used in the production of high-fructose syrups (HFSs) containing approximately 53–54% glucose and 42% fructose (Vermelho et al., 2016).

Glucose isomerase production can be obtained industrially using microorganisms such as *Anoxybacillus gonensis*, *Clostridium thermosulphurogenes*, *Thermus thermophilus*, *B. subtilis*, *Bacillus thermoantarcticus*, *Actinoplanes missouriensis*, *Streptomyces rubiginosus*, and *Streptomyces murinus* (Vermelho et al., 2016).

Invertases

Invertases (EC 3.2.1.26), also known as saccharases, invertins, sucrases, β-fructofuranosidases, and β-D-fructofuranosidase fructohydrolases, catalyze the formation of two monosaccharides (i.e., fructose and glucose) by hydrolyzing terminal nonreducing β-fructofuranoside (Singh et al., 2019; Vieira & Delerue-

Matos, 2020). Invertases like glucose isomerase are used in HFS preparation by sucrose inversion (Vermelho et al., 2016).

Penicillium sp. and *Aspergillus* sp. are among the fungal strains with high potential for producing invertases. They include *P. purpurogenum*, *Penicillium pinophilum*, *Penicillium citrinum*, and *Penicillium chrysogenum* among the Penicilli and *A. flavus*, *Aspergillus ochraceus*, *A. japonicas*, *Aspergillus caespitosus*, *A. oryzae*, *Aspergillus casiellus*, *Aspergillus fumigatus*, *Aspergillus parasiticus*, and *A. niger* among the Aspergilli. Other strains of fungi that are invertase producers include *Thermomyces lanuginosus*, *Rhodotorula glutinis*, *Cladosporium cladosporiodes*, and *Sclerotinia sclerotiorum*. Strains of bacteria with the capacity to produce invertase include *Bifidobacterium breve*, *B. cereus*, and *Arthrobacter globiformis*, whereas strains of yeasts with high invertase yield include *Hansenula polymorpha*, *Leucosporidium antarcticum*, and *Kluyveromyces marxianus* (Vermelho et al., 2016; Singh et al., 2019).

Keratinases

Keratinases (EC 3.4.21) are a group of enzymes that belong to metallo/serine peptidases, which are responsible for the hydrolysis of keratins (Vermelho et al., 2016, Singh et al., 2019). Keratinases have specificities toward a wide range of protein substrates that are proteinase K-resistant, indicating their differences from the conventional proteases. The structure of keratin is highly resistant to natural degradation as well as degradation enhanced by proteases and, as such, requires keratinases, which are a specific class of enzymes that are proteolytic in action (Singh et al., 2019).

Keratinases are produced by several bacteria as their extracellular material. A majority of these bacteria belong to the *Bacillus* genus such as *B. licheniformis* and *B. subtilis*, whereas *Aspergillus fumigates*, a fungus that has been demonstrated in the past to utilize flour of chicken feathers as a source of nitrogen and carbon, belongs to the *Aspergillus* genus (Vermelho et al., 2016).

Transglutaminases

Transglutaminase (TGase) (EC 2.3.2.13) is an enzyme that catalyzes the reaction involving the transfer of acyl between groups of primary amines and peptide-bound carboxyamide (Vermelho et al., 2016). As a result of the aforementioned reaction, there is isopeptide linkage formation in the matured protein (Sonali & Arora, 2020).

Three approaches exist for TGase industrial production. They are (1) extraction and purification of the enzyme from animals such as swine, cattle, or fish; (2) use of microorganisms such as *Streptoverticilium* sp., *Aspergillus* sp., *Bacillus* sp., and *Escherichia coli* for enzyme production, where the quantity produced can be upregulated through microbial genetic manipulation; and (3) discovery of new microbial strains with high potential to produce TGase. TGase is currently produced commercially using *Streptoverticilium* sp. Several research works have been put in place to increase the availability of TGase with the use of DNA recombinant technology to induce TGase production in *E. coli* (Vermelho et al., 2016).

3 Microbial Enzyme Applications in Industry

The global production of industrial enzymes is majorly dominated by enzymes of microbial origin, since 15% are obtained from plants, whereas bacteria alone hold a record share of 35% of the total industrial enzymes with yeasts and fungi having a combined 50% share (Kapoor et al., 2020). The importance of microbial enzymes in industrial processes cannot be overemphasized. They are more stable, efficient, and can be used under extreme conditions (i.e., temperature and pH). The use of microbial enzymes in various industries helps in eliminating the use of harsh chemicals, organic solvents, and extreme temperatures and pH in the process and produces high-quality products, often of high purity, with low production cost and risks (Kapoor et al., 2020). The wastes generated have low toxicity, thus reducing environmental impact and other risks associated with chemical catalysts (Sanchez & Demain, 2017). Furthermore, production of enzymes using microorganisms provides an added advantage since these microorganisms can be upregulated and cultured in large amounts within a short fermentation period to produce large quantities of the enzymes (Kapoor et al., 2020). Microbial enzymes are essential and needed in various bioprocesses at the industrial level including, but not limited to, the biofuel, animal feed, chemical, beverage, food, pharmaceutical, textile, detergent, leather, cosmetic, and paper and pulp industries among many others (Avendaño et al., 2016; Singh et al., 2016, 2019). Below are some of the applications of enzymes in various industries.

3.1 Food Industry

For centuries, humans have relied on enzymes and microorganisms for the production of various food products (Fernandes & Carvalho, 2017). Currently, there is a surge in the world's population, and, in order to meet the increase in demand for quality food, humans have sought out enzymes that can be used in the biotransformation of raw food materials into good, nutritive food products. These enzymes, also known as biomolecules, help in enhancing the shelf-life, texture, color, aroma, flavor, and nutritive value of food (Avendaño et al., 2016). However, safety issues concerning the involvement of enzymes in food and ingredient production have been raised. A typical example is seen in sweetener technology as well as in fat alteration, which has increased enzyme proliferation in the respective applications. The food industry enzyme application is bifurcated into various sectors such as breweries, juice, baking, and dairy (Sonali & Arora, 2020).

Fruit Juice Production

Enzymes have been employed to play important roles in processing and preparing various vegetable and fruit juices such as lemon, carrot, pineapple, grapefruit, orange, and apple among many others (Liu & Kokare, 2017). Some of these enzymes used in fruit juice preparation aid in increasing the efficiency of certain operations, such as clarification, juicing, extraction, peeling, and extraction, and in improving the stability and quality of the final product (Singh et al., 2016). The involvement of enzymes in vegetable and fruit juice production helps in digesting the starch, pectin, cellulose, and proteins that they contain. Thus, the processing time is reduced with an improved yield as well as enhanced sensory properties (Singh et al., 2016).

Pectin is a hydrocolloid that has a high affinity to water, which is capable of easily forming a gel under certain conditions. Pectin is abundant in vegetables and fruits. The addition of polygalacturonase, pectin esterase, pectin, or pectinase helps in collapsing the pectin gel often formed in fruit juice, thereby reducing the viscosity in the process and improving pressability (Liu & Kokare, 2017). The juice present in the pulp of fruits can be liberated through the addition of cellulase, xylanase, and pectinase (Patel et al., 2017). For enhanced process yield and performances, cellulase in synergy with other enzymes (i.e., amylase) that enhance maceration is used. Cellulases also play an important role in fruit juice extraction as well as in the stabilization and clarification of fruit juice (Singh et al., 2016). Sindhu et al. (2018) reported the use of cellulase in flavonoid extraction from seeds and flowers of fruits. The reason why cellulase is preferred over conventional methods of extraction purposes is due to the low processing time required as well as the use of low heat, which prevents damages usually encountered with conventional methods. In addition, cellulase promotes high-yield products (Sindhu et al., 2018). Cellulases are also used for extracting phenolic compounds present in grape pomace (Sindhu et al., 2018).

Cellulases and amylases are utilized to reduce the viscosity of purees and nectars in fruits such as peach, pear, papaya, plum, mango, and apricot among others (Singh et al., 2016; Sindhu et al., 2018). Cellulase has been reported to be used in reducing citrus fruit bitterness as well as enhancing its taste and aroma (Sindhu et al., 2018). Similarly, limonene and naringinase (EC 3.2.1.40) are employed as debittering enzymes to hydrolyze the bitter components of citrus juice (Singh et al., 2016). The combined action of pectinases and β-glucosidases causes alterations to the aroma, flavor, and structure of vegetables and fruits. The scavenging of oxygen present in fruit juice is carried out by flavoprotein glucose oxidase in order to prevent unwanted changes in taste and color that usually occur during storage (Liu & Kokare, 2017). Liquefaction of vegetables and fruits is completely carried out by the actions of amylases and hemicellulases together with pectinases (Singh et al., 2016).

Dairy Products

Dairy products occupy a vast area in the food industry today. Dairy products include yogurt, ice cream, butter, cheese, cream, and milk among many others. The major functions of microbial enzymes in the aforementioned dairy products are to improve the processes involved or enhance the features of the final products, including structure, texture, appearance, composition, consistency, aroma, color, and flavor (Avendaño et al., 2016; Singh et al., 2016).

In the digestion of fat, lipases are the major enzymes employed to catalyze triacylglycerol reactions to di- or monoglycerides, glycerol, and fatty acids (Avendaño et al., 2016; Al-Manhel, 2018). Most researches have indicated the application of lipases in the dairy industry, especially in accelerating the ripening of cheese and in the hydrolysis of fat present in milk. Lipase is also used for enhancing flavors in cheese (Chandra et al., 2020). A variety of products can be produced via biochemical pathways, which could be either primary or secondary with different characteristics of the end products (Al-Manhel, 2018). The type of lipase source determines its application in cheese production (Avendaño et al., 2016; Sindhu et al., 2018). For instance, Romano cheese is produced with pregastric lipases obtained from lambs/kid goats. Camembert cheese, however, is produced with lipases obtained from *Penicillium camemberti*; lipases from *A. oryzae* or *A. niger* are used in Cheddar cheese production (Sindhu et al., 2018). The texture and softness of dairy products can be enhanced via lipase catalyzation. In margarine and butter, lipases are used to enhance the flavor (Sindhu et al., 2018).

β-Galactosidase or lactase is another essential enzyme within the confinement of the dairy industry (Fernandes & Carvalho, 2017). The hydrolysis of galactose and glucose from lactose is made possible through the catalytic actions of lactase. Lactase acts as a digestive enzyme, which improves the sweetness and solubility of dairy products. People who are lactose-intolerant can now consume dairy products that have been hydrolyzed with lactose without having any negative reactions to their bowel (Liu & Kokare, 2017). Lactose intolerance often occurs as a result of an individual's inability to produce the lactase digestive enzyme (Fernandes & Carvalho, 2017). The lactose content of milk and other milk products needs to be removed or minimized through the actions of lactase in order to prevent diarrhea, severe dehydration of tissues, and other fatal consequences in lactose-intolerant individuals (Singh et al., 2016). Dairy wastes can also be treated using lactase before disposal; moreover, these wastes can be transformed into other useful substances (Liu & Kokare, 2017). Lactase can also be used to enhance sweetness in milk and its by-products through the hydrolysis of lactose into glucose and galactose. As such, addition of sugar is minimized or completely eliminated in milk drinks.

Whey, known as the liquid/watery part of milk that is generated during the processing of milk to curd, is treated using lactase. Traditionally, the large amounts of whey generated after curd production have been used as animal feeds or fertilizers; sometimes, they are simply dumped in watercourses or sewers, which is of course a serious concern to environmental experts owing to the deleterious effects they have on the ecosystem (Fernandes & Carvalho, 2017). Whey can be used in the formation of sweet syrup by treating it with lactase to hydrolyze the lactose in it. This sweet syrup eventually formed has been employed as a source of sugar in feedstuff, ice creams, dairy desserts, and other confectioneries (Fernandes & Carvalho, 2017).

About 20–30% of coagulants used across the globe in milk production are represented by chymosin. Like chymosin, enzymes such as proteases, proteinases, and peptidases are utilized as coagulants in the production of cheese (Singh et al., 2016). Cheese ripening is favored by proteinases; they accelerate the hydrolysis of protein, which is one of the most vital biochemical events encountered at this stage with a resultant impact on the flavor and texture of the final product (Liu & Kokare, 2017). Peptidases, on the other hand, cause the removal of bitterness, usually produced during ripening by proteinases (Singh et al., 2016). The functional properties of dairy products are usually improved as a result of polymerization reactions catalyzed by transglutaminase (Singh et al., 2016).

Baking

Microbial enzymes applied in the baking industry provide dough stability, flour enhancement, prolonged softness of crumb and freshness, and increased volume, improve color and texture, and maintain uniform structure of crumbs (Avendaño et al., 2016, Singh et al., 2016).

Traditionally, bread and baked food products have been produced with the aid of proteases. The proteolytic properties of proteases have been advantageously used on dough and gluten. As a result, the time required for mixing is reduced with improved uniformity throughout the dough (Fernandes & Carvalho, 2017; Patel et al., 2017). Aside from that, the presence of proteases in dough helps in regulating the strength of gluten, thereby making kneading and pulling easier. The use of endopeptidases is, however, more pronounced as their actions positively affect the dough rheology and network of gluten formed (Fernandes & Carvalho, 2017). Exopeptidases, on the other hand, impact the color and flavor of dough due to the Maillard reactions they catalyze with a resultant release of sugars and amino acids present. Owing to the environmental friendliness of proteases, they are now used in substituting sodium metabisulfite in the conditioning of dough (Fernandes & Carvalho, 2017).

Amylases (α-amylases and β-amylases) could be used singly or synergistically in the baking industry (Fernandes & Carvalho, 2017). The hydrolysis of starch to dextrin is catalyzed by α-amylases (releasing dextrins that are low in the molecular chain); a further hydrolysis to maltose is catalyzed by β-amylases, which can be utilized by yeasts as a fermentable sugar (Fernandes & Carvalho, 2017). The actions of these amylases greatly enhance the shelf-life, bread volume, freshness, and softness and reduce the viscosity of dough (Singh et al., 2016; Sonali & Arora, 2020). Furthermore, the formation of reducing sugars promotes the Maillard reactions, as evidenced by the browning of crust that intensified the pleasant flavor. The synergistic actions of glucoamylases and maltogenic amylases prevent or minimize staling during storage. Staling is a physical and chemical process that decreases the palatability of baked food products (Sonali & Arora, 2020). Staling results in a noticeable increase in crumb elasticity and in the leathery and tough crust appearance (Fernandes & Carvalho, 2017; Sonali & Arora, 2020).

Other enzymes such as xylanases, transglutaminases, and laccases enhance the quality and elasticity of dough by forming a fine and homogeneous structure (Singh et al., 2016). The enzymes also add to the freshness, softness, and color and increase volume while decreasing the off-flavor at the end of production (Avendaño et al., 2016; Sonali & Arora, 2020). During production, the flour is whitened by the actions of lipoxygenases; likewise, the viscoelastic properties of the dough are also improved (Avendaño et al., 2016). The flavors of bakery products are also improved by lipases by releasing short-chain fatty acids, which aid in preserving as well as increasing the shelf-life of baked products. The use of lipases in synergy with emulsifiers (i.e., insulin) provides some effects that are beneficiary in terms of rheological modifications such as modification of cake crumbs with high cell structures that are homogeneous (Singh et al., 2016; Chandra et al., 2020; Sonali & Arora, 2020). During baking, the formation of acrylamide is inhibited by asparaginase through free asparagine hydrolysis to aspartic acid. The concentration of acrylamide was reported to be reduced by 95% with no effects on the sensory properties of the final products (Avendaño et al., 2016). The evaluation of asparaginase was made by the Food and Agriculture Organization (FAO); in their observations, they found that asparaginase activity depends on some factors such as the type of ingredient used, dose, and reaction conditions utilized during processing (Avendaño et al., 2016).

Beverages

For production of fermented beverages like beer and wine, the primary focus is on the improvement of yield, optimization of processes, and enhancement or maintenance of flavors and colors. By involving enzymes in beverage production, the aforementioned objectives are achieved. Aside from that, the calorie levels of sulfur and beer are reduced in the process of clarifying wine (Avendaño et al., 2016).

Pectinases are essential in beer and wine production; they facilitate the extraction processes, filtration, and aid in improving the yield of juice, odor, flavor, and clarity (Avendaño et al., 2016; Fernandes & Carvalho, 2017). Grapes used for wine production contain pectinases; however, these pectinases have low activity. As such, pectinases from microbial enzymes are used due to their stability and ability to withstand high fermentation conditions. Producers of flavored beverages and vegetable and fruit juices are more interested in tackling problems related to viscosity, clarification, quality, stability, and yield, which are important features/conditions that influence consumer acceptance of the particular product (Avendaño et al., 2016).

One of the problems encountered in the brewing industry is the formation of haze on products. However, this can be avoided through the use of laccases, as demonstrated in the past in the oxidation of polyphenol compounds. In addition, laccases improve the lifespan of beer through the removal of oxygen in the last step involved in the production of beer (Sindhu et al., 2018). Off-flavor, encountered in beverages as a result of the presence of phenolic compounds, can be reduced, if not removed completely, through the actions of laccases (Avendaño et al., 2016). Polyphenols are traditionally removed from white wines through the application of sulfur at a high dose or the use of polyvinylpolypyrrolidone, which usually affects the organoleptic properties of wine. However, the use of laccases can effectively remove these polyphenolic compounds without interfering with the taste since it selectively removes the polyphenolic compounds (Avendaño et al., 2016, Fernandes & Carvalho, 2017). Flavourstar is a commercial laccase produced by Novozymes and has been employed industrially for the removal of compounds that cause off-flavor in the brewing industry (Sindhu et al., 2018).

In the distillation of alcoholic beverages, microbial amylases are utilized in the hydrolysis of starch into fermentable sugar prior to the fermentation proper; at the end, turbidity is minimized, giving the beverage a pleasant look (Fernandes & Carvalho, 2017; Patel et al., 2017). The usage of enzymes in the hydrolysis of starch adjuncts and other unmalted barley adjuncts facilitates production in the long run and reduces the entire production cost (Singh et al., 2016). Proteases are used in controlling chill haze in beer production (Singh et al., 2016; Patel et al., 2017). Glucanases are another group of enzymes that is extremely important in the brewing industry; they are essential in breaking the cell walls of cereals and grains. For example, the cell wall of barley is made up of glucans (70%), which are not easily degraded. However, the use of glucanase will help hasten the whole production process, thereby reducing processing time as well as cost. Similarly, xylanase contributes greatly to the degradation of plant cell walls, since they have been demonstrated to be effective in the breakdown of nonstarch polysaccharides such as arabinoxylans, which are also present in significant amounts (Liu & Kokare, 2017). Proteases (exo- and endopeptidases), on the other hand, hydrolyze the large-chain protein molecules present in the cell walls of cereals, thereby facilitating the accessibility of amylolytic enzymes to starch; in the process, amino acids and other small peptides are made available in the fermentation medium with an ultimate influence on the final flavor (Fernandes & Carvalho, 2017; Liu & Kokare, 2017). It is important to state that excess of proteolysis has negative effects on the foam stability of beers, whereas deficiency in proteolysis also affects the stability of colloids in beer. As such, proteolysis needs to be controlled to avoid its negative impacts on beer (Fernandes & Carvalho, 2017). Other enzymes such as α-glucosidases, pullulanases, amyloglucosidases, and α- and β-amylases are also required in processes involved in the hydrolysis of starch to units of glucose (Fernandes & Carvalho, 2017).

3.2 Textile Industry

A lot of waste is being generated from the textile industry due to some activities such as fabric desizing and bleaching, which involves the use of dyes and chemicals often resulting in environmental pollution (Singh et al., 2020). As a sequel to that, technologies that are friendly to the environment with a relative production of quality products are preferred (Singh et al., 2016). One of the technologies greatly used is in the use of enzymes in processing fibers (Singh et al., 2016). Oxidoreductases and hydrolases are groups of enzymes utilized in the pretreatment as well as finishing of cotton materials (Singh et al., 2016). Hydrolases such as esterases/lipases, pectinases, proteases, cutinases, cellulases, and amylases are all involved in wool finishing, desizing, denim finishing, cotton softening, wool anti-felting, synthetic fiber modification, bioscouring, and polishing of fabrics (Patel et al., 2017). Oxido-reductases as well as other groups of enzymes such as ligninases, peroxidases, laccases, and catalases are all involved in wool finishing, dye decolorization, bleach termination, and biobleaching among others (Singh et al., 2016, Patel et al., 2017).

3.3 Detergent Industry

The contribution of enzymes to the growth and development of the detergent industry cannot be overemphasized, since it is practically involved in the efficiency and satisfaction we derive today from its usage. The use of detergents, such as in laundering, dishwashing, and institutional, industrial, and domestic cleaning, is vast (Sonali & Arora, 2020). Enzymes present in detergents function in the removal of stains such as starch, proteins, fats, and oils from materials (Singh et al., 2016). Proteolytic enzymes mostly produced by *Bacillus* sp., such as *Bacillus brevis* and *B. cereus*, are used by detergent companies in the making of detergents (Sonali & Arora, 2020).

3.4 Paper and Pulp Industry

The past two decades have witnessed intensive studies on the applications of varying enzymes in the paper and pulp industry. The use of microbial enzymes in this industry has been able to alleviate sustainability issues exhibited toward the safety of the ecosystem. As a result, utilization of efficient enzymes has been able to reduce the usage of harsh chemicals and energy as well as processing time through improved bleaching and deinking (Sonali & Arora, 2020). Wastes generated from this industry are also treated using enzymes, thereby increasing the chemical oxygen demand (COD) similar to the biological oxygen demand (BOD) (Singh et al., 2016). Kraft pulp prebleaching is currently the major application of enzymes. The increment in water preservation and pulp fibrillation in virgin pulps reduces the processing time and cost of paper production (Singh et al., 2016).

In the production of paper, lignin needs to be degraded and separated from wood pulp. Before now, lignin was traditionally removed by reagents containing chlorine, which is a potential pollutant to the environment. However, this could be overcome using laccases. Laccases are also important in the decolorization and degradation of chlorophenol, chlorolignin, and waste effluents generated from paper and pulp mills (Sonali & Arora, 2020). Paper and pulp mill effluents usually contain pigments from elementary mixtures together with extracts from plants with high presence of cellulose and other substances such as heavy metals (Sonali & Arora, 2020). Lignin is the major component of phenolic compounds that is solubilized and eradicated during the pulping process (Sonali & Arora, 2020). Filamentous fungi can be used in the production of various classes of laccases, which can be utilized in the deterioration of wastewater. Other areas of interest in the use of laccases include biopulping (Singh et al., 2016; Sonali & Arora, 2020).

Another group of important enzymes used in the paper and pulp industry is xylanases. Xylanases are applied in pulp bleaching with a resultant liberation of fragments of lignin through hydrolysis of xylan. As such, the need for chemicals that are chlorine-based with bleaching potential is greatly reduced (Patel et al., 2017). Pretreatment of wood with xylanases improves sodium hydroxide distribution, which enhances the traditional processes involved in pulping both softwood and hardwood. For bioleaching processes, the enzyme needs to function at a higher temperature (i.e., thermostable), be alkalophilic, and free from cellulases (Singh et al., 2016). Aside from the use of xylanases in bleaching, they are also useful in increasing the fibrillation of pulp by reducing the time required for thrashing of pulps, thereby prolonging the freeness of reused fibers (Sonali & Arora, 2020).

Other enzymes such as lipases have been employed in enhancing deinking as well as pitch control, whereas cellulases are utilized in developing bioprocesses, which can be utilized in recycling printed papers that have already been used (Singh et al., 2016). In addition, the aforementioned enzymes play important roles in the manufacture of cardboard that can be recycled, separating/removing papers that are adhered together, and in the production of soft papers such as sanitary and towel papers (Sonali & Arora, 2020).

3.5 Leather Industry

The leather industry generates a lot of wastes at varying stages of processing; most of these wastes generated have deleterious effects on the environment. This event can, however, be alleviated with the application of enzymes in some of the processes involved in the conversion of hides and skin to leather. The enzymes not only enhance the conversion rate of hides and skin but also improve the quality of leather produced as well as shrink the amount of wastes that was supposed to be generated without the application of enzymes (Singh et al., 2016). Some of the stages employed in the processing of leather include tanning, degreasing, picking, bating, dehairing, liming, soaking, and curing (Singh et al., 2016).

Enzymes utilized in various processes in the leather industry are lipolytic and proteolytic in nature; examples are lipases and neutral and alkaline proteases (Singh et al., 2016). These enzymes are important and are widely used as a result of the structure of the substrate (i.e., the animal skin) as unwanted parts are easily removed (Patel et al., 2017). For example, alkaline proteases are normally added to the animal skin in the soaking stage. By so doing, the dry animal skin takes up water quickly, which promotes the degradation and removal of fats, dirt, and proteins with an ultimate reduction in the processing time (Patel et al., 2017). The availability of microbial alkaline proteases makes them a better option than the use of pancreatic trypsin in the soaking phase (Patel et al., 2017). Proteases are also used in dewooling and dehairing the animal skin, thereby producing quality leathers (i.e., fewer spots, softer leather, cleaner and stronger surface) at the end. Lipases are, however, used in the bating phase specifically in the removal of grease (Patel et al., 2017). In liming, some of the advantages of using microbial enzymes against chemicals include production of stainless pelt, improved recovery of hair, reduced odor, and low COD and BOD (Singh et al., 2016).

3.6 Animal Feed Industry

There is a continuous increase in global population with a proportional increase in the demand for quality food products such as milk and meat. In order to meet this global demand, animal feeds are being incorporated with enzymes to aid digestibility and optimum feed utilization to provide necessary nutrients for animals' growth and reproduction (Patel et al., 2017). Although the addition of enzymes in animal feeds exploded in the 1990s, it was first initiated back in the 1980s. Currently, this practice is known across the globe since it is sustainable and ensures availability of food (Singh et al., 2016, Patel et al., 2017).

Among the numerous enzymes used in animal feeds, phytase is the largest and the most common enzyme utilized in the animal feed industry. This is because, in a cereal-based meal, phytase helps in utilizing phosphorous that is found to be naturally bonded with phytic acid (Singh et al., 2016). As a result, the supplementation of animal feeds with phosphorous is unnecessary since it is made available by phytase. The *Aspergillus* sp. are potent producers of phytase, which are commercially used in animal feed formulations (Patel et al., 2017).

Monogastric animals find it difficult to digest feeds that are plant-based with abundant hemicellulose and cellulose content. However, the addition of enzymes such as β-glucanase and xylanase aids in completely digesting such plant materials into simple monosaccharides (Singh et al., 2016). Increase in animal weight and growth has been demonstrated as a net effect of the actions of enzymes; increase in the feed conversion rate has also been demonstrated by animal feeds with meals augmented by enzymes. Patel et al. (2017) documented an increase (7–10%) in the availability of metabolized energy in wheat-based broiler feeds augmented by xylanases. Proteases are added to animal feeds in order to make proteins available for animal absorption through the degradation of peptide bonds into amino acid constituents. As a result, the antinutritional factors are limited. Aside from that, enzymes in feeds help in reducing the cost of producing animal feeds while maintaining the quality of animal products (Singh et al., 2016).

3.7 Cosmetic Industry

The utilization of enzymes in the cosmetic industry has drastically increased in recent times. Enzymes have been used as eliminators of free radicals as well as some other functions in commercially available products such as toothpastes, hair dyes, hair sprays, mouthwashes, and creams (Singh et al., 2016).

Lipases have great roles to play in cosmetics especially in surfactants and in the manufacturing of perfumes. Through glycerol esterification, two surfactants, namely, diacylglycerols and monoacylglycerols are yielded and used in the production of perfumes as well as in other cosmetics (Sonali & Arora, 2020). The major lipase-producing microorganisms used by the cosmetic industry are *Rhizomucor miehei* and *Candida cylindracea* (Sonali & Arora, 2020).

Levansucrase (EC 2.4.1.10) produced by *Zymomonas mobilis* has been employed in the production of levan compounds with great applications as cosmeceuticals (Srikanth et al., 2015; Tezgel et al., 2020). Polymers of levan possess transepidermal water loss (TEWL) properties and can be used as perfect moisturizing agents in place of hyaluronic acid (Tezgel et al., 2020). Biopolymers of levan have great applications in other beauty products such as in hair sprays, shampoos, and in moisturizers (Tezgel et al., 2020).

In mouth washes and toothpastes, papain and endoglycosidase (EC 3.2.1.96) are the major enzymes used, since their actions cause teeth whitening as well as removal of patches, gum tissues, odor, and teeth deposits. These enzymes also have the potential to remove fatty alcohol, precursors of vitamins, and few enzymes, which are sometimes attached to polymeric molecules (Singh et al., 2016). In addition to the aforementioned uses of papain and endoglycosidase, they are also used in cleaners utilized in cleaning protein films from eye contact lenses (Sonali & Arora, 2020). Microorganisms such as *Rhodococcus* sp. and *Mucor hiemalis* are responsible for endoglycosidase production (Sonali & Arora, 2020).

Another enzyme specifically utilized for scavenging free radicals in cosmetic products is superoxide dismutase (SOD) (EC 1.15.1.1). SOD functions in preventing skin damage, which usually occurs as a result of toxic wastes, environmental pollution, pathogenic bacteria, and some other factors that are deleterious to the human skin. The combination of peroxidase and SOD as eradicators of free radicals helps in reducing the erythema induced by ultraviolet (UV) radiation such as skin redness in sunscreen creams (Sonali & Arora, 2020). Many microorganisms are

SOD producers; examples are *Thermomyces lanuginosus*, *Anoxybacillus gonensis*, and *Thermoascus aurantiacus* (Singh et al., 2016, Sonali & Arora, 2020).

4 Enzymes in Medicine

In modern-day medicine, microbial enzymes are used as prodrugs, supplements, and biomarkers or in the synthesis of drugs (metabolic or digestive) (Das & Goyal, 2014). Enzymes mostly used in medicine have these two unique features: (a) they are highly specific to a particular substrate and, as such, bind and act with high affinities and (b) they can convert numerous targeted molecules to a particular desired product owing to their catalytic properties (Das & Goyal, 2014; Mane & Tale, 2015). The aforementioned features are now used in making potent drugs for treating numerous health disorders in humans (Mane & Tale, 2015). For example, lipases produced using *Candida rugosa* are used in the production of lovastatin. Lovastatin is a common drug that aids in reducing cholesterol intensity in serum (Sonali & Arora, 2020). Lipases from *Serratia marcescens* are used in diltiazem hydrochloride production, for asymmetric hydrolysis of some key intermediates such as esters of 3-phenylglycidic acid. Lipases have also been utilized in producing numerous enantiopure molecules such as esters, amides, alcohols, and carboxylic acids, which have been utilized in formulations of antiviral, anticancer, anticholesterol, and antihypertensive drugs as well as in drugs against Alzheimer's disease (Sonali & Arora, 2020).

Enzymes in medicine are used for treating enzyme-related deficiencies as well as some health disorders in a process known as enzyme therapy. In humans, enzymes assist in body detoxification, muscle contraction, food digestion, and immune system fortification and reduce stress on some vital organs such as the pancreas among others (Meghwanshi et al., 2020). In this aspect (i.e., therapeutic usage), there are numerous applications of enzyme therapy in medicine such as treatment of cystic fibrosis (CF) and pancreatic insufficiency, lactose intolerance, removal of dead tissues, metabolic disorders, cancers, and treatment of genetic diseases such as Gaucher, Fabry, and phenylketonuria (PKU) among many others (Fig. 1) (Gurung et al., 2013). This enzyme therapy could be nonsystematic or systemic, and the administration routes are multiple such as topical, intravenous, or oral (Mane & Tale, 2015).

4.1 Enzymes for the Treatment of Damaged Tissues

Proteolytic enzymes from microbial sources have been investigated for their use in the debridement of burnt and damaged skin. At first, it seems impossible with lots of inconsistent results. However, the formulation of these enzymes, known as Debrase Gel Dressing, has been made possible by advancements in DNA recombinant

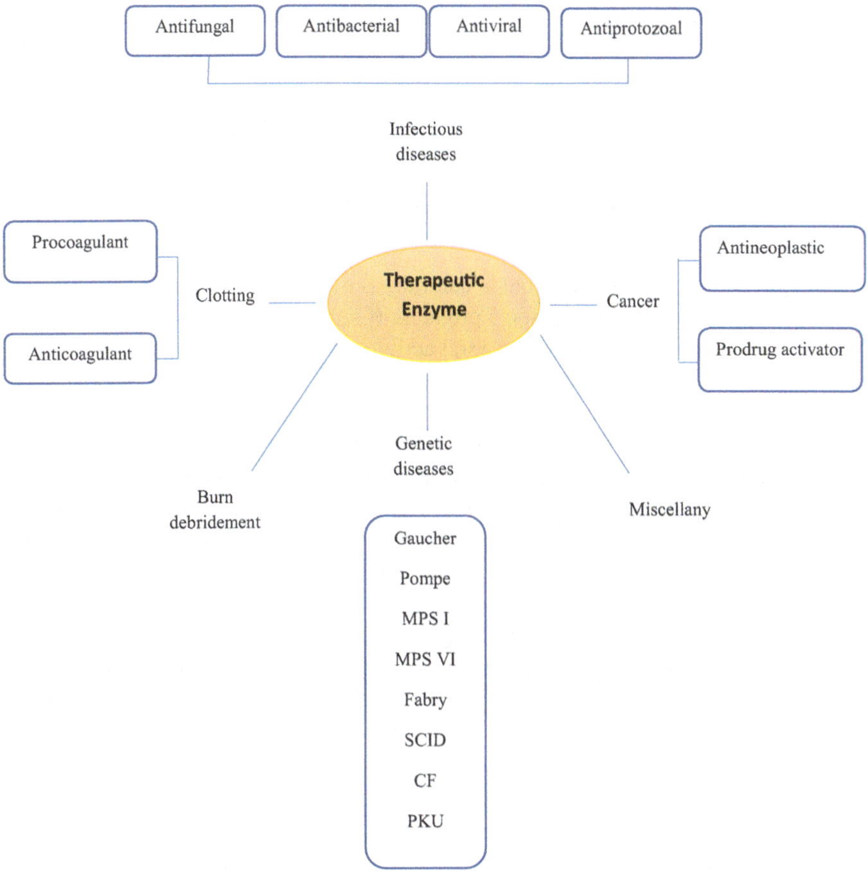

Fig. 1 Therapeutic enzymes applications in various diseases and disorders (Gurung et al., 2013)

technology (Mane & Tale, 2015). It was first formulated with enzymes extracted from the stem of pineapples and got approval from the United States Food and Drug Admintration (USFDA) in 2002, and it has since then undergone various clinical trials in patients with full thickness of varying thermal burns in the United States and across Europe (Meghwanshi et al., 2020). A proteolytic enzyme produced by *Vibrio proteolyticus* was reported by Reshma (2019) is effective as a debridement agent, and is used in treating severe secondary burns. It is currently in circulation with Vibrilase™ as its trade name. Another enzyme used in the treatment of damaged skin as a result of skin ulcers of burns is collagenase (Mane & Tale, 2015). Collagenase aids in breaking and removing dead skin in the form of a repair mechanism. Chondroitinase is another important enzyme that has been utilized in the regeneration of a spinal cord that has been injured. Chondroitinases function in the removal of glial scars with a resultant accumulation of chondroitin sulfate, which stops the growth of axons (Gurung et al., 2013). Similarly, hyaluronidase has been reported to possess hydrolytic activity that is similar to those of chondroitinases on chondroitin sulfate, which aids damaged nerve tissue regeneration (Gurung et al., 2013).

4.2 Enzymes for the Treatment of Infectious Diseases

Lysozyme, a bactericidal agent that is naturally produced in the human body, is now added to the list of numerous food products (Gurung et al., 2013). Research has shown that this enzyme exhibits antiviral activity against human immunodeficiency virus (HIV), which is similar to the activity of RNase U and RNase A, and, as such, can be formulated for treating viral infections such as HIV (Gurung et al., 2013; Meghwanshi et al., 2020). Interestingly, lytic enzymes can be obtained from bacteriophage as it has been demonstrated in killing varying bacteria such as *Bacillus anthracis*, *Clostridium perfringens*, and *Streptococcus pneumonia* (Gurung et al., 2013, Meghwanshi et al., 2020). Another example of enzymes used as an antimicrobial is chitinase, which is responsible for hydrolyzing chitin. Most antimicrobials target chitin; this is due to its abundant presence in the cell wall of varying pathogens like protozoa, helminths, and fungi (Meghwanshi et al., 2020). Examples of microorganisms reported to produce chitinase are *P. cepacia*, *Serratia marcescens*, *Micromonospora carbonacea*, *Paenibacillus ehimensis*, *Streptomyces viridodiasticus*, *Pseudomonas fluorescens*, and *Bacillus aerius* among many others (Mishra et al., 2020).

4.3 Enzymes for the Treatment of Cancer

Cancer is one of the leading causes of mortality across the globe. As a result, lots of efforts have been put in place to tackle this menace. In the process, enzyme technology has been explored as a possible means of cancer treatment. Recent reports by Meghwanshi et al. (2020) have documented the use of an arginine-degrading enzyme known as polyethylene glycol (PEG)/PEGylated arginine deaminase in the inhibition of hepatocellular carcinomas and melanomas in humans. Pegaspargase is another PEGylated enzyme in circulation with the name "Oncaspar 1." These enzymes have also been demonstrated in treating children diagnosed with acute lymphoblastic leukemia and are currently available for clinical use (Meghwanshi et al., 2020). PEGylated enzymes are efficient in treating cancerous cells since normal cells have the capacity to synthesize asparagine, which is not the case with cancerous cells and, as such, these die in the presence of the enzyme (Reshma, 2019). PEG-asparaginase and asparaginase are among the adjuncts used for an effective standard chemotherapy (Gurung et al., 2013, Meghwanshi et al., 2020). An extremely essential property of oncogenesis is the ability to proliferate.

Chondroitin sulfate proteoglycan removal by chondroitinases AC and chondroitinase B (to a lesser extent) prevents the growth of tumors, neovascularization, and metastasis. Another application of enzymes in cancer therapy is described by antibody-directed enzyme prodrug therapy (ADEPT). ADEPT is distinctive and effective in cancer therapy. It involves the movement of monoclonal antibodies carrying specific enzymes targeted at cancerous cells. Once this enzyme comes in contact with cancerous cells, it activates the prodrug, which goes on to destroy the cancerous cells without touching healthy cells (Gurung et al., 2013). This particular approach is now utilized for discovering and developing therapeutics drugs, which are targeted on enzymes associated with tumor cells and are responsible for activation of prodrugs. This method is known as targeted enzyme prodrug therapy (TEPT); it is a platform that involves enzymes with targeting domains that are antibody-like (Gurung et al., 2013). Another enzyme with the potential to be used for treating malignant tumors is lipase; lipase causes the activation of tumor necrosis factor (Mane & Tale, 2015).

4.4 Treatment of Exocrine Pancreatic Insufficiency

Exocrine pancreatic insufficiency is a severe disease condition that occurs alongside other diseases such as pancreatic cancer, post-pancreatic surgery, cystic fibrosis, and chronic pancreatitis. Pancreatic enzymes are important for digesting food substances in the gut, facilitating the absorption of nutrients (Das & Goyal, 2014). In situations of pancreatic cancer or chronic pancreatitis, the pancreas usually lacks the ability to secrete enough enzymes to completely digest the food particles ingested. Due to this, the food is not be completely absorbed from the intestine, which leads to the phenomenon known as malabsorption (Das & Goyal, 2014). The major problem in malabsorption comes from incomplete digestion of fat. However, this disorder can be treated utilizing enzyme replacement therapy with pancreatic enzymes or pancrelipase (a mixture of amylase, protease, and lipase). The amylase is responsible for digesting starch/carbohydrate, protease for digesting protein, and lipase for digesting fat (Das & Goyal, 2014). While using exogenous pancrelipase, the dosage needs to be taken into consideration as an overdose can cause an upset stomach or diarrhea. Pancrelipase drugs are currently in circulation with brand names such as Pancrex, Pancrease HL, Nutrizym, and Creon (Das & Goyal, 2014).

4.5 Treatment of Dupuytren's Disease

A fibroproliferative disorder that affects the palmar fascia limiting the functions of the hand is known as Dupuytren's disease. This disorder disables the hands' functions, with an ultimate reduction in the sufferer's quality of living (Reshma, 2019). If it is not arrested on time, Dupuytren's disease can progress into a permanent symptomatic flexion contracture affecting the digits (Reshma, 2019). However, microbial enzymes such as collagenase (EC 3.4.24.3) from *Clostridium* can be utilized in treating Dupuytren's disease. Although effective, injecting collagenase from *Clostridium histolyticum* usually gives rise to some side effects such as pain at the site of injection, bruising, hemorrhage, edema, skin lacerations, and, less frequently, rupture of flexor tendon and pulley. Collagenases have been proven to be effective in numerous in vivo studies, recording more than 60% reduction in contracture of infected individuals, who have been injected with clostridial collagenases (Reshma, 2019).

4.6 Treatment of Inflammation and Pain

For a decade now, serratiopeptidase or serrapeptase (EC 3.4.24.40) has been available for use by clinicians (Reshma, 2019). α-2-macroglobulin binds with serratiopeptidase in blood in a ratio of 1:1, thereby helping mask its antigenicity while retaining its enzymatic functions during its slow transfer to the inflammation site (Reshma, 2019). Once it gets there, serrapeptase breaks down the by-products formed as a result of blood coagulation at the injury site where the inflammation is formed; these by-products are insoluble proteins known as fibrin. This enzyme also enhances the drainage of the pus formed with a resultant increase in the speed at which damaged tissues are repaired (Mane & Tale, 2015). Serratiopeptidase causes the hydrolysis of serotonin, histamine, and bradykinin, which are responsible for its edematous status. Other functions of serratiopeptidase include facilitating microcirculation, reducing swelling, and causing sputum expectoration. Among serratiopeptidase-producing microorganisms, *Serratia marcescens* has been reported to be used efficiently in the commercial production of serratiopeptidase, which is now employed as an anti-inflammatory agent against fibrocystic breast disease as well as carpel tunnel syndrome (Reshma, 2019).

4.7 Prevention of Blood Clotting

Nattokinase (EC 3.4.21.62) is most times isolated from *B. subtilis*. It is a serine proteinase that is often employed in reducing some factors present in the blood, which are responsible for blood clotting, as well as lipids that participate greatly in causing/increasing the risk of cardiovascular diseases (CVDs) (Mane & Tale, 2015). Nattokinases are sometimes formulated into drugs taken orally known as CVD nutraceutical. CVD prevents/decreases the levels of fibrinogen in the plasma as well as factors VII and VIII (Mane & Tale, 2015). Nattokinases also have the potential for prolonging the actions of blood clotting factors, thereby preventing blood coagulation as well as dissolving the thrombus (Mane & Tale, 2015). Aside from nattokinase, staphylokinase (EC 3.4.99.22) also helps in preventing the clotting of blood through its fibrolytic actions. Staphylokinase causes the conversion of plasminogen, a proteolytic enzyme that is inactive, to its active state known as plasmin. Plasmin is often employed in treating myocardial infarction. Plasmin can also stimulate lysis of platelet- and erythrocyte-rich clots (Reshma, 2019).

4.8 Genetic Correction of Gaucher Disease

β-Glucocerebrosidase (EC 3.2.1.45), also known as D-glucosyl-*N*-acylsphingosine glucohydrolase, is an enzyme exhibiting glucosylceramidase activity (Meghwanshi et al., 2020). β-Glucocerebrosidase possesses the capability to hydrolyze the linkages and bonds present in chemical glucocerebroside, which is an intermediate in the metabolisms of glycolipids. Gaucher disease arises as a result of glucocerebrosidase gene mutation affecting lysosomal storage function, which is characterized by glucocerebroside accumulation (Meghwanshi et al., 2020). Parkinson's disease is also associated with mutations in genes expressing glucocerebrosidase production. β-Glucocerebrosidase is currently employed in treating Gaucher disease through enzyme replacement therapy. In this treatment method, modified exogenous placental glucocerebrosidase is targeted within the human body to its correct position or compartment. Commercially, the enzyme is available in clinics as alglucerase injection, with Ceredase as its brand name (Meghwanshi et al., 2020).

4.9 Enzymes Used as Digestive Aids

Today, several enzymes are employed in treating some problems associated with the inability of the body to digest sugars. For example, the enzyme α-galactosidase is administered to people having problems with digestion such as diarrhea, gas, and bloating whenever they ingest food like *Brassica* vegetables (i.e., Brussels sprout, broccoli, and cabbage) and proteins (Meghwanshi et al., 2020). The sugar-terminal hydrolysis of α-galactosidic residues present in the aforementioned food is catalyzed by α-galactosidase, which causes discomfort if undigested since the sugars can be fermented in the gut of the consumer causing the undesirable events. These days, lactase and α-galactosidase are very much available in the form of supplements, which are ready to be used. Lactose intolerance, as mentioned earlier, is a condition in which affected individuals are unable to utilize the sugar present in food substances as a result of inefficient production of the enzyme lactase. For example, milk contains varying nutrients including the sugar lactose, which, if not digested completely, can cause various forms of stomach problems (Fernandes & Carvalho, 2017; Liu & Kokare, 2017). However, this condition can be overcome through supplementing milk products with lactase to facilitate complete digestion of the lactose present in milk.

The addition of lactase aids in breaking down lactose into its corresponding monomers, which are galactose and glucose. By so doing, the effects of lactose intolerance such as diarrhea, gas, and bloating are alleviated (Singh et al., 2016). The microorganisms commonly involved in lactase production are *K. lactis* and *A. niger* (Fernandes & Carvalho, 2017). Lipase is also used to facilitate digestion; it is used in treating disorders such as cutaneous manifestations as a result of digestive allergies, treatment of disturbances of the gastrointestinal system, and dyspepsia (Mane & Tale, 2015).

Strict compliance with a particular diet is encountered by individuals with phenylketonuria (PKU). PKU is a genetic disorder transferable between generations; this disorder occurs as a result of the absence or insufficient presence of the phenylalanine hydroxylase enzyme (Meghwanshi et al., 2020). The aforementioned enzyme enhances the formation of tyrosine through the conversion of phenylalanine and, as a result, maintains the tyrosine levels in the body. Oral treatment for PUK has been developed with the help of phenylalanine ammonia lyase (PAL) from plants, which is overexpressed by recombinant yeast. PAL is currently in circulation in the market with Phenylase™ as its trade name (Meghwanshi et al., 2020). Hydrolysis of phenylalanine by PAL in the gastrointestinal tract has been demonstrated. Individuals with weak immune systems such as those infected with HIV usually encounter malabsorption of fat. However, this can be overcome by supplementing their diet with a pancreatic enzyme cocktail containing enzymes such as lipases, proteases, and amylases (Meghwanshi et al., 2020). Pancreatic insufficiency usually encountered by individuals with cystic fibrosis can be overcome with the help of a pancreatic enzyme cocktail such as "TheraCLEC™—Total," which is commercially available (Meghwanshi et al., 2020).

5 Conclusions and Future Perspective

Microorganisms are efficient in the provision of enzymes for various industrial processes in industries such as food, leather, chemical, textile, animal feed, agricultural, and medicine among many others. Their ability as well as versatility in the conversion of cheap substrates into valuable enzymes make them an important tool and a source of a wide array of enzymes for production. Enzymes obtained from microbial sources are efficient, cheap, and stable with high activities. Even when used under extreme conditions such as pH or temperature, they still yield quality products and can be recovered at the end of their applications. Furthermore, microbial enzymes reduce processing time and, as such, reduce the overall production cost. Microbial enzymes are the preferred alternative to chemical catalysts, owing to the fact that they have low activation energy and do not pose any threat to environmental safety. Enzymes from microbial sources provide an excellent avenue for enzyme improvement as well as upregulation through genetic editing and manipulation of the microbial genome. Since microorganisms are ubiquitous, their presence in extreme environments such as the Polar Regions, volcanoes, extremely arid deserts, deep oceanic floors, and trenches among others have been reported by various scientists across the globe. Some of these microbes possess great potential for the production and use of enzymes under extremely high or low temperature, pH, or pressure. The major challenges, however, are the difficulty in isolating and culturing microbes from extreme environments in their pure form. As such, new approaches such as those utilizing genomics and metagenomics should be developed together with some classical techniques in the screening and isolation of microorganisms with the potential for producing enzymes with advanced catalytic properties.

References

Al-Manhel, A. J. A. (2018). Application of microbial enzymes in dairy products: A review. *Basrah Journal of Agricultural Sciences, 31*(1), 20–30. https://doi.org/10.21276/basjas

Avendaño, K. A., Anguiano, M., López, C. E., Montañez, L. E., Sifuentes, L., & Balagurusamy, N. (2016). Microbial enzymes applications in food processing. *Agro Food Industry Hi-Tech, 27*(4), 63–67.

Bansal, N., Tewari, R., Soni, R., & Soni, S. K. (2012). Production of cellulases from Aspergillus niger NS-2 in solid state fermentation on agricultural and kitchen waste residues. *Waste Management, 32*, 1341–1346.

Bhandari, S., Poudel, D. K., Marahatha, R., Dawadi, S., Khadayat, K., Phuyal, S., Shrestha, S., Gaire, S., Basnet, K., Khadka, U., & Parajuli, N. (2021). Microbial enzymes used in bioremediation. *Journal of Chemistry, 2021*, 8849512. https://doi.org/10.1155/2021/8849512

Bhatt, B. M., Trivedi, U. B., & Patel, K. C. (2020). Extremophilic amylases: Microbial production and applications. In N. K. Arora, J. Mishra, & V. Mishra (Eds.), *Microbial enzymes: Roles and applications in industries* (pp. 185–206). Springer Nature.

Chandra, P., Enespa Singh, R., & Arora, P. K. (2020). Microbial lipases and their industrial applications: A comprehensive review. *Microbial Cell Factories, 19*, 169. https://doi.org/10.1186/s12934-020-01428-8

Das, D., & Goyal, A. (2014). Pharmaceutical enzymes. In S. K. Brar, G. S. Dhillon, & C. R. Soccol (Eds.), *Biotransformation of waste biomass into high value biochemicals* (pp. 367–389). Springer. https://doi.org/10.1007/978-1-4614-8005-1

Fernandes, P., & Carvalho, F. (2017). Microbial enzymes for the food industry. In *Biotechnology of microbial enzymes* (pp. 513–544). Elsevier. https://doi.org/10.1016/B978-0-12-803725-6.00019-4

Gurung, N., Ray, S., Bose, S., & Rai, V. (2013). A broader view: Microbial enzymes and their relevance in industries, medicine, and beyond. *BioMed Research International, 2013*, 329121. https://doi.org/10.1155/2013/329121

Kapoor, D., Sharma, P., Sharma, M. M. M., Kumari, A., & Kumar, R. (2020). Microbes in pharmaceutical industry. In S. G. Sharma, N. R. Sharma, & M. Sharma (Eds.), *Microbial diversity, interventions and scope* (pp. 259–300). Springer Nature.

Liu, X., & Kokare, C. (2017). Microbial enzymes of use in industry. In *Biotechnology of microbial enzymes* (pp. 267–298). Elsevier. https://doi.org/10.1016/B978-0-12-803725-6.00011-X

Mane, P., & Tale, V. (2015). Overview of microbial therapeutic enzymes. *International Journal of Current Microbiology and Applied Sciences, 4*(4), 17–26.

Meghwanshi, K. G., Kaur, N., Verma, S., Dabi, N. K., Vashishtha, B., Charan, P. D., Purohit, P., Bhandari, H. S., Bhojak, N., & Kumar, R. (2020). Enzymes for pharmaceutical and therapeutic applications. *Biotechnology and Applied Biochemistry, 67*, 586–601. https://doi.org/10.1002/bab.1919

Mishra, P., Mishra, J., Dwivedi, S. K., & Arora, N. K. (2020). Microbial enzymes in biocontrol of phytopathogens. In P. Mishra, S. K. Dwivedi, & N. K. Arora (Eds.), *Microbial enzymes: Roles and applications in industries, microorganisms for sustainability* (pp. 259–289). Springer Nature. https://doi.org/10.1007/978-981-15-1710-5_1

Naureen, I., Bashir, M. J., Ahmed, W., Hussain, S. A., Naveed, M. H., Ijaz, A., & Nadeem, F. (2021). A review on microbial enzymes, synthesis, biological role, current applications and future perspectives. *Schizophrenia Bulletin, 7*(3), 44–48. https://doi.org/10.36348/sb.2021.v07i03.005

Niyonzima, F. N., Veena, S. M., & More, S. S. (2020). Industrial production and optimization of microbial enzymes. In N. K. Arora, J. Mishra, & V. Mishra (Eds.), *Microbial enzymes: Roles and applications in industries* (pp. 115–137). Springer Nature.

Pandey, K., Singh, B., Pandey, A. K., Badruddin, I. J., Pandey, S., Mishra, V. K., & Jain, P. A. (2017). Application of microbial enzymes in industrial waste water treatment. *International Journal of Current Microbiology and Applied Sciences, 6*(8), 1243–1254. https://doi.org/10.20546/ijcmas.2017.608.151

Patel, A. K., Singhania, R. R., & Pandey, A. (2017). Production, purification, and application of microbial enzymes. In *Biotechnology of microbial enzymes* (pp. 13–41). Elsevier. https://doi.org/10.1016/B978-0-12-803725-6.00002-9

Preethi, P. S., Gomathi, A., Srinivasan, R., Kumar, J. G. P. S., Murugesan, K., & Muthukailannan, G. K. (2020). Laccase: Recombinant expression, engineering and its promising applications. In N. K. Arora, J. Mishra, & V. Mishra (Eds.), *Microbial enzymes: Roles and applications in industries* (pp. 63–86). Springer Nature.

Reshma, C. V. (2019). Microbial enzymes: Therapeutic applications. *Microbiology Research Journal International, 27*(2), 1–8.

Sanchez, S., & Demain, A. L. (2017). Useful microbial enzymes – An introduction. In G. Brahmachari (Ed.), *Biotechnology of microbial enzymes, production, biocatalyst and industrial application* (pp. 1–11). Elsevier.

Sindhu, R., Binod, P., Beevi, U. S., Abraham, A., Mathew, A. K., Madhavan, A., Rebello, S., & Pandey, A. (2018). Applications of microbial enzymes in food industry. *Food Technology and Biotechnology, 56*(1), 16. https://doi.org/10.17113/ftb.56.01.18.5491

Singh, R., Kumar, M., Mittal, A., & Mehta, P. K. (2016). Microbial enzymes: Industrial progress in 21st century. *3 Biotech, 6*, 174. https://doi.org/10.1007/s13205-016-0485-8

Singh, R. S., Singh, T., & Pandey, A. (2019). Microbial enzymes - An overview. In *Advances in enzyme technology*. Elsevier. https://doi.org/10.1016/B978-0-444-64114-4.00001-7

Singh, G., Dwivedi, S. K., & Mishra, J. (2020). Role of fungal enzymes in the removal of azo dyes. In N. K. Arora, J. Mishra, & V. Mishra (Eds.), *Microbial enzymes: Roles and applications in industries* (pp. 231–258). Springer Nature.

Sonali, & Arora, R. (2020). Industrial potential of microbial enzymes. In S. G. Sharma, N. R. Sharma, & M. Sharma (Eds.), *Microbial diversity, interventions and scope* (pp. 301–320). Springer Nature.

Srikanth, R., Siddartha, G., Sundhar, R. C. H., Harish, B. S., Janaki, R. M., & Uppuluri, K. B. (2015). Antioxidant and anti-inflammatory levan produced from *Acetobacter xylinum* NCIM 2526 and its statistical optimization. *Carbohydrate Polymers, 123*, 8–16.

Tezgel, N., Kırtel, O., Van den Ende, W., & Oner, E. T. (2020). Fructosyltransferase enzymes for microbial fructan production. In N. K. Arora, J. Mishra, & V. Mishra (Eds.), *Microbial enzymes: Roles and applications in industries* (pp. 1–40). Springer Nature.

Thakur, A., Putatunda, C., Sharma, R., Mehta, R., Solanki, P., & Bhatia, K. (2020). Innovative techniques for improving microbial enzyme production. In S. G. Sharma, N. R. Sharma, & M. Sharma (Eds.), *Microbial diversity, interventions and scope* (pp. 157–184). Springer Nature.

Vandenberghe, L. P. S., de Carvalho, J. C., Libardi, N., Rodrigues, C., & Soccol, C. R. (2016). Microbial enzyme factories: current trends in production processes and commercial aspects. In *Agro-industrial wastes as feedstock for enzyme production*. Elsevier. https://doi.org/10.1016/B978-0-12-802392-1.00001-0

Vermelho, A. B., Cardoso, V., Nascimento, R. P., Pinheiro, A. S., & Rodrigues, I. A. (2016). Application of microbial enzymes in the food industry. In R. V. Ravishankar (Ed.), *Advances in food biotechnology* (pp. 105–131). John Wiley & Sons.

Vieira, E. F., & Delerue-Matos, C. (2020). Exploitation of *Saccharomyces cerevisiae* enzymes in food processing and preparation of nutraceuticals and pharmaceuticals. In N. K. Arora, J. Mishra, & V. Mishra (Eds.), *Microbial enzymes: Roles and applications in industries* (pp. 41–62). Springer Nature.

Vishnoi, N., Dixit, S., & Mishra, J. (2020). Microbial lipases and their versatile applications. In N. K. Arora, J. Mishra, & V. Mishra (Eds.), *Microbial enzymes: Roles and applications in industries* (pp. 207–230). Springer Nature.

Vyas, A., & Yakubu, A. (2020). Microbial enzymes from in vitro to market. In S. G. Sharma, N. R. Sharma, & M. Sharma (Eds.), *Microbial diversity, interventions and scope* (pp. 185–215). Springer Nature.

Microbial Enzymes in the Biosynthesis of Metal Nanoparticles

Kondakindi Venkateswar Reddy, Nalam Renuka Satya Sree, Palakeerti Srinivas Kumar, and Pabbati Ranjit

Abstract Metal-based nanoparticles (MtNPs) are at the cutting edge of green nanotechnology and have gained tremendous attention across the globe due to their unique electronic as well as optical properties and their potential applications in different areas. In the past few years, besides chemical and physical methods, biological means to synthesize MtNPs have been largely preferred as microbial enzyme-mediated MtNPs are benign and cost-effective. This chapter focusses on discussing various microbial means including bacteria, actinomycetes, fungi, yeasts, and algae, which are utilized to synthesize MtNPs, and the detailed antimicrobial mechanism of the key microbial enzyme-mediated synthesis of metal nanoparticles. This chapter also elaborates on the wide range of MtNP applications and the recent advancements and challenges faced in microbial enzyme-mediated nanoparticle synthesis.

1 Introduction

Nowadays, research attention on metal nanoparticles (MtNPs) and their synthesis has gained significance due to MtNPs' groundbreaking applications in diverse industrial realms. Nanoparticles, which are solid particulate dispersion, have dimension ranges between 1 and 100 nm. Nanoparticles (NPs) have provided an opportunity to discover several design patterns of advanced materials and thus have opened the path for the evolution of nanoparticle properties by modulating the morphology, size, and distribution. The most prominent property of nanoparticles is the high ratio of surface area to volume, which enhances molecular interaction (Gahlawat & Choudhury, 2019). The word "nano" is drawn from "*nanos*," a Greek word meaning "dwarf," a prefix used to describe one billionth the size of things. Richard Feynman first presented nanotechnology in his lecture at the American Institute of

K. V. Reddy · N. R. S. Sree · P. S. Kumar · P. Ranjit (✉)
Center for Biotechnology, Institute of science and Technology, JNTUH, Hyderabad, Telangana, India

N. R. Maddela et al. (eds.), *Ecological Interplays in Microbial Enzymology*, Environmental and Microbial Biotechnology,
https://doi.org/10.1007/978-981-19-0155-3_15

Technology. Nanoparticles are considered to be better than bulk materials as they possess enhanced Rayleigh scattering, better surface plasmon resonance (SPR), and surface-enhanced Raman scattering (SERS) in MtNPs. Hence, NPs are deemed to be the next-generation building blocks of electronics, optoelectronics, and biosensors. While discovering microbial synthesis of NPs, scientists found siliceous materials by diatoms, magnetite particles by magnetotactic bacteria, calcium and gypsum lawyers by S-layers bacteria were involved in the synthesis of NPs. These interactions among microbes and metals have laid an interesting research path toward potential applications in the areas of biomineralization, biocorrosion, bioremediation, and bioleaching (Navia-Mendoza et al., 2021; Roy et al., 2020). The biosynthesis of NPs has evolved as a promising field of research, interconnecting nanotechnology and biotechnology (Narayanan & Sakthivel, 2010). NPs are excellent catalysts and absorbents due to the large number of allied adsorption sites (Saravanan et al., 2021). Moreover, there is immense significance in NP synthesis due to their remarkable chemical, electronic, and photoelectrochemical properties. Striking progress has been made in nanoparticles, and the current advances in organizing nanoscale structures into superstructures ensure their imperative role in the key technologies of the new era in the field of science. NPs are achieving prominence in biomedical sciences, energy sciences, magnetics, optics, and catalysis. Techniques generally used to manufacture NPs include molecular, atomistic, and particulate processing with the help of vacuum or a liquid medium, but most of them are expensive and ineffective. Therefore, there is an urge to upgrade to nontoxic, clean, and eco-friendly procedures, which has encouraged researchers to focus on biological systems (Mandal et al., 2006). The inimitable nature of NPs makes them exceptional materials for use in the innovative design and advancement of tools for assessing agriculture and industries. Nanotechnology can enhance agricultural practices and product quality by employing nanoparticle-centered fertilizers and pesticides. Microbes are considered as significant nanofactories that have the ability to hoard and detoxify metal salts with the help of several reductase enzymes that have the ability to reduce heavy metals to MtNPs. The latest research has demonstrated that bacteria belonging to the genera including *Bacillus*, *Pseudomonas*, *Klebsiella*, *Escherichia*, *Corynebacterium*, *Lactobacillus*, *Trichoderma*, *Rhodobacter*, *Weissella*, *Brevibacterium*, *Sargassum*, *Desulfovibrio*, *Shewanella*, *Pyrobaculum*, *Plectonema boryanum*, *Aeromonas*, and *Rhodopseudomonas* have the ability to synthesize nanoparticles (Bahrulolum et al., 2021). MtNPs have a major role in drug delivery and treatment procedures. A drug carrier, which is a nanoscale material, acts as a single unit during its transport in treatment. The size of these nanoclusters ranges between 1 and 10 nm. An assemblage of nanoparticles or nanoclusters is known as "nanopowders," whereas NPs in the crystal form are termed "nanocrystals." The synthesis, assemblage, and organization of these nanocrystals and nanopowders are controlled by biological entities as patterns (Zhang et al., 2020). Nanoparticle production can be carried out either by a top-down strategy, where the bulk material is fragmented into small particles using physicochemical techniques, or by a bottom-up strategy, where MtNP synthesis happens by atoms through nuclei self-assembly, which includes chemical and

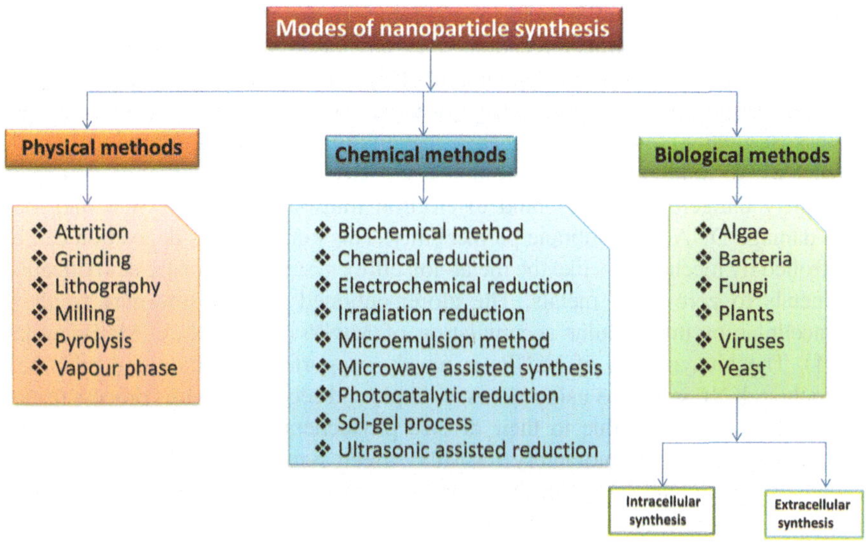

Fig. 1 Synthesis of NPs by different methods. (Modified from Koul et al., 2021)

biological processes as indicated in Fig. 1. Physical procedures require high energy, which is capital-intensive, and, also, the yield rate is low. Chemical procedures are mostly preferred as they require less energy and involve toxic chemicals that are carcinogenic, genotoxic, and cytotoxic in nature. Hence, there is a need for an eco-friendly approach. The synthesis of MtNPs with the help of microbes has advanced as a promising strategy, which is a green approach.

The microbial path facilitates inexpensive, reliable, and nontoxic MtNPs and is also easy to scale-up. Another distinctive feature of biosynthesized NPs is their ability to serve as a template for the production and arrangement of nanorange particles into precise structures (Gahlawat & Choudhury, 2019). In addition to the above, microbes have also been effectively applied to generate nanohybrid systems with high catalytic properties (Palomo & Filice, 2016). The production of nanomaterials like metals, silica, and metal alloys using biological systems is termed "biomineralization." The finest example of the biosynthesis of nanostructured complexes is magnetotactic bacteria, a heterogeneous group of bacteria located in aqueous environments. Through intracellular mechanisms, they involve in the production of magnetic nanocrystals in magnetosomes that are made up of gregite (Fe_3S_4) or magnetite (Fe_3O_4). Their size ranges from 40 to 140 nm, and they are covered by a membrane. Magnetotactic bacteria have the ability to travel alongside Earth's magnetic lines, which helps them in migration. Another example includes unicellular eukaryotic organisms like diatoms, which are found in brackish water, seas, freshwater, and oceans. They have a unique ornamented cell wall composed of polysialic acid. Similarly, sponges also produce intricate skeletal elements. In this manner, metal and metal alloy NPs are produced by microbes, and these NPs are the result of detox pathways. More precisely, metals such as silver, gold, copper, manganese, and cadmium, besides their presence in the environment, play a crucial role in several biochemical reactions that are based on heavy metal complex formation mechanisms like oxygenic photosynthesis involving water cleavage, nitrogen fixation, respiration, cleavage of urea, and hydrogen assimilation. Nevertheless, at greater concentrations, heavy metal ions are involved in producing toxic effects by forming nonspecific complexes. Heavy metals are highly toxic and extremely dangerous as they bind to electron transport chain enzymes and may also damage DNA or membrane permeability. This kind of toxic deposition can be controlled by mechanisms like the metal–ion efflux system, enzymatic detoxification of metals to zerovalent metals. The aforementioned mechanisms may result in extracellular or intracellular accumulation of MtNPs in microbes (Maliszewska, 2011). The biosynthesis of MtNPs is mostly supported by plants and microbes. Even though NP synthesis using plants is simple and economical, this process results in polydispersed NPs, due to their diverse phytochemistry. Hence, microbes are regarded as a probable biofactory for MtNP green synthesis. Several microbes are capable of producing inorganic materials by intracellular and extracellular mechanisms. The intracellular mechanism of a cell in a microbe has an incredible ion transport system. The bacterial cell wall, due to its electrostatic forces, attracts metal ions. Besides, bacterial cell walls also comprise enzymes that reduce metal ions. On the other hand, the extracellular mechanism completely relies on the reductase enzymes present in the microbial cell wall involved in metal reduction (Ovais et al., 2018a, b).

Based on their origin, NPs can be categorized into three forms: Natural NPs are considered to exist in Earth's environment in the form of lunar dust, volcanic dust, mineral composites, etc. Incidental NPs are considered as man-made waste particles generated through industrial processes including coal combustion, diesel exhaust, etc. Engineered NPs are further categorized into four types:

1. NPs composed of carbon compounds like single- and multiwalled carbon nanotubes and fullerene
2. NPs combined with nanogold, quantum dots, nanoaluminum, nanozinc, and other metal-based particles
3. NPs with dendrimers, nanosized polymers that have the advantage of being customized to achieve a fixed chemical reaction
4. Composites that help one NP to bind to another NP

Engineered NPs, due to their performance, have a huge progressive impact on enhancing areas including economy, pharmaceuticals, consumer products, agriculture, and energy. Due to their surface tailorability, size, multifunctionality, and enhanced solubility, NPs are attracting several research opportunities for biologists. Their novel characteristics help them interact with complex biological processes in innovative approaches. This swift mount in nanotechnology offers cross disciplinary researchers to collaborate with each other, which undoubtedly rise break throughs. (Sardar et al., 2014). Microbes can be simply modified by genetic engineering that helps cells overexpress a particular enzyme used in the green synthesis of MtNPs. During production, aggregated NPs do not directly come in contact with one another; this is prevented by capping proteins like cytochrome c secreted by microbes (Talebi et al., 2010).

Once designed, NPs are released into the environment, and it is tedious to detect where they are present. If they are not secured properly, they may cause environmental risks. Hence, several concerns like environmental deposition, increased expenses, low detection limits, environment soundness, and regeneration must be handled primarily before synthesis (Saravanan et al., 2021). Due to numerous complexities in detection of the specific chemical compounds responsible for MtNP synthesis, the green synthesis of NPs still remains a challenge (Ovais et al., 2018a, b).

The following sections of this chapter will encompass an elaborate discussion on MtNP green synthesis by utilizing diverse microbes, the mechanisms involved, enzymes secreted by microbes, and the strategies used by microbes to synthesize NPs. Furthermore, the recent mechanistic approaches of MtNP biosynthesis, their potential applications, challenges, and future prospects have been briefed.

2 Microbial Enzyme-Mediated Synthesis of MtNPs

The metal and microbial interaction is well-acknowledged in biotechnological processes. The biodiversity of microbes and their simple cultivation under biochemical, molecular, and cellular mechanisms ensure that MtNP synthesis can be achieved (Das et al., 2017). The synthesis of NPs by microbes is mostly performed using a bottom-up technique by oxidation or reduction (Kapoor et al., 2021). In recent decades, MtNP biosynthesis using microbes such as bacteria, fungi, actinomycetes, yeasts, marine algae, and virus has gained great attention in the area of green technology (Gahlawat & Choudhury, 2019) and has been extensively used for the synthesis of gold, copper, silver, cadmium, iron, platinum, etc. (Kapoor et al., 2021), as listed in Table 1 and as described below.TEM: Transmission Electron Microscope, EDX: Energy Dispersive X-ray, FTIR: Fourier Transform Infrared Spectroscopy, DLS: Dynamic Light Scattering, XRD: X-Ray Diffraction, SEM: Scanning Electron Microscope, ATR-FTIR: Attenuated Total Reflectance--FTIR, ESI: Electrospray Ionization, HRTEM: High Resolution TEM.

2.1 *Bacteria- and Actinomycete-Mediated MtNP Synthesis*

In microbes, bacteria, compared to others, can be easily manipulated and modified genetically for MtNP green synthesis. As MtNPs have to be synthesized from heavy metals, which creates a harsh and toxic environment, bacteria naturally evolved defense mechanisms including efflux pumps, intracellular sequestration, extracellular precipitation, and altered metal–ion concentrations to manage the aforementioned stress conditions (Gahlawat & Choudhury, 2019). The cell wall of bacteria plays a

Table 1 Microbial green synthesis of MtNPs (modified from Kapoor et al., 2021)

Source	Synthesis methods	Characterization	Types of MtNPs	Morphology	Size (nm)
Bacteria					
Bacillus methylotrophicus	Extracellular	TEM, EDX	Silver	Spherical	10–30
Brevibacillus formosus	Cell culture	FTIR, TEM, DLS	Gold	Spherical	5–12
Escherichia coli	Extracellular	TEM	Silver	Spherical	10–100
Lactobacillus sp.	Cell culture	XRD, TEM	Titanium	Spherical	40–60
Novosphingobium sp.	Extracellular	XRD, TEM	Silver	Spherical Crystalline	8–25
Pseudomonas fluorescens	Extracellular	FTIR, XRD, EDS, TEM	Silver	Cubic Spherical Oval	10–60
Pseudomonas putida	Extracellular	FTIR, SEM	Silver	Spherical	70
Serratia nematodiphila	Extracellular	TEM, XRD	Silver	Crystalline	10–31
Actinomycetes					
Streptomyces capillispiralis	Extracellular	TEM, XRD	Copper	Spherical	10–31
Streptomyces hygroscopicus	Intracellular	TEM	Gold	Spherical	3.69–59
Streptomyces kasugaensis	Cell filtrate	TEM, FTIR	Silver	Rounded	4.2–65
Fungi					
Arthroderma fulvum	Cell filtrate	TEM, XRD	Silver	Spherical	15.5
Aspergillus niger	Extracellular	TEM, ESI	Silver	Spherical	20
Aspergillus fumigatus	Cell-free filtrate	ATR-FTIR, XRD, SEM	Silver	Spherical	322.8
Aspergillus oryzae	Cell filtrate	XRD, TEM, FTIR	Silver	Spherical	7–27
Colletotrichum sp.	Cell-free extract	XRD, TEM, FTIR	Silver	Myriad	5–60
Coriolus versicolor	Extracellular and intracellular	XRD, TEM, FTIR	Silver	Spherical	25–75
Epicoccum nigrum	Extracellular	XRD, TEM	Silver	Spherical	1–22
Fusarium keratoplaticum	Culture filtrate	XRD, FTIR	Silver	Spherical	6–36
Fusarium oxysporum	Extracellular	FTIR, TEM	Silver	Spherical	5–13
Penicillium oxalicum	Extracellular	XRD, FESEM	Silver	Spherical	10–40
Rhizopus stolonifer	Culture filtrate	XRD, TEM, FTIR	Gold	Spherical	9.47

(continued)

Table 1 (continued)

Source	Synthesis methods	Characterization	Types of MtNPs	Morphology	Size (nm)
Yeast					
Candida albicans	Cell-free extract	TEM	Gold	Spherical	5–13
Candida glabrata	Intracellular and extracellular	TEM	CdS	Hexamer	20–29
Caulerpa racemosa	Cell extract	TEM, XRD	Silver	Spherical Triangular	5–25
Pichia jadinii	Intracellular	TEM	Gold	Various	–
Saccharomyces cerevisiae	Cell wall cytoplasm	TEM	Gold	Spherical	15
Sargassum bovinum	Cell extract	TEM, XRD, EDX	Palladium	Octahedral	5–10
Sargassum longifolium	Cell extract	TEM, SEM	Silver	Spherical	40–85
Sargassum tenerrimum	Cell extract	HRTEM, FTIR	Gold	Rounded	5–45

TEM: Transmission Electron Microscope, EDX: Energy Dispersive X-ray, FTIR: Fourier Transform Infrared Spectroscopy, DLS: Dynamic Light Scattering, XRD: X-Ray Diffraction, SEM: Scanning Electron Microscope, ATR-FTIR: Attenuated Total Reflectance -FTIR, ESI: Electrospray Ionization, HRTEM: High Resolution TEM

crucial role as metals must penetrate through the cell wall before entering into the cytoplasm and reverse back through the same wall for liberating extracellularly. The peptidoglycan in the cell wall presents polyanions for metals and chemical reactive group stoichiometric interaction, followed by inorganic metal accumulation (Das et al., 2017).

Various bacterial species are readily employed as green nanofactories for MtNPs. For instance, *Pseudomonas stutzeri* AG 259 is involved in silver NP synthesis in crystal form observed under transmission electron microscopy (TEM). *Rhodopseudomonas capsulata* is observed for extracellular production of gold NPs. The *Morganella* species have been reported to produce silver NPs extracellularly. *Klebsiella aerogenes*, when subjected to Cd^{2+} ions, obtained CdS NPs synthesized intracellularly (Sardar et al., 2014). Pure gold NPs are obtained from *Delftia acidovorans* by inducing resistance against gold ions and forming inert gold NPs (AuNPs). *Bacillus sphaericus* JG-A12 has been reported to accumulate excess concentrations of Cd, Al, Cu, U, and Pb (Zhang et al., 2020). Magnetite NPs are formed by *Aquaspirillum magnetotacticum*, sulfate-reducing bacterium MV-1, *Magnetospirillum Gryphiswaldense*, and *Magnetospirillum magnetotactum*. *Desulfovibrio desulfuricans* has been observed to synthesize palladium NPs extracellularly. Gram-positive bacteria isolated from sediments have the ability to reduce sulfate, identified as *Desulfosporosinus* sp., and synthesize ZnS and uranium NPs. *Lactobacillus* has also been observed to assist silver and gold NP crystal synthesis. While $AgNO_3$ is subjected to *Staphylococcus aureus*, AgNP formation was observed extracellularly (Talebi et al., 2010). Other bacteria like *Klebsiella pseudomonas*, *Alteromonas*, *Bacillus subtilis*, *Escherichia coli*, *Bacillus megaterium*, *Ochrobacterum*, etc., were also reported to synthesize MtNPs.

Actinomycetes are microbes that have characteristics similar to those of fungi and bacteria. Actinomycetes including *Streptomyces griseoruber* and *Streptomyces capillispiralis* are known for their extensive production of copper and gold NPs (Gahlawat & Choudhury, 2019). When subjected to gold ions, *Thermomonospora* sp. has been observed to reduce metal ions to gold NPs extracellularly. *Rhodococcus* sp., which is an alkalotolerant actinomycete, has been reported to synthesize gold NPs intracellularly, which are concentrated on the cell membrane (Talebi et al., 2010).

2.2 Fungi- and Yeast-Mediated MtNP Synthesis

Fungi-mediated MtNP synthesis is an alternative, simple, and basic process, which has been discovered for NP production. When compared to bacteria, fungi have a higher yield of NPs and have greater tolerance to heavy metals (Gahlawat & Choudhury, 2019). Mycosynthesis of nanoparticles has been successfully employed for the large-scale production of various MtNPs. In intracellular synthesis, the metal ions are transformed into low-toxic substances in the mycelia, which are utilized by the fungi themselves. Extracellular NP synthesis includes the utilization of fungal extracts. *Rhizopus stolonifer* extracts were observed to mediate the production of monodispersed AgNPs. *Candida glabrata* was stated to extracellularly synthesize AgNPs (Ovais et al., 2018a, b). *Fusarium oxysporum* was reported to mediate gold NPs in the presence of $AuCl_4^-$ ions and nicotinamide adenine dinucleotide (NADH) enzymes (Kapoor et al., 2021). It was also reported for its extensive use in synthesizing AgNPs and for capping with fungal proteins. *Aspergillus fumigatus* was observed to produce AgNPs of size 5–25 nm extracellularly. AuNPs are synthesized using *Verticillium* sp. by reducing $AuCl_4$ located on the mycelia surface. *Neurospora crassa*, a well-known microbe, was used for synthesizing platinum NPs intracellularly (Zhang et al., 2020). AgNPs have also been observed to be synthesized by *Penicillium* sp. and *Trichoderma asperellum*. *Hormoconis resinae* can synthesize AgNPs (Talebi et al., 2010). *Caribena versicolor*, which is a white rot fungus, yielded AgNPs via extracellular and intracellular modes. *Aspergillus flavus* was observed to synthesize monodispersed AgNPs along with the *sil* gene in plasmids, which are involved in silver ion reduction on a large scale (Das et al., 2017).

Apart from fungi, yeasts are also observed to synthesize MtNPs. Yeasts have the inbuilt ability to absorb and gather high amounts of toxic metal salts from their environment. They can adapt to harsh metal toxic conditions using diverse detoxification mechanisms such as chelation, intracellular sequestration, and bioprecipitation. These properties of yeasts make them useful for synthesizing MtNPs. A marine strain *Yarrowia lipolytica* was mediated to synthesize AgNPs in a cell-associated manner. Extracellular AgNP synthesis was observed in *Candida utilis* (Gahlawat & Choudhury, 2019). *Schizosaccharomyces pombe* and *Candida glabrata* lead to the synthesis of CdS quantum dots intracellularly when subjected to Cd^{2+} ions (Talebi et al., 2010). *Pichia jadinii* reduced gold ions into gold NPs with the help of enzymes present in the cell wall or cytoplasm (Kapoor et al., 2021).

2.3 *Algae-Mediated MtNP Synthesis*

Similar to yeasts, there are various reports on algae-mediated MtNP synthesis. Algae, which are aquatic oxygenic photoautotrophs, can be utilized for MtNP production. In algae, due to the electrostatic forces between ions and carboxylate groups that are negatively charged, the metal ions affix to the surface of the algal cell. Later, with the help of enzymes, these metal ions are reduced to nuclei formation, followed by metal–ion reduction (Kapoor et al., 2021). *Chlorella vulgaris*, a unicellular microalga, synthesized AgNPs with a size range from 5.7 to 9.8 nm. *Sargassum bovinum*, a marine alga, synthesized palladium NPs. Another marine alga—*Sargassum plagiophyllum*—synthesized silver chloride NPs. A marine green alga—*Caulerpa racemosa*—was also involved in AgNP synthesis. Gold NPs were produced by brown algae such as *Turbinaria* and *Sargassum tenerrimum* (Gahlawat & Choudhury, 2019). *Gelidium amansii* synthesized AgNPs, which have shown antimicrobial properties. A microalga, which is a sea weed—*Sargassum crassifolium*—has been extensively used to biosynthesize AuNPs. (Au) shell (Ag) NP, which is a novel NP, has also been synthesized from *Spirulina platensis* (Ovais et al., 2018a, b). *Cylindrospermum stagnale*, a filamentous heterocystous strain, and *Nostoc linckia* have been observed to synthesize AgNPs (Zhang et al., 2020).

3 Mechanisms Involved in the Microbial Enzyme-Mediated Synthesis of MtNPs

Mostly, microbes carry out the biosynthesis of MtNPs by trapping metal ions from their surroundings and converting them into elemental forms with the help of enzymes. However, not all microbes have this ability, as MtNP synthesis happens through cellular enzymes and metabolic pathways, which may not take place in few microbes. The biosynthesis of MtNPs also depends on the microbial ability to tolerate toxic heavy metals. Generally, microbes that dwell in metal-rich environments exhibit high resistance to heavy metals as they exhibit mechanisms like chelation, both extracellularly and intracellularly (Bahrulolum et al., 2021). The biosynthesis of MtNPs can be classified into two types.

Biosorption

The process of metal cation interaction with the microbial cell wall is known as biosorption of metals, and it mainly encompasses mechanisms such as precipitation, complexation, ion exchange, and physisorption. Microbes generally release extra

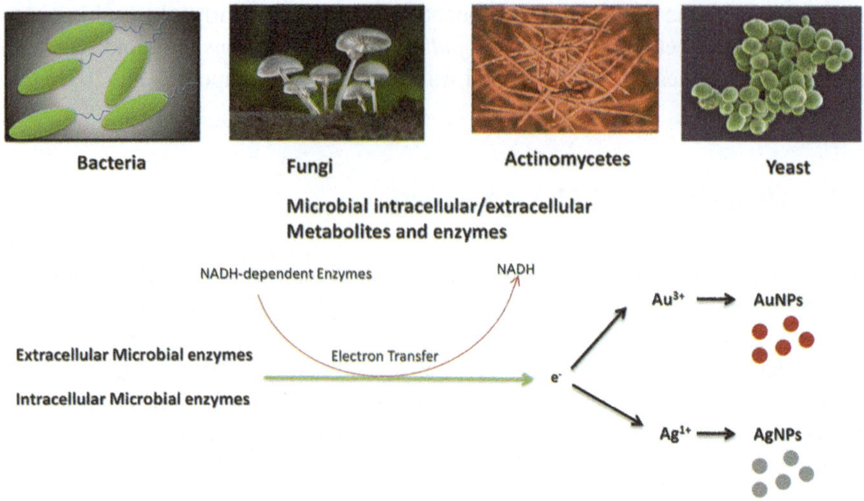

Fig. 2 Mechanism involved in MtNP synthesis via NADH- and NADH-reliant microbial enzymes. (Modified from Ovais et al., 2018a, b)

polysaccharide substances comprising lipopolysaccharides (LPSs), glycoproteins, etc., which mostly contain anionic functional groups, which attract cations from pollutants and toxic substances. In the case of bacteria, cell components like lipopolysaccharides, peptidoglycans, phospholipids, and teichoic acids accomplish positive metal binding to the cell wall with negative charge. In fungi, the main component of the cell wall was observed to be chitin, which is responsible for heavy metal complexation that results in MtNPs.

Bioreduction

Bioreduction involves the chemical reduction of metal salts into stable forms with the help of microbial enzymes. MtNP synthesis is triggered by various compounds like amides, amines, proteins, carbonyl groups, alkaloids, pigments, and other reducing substances, which exist in microbial cells (Saravanan et al., 2021). Based on other reducing substances—enzymes—and the site where the biosynthesis of MtNPs happens, synthesis can be extracellular or intracellular as mentioned below.

Extracellular Biosynthesis of MtNPs

In this mechanism, extracellular microbial enzymes play a crucial role as reducing substances in MtNP green synthesis. In fungi, extracellular enzymes like cellobiohydrolase D, acetyl xylan esterase, and glucosidase participate in MtNP synthesis. Cofactors like nicotinamide adenine dinucleotide (NADH) and the reduced form of nicotinamide adenine dinucleotide phosphate (NADPH)-dependent enzymes play a crucial part as reducing agents by transferring electrons from NADH by NADH-dependent enzymes that act as electron carriers as mentioned in Fig. 2. *Rhodopseudomonas capsulata* extracellularly synthesizes AuNPs by transferring electrons from NADH with the help of NADH-dependent reductase enzymes,

Table 2 Enzymes utilized for MtNP biosynthesis (modified from Khan et al., 2016)

Source	Enzyme	Types of NPs	Size (nm)
Anabaena flos-aquae	Nitrogenase	Palladium, Gold, Silver, Platinum	3.5–40
Aspergillus niger	Nitrate-dependent reductase	Silver	1–20
Calothrix pulvinata	Nitrogenase	Palladium, Gold, Silver, Platinum	3.5–40
Enterobacter cloacae	Nitro reductase	Silver	28.2–122
Escherichia coli	Nitro reductase	Silver	28.2–122
Fusarium oxysporum	α-NADPH-dependent sulfite reductase	Gold	7–20
	Hydrogenase	Platinum	100–180
	Nitrate-dependent reductase	Silver	20–50
	Nitrate reductase	Silver	10–25
	Sulfate reductase	Cadmium	5–20
Klebsiella pneumonia	Nitro reductase	Silver	28.2–122
Leptolyngbya foveolarum	Nitrogenase	Palladium, Gold, Silver, Platinum	3.5–40
Pleurotus ostreatus	Laccase	Gold	22–39
Rhodopseudomonas capsulata	NADH-dependent enzymes	Gold	10–20
Sclerotium rolfsii	NADPH-dependent reductases	Gold	25
Tethya aurantia	Hydrolase	Gallium	–
Thermomonospora sp.	Sulfite reductase enzyme	Gold	2–6

followed by the conversion of gold ions to AuNPs. The shuttle system with quinone- and nitrate-based reductases present in *Fusarium oxysporum* reduces gold and silver ions and synthesizes AuNPs and AgNPs extracellularly (Kapoor et al., 2021). *Penicillium brevicompactum* reduces silver ions with the help of nitrate reductase, an NADH-based enzyme (Ovais et al., 2018a, b).

Intracellular Biosynthesis of MtNPs

In the intracellular mechanism, interactions occur mainly between intracellular enzymes and the positive charge of metal ions, which ultimately lead to reduction as mentioned in Table 2. When MtNPs synthesized by this process were observed under a microscope, accumulation of MtNPs in the cell wall, cytoplasmic membrane, and the periplasmic space was noted. MtNP accumulation occurs due to the metal–ion diffusion across the plasma membrane and reduction by enzymes. *Rhodococcus* sp., which is an alkalotolerant actinobacteria, intracellularly synthesize AuNPs when subjected to $AuCl_4$ ions. Enzymes present on the mycelial surface and in the cytoplasmic membrane effectively mediate Au^{3+} reduction. Similarly, in *Verticillium*, with the help of reductase enzymes, AuNPs were trapped in the cell wall and cytoplasmic membrane by reducing Au^{3+} ions intracellularly (Ovais et al.,

2018a, b). In yeasts, the interaction between metal cations and amide groups in the yeast cell wall occurs intracellularly, followed by enzyme-mediated metals ion reduction and MtNP formation (Kapoor et al., 2021).

3.1 Mechanisms in Gold NP Green Synthesis

Gold ion reduction to gold atom involves atomic interaction with the cell surface and accumulation of atoms, leading to the formation of AuNPs. While in bulk quantity, Au exists as a nonreactive inert metal for many chemical interactions; but, when MtNPs are synthesized, gold exhibits unique properties like quantum size effects, localized energy-level changes, electronic properties, and localized surface plasmon resonance (LSPR), which have extreme chemical modifications while transforming from bulk to MtNPs. The mechanisms employed by microbes for detoxifying heavy metals include metal binding, vacuole compartmentalization, and volatilization. With the help of Fourier transform infrared (FTIR) analysis, the presence of the NH stretch and carbonyl stretch was observed to possess stronger bonding with MtNPs, which helps in coat formation that ultimately prevents AuNP agglomeration mediated by the alga *Galaxaura elongata* (Menon et al., 2017). AuNPs synthesized using α-amylase readily stabilized NPs by capping, and this mechanism is mediated by reducing tetrachloroaurate (Khan et al., 2016). In the case of *Stenotrophomonas maltophilia*, the AuNPs resulted due to the action of NADPH-dependent enzymes. They reduced gold ions (Au^{3+}) to gold NPs (Au), mediated by the electron shuttling system (Nangia et al., 2009).

3.2 Mechanisms in Silver NP Green Synthesis

AgNPs exhibit three strategic mechanisms to combat bacteria—damage to the cell membrane and cell wall, damage by intracellular penetration, and damage by oxidative stress.

Damage to the Cell Wall and Cell Membrane

The cell wall components and cell membrane exhibit diverse adhesion pathways for NPs. The cell wall and cell membrane mainly function as a protective barrier for microbes to combat unfavorable environmental conditions and also to balance homeostasis during nutrient transportation. AgNPs mostly show greater antibacterial activity against Gram-negative bacteria, which possess LPSs, and act as a major attractive factor to abide as they have a negative charge. AgNP interaction with microbes starts with AgNPs binding to the bacterial cell wall and cell membrane. Here, the interaction occurs due to the electrostatic force between the bacterial membrane's negative charge and positively charged AgNPs or it may be due to the low zeta potential on the surface of the bacterial cell. Due to this interaction, morphological change occurs, which is induced by NPs, leading to cell integrity disruption, which ultimately leads to cell content leakage into the medium along with DNA, metabolites, proteins, etc., finally resulting in cell death. Thus, cell wall damage with loss of cell integrity remains a primary antimicrobial mechanism.

Damage by Intracellular Penetration

Based on the level of membrane damage, AgNPs can invade the cell and damage vital cell functions by reacting with proteins and DNA. The silver of AgNPs transforms the natural relaxed state of DNA to the condensed state with low replication ability. Besides structural change, AgNPs also degrade or denature DNA, and recent proteomics studies have mentioned the effects of AgNPs on proteins and their synthesis. They also react with the thiol group in proteins, which is a functional group in cysteine. AgNPs have the ability to suppress hydroxylamine oxidoreductase and monooxygenase production, which catalyzes crucial reactions in nitrification, leading to inhibition of nitrification and protein synthesis. AgNPs also affect ribosomal subunits, leading to protein biosynthesis termination.

Damage by Oxidative Stress

Oxidative stress is mainly caused by reactive oxygen species (ROS), which have a high redox potential. Generally, in cells, antioxidants and ROS are in a balanced state. If an imbalance occurs, then the cell redox balance favors ROS, leading to oxidative stress. When oxidative stress escapes the cell defense mechanisms, cell components will be exposed to ROS and free radicals including hydrogen peroxide, hypochlorous acid, a singlet oxygen, and a superoxide anion. Damage to DNA comprises single- and double-stranded breaks, mutations, deletions, protein cross-linking, and adduct formation. ROS functions mediated by AgNPs may damage bacterial oxidative phosphorylation by inhibiting related enzymes, and protein leakage may occur, resulting cell death. Besides cell wall and intracellular damage, oxidative stress-mediated AgNPs also modify gene expression or may damage cell signaling and transduction pathways (Roy et al., 2019).

Microbes that generally exhibit resistance against Ag ions tend to synthesize AgNPs. The defense mechanism exhibited by microbes varies with the organism. *Bacillus lichenformis* synthesizes AgNPs without cell death at 1 mM concentration, but, when the gradient is raised to 10 mM, the microbe faces cell death. A microbe like *Fusarium oxysporum* is considered to accumulate AgNPs in its periplasmic space, due to the presence of the "nitrate reductase" enzyme NO_3^- (nitrate) that converts to NO_2^- (nitrite). While this reduction mechanism occurs, the electrons from NO_3^- are transferred to silver ions (Ag^+), which, in turn, reduce to silver metal Ag (Roy et al., 2019), as mentioned in Fig. 3.

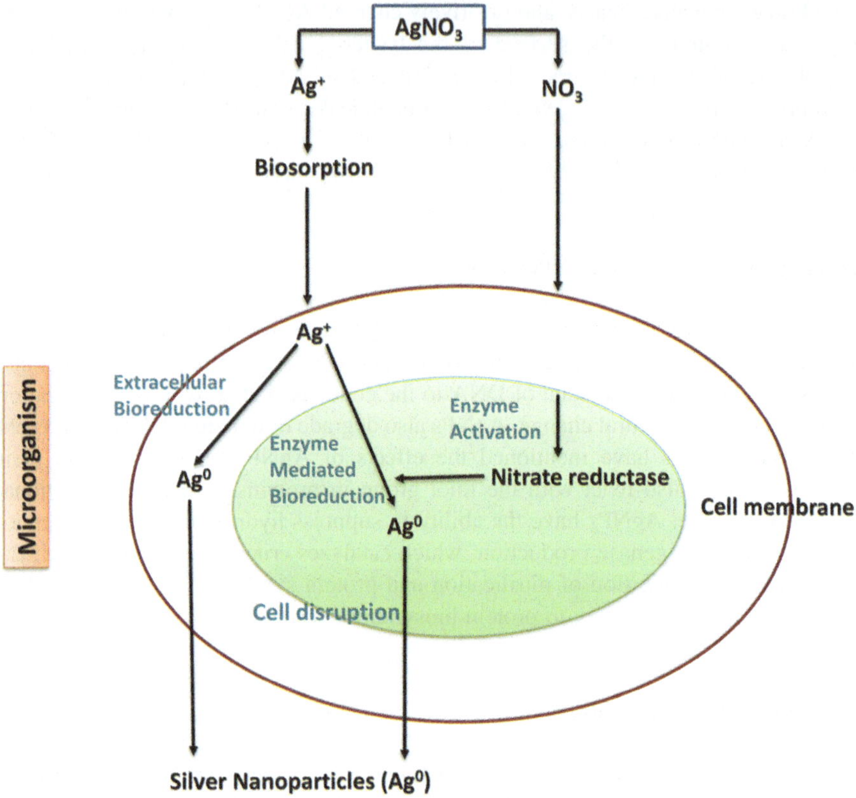

Fig. 3 Schematic representation of silver ion reduction using nitrate reductase. (Modified from Saravanan et al., 2021)

In nature, lysozymes present in higher organisms as a part of innate immunity and mainly attack bacteria by degrading the peptidoglycan layer via enzymatic hydrolysis. Using lysozymes as a catalytic factor, AgNPs are synthesized. Lysozymes are involved in AgNP synthesis by nucleating and reducing. Lysozymes and silver acetate at saturated concentrations are dissolved in methanol. AgNP formation occurs when this Ag–Lyso mixture in methanol is exposed to light. The unique amphipathic form of lysozymes in methanol leads to Ag metal reduction, resulting in stable silver particles.

Even though methanol affects lysozyme conformation, one hypothesis states that lysozyme-altered conformation only favors AgNP synthesis. The hydrophobicity of methanol plays a major role in stabilizing Ag colloids by reducing NP adsorption and accumulation. Lysozyme is a protein with a high cationic nature and in a globular form; it exhibits an amphiphilic form, which is involved in positive–positive or hydrophobic-induced revulsion between NPs, and the reaction is completed only when subjected to water in the final step as mentioned in Fig. 4.

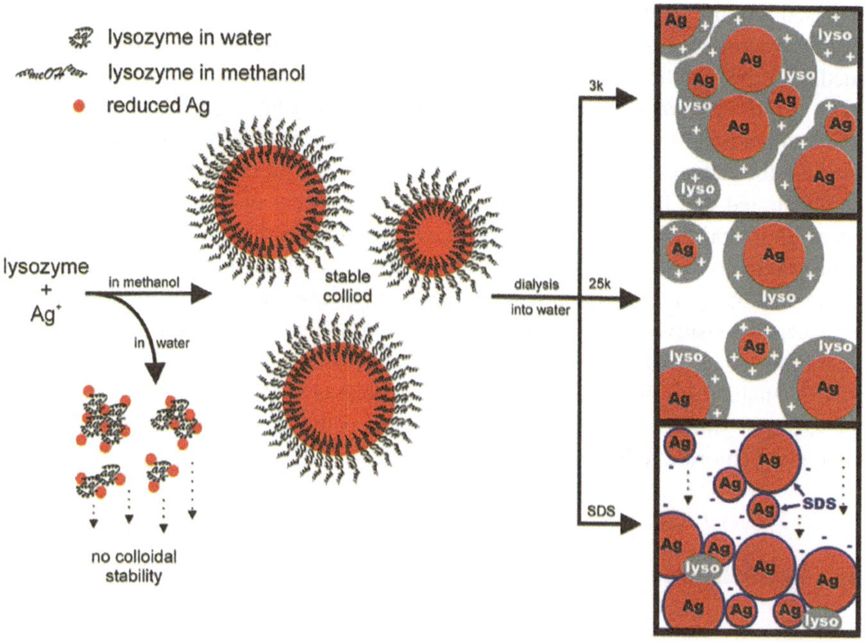

Fig. 4 Mechanism involved in Ag-Lyso NP formation. (Adopted from Matthew Eby et al., 2009)

Lysozyme-mediated AgNP formation exhibits numerous advantages when compared to inorganic synthesis. This process represents an inexpensive, passive, and simple methodology to synthesize large amounts of stable AgNPs (Matthew Eby et al., 2009).

3.3 Mechanisms in Platinum NP Green Synthesis

Several mechanisms have been demonstrated for PtNP bacterial synthesis. Using bacterial cellulase obtained from *Acetobacter xylinum* PtNPs, here hydrogen gas actas reducing agents. K_2PtCl_4 is converted to $PtCl_2$ $(H_2O_2)_2$ by solvolysis. *Desulfovibrio alaskensis* G20, an anaerobic sulfate-reducing bacterium, synthesizes PtNPs extracellularly, mediated by cytochromes and hydrogenases. PtNPs, in this mechanism, accumulate on the outer surface of the cell. In *Desulfovibrio vulgaris*, seven Hases enzymes are secreted, and, out of them, four are present in the periplasmic space and three are located in the cytoplasm. Hases play a major role in PtNP synthesis by the reduction mechanism. Several hydrogenases present in *E. coli* MC4100 catalyze hydrogen to protons and electrons, which facilitates PtNP production. In cyanobacteria, PtNP synthesis occurs by reduction of metal salts, which takes place in vegetable cells and heterocysts. After synthesis, PtNPs are released into the medium. This mechanism is favored by nitrogenases. In few other cyanobacteria, two classes of reducing enzymes—nitrogenases and hydrogenases—mediate PtNP synthesis. *Calothrix* and *Anabaena* have nitrogenases in their heterocysts. *Anabaena variabilis* secretes two kinds of nitrogenases, which can be active under either aerobic or anaerobic conditions that occur in heterocysts. Hydrogenases mediate the reduction process to obtain molecular hydrogen. In few reports, it was demonstrated that PtNP synthesis happens with the help of two kinds of hydrogenases. In the primary stage, platinum-IV was converted into platinum-II, which was mediated by a two-electron bioreduction process carried out by cytoplasmic hydrogenase—an oxygen-tolerant enzyme. In the second stage, periplasmic hydrogenase—an oxygen-sensitive enzyme—mediated the conversion of platinum-II ion into platinum metal. Alternatively, the *Streptomyces* sp. extracellularly synthesize PtNPs, enabled by the chloride reductase enzyme. PtNPs are also synthesized by the electron shuttle–enzymatic metal reduction pathway mediated by the NADP-dependent chloride reductase (Bloch et al., 2021).

4 Applications

MtNPs have diverse applications in both physicochemical and biomedical fields. They can be utilized for biosensing, drug delivery, biomolecular recognition, and bioimaging. MtNPs are combined with different materials, which are used in daily life such as toothpastes, water purification systems, deodorants, humidifiers, and cosmetics, as they possess antimicrobial properties. They also play a crucial role in agricultural technologies that include detection and diminution of plant diseases and curtailing nutrient leakage to boost crop yield (Zhang et al., 2020).

Gold NP Applications The utilization of AuNPs started in the sixteenth century for staining and medical purposes. AuNPs can be also used in colorimetric techniques for detecting heavy metal salts in an aqueous medium. AuNPs are also helpful in treating B-cell chronic lymphocytic leukemia (CLL). Gold NPs are also used in the Carter–Wallace home pregnancy test (Sardar et al., 2014). AuNPs from rattle bush—*Sesbania drummondii*—have revealed the catalytic action that reduces the aromatic nitro components in waste cleansing (Zhang et al., 2020).

Silver NP Applications AgNPs are extensively applied as antimicrobial substances in medical, commercial, and consumer products. They exhibit larvicidal activity against malaria and filariasis vectors. They are mostly combined with topical creams and ointments, which are helpful in preventing burn infections and open wounds (Sardar et al., 2014). Besides topical ointments and creams, AgNPs have become a part of clothing, which guards the wearer from emanating body odor. AgNP-based products are also approved by organizations like United States Environmanetal Protection Agency (US-EPA), the United States Food and Drug Administration (US-FDA), SIAA (Society of industrial technology for antimicrobial agents) of Japan, Koreas Testing, and FITI Testing and Research Institute Korea. They also exhibit antitumorigenic activity as they comprise cytotoxic activity against different tumor cells (Zhang et al., 2020).

Platinum NP Applications PtNPs have been observed to have anticancer and catalytic activities, antibacterial and antioxidant properties, and can be used in drug delivery and bioimaging (Bloch et al., 2021).

MtNPs exhibit outstanding antifungal properties. In few reports, it has been mentioned that CuNPs obtained from *Streptomyces* sp. exhibited antifungal activity against pathogenic fungi like *Pythium ultimum*, *Fusarium oxysporum*, *Alternaria alternata*, and *Aspergillus niger*. AgNPs have also found to be active against *Fusarium oxysporum*.

In recent reports, it has been observed that nanoclay crystals or minerals can be also utilized as fertilizers. NPs also participate in nanopesticide production. Diseases caused by the Baculovirus, *Sitophilus oryzae*, and nuclear polyhedrosis in silkworms can be regulated by ZnNPs and AgNPs (Koul et al., 2021).

Due to their large surface area with numerous active sites, MtNPs exhibit catalytic activity. CuNPs degrade different azo dyes like Congo red, malachite green, Reactive Black-5, and Direct Blue-1. Excessive use of aromatic chlorinated compounds results in air, water, and soil pollution. Dehalogenation of these compounds can be carried out by Pd-based NPs, which are synthesized by *Desulfovibrio vulgaris* and *Desulfuricans* (Kapoor et al., 2021). NPs have also been utilized to enhance the reaction rates of microbiological reactions. Nanoparticles, due to their small size, can pass through epithelial junctions of the skin and blood barriers. They enhance the solubility of hydrophobic substances and make them favorable for application. It is considered that NP-mediated drug delivery could notably decrease anticancer drug dosage, enhance specificity, lower toxicity, and enhance efficacy. Magnetic NPs play a crucial role in the treatment of hyperthermia in cancer by subjecting cancer tissue sites to local heating with an external magnetic field at the specific target areas (Li et al., 2011). In the food industry, NPs are widely used in food processing and food packaging. AgNPs, with their unique ability to penetrate bacterial biofilms, prevent packaging material contamination and help in the cleansing process while packaging food (Koul et al., 2021).

5 Large-Scale Production of MtNPS

Microbial fermentation symbolizes a state-of-the-art tactic for the production of nanoscale structures on a large scale. Recently, researchers have discovered large-scale production of NPs by utilizing biogenic pathways with narrow size. While producing tailor-made MtNPs on a large scale, factors like dosage, basal medium composition, biomass concentration, and types of precursors utilized play a crucial role, which need to be taken care of while production. The first large-scale production reported was synthesis of metal-substituted magnetic NPs by the *Thermoanaerobacter* sp. TOR 39. Zn-substituted magnetites were produced from this strain in enormous quantities (1 kg/30 L) with less expenditure. Magnetic NPs have attracted huge attention in the recent decades as they have potential applications in bioremediation, magnetic resonance imaging (MRI), catalysis, biosensor development, and data storage and they can be easily handled in the magnetic field. The same train of thought was also employed for CdSNP extracellular production on a large scale. The obtained CdS crystallites have a size of less than 10 nm, and the process was effortlessly scalable. Using an anaerobic metal-reducing thermophilic bacterium—*Thermoanaerobacter* sp.—ZnS NPs were reported to effectively synthesize on a large scale in a controlled and reproducible manner, ranging from 10 mL to 24 L, which yielded 5 g/L every month (Gahlawat & Choudhury, 2019). For large-scale production of MtNPs, only few reports have focused on cultivation methods. These strategies will be extremely helpful to attain a higher yield of MtNPs and thus need to be studied further.

6 Regulation of NP Morphology

The electronic and optical parameters of NPs are generally dependent on their size and shape. Microbes are capable of regulating their own shape and size (Abada et al., 2017). The enzymatic production of MtNPs is highly based on the nature of the enzyme, pH of the mixture, and metal salt. The interaction of proteins with MtNPs and the nature of capping proteins also affect the size and shape. All these factors lead to different morphologies, monodispersity index, and size control. An extract from *Sclerotium rolfsii*—a fungus—contains an NADPH-dependent enzyme, which is used to develop AuNPs rapidly. As mentioned in few reports, the size and shape of AuNPs can be controlled by altering the cell extract and salt ratios without affecting enzyme activity. *Rhodopseudomonas capsulata* is used to synthesize AuNPs with different shapes and sizes. In regulating AuNP morphology, pH plays a crucial role (Khan et al., 2016). It has been reported that AuNPs exhibit several kinds of morphologies and size. This can be achieved by controlling factors like gold concentration, pH, and temperature. Mms6, an associated protein, possesses a sequence motif, which has greater affinity to iron ions. These components also exhibited similar morphologies—cuboctahedral—and size—20 nm. This demonstrates that Mms6 has greater effects on AgNP morphology (Abada et al., 2017).

7 Advances in MtNP Microbial Biosynthesis

As described earlier, wild-type microbial strains have succeeded in synthesizing various MtNPs. However, the efficiency, composition, and size of NPs need to be regulated for better yield and applications. Recently, researchers have been focusing on MtNP biosynthesis using engineered microbes like recombinant *E. coli*. As mentioned in few reports, recombinant *E. coli* cells were allowed to grow until specific cell density and then incubated with metal–ion solutions to synthesize desired MtNPs. It has been mentioned that phytochelatins (PCs) form metal complexes with Cu, Cd, Ag, Hg, and Pb, whereas MTs complex with Cd, Cu, and Zn ions. To increase the potential applications of MtNPs, a recombinant *E. coli* strain was modified to express both MT genes and PC synthases. PC synthases and MT coexpression can develop an exceptional cell environment for synthesizing bimetallic, trimetallic, noble, semiconducting, rare-earth, and magnetic MtNPs like CdSe, CdSeZn, FeCoMn, FeCoNi, FeAg, and AuCdSeZn. The sizes of these MtNPs along with quantum dots can be regulated by altering the concentrations of treated MtNPs by up to 5 mM. Nevertheless, the biosynthesis of MtNPs is problematic to some extent due to the effects of metal toxicity on cell viability and heterogeneity of cells. To overcome the effects of metal toxicity, a microfluidic system was developed, which generates microdroplets. Metal ions and recombinant *E. coli* were embedded in the microdroplets dosing in nanoliters. This biosynthetic in vitro system acts as an individual artificial chemical bioreactor with hydrogel, a microfluidic device, and recombinant *E. coli* extracts. By this mechanism, the cell size can be regulated and easily separated due to unique color differences. Using this system, various types of MtNPs and combinations of metal NPs can be developed on a large scale (Park et al., 2015).

8 Challenges and Limitations of Biosynthesized MtNPs

- Nanotechnology is an interdisciplinary field, which provides coordination between chemists, engineers, and biologists to develop novel therapeutic MtNPs. Microbial synthesis of MtNPs is well-focused by researchers, where metal ions are converted to MtNPs by the reduction mechanism as a defense mechanism exhibited by microbes.
- The primary challenge is the selection of the best and suitable microbial strains by considering several intrinsic factors like replication, growth rate, biochemical activity, etc.
- Other crucial criteria include controlling the size, shape, and monodispersity of MtNPs to achieve proper drug delivery and desired therapeutic effects. Among the huge pool of enzymes, finding the key enzymes responsible for detoxification of metals to MtNPs remains as challenge.
- Maintaining the desired conditions for enzyme activity including nutrient supplements, pH, temperature, inoculum size, and amount of light required for synthesis is crucial.
- In addition, there is a huge difference in the production of MtNPs in a laboratory and in industries on a large scale, which primarily needs trial and error methodology to scale up, which should be done precisely.
- Knowing the microbial source is extremely important, as, rarely, the use of toxic microbes as a source may lead to pathogenicity (Ovais et al., 2018a, b).

Apart from optimizing desired environmental conditions, biofilm application is another crucial approach for effective MtNP synthesis, but only few reports have demonstrated the role of biofilms in MtNP synthesis. Numerous reports have mentioned that MtNPs exhibit toxicity to some extent. The cytotoxicity of biosynthesized MtNPs depends on several factors like shape, size, capping agent, NP density, and pathogen type that is evaluated for toxicity. Cytotoxicity is considered to be generated due to ROS, which lowers glutathione levels and increases free radicals. However, this toxicity can be controlled by coating MtNPs to some extent (Gahlawat & Choudhury, 2019). However, even low toxicity influences MtNP utilization as the metal ions may penetrate into the cells of organisms, which leads to oxidative stress (Saravanan et al., 2021).

9 Conclusions

Nowadays, researchers are focusing more on the large-scale production of MtNPs with narrow size and on unexplored downstream processing. Large-scale production of MtNPs is generally halted by factors including low yield, high energy requirement, high cost, and polydispersity. Combined research on MtNP fermentation along with understanding the mechanisms involved could increase the chances for tailor-made, cost-effective MtNP synthesis. Recently, the microdroplet-based approach to synthesize MtNPs in vitro from engineered microbial cells has gained huge attention as it has the ability to regulate the polydispersity, size, chemical composition, and morphology, but there is a need for further investigation. The toxicity of MtNPs is enhanced when present as fine-sized particles. Future research must focus on NP stability and reactive parameters. The development of novel and efficient MtNPs with no toxicity can be achieved by tailored, engineered microbial enzyme-mediated synthesis of MtNPs. Therefore, it can be concluded that MtNPs have a bright future due to their potential applications and eco-friendliness. If properly employed and further investigated, MtNPs can become a game changer in future.

References

Abada, E., Al-Faifi, Z., & Osman, M. (2017). Enzymes and nanoparticles produced by microorganisms and their applications in biotechnology. In *Fungal biology: Fungal nanotechnology*. Springer. https://doi.org/10.1007/978-3-319-68424-6_7

Bahrulolum, H., Nooraei, S., Javanshir, N., Tarrahimofrad, H., Mirbagheri, V. S., Easton, A. J., & Ahmadian, G. (2021). Green synthesis of metal nanoparticles using microorganisms and their application in the agrifood sector. *Journal of Nanobiotechnolgy, 19*, 86. https://doi.org/10.1186/s12951-021-00834-3

Bloch, K., Pardesi, K., Satriano, C., & Ghosh, S. (2021). Bacteriogenic platinum nanoparticles for application in nanomedicine. *Frontiers in Chemistry, 9*, 624344. https://doi.org/10.3389/fchem.2021.624344

Das, R. K., Pachapur, V. L., Lonappan, L., Naghdi, M., Pulicharla, R., Maiti, S., Cledon, M., Dalila, L. M. A., Sarma, S. J., & Brar, S. K. (2017). Biological synthesis of metallic nanoparticles: Plants, animals and microbial aspects. *Nanotechnology for Environmental Engineering, 2*, 18. https://doi.org/10.1007/s41204-017-0029-4

Gahlawat, G., & Choudhury, A. R. (2019). A review on the biosynthesis of metal and metal salt nanoparticles by microbes. *RSC Advances, 9*, 12944. https://doi.org/10.1039/c8ra10483b

Kapoor, R. T., Salvadori, M. R., Rafatullah, M., Siddiqui, M. R., Khan, M. A., & Alshareef, S. A. (2021). Exploration of microbial factories for synthesis of nanoparticles – A sustainable approach for bioremediation of environmental contaminants. *Frontiers in Microbiology, 12*, 658294. https://doi.org/10.3389/fmicb.2021.658294

Khan, I., Dutta, J. R., & Ganesan, R. (2016). Enzymes action on materials: Recent trends. *Journal of Cellular Biotechnology, 1*(2015/2016), 131–144. https://doi.org/10.3233/JCB-15013

Koul, B., Poonia, A. K., Yadav, D., & Jin, J.-O. (2021). Microbe-mediated biosynthesis of nanoparticles: Applications and future prospects. *Biomolecules, 11*, 886. https://doi.org/10.3390/biom11060886

Li, X., Huizhong, X., Chen, Z.-S., & Chen, G. (2011). Biosynthesis of nanoparticles by microorganisms and their applications. *Journal of Nanomaterials, 2011*, 270974. https://doi.org/10.1155/2011/270974

Maliszewska, I. (2011). Microbial synthesis of metal nanoparticles. In *Metal nanoparticles in microbiology*. Springer. https://doi.org/10.1007/978-3-642-18312-6_7

Mandal, D., Bolander, M. E., Mukhopadhyay, D., Sarkar, G., & Mukherjee, P. (2006). The use of microorganisms for the formation of metal nanoparticles and their application. *Applied Microbiology and Biotechnology, 69*, 485–492. https://doi.org/10.1007/s00253-005-0179-3

Matthew Eby, D., Schaeublin, N. M., Farrington, K. E., Hussain, S. M., & Johnson, G. R. (2009). Lysozyme catalyzes the formation of antimicrobial silver nanoparticles. *ACS Nano, 3*, 984.

Menon, S., Rajeshkumar, S., & Venkat Kumar, S. (2017). A review on biogenic synthesis of gold nanoparticles, characterization, and its applications. *Resource-Efficient Technologies, 3*, 516. https://doi.org/10.1016/j.reffit.2017.08.002

Nangia, Y., Wangoo, N., Goyal, N., Shekhawat, G., & Suri, C. R. (2009). A novel bacterial isolate *Stenotrophomonas maltophilia* as living factory for synthesis of gold nanoparticles. *Microbial Cell Factories, 8*(1), 1–7. https://doi.org/10.1186/1475-2859-8-39

Narayanan, K. B., & Sakthivel, N. (2010). Biological synthesis of metal nanoparticles by microbes. *Advances in Colloid and Interface Science, 156*, 1–13. https://doi.org/10.1016/j.cis.2010.02.001

Navia-Mendoza, J. M., Zambrano-Intriago, L. A., Maddela, N. R., Duarte, M. M. M. B., Quiroz-Fernández, L. S., Baquerizo-Crespo, R. J., & Rodríguez-Díaz, J. M. (2021). Advances in the application of nanocatalysts in photocatalytic processes for the treatment of food dyes: A review. *Sustainability, 13*(21), 11676.

Ovais, M., Khalil, A. T., Ayaz, M., Ahmad, I., Nethi, S. K., & Mukherjee, S. (2018a). Biosynthesis of metal nanoparticles via microbial enzymes: A mechanistic approach. *International Journal of Molecular Sciences, 19*, 4100. https://doi.org/10.1007/s00253-018-9146-7

Ovais, M., Khalil, A. T., Islam, N. U., Ahmad, I., Ayaz, M., Saravanan, M., Shinwari, Z. K., & Mukherjee, S. (2018b). Role of plant phytochemicals and microbial enzymes in biosynthesis of metallic nanoparticles. *Applied Microbiology and Biotechnology, 102*(16), 6799–6814. https://doi.org/10.1007/s00253-018-9146-7

Palomo, J. M., & Filice, M. (2016). Biosynthesis of metal nanoparticles: Novel efficient heterogeneous nanocatalysts. *Nanomaterials, 6*, 84. https://doi.org/10.3390/nano6050084

Park, T. J., Lee, K. G., & Lee, S. Y. (2015). Advances in microbial biosynthesis of metal nanoparticles. *Applied Microbiology and Biotechnology, 100*(2), 521–534. https://doi.org/10.1007/s00253-015-6904-7

Roy, A., Bulut, O., Some, S., Mandal, A. K., & Deniz Yilmaz, M. (2019). Green synthesis of silver nanoparticles: Biomolecule-nanoparticle organizations targeting antimicrobial activity. *RSC Advances, 9*, 2673. https://doi.org/10.1039/c8ra08982e

Roy, K., Mukherjee, A., Maddela, N. R., Chakraborty, S., Shen, B., Li, M., Du, D., Peng, Y., Lu, F., & Cruzatty, L. C. G. (2020). Outlook on the bottleneck of carbon nanotube in desalination and membrane-based water treatment—A review. *Journal of Environmental Chemical Engineering, 8*(1), 103572.

Saravanan, A., Senthil Kumar, P., Karishma, S., Vo, D.-V. N., Jeevanantham, S., Yaashikaa, P. R., & George, C. S. (2021). A review on biosynthesis of metal nanoparticles and its environmental applications. *Chemosphere, 264*, 128580. https://doi.org/10.1016/j.chemosphere

Sardar, M., Mishra, A., & Ahmad, R. (2014). Biosynthesis of metal nanoparticles and their applications. In *Biosensors nanotechnology* (pp. 239–266). Wiley. https://doi.org/10.1002/9781118773826.ch8

Talebi, S., Ramezani, F., & Ramezani, M. (2010). *Biosynthesis of metal nanoparticles by micro-organisms* (Vol. 10, pp. 12–18). EU.

Zhang, D., Ma, X.-L., Gu, Y., Huang, H., & Zhang, G.-W. (2020). Green synthesis of metallic nanoparticles and their potential applications to treat cancer. *Frontiers in Chemistry, 8*, 799. https://doi.org/10.3389/fchem.2020.00799

Part III
Impact of Environmental Disturbances on Microbial Enzymes

Effects of Agrochemicals on Soil Microbial Enzymes

**Babafemi Raphael Babaniyi, Samuel O. Thompson,
Olusola David Ogundele, and Obagboye Fredrick Oluwole**

Abstract This chapter considers the effects of agrochemicals (pesticides and fertilizers) on microbial enzymes (fluorescein diacetate hydrolases, acid phosphatases, alkaline phosphatases, phosphatases, β-glucosidases, cellulases, ureases, and arylsulfatases). The pesticides considered include fungicides, insecticides, and herbicides. Soil is not a mass of dead debris, arising from physical and chemical processes of soil formation, but is a mixture of decomposed plant and animal remains. Microbial enzymes in the soil aid in the recycling of carbon and nutrient assimilation. The cell control mechanisms of nutrients, coupled with carbon, nitrogen (N), and phosphorous (P) uptake, trigger biomass growth and increase the rate of enzyme synthesis and secretion. The impacts of agrochemicals on microbes and their extracellular enzymes are generally known to be unpleasant. These impacts include, but are not limited to, destruction of microbial habitats, ecological succession, reduction of microbial communities, development of new strains, and multiple drug-resistant microbes. These effects may result in increased pathogenic activities, reduction in soil fertility, high soil acidity, eradication or reduction of the natural flora of a particular ecology (both flora and fauna), low crop yield, etc.

1 Introduction to Agrochemicals

Agrochemicals, in general are referred to as products that include fertilizers, fungicides, insecticides, nematicides, etc., which enhances plant growth (Biswas et al., 2014). Over the last few decades, a large amount of chemicals have been used in agriculture to increase the production of crops in both developed and developing

B. R. Babaniyi
Bioresources Development Centre, National Biotechnology Development Agency, Nigeria

S. O. Thompson (✉) · O. F. Oluwole
Chemistry Department, Federal University of Technology Akure, Akure, Nigeria

O. D. Ogundele
Chemical Sciences department, Achievers University, Owo, Nigeria

N. R. Maddela et al. (eds.), *Ecological Interplays in Microbial Enzymology*,
Environmental and Microbial Biotechnology,
https://doi.org/10.1007/978-981-19-0155-3_16

countries (Tan et al., 2020). Crops tend to grow, slowing with inadequate provision of the right nutrient. Hence, to overcome these problems, agrochemicals are applied with their best modified and oriented results. These are chemicals used mainly in agriculture to aid crop growth and safety. They are applied in various practices of the farming sector such as crop shifting, poultry, dairy farming, commercial farming, horticulture, etc. (Princy & Prabagaran, 2020).

Agrochemicals are produced to protect agricultural crops from pests and for increasing crop yields. They are inorganic fertilizers and pesticides that provide benefits and manage the agricultural ecosystem. The continuous use of pesticides has affected the entire ecosystem and also the microorganisms in soil (Onder et al., 2011). Weeds and insects are the main reducing biotic factors in agriculture; they reduce crop yield, resource use efficiency, and productivity (Oliveira et al., 2014). Agrochemicals are usually harmful and may cause major environmental risks.

Different researchers have proven the adverse effects of agrochemicals on soils and ecosystems at large and consider them a matter of major concern that needs attention especially because of their studied impacts on pathogens, fertility, microorganisms, and enzymes (Mergel et al., 1998; Nannipieri et al., 2008; Steinauer et al., 2016; de Vries et al., 2019; Perucci et al., 2000; Vischetti et al., 2000, 2002; Puglisi et al., 2005, 2012; Nannipieri et al., 2012; Sofo et al., 2012; Suciu et al., 2019). Although the results varied in some aspects, the major negative impacts were clearly stated and explained. According to Boivin and Poulsen (2017), it has become mandatory that in most countries any pesticide must be authorized before use, in which case, before authorization, a risk assessment procedure must have been conducted to ascertain its safety for nontarget organisms. The reason for the risk assessment comes from the high rate of the adverse effects of pesticides and other agrochemicals on the ecosystem, which is quantifiably related to different concentrations of their use in a particular environment (Desneux et al., 2007; Beketov et al., 2013; Brühl et al., 2013; Wood & Goulson, 2017).

2 Types of Agrochemicals

Agrochemicals are widely used in farming activities; they are known as pesticides, which include herbicides, insecticides, fungicides, nematicides, rodenticides, and molluscicides. Agrochemicals also include fertilizers and soil conditioners.

2.1 Pesticides

Pesticides are substances used for preventing, repelling, destroying, reducing, or eliminating damages caused by pests (Eldridge, 2008). They are used to control some types of organisms known as pests, which are harmful to cultivated plants and animals. They mostly work through poisoning of pests. Pests can be insects, plant pathogens, weeds, and microbes that compete with humans for food, destroy properties, and carry or help spread diseases. Most commonly, they are used in health sectors and for agricultural crops (Yadav et al., 2015). Naturally, pesticides may generally become harmful to other nontarget organisms, including humans. Therefore, it is important to be careful when handling them and they must be safely disposed.

2.2 *Insecticides*

Insecticides are commonly used to protect households, restaurants, hospitals, farms, forest plantations, etc. from insects. These substances offer protection from harmful insect-borne diseases, insect pests in warehouses, and agricultural and forest pests (Cardoso & Alves, 2012). In general, they are used to destroy insects. Insecticides can be ovicides that kill eggs or larvicides that kill larvae. They are categorized based on their mode of action and structure. Many insecticides act on the insects' nervous system (e.g., cholinesterase inhibition), whereas others act as growth regulators or endotoxins (Relyea, 2005).

2.3 *Herbicides*

Weeds have been known to affect human activities, especially in agriculture, since ages. The growth of these weeds can be controlled with the use of pesticides. Herbicides are chemicals used to manipulate or control undesirable vegetation (Belden & Lydy, 2000). They are generally applied to control or kill plants, weeds, and herbs. Their application occurs more frequently in row crop farming where they are applied before or during planting to maximize crop productivity by minimizing other vegetation. Herbicides can act by inhibiting cell division, photosynthesis, or amino acid production by mimicking natural plant growth hormones that cause deformities (Ross & Childs, 1996).

2.4 *Fertilizers*

Fertilizers are materials of synthetic or natural origin that are applied to plant tissues or soil with the aim of supplying the needed nutrients. Many sources of fertilizers exist naturally or are industrially produced (Scherer et al., 2009). These are compounds used for enhancing plant development; they add the needed nutrients to the soil and eliminate nutrient deficiency. For most modern agricultural practices, fertilization focuses on three major macronutrients, namely, nitrogen, phosphorous, and potassium, with the occasional addition of supplements for micronutrients (Scherer et al., 2009). Fertilizers can be categorized into two types: organic and

Table 1 Agrochemicals and active ingredients (Lamberth et al., 2013; Jeschke 2016; Hamilton 2001)

Agrochemicals	Active ingredients
Insecticides	Abamectin, cyfluthrin, fipronil, deltamethrin, permethrin, bifenthrin, and pyrethrum
Herbicides	Atrazine, butachlor, dithiopyr, flufenacet, isoproturon, and chlorimuron
Fungicides	Captan, dinocap, pyrimethanil, quinoxyfen, iprodione, fenarimol, and azoxystrobin
Nematicides	Chloropicrin, 1,3-dichloropropene, dimethyl disulfide, allyl isothiocyanate, and oxamyl
Fertilizers	Nitrogen, phosphorus, potassium, magnesium, and calcium

inorganic fertilizers. Organic fertilizers are naturally existing substances prepared through natural processes. Inorganic fertilizers, also called synthetic fertilizers, are manufactured artificially using chemical processes (Table 1).

2.5 *Soil Conditioners*

To keep all soils in good conditions, the best thing to do is to add things that help keep it in good conditions. These good things are called soil conditioners that include manures, composts, peats, livestock manures, and leaves. Conditioners are products applied to the soil to improve soil properties and to control erosion (Baumhardt & Blanco Canqui, 2014). Soil conditioners boost the water holding capacity and aeration of the soil. Some of the conditioners used to reduce water erosion include polyacrylamide (PAM), phosphogypsum, flue gas desulfurization (FGD) gypsum, etc.; all these conditioners are laid on the soil and then mixed. Conditioners are not a substitute to soil conservation practices, but they should be used as companions to other practices (Baumhardt & Blanco Canqui, 2014).

3 Importance of Agrochemicals

If agrochemicals are handled with care, they will produce fruitful results. Crop protection solutions allow growers in crop production processes to increase output and crop yield. As weeds, pests, and diseases have an impact of up to 30% on the future crop production worldwide, food production will deteriorate without crop protection chemicals (Princy & Prabagaran, 2020). The benefits of agrochemicals are not limited to growing crop yields. Agrochemicals are also used to prevent the negative impacts caused to society in many ways; for example, trees and weeds growing under power lines when left unchecked would result in power outages (Sharma et al., 2019). Herbicides are also widely used to control unwanted vegetation along national highways, roadsides, in parks, and in other public areas to ensure public safety and convenience. In food processing, insecticides are used in permissible levels to protect raw commodities and packaged groceries from insects infesting during the processing, manufacturing, and packaging stages. Pesticides are also used in homes for controlling insects and pests (Sharma et al., 2019).

4 Environmental Impacts of Agrochemicals

Along with having positive impacts, the negative impacts of agrochemicals are becoming clear. The uses of agrochemicals pose threats and cause harm to the ecological balance and environment. These agrochemicals cause pollution; they enter water bodies and kill many fishes (Aktar et al., 2009). During many uses of pesticides in agriculture, their exposure to other organisms, including humans, is not well controlled, which then causes several problems. Pesticides keep accumulating in soil residues and cause biomagnification in plant and animal tissues; this is dangerous to humans and can cause health problems (Hans & Faroq, 2000). Microorganisms become resistant to pesticides, which is a serious issue. In general, the effects of pesticides will vary depending on the chemical dosage, various environmental factors, and the properties of the soil.

Agricultural runoffs often contain developed levels of heavy metals from fertilizers and other agricultural chemicals applied to the fields. These chemicals are washed away with rainfall runoffs into rivers, streams, and reservoirs, thus polluting water bodies and modifying aquatic habitats (Ogbodo & Onwa, 2013). There could be potential damage to soil organisms from high concentrations of agrochemicals. The effects of agrochemicals can be either direct (immediate or short-term impacts), due to the harm to organisms that come in contact with the chemicals, or indirect due to changes caused by the chemicals to the environment or food source of the organisms (Ogbodo & Onwa, 2013). The direct effects of these chemicals can be short, obvious in the first season after application of the fertilizer or in the long term if repeated addition has taken place. The indirect effects may be long term; they may take up to one season or more to build up due to soil organic matter levels, changes in productivity or pH, and residue inputs (Bunneman & McNeil, 2004). Nitrate pollution has been reported to be a result of excessive use of fertilizers. Nitrate is a chemical compound that is toxic to animals and humans if exposed to high concentrations (Princy & Prabagaran, 2020).

5 Soil Microbial Enzymes

Soils home a vast majority of microbes that are accountable for the disintegration of organic matter and the mobilization of nutrients. Microbes in soil have the highest genetic diversity, and they participate in maintaining the functionality of plant diversity and other various important processes in the ecosystem (Zhang et al.,

2018). Living organisms in the soil are grouped into two types, viz. soil and soil fauna. Soil is not an inert stable material but is a medium that supports life. Soil is dynamic in nature; it is composed of a mass of dead debris of plant and animal remains. Soil structure and fertility are aided by soil microorganisms; this is one of the major microbial activities that take place in the formation of soil. Microorganisms in soil can be grouped as bacteria, actinomycetes, fungi, algae, and protozoa. Each of these groups possesses characteristics and functions that determine the group they belong to in soil.

5.1 Soil Faunas

These include invertebrates that contribute to the breaking of organic matter and the presupplying of nutrients to microorganisms by reducing the size of the organic matter in the process of feeding. Apart from increasing the surface area, faunas promote bioturbation of litters and also enhance formation of soil enzymes (Rao et al., 2017). Microbial communities in soil correspond to soil biogeochemical processes and play a vital role in soil nutrient cycles and turnover (Zeng et al., 2016). Biochemical processes contribute to direct changes in the soil microbial community structure, which may affect microbial functions and population (Sekaran et al., 2019b).

5.2 Soil Enzymes

These are responsible for the biochemical activities of organic matter transformation in the soil processes, such as soil physical properties, microbial activity, and nature of biomass. Enzymes can be extracellular or intracellular. Intracellular enzymes are bound to the cell walls of living and metabolically viable cells, such as spores. Extracellular enzymes are discharged into the soil and "permanently" stick to clay and humic colloids through ionic interplay, hydrogen bonding, and covalent bond immobilization. Soil enzymes aid in catalytic decomposition of organic matter and production of nutrients and vehemently enhance transformation of energy, environmental quality, and agronomic productivity. Nonetheless, tillage, monoculture, and removal of residues adversely affect the enzymatic processes and availability of nutrients to plants. Enzymatic activity reduces due to an increase in soil depth. Moreover, soil enzymes reveal early changes in soil health due to quick response to changes in soil management and environmental factors such as soil quality. Meanwhile, understanding the relationship between various forms of enzymes in relation to biotic and abiotic factors will be a panacea for determining the potential effects of soil management, functionality and productivity of an ecosystem, and changes in the environment (Rao et al., 2017).

6 Production of Soil Enzymes

Microbial enzyme secretion in the soil is favored by natural selection processes, which control the intake of carbon and nutrients to the cell. Production of enzymes requires cellular management of the available minerals to produce enzymes with the advantage of increasing assimilation of nutrients, energy production, and low molecular mass of organic compounds. Carbon, nitrogen, phosphorous, and nutrients are needed for energy in the form of adenosine triphosphate (ATP), enzyme (protein) synthesis and secretion, production and arrangement of membrane transporters for the uptake of nutrients to enhance formation, and discovery of efficient surfaces for microbial interaction. For instance, enzyme production by *Bacillus licheniformis* requires approximately 1–5% of carbon and nitrogen intake. In addition, the *Escherichia coli* synthesis of ATP costs of protein per unit mass of the enzymes that are secreted is significantly reduced compared to protein retained within the cell (Fig. 1) (Burns et al., 2013).

Clearly, extracellular enzymes are responsible for the microbial recycling of energy and carbon. Elevated concentrations of N and P in plants trigger production of enzymes, leading to decomposition and recycling of nutrients. An increase in enzyme activity in response to the available resource contributes to excess release of product reaction; hence, a possible synergy between enzyme activity and resource availability is envisioned. Normally, enzyme synthesis and secretion is aided by substrate availability, but the substrate may not be the main facilitator of enzymes. More so, adequate density with the right aggregate microbial degraders is a factor for successful catalysis and subsequent microbial proliferation (Franklin et al., 2011).

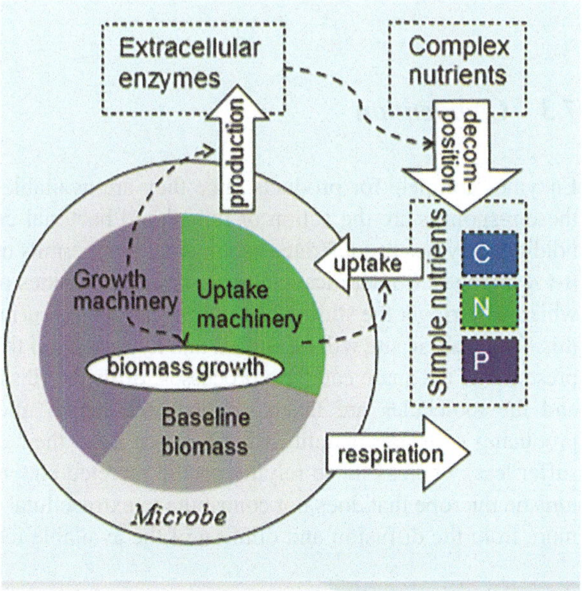

Fig. 1 Extracellular production of enzymes (Franklin et al., 2011)

7 Groups of Cellular Enzymes and Their Activities

7.1 *Mobile Extracellular Enzymes*

Nearly all extracellular enzymes move (diffuse) away from their parent cells because they are more active than intracellular enzymes due to possession of disulfide bonds and they are glycosylated. Extracellular enzymes have modified structures, which make them stable, with the ability to resist proteases and modulate cell adhesion. An increase in the gap between extracellular enzymes and a cell, leads to a reduction reaction on the sum of products trapped by the cell per unit of enzyme yield due to loss in product reaction, concentration of the substrate and enzymes, and diffusional environment (Burns et al., 2013).

7.2 *Immobile Extracellular Enzymes*

Some of the extracellular enzymes are immobilized; they stick to clay, humic acids, and particulate organic matter, which make them comfortable, active, and persistent for a longer time. The activity of static enzymes is low compared to that of their diffusible counterparts; they are confined to a position and so are unable to access the substrate oftentimes. Soil entrapment of enzymes serves as housing for the enzymes toward effective catalytic process in soils and also, provide energy for microbes when they are stressed out or during low accessibility of biomass (Feketeová et al., 2021; Quiquampoix & Burns, 2007).

7.3 *Competition*

Enzymes compete for products once they are available in different forms. Among these responses are the action of fungal and bacterial celluloses engulfing moieties holding enzymes to substrates in a manner that permits the catalytic site to cleave the β-1,4 linkages. Sometimes, cellulose-holding moieties may split from the substrate, which will trigger the sliding of enzymes across the surface of fibrillary cellulose. By this, the catalytic site will be shifted and hydrolysis of the substrate will occur. In the presence of adequate catalytic processes, diffusible dissolved products are emitted, and the molecules are taken up by some active microbes before the enzyme-producing cells can benefit. Microbes that keep their extracellular enzymes intact suffer less, whereas those relying on the secreted enzymes are affected. An opportunistic microbe that does not contribute to extracellular enzyme production benefits more from the diffusion and dilution of the available resources (Allison, 2005).

7.4 Cells Engulfed by Extracellular Enzymes

Diffusional losses are minimal in enzymes secreted from the cytoplasmic membrane. The enzyme is structured in a manner that its viable site is exposed, which makes it vulnerable to microbial attacks, but the proteases are protected. Proteases give protection to enzymes, provide strength to scavenge, prompt responses to the substrate distal to the cell, and aid in unavoidable reductions in freely diffusible enzymes. Apart from substrate diffusion and convection, cell possession of the enzyme holds onto the principle of Brownian motion, which aids in the collection of substrates through signals and chemical gradients to initiate and control movement toward efficient energy sources. This is possible due to the chemotaxis process, which empowers microorganisms to find gradients and enhance migration to elevated concentrations of the substrate (Centler et al., 2011). Possession of extracellular enzymes within the periplasm of some Gram-negative bacteria accounts for the survival of periplasmic enzymes through metabolic synthesis of protein as a result of shut down of cells due to starvation. The adhesive nature of the polymeric material (biofilm) enhances the attachment of microbes, thus producing enzymes to directly bind to insoluble substrates. Dissolution of substrates by extracellular enzymes betides at the interface of reaction products entering the biofilm, leading to reduction of diffusional and convective effects associated with the unavailability of the biofilm (Van Horn et al., 2011). Polysomes are associated with the anaerobic thermophile *Clostridium thermocellum*. A large number of extracellular enzymes secreted by *C. thermocellum* are polygalacturonate hydrolases, endoglucanases, exoglucanases, β-glucosidases, lichenases, laminarinases, xylosidases, galactosidases, mannosidases, pectin lyases, pectin methylesterases, cellobiose phosphorylases, cellodextrin phosphorylases, and xylenes (Burns et al., 2013) (Table 2).

8 Significance of Microbial Enzymes in Soil

Soils are the naturally occurring physical covering of Earth's surface and represent the interface of the three material states, namely, solids, liquids, and gases. Soil is an excellent culture medium for the growth and development of various microorganisms. Soil is not an inert static material; it is a medium pulsating with life (Eilers et al., 2012).

Soils are the foundation of all terrestrial ecosystems and are home to a vast diversity of bacteria, archaea, fungi, insects, annelids, and other invertebrates as well as plants and algae. These soil dwellers are referred to as microbes, and they play a major role in the human society. We depend on soils for the basis on which we and our buildings stand and for the production of food and other materials. Indeed, soils influence most ecosystem services on which we depend (Dominati et al., 2010).

Soil microbes, bacteria, archaea, fungi, and all others play diverse and often critical roles in these ecosystem services. The vast metabolic diversity of soil

Table 2 Some soil enzymes, their sources, and their functions (Rao et al., 2017)

Enzymes	Sources	Functions	Influencing agents	Catalytic reactions	Products
Arylsulfatase	Secreted by plants, animals, and microorganisms	Cycling of sulfur	pH, organic matter composition, heavy metal pollution, and the presence of organic sulfate esters	Hydrolysis of sulfate esters	Sulfate (SO_4^{2-})
Phenol oxidase	Plants and microorganisms	Carbon cycling	Soil pH, rainfall, temperature, excess nitrogen, and soil organic materials	Hydrolysis of lignin	Humic materials
Urease	Invertebrates, plants, and microorganisms	Nitrogen cycling	Organic matter content, management practices, soil depth, heavy metals, temperature, pH, and cropping history	Urea hydrolysis	Ammonia (NH_3) and CO_2
α-Amylase	Plants and microorganisms	Carbon cycling	Environment, soil and vegetation types, and management practices	Starch hydrolysis	Glucose
Dehydrogenase	Microorganisms	Microbial oxidative actions and carbon cycling	Content of water in the soil, temperature, pesticides, trace elements, and pollution	Organic compound oxidation	Hydrogen transfer to NAD
β-Amylase	Plants	Carbon cycling	Vegetation types and management practices	Starch hydrolysis	Maltose
Protease	Microorganisms and plants	Nitrogen cycling	Humic acid concentration and excess of carbon and nitrogen	Mineralization of nitrogen	Nitrogen availability in plants
Alkaline phosphatase	Bacteria	Phosphorus cycling	Organic content, pH, management practices, pollution, and crop species	Hydrolysis of esters and anhydrides of phosphoric acid	Phosphate (PO_4)
Chitinase	Microorganisms and plants	Carbon and nitrogen cycling	Level of atmospheric CO_2 and soil depth	Degradation and hydrolysis of chitin	Carbohydrates and inorganic nitrogen

Endo-1,4-β-glucanase	Microorganisms, protozoa, and termites	Cellulose endohydrolysis	Temperature, pH, water, oxygen contents, quality of organic matter, fungicides, and mineral elements	Cellulose endohydrolysis	Oligosaccharides
Xylanase	Plant cell	Degradation of the linear polysaccharide xylan	Temperature and mineral content	Breaking of hemicellulose	Xylose

microbes means that their activities drive or contribute to the cycling of all major elements and this cycling affects the structure and the functions of soil ecosystems as well as the ability of soils to provide services to people. Collectively, soil microbes play an essential role in nutrient cycling, recycling of wastes and detoxification, decomposing organic matter, and biogenic element circulation, which makes nutrients available to plants; they are also important for the development of healthy soil structures (Di et al., 2010).

Microbes are the smallest organisms (<0.1 mm in diameter) and are extremely abundant and diverse. They include protozoa, bacteria, nematodes, fungi, and actinomycetes. Most of them are able to decompose almost any existing natural material. Microorganisms transform organic matter into plant nutrients that are assimilated by plants. Soil microbes represent a large fraction of the global terrestrial biodiversity. Microbes include:

1. Bacteria: Bacteria are the crucial workforce of soils. They are the final stage of breaking down nutrients and releasing them into the root zone of a plant. In fact, the Food and Agriculture Organization once stated "Bacteria may well be the most valuable of life forms in the soil" (Hobbie, 2006).
2. Actinomycetes: Actinomycetes were once classified as fungi and act similarly in the soil. However, some actinomycetes are predators and will harm the plants, whereas others living in the soil can act as antibiotics for the plants.
3. Fungi: Like bacteria, fungi also live in the root zone and help make nutrients available to plants. For example, mycorrhizae, which is the association between roots and fungi, facilitates water and nutrient uptake by the roots and plants to provide sugars, amino acids, and other nutrients (Hibbett et al., 2007).
4. Protozoa: Protozoa are larger microbes that ingest bacteria and are surrounded by them. In fact, nutrients that are consumed by bacteria are released when protozoa, in turn, ingest the bacteria.
5. Nematodes: Nematodes are microscopic worms that live around or inside plants. Some nematodes are predators, whereas others are beneficial as they consume pathogenic nematodes and secrete nutrients to the plants.

Although there are several other soil microbes, the ones listed above are the most abundant. Microbes play a pivotal role in the cycling of nutrients essential for life; they exclusively mediate nitrogen fixation, denitrification, and nitrification. For example, soil microbes play major roles in cycling carbon, nitrogen, and phosphorus, which are essential for producing biomolecules such as amino acid, proteins, DNA, and RNA—the fundamental compounds of life. Many plant nutrients are ultimately derived from weathering of minerals. Mineral weathering by soil bacteria and fungi plays a significant role in ion cycling and plant nutrition (Philippot et al., 2007).

Carbon Cycling Microbes play major roles in the cycling of carbon—the key constituents of all living organisms. Primary producers fix carbon dioxide and convert it into organic materials. In terrestrial ecosystems, the primary producers of organic materials are plants, although surface-dwelling algae and cyanobacteria, both free-living and symbiotic as lichens, can significantly contribute to carbon fixation in some ecosystems. Within soil, autotrophic microbes can also fix carbon dioxide (Eilers et al., 2010).

Nitrogen Cycling All organisms require nitrogen because it is an essential element in protein and nucleic acids. Animals derive nitrogen from organic sources, whereas plants require inorganic nitrogen sources such as ammonium and nitrate or relatively depolymerized nitrogen sources such as single amino acids. Microbes play an important role in the nitrogen cycle; they carry out processes not carried out by other organisms, namely, nitrogen fixation, dissimilatory nitrate reduction to ammonia (DNRA), anaerobic ammonium oxidation (anammox), etc. Because nitrogen is the major limiting nutrient for plant biomass production in terrestrial habitats, the rates of this microbial process often limit ecosystem productivity (Philippot et al., 2007).

Biodegradation Many years of laboratory studies have provided a wealth of information about how microbes biodegrade or detoxify organic contaminants. It describes the establishment of enrichment cultures for detection of biotransformation of contaminants under a range of environmental conditions, for example, pH or nutrient or oxygen availability. The source of microbes for the enrichment cultures are typically soils contaminated with the compound of interest. Where possible, pure cultures that can degrade the contaminants are obtained and have been used for biochemical and molecular characterization of the degradation pathways (Dominati et al., 2010).

Heterotrophic bacteria in soil—for example, *Pseudomonas*, *Sphingomonas*, and *Mycobacterium*—have often been implicated in oil degradation. *Pseudomonas*, for example, has been well studied, and the genes and enzymes responsible for degrading alkanes, monoaromatics, naphthalene, and phenanthrene as a sole carbon source under aerobic conditions are well understood. Knowledge of the mechanisms that microbes use to degrade oil has been applied in situ. For example, enhancing oil degradation in soil typically involves addition of nutrients (N and P) and sometimes oxygen and water (Fierer et al., 2007).

There is usually no need to add hydrocarbon-degrading bacteria to oil-contaminated sites because they are ubiquitous in soil, and, when oil is spilled, they increase in numbers. However, high concentrations of hydrocarbons can deplete the available nitrogen and phosphorus because these elements are assimilated during biodegradation; consequently, the activity of the hydrocarbon degraders may become limited by these nutrients. They are also responsible for the chemical degradation of pesticides; examples include bacteria and fungi (Philippot et al., 2007).

Soil microbes are responsible for maintaining soil quality and health; they are also involved in disease transmission and control and increase soil aeration and penetrability (Dominati et al., 2010).

Generally, microbes play the foremost role in soil formation and ecology because they, as "natural soil engineers," regulate the flux of nutrients to plants and pop up nitrogen fixation and detoxification and ultimately promote detoxification of naturally occurring inorganic and organic pollutants in soil (Fierer et al., 2007).

The quantitative composition of the population and its qualitative nature depend largely on the origin and nature of the soil and the relative composition of its inorganic and organic constituents. The prevailing climate and growing vegetation also greatly influence the nature and abundance of microbes that inhabit the particular soil. Soil microbes play a crucial role in returning nutrients to their mineral forms, which plants can take up again (Hobbie, 2006).

This process is known as mineralization. Biological nitrogen fixation contributes about 60% of the nitrogen fixed on Earth. Some soil microbes yield numerous substances that boost plant growth. They break down organic matter, create humus, and also promote plant growth (Dominati et al., 2010).

Furthermore, soil microbes produce antimicrobial agents and enzymes used for biotechnological purposes. They also mobilize nutrients from insoluble minerals to support plant growth. Macropores are formed by plant roots, earthworms, and other soil biota, which may depend on soil microbes as food or for nutrients. In concert with the organic matter and clay content of soils, microbial products add to both the wettability and the hydrophobicity of soils, impacting the property of the soil to filter contaminants (Hobbie, 2006).

Soil bacteria, fungi, and archaea comprise the vast majority of the biological variety on Earth. They also make up the foundation of soil food networks, thereby sustaining the variety of higher trophic intensities. Interactions between plants and soil microbes often decide plant biodiversity. Beneficial species include fungi, archaea, and bacteria that promote plant development by outcompeting invading pathogens and increasing nutrient availability (Eilers et al., 2012). By mineralizing soil carbon and nutrients, microbes are major determinants of the carbon storage capacity of soils.

9 Effects of Herbicides, Fungicides, and Insecticides on Microbial Enzymes

9.1 Effects on Dehydrogenase Activity

Dehydrogenase occurs in all living microbial cells, and it is linked to microbial respiratory processes (Bolton et al., 1985). Author findings showed that all fungicides except Prochloraz at a recommended field application dose between pH 4.4 and 7.5 have both negative and positive effects on dehydrogenase enzyme activities and population (Chen et al., 2001; Burrows & Edwards, 2004; Bending et al., 2007; Bello et al., 2008; Rasool & Reshi, 2010; Ataikiru et al., 2019; Małgorzata et al., 2021). Most insecticides have no effects or a slight inhibition effect (Caceres et al., 2009; Beulke & Malkomes, 2001; Kalam et al., 2004; Yao et al., 2006; Jastrzebska, 2011; Gangan et al., 2015; Nataraj et al., 2017; Madhavi et al., 2019). Similarly, herbicides also cause inhibition of the activity of dehydrogenase enzymes not minding the application dose or pH (Sebiomo et al., 2012; Filimon et al., 2021), except butachlor (Min et al., 2002; Xia et al., 2011). In summary, pesticides (fungicides, herbicides, and insecticides) may have no effect, inhibitory effects, or sometimes enhance the activities depending on the pesticide and conditions involved in their applications.

9.2 Fluorescein Diacetate Hydrolase

The influence of insecticides on fluorescein diacetate hydrolase is not much; however, Das et al. (2007) and Bishnu et al. (2012) conducted some research on it explaining that its activities could be enhanced by the imidazolines (Imazethapyr) and organochlorines (endosulfan) families (Perucci et al., 2000; Kalyani et al., 2010; Riah et al., 2014; Mariane et al., 2020). Authors noted that application doses have similar or same effects on its activities (Bishnu et al., 2012). Fluorescein diacetate hydrolase activity in soil is poorly influenced by herbicide or insecticide applications, except endosulfan applications, which seem to stimulate this activity (Wassila et al., 2014).

9.3 Cellulase and β-Glucosidase

The effects of fungicides and herbicides were tested by different authors and they were discovered to have no solid impact on the activity of cellulase (Bishnu et al., 2012; Tejada et al., 2011; Niemi et al., 2009; Gundi et al., 2007; Omar & Abdel-Sater, 2001). However, Gundi et al. (2007) went further to show that there is a valid relationship between some insecticides (monocrotophos, quinalphos, and profenofos) and cellulolytic bacteria population growth. Similarly, Tejada (2009) noted the inhibition of the β-glucosidase activity by glyphosate and diflufenican combination. Among the various insecticides, Defo et al. (2011) observed an enhancement of β-glucosidase activity by endosulfan at high concentrations above the normal dose. Wassila et al. (2014) were able to support the claim that the effects of the endosulfan insecticide may be related to the strong functional redundancy of β-glucosidase activity.

9.4 Effects on Phosphomonoesterase Enzymes

The effects of pesticides on enzymes have been studied by many researchers who have come to the conclusion that pesticides either decrease enzyme activity or, in some cases, have no effect on them (Schneider et al., 2001. Kalam et al., 2004; Yan et al., 2011; Dick et al., 2000), depending on some conditions like doses, soil pH, and other physiochemical properties of soil (Min et al., 2002; Tejada, 2009). For the sake of differentiation, Rasool and Reshi (2010) noted an inhibition of the activity of the alkaline form of the enzyme when fungicides are used, which was also confirmed by Sharma et al. (2010), but an enhancement of the activity of the acid phosphatase. The different responses between the alkaline and acidic forms of the enzyme can be attributed to their sensitivity (Klose et al., 2006) Monkiedje et al. (2002) furthered this research and discovered that fungicides at basic pH will inhibit alkaline phosphate activity; this was also confirmed by other authors (Bello et al., 2008; Tejada et al., 2011; Yan et al., 2011). Studies by Perucci et al. (2000), Omar and Abdel-Sater (2001), and Bacmaga et al. (2012) showed that the type of insecticide has to do with their reaction to it. For example, Xia et al. (2011) discovered that butachlor enhances the activity of the enzyme, especially the alkaline type. Similar to the responses with herbicides, insecticides may inhibit acid phosphatase and enhance alkaline phosphatase activity, and vice versa (Omar & Abdel-Sater, 2001; Cycoń et al., 2010; Defo et al., 2011; Jastrzebska, 2011; Madhuri & Rangaswamy, 2002; Yao et al., 2006).

9.5 Nitrogen Cycle and Enzymatic Activity of Urease

Antonious (2003) explained that urease is generally beneficial because it helps maintain nitrogen availability to plants. The study summaries of certain authors observe that herbicides and fungicides do not have any effects on urease activities (Cycoń et al., 2010; Romero et al., 2010; Tejada et al., 2011; Yan et al., 2011; Bacmaga et al., 2012), but some studies have recorded a decrease in urease activity, e.g., carbendazim and validamycin (Sukul, 2006; Caceres et al., 2009; Tejada, 2009). Generally, pesticides do not seem to affect the activity of this enzyme (Niemi et al., 2009; Tejada, 2009; Vavoulidou et al., 2009). It is difficult to identify a clear response of the activity of this enzyme to pesticides as it has received only a few mentions in the literature in past years.

10 Effects of Application of Fertilizers on Enzymatic Activities

Wang et al. (2020) used organic fertilizers on four types of soil enzymes (ureases, sucrases, alkaline phosphatases, and catalases), which did not significantly respond to the addition of vermicompost and mushroom residue fertilizers. Urease activities declined as a result of vermicompost and mushroom residue applications. However, sucrase, alkaline phosphatase, and catalase activities increased to varying degrees under the different levels of treatment of vermicompost and mushroom residue fertilizers. Sawicka et al. (2020), using nitrogen fertilizers, observed that the activity of dehydrogenases, phosphatases, and ureases changed as the nitrogen dose increased. The polynomial regression analysis enabled a better understanding of those dependences. However, soil acidity did not have a significant influence on either the enzymatic activity or the physicochemical characteristics of soil under the cultivation of sweet potatoes. Ye and Peng (2019) discovered that NPK fertilizers improve soil enzyme activity. The long-term effects of fertilizers were considered by Chew et al. (2019), using a combination of inorganic and organic fertilizers, who discovered that they enhanced dehydrogenase, urease, alkaline phosphatase, invertase, and glomalin enzymes. From different authors and the literature, it was discovered that factors including time, type of fertilizer (inorganic or organic), dose of application, and soil parameters are responsible for the response of soil enzymes (Gostkowska et al., 1998; Lü et al., 2018; Sekaran et al., 2019a).

11 Relationships Between Pesticide Mechanisms of Action and Enzymatic Responses

11.1 Pesticides

Gianfreda and Rao (2008) noted that the relationships between pesticide action and enzymatic responses have been known to be direct and indirect, which could include active site binding or a nutrition source for the enzymes (Tabatabai, 1994); the former could cause a change in the catalytic reaction, and the latter could cause a biosynthesis of the enzymes by induction (Cycoń et al., 2006; Tejada, 2009; Zabaloy et al., 2012; Chishti et al., 2013). The relationships are strongly related to the functionality power or resistance of the target (Chaer et al., 2009; Griffiths & Philippot, 2013; Puglisi et al., 2012) and to the physicochemical properties of soil, pH, humus, clay content, or organic matter, which have been known to affect the expression and proper function of the pesticide in soil (Chen et al., 2001; Gundi et al., 2007; Defo et al., 2011; Muñoz-Leoz et al., 2013).

11.2 Fungicides

It has been noted that the high application of fungicides has destructive effects on the fungal population but enhances the bacterial population (Monkiedje & Spiteller, 2002; Moharram et al., 2004; Cycoń et al., 2006; Bending et al., 2007; Cycoń et al., 2010), which explains why bacteria use dead fungi as a source of nutrients and energy for their population increase (Cycoń et al., 2006; Tejada et al., 2011). According to Muñoz-Leoz et al. (2013), microbial biomass decrease is parallel to the decrease in enzymatic activities after the use of fungicides, which may lead to a global unpleasant response within 28–50 days of incubation. The effect on the field is not far apart because even at a recommended standard of application, the field is still found to be under the negative influence even after 3 years (Niewiadomska, 2004; Niewiadomska & Klama, 2005).

11.3 Insecticides

Endosulfan (one of the commonly used insecticides in the world) has been noted to cause an increase in microbial biomass carbon (Kalyani et al., 2010; Joseph et al., 2010; Xie et al., 2011), but the opposite action has been noted for organophosphate. This claim has been affirmed using chlorpyrifos and monocrotophos, which are two major molecules in that class; it showed that they caused a decrease in microbial biomass carbon in soil and also have adverse effects on soil bacterial and fungal counts (Shan et al., 2006; Vischetti et al., 2007; Zayed et al., 2008). This does not enable all molecules of organophosphate to function similarly (Martinez-Toledo et al., 1992; Tejada, 2009).

11.4 Herbicides

Herbicides that inhibit the acetolactate synthase enzyme and photosynthesis process have predominately neutral effects on soil enzymatic activities. Radivojevic et al. (2008) noted that the addition of atrazine had no effect on soil microbial activity, bacterial density, and functional richness, whereas metsulfuron-methyl herbicides had a little effect (Zabaloy et al., 2008). Researchers have noted that the recommended field rate of glyphosate had a benign effect (Barriuso & Mellado, 2012; Hart et al., 2009), whereas above the concentration, enhancement of bacteria was discovered (Ratcliff et al., 2006; Weaver et al., 2007).

12 Conclusions

Adequate secretion of microbial enzymes is a significant factor in enriching soil for profitable agricultural practices. Soil enzymes improve the soil biogeochemical processes, soil health, and quality. Soil enzymes are influenced by the physical, chemical, and biological properties of soil, which are the functions of biomass content and nutrient quality, resulting in the synthesis and secretion of enzymes in the soil. Unfortunately, modern-day agricultural practices coupled with other factors pose threats to the microbial community, most especially application of chemicals, soil management practices, and environmental factors. Agrochemical application reduces microbial community and ecological niche and hampers the response of microbes toward nutrients. For mediating the effects of agrochemicals on the microbial community for effective secretion of soil enzymes, natural attenuation of soil will be the leading option. In addition, the use of harmful agrochemicals such weed killers should be stopped. It is a known fact that the total stoppage of agrochemicals is not possible, but the mode of action can be selective, most especially when dealing with pests. Culturing and multiplication of bacteria-producing enzymes for desired purposes will also serve as a meditative approach through biotechnological means.

References

Aktar, W., Sengupta, D., & Chowdury, A. (2009). Impact of pesticides use in agriculture; Their benefits and hazard. *Interdisciplinary Toxicology, 2*, 1–12.

Allison, S. D. (2005). Cheaters, diffusion and nutrients constrain decomposition by microbial enzymes in spatially structured environments. *Ecology Letters, 8*, 626–635.

Antonious, G. F. (2003). Impact of soil management and two botanical insecticides on urease and invertase activity. *Journal of Environmental Science and Health. Part. B, 38*, 479–488.

Ataikiru, T. L., Okpokwasili, G. S. C., & Okerentugba, P. O. (2019). Impact of pesticides on microbial diversity and enzymes in soil. *South Asian Journal of Research in Microbiology, 4*, 1–16.

Bacmaga, M., Boros, E., Kucharski, J., & Wyszkowska, J. (2012). Enzymatic activity in soil contaminated with the Aurora 40 Wg herbicide. *Environment Protection Engineering, 38*, 91–102.

Barriuso, J., & Mellado, R. P. (2012). Glyphosate affects the rhizobacterial communities in glyphosate-tolerant cotton. *Applied Soil Ecology, 55*, 20–26.

Baumhardt, R. L., & Blanco Canqui, H. (2014). Soil conservation practices. In *Encyclopedia of agriculture and food systems*. Elsevier.

Beketov, M. A., Kefford, B. J., Schäfer, R. B., & Liess, M. (2013). Pesticides reduce regional biodiversity of stream invertebrates. *Proceedings. National Academy of Sciences. United States of America, 110*, 11039–11043.

Belden, J., & Lydy, M. J. (2000). Impact of atrazine organophosphate insecticide toxicity. *Environmental Toxicology and Chemistry, 19*, 2266–2274.

Bello, D., Trasar-Cepeda, C., Leiros, M. C., & Gil-Sotres, F. (2008). Evaluation of various tests for the diagnosis of soil contamination by 2,4,5-trichlorophenol (2,4,5-TCP). *Environmental Pollution, 156*, 611–617.

Bending, G. D., Rodriguez-Cruz, M. S., & Lincoln, S. D. (2007). Fungicide impacts on microbial communities in soils with contrasting management histories. *Chemosphere, 69*, 82–88.

Beulke, S., & Malkomes, H. P. (2001). Effects of the herbicides metazachlor and dinoterb on the soil microflora and the degradation and sorption of metazachlor under different environmental conditions. *Biology and Fertility of Soils, 33*, 467–471.

Bishnu, A., Chakraborty, A., Chakrabarti, K., & Saha, T. (2012). Ethion degradation and its correlation with microbial and biochemical parameters of tea soils. *Biology and Fertility of Soils, 48*, 19–29.

Biswas, S. K., Rahman, S., Kobir, S. A., Ferodus, T., & Banu, N. A. (2014). A review on impact of agrochemicals on human health and environment. *Plant Environment Development, 3*, 31–35.

Boivin, A., & Poulsen, V. (2017). Environmental risk assessment of pesticides: State of the art and prospective improvement from science. *Environmental Science and Pollution Research, 24*, 6889–6894.

Bolton, H., Elliott, L. F., Papendick, R. I., & Bezdicek, D. F. (1985). Soil microbial biomass and selected soil enzyme-activities: Effect of fertilization and cropping practices. *Soil Biology and Biochemistry, 17*, 297–302.

Brühl, C. A., Schmidt, T., Pieper, S., & Alscher, A. (2013). Terrestrial pesticide exposure of amphibians: An underestimated cause of global decline? *Scientific Reports, 3*, 1135.

Bunneman, E. K., & McNeil, A. (2004). Impact of fertilizers on soil biota. In R. Lines-Kelly (Ed.), *Soil biology in agriculture. Proceedings of a workshop on current research into soil biology in agriculture* (pp. 63–71). NSW Department of Primary Industries..

Burns, R. G., De Forest, J. L., Marxsen, J., Sinsabaugh, R. L., Stromberger, M. E., Wallenstein, M. D., Weintraub, M. N., & Zoppini, A. (2013). Soil enzymes in a changing environment: Current knowledge and future directions. *Soil Biology and Biochemistry, 58*, 216–234.

Burrows, L. A., & Edwards, C. A. (2004). The use of integrated soil microcosms to assess the impact of carbendazim on soil ecosystems. *Ecotoxicology, 13*(1), 143–161.

Caceres, T. P., He, W. X., Megharaj, M., & Naidu, R. (2009). Effect of insecticide fenamiphos on soil microbial activities in Australian and Ecuadorean soils. *Journal of Environmental Science and Health. Part. B, 44*, 13–17.

Cardoso, E. J. B. N., & Alves, P. R. N. (2012). *Ecotoxicology: Soil ecotoxicology*. InTechOpen.

Centler, F., Fetzer, I., & Thullner, M. (2011). Modeling population patterns of chemotactic bacteria in homogeneous porous media. *Journal of Theoretical Biology, 287*, 82–91.

Chaer, G., Fernandes, M., Myrold, D., & Bottomley, P. (2009). Comparative resistance and resilience of soil microbial communities and enzyme activities in adjacent native forest and agricultural soils. *Microbial Ecology, 58*, 414–424.

Chen, S. K., Edwards, C. A., & Subler, S. (2001). Effects of the fungicides benomyl, captan and chlorothalonil on soil microbial activity and nitrogen dynamics in laboratory incubations. *Soil Biology and Biochemistry, 33*, 1971–1980.

Chew, K. W., Chia, S. R., Yen, H. W., Nomanbhay, S., Ho, Y. C., & Show, P. L. (2019). Transformation of biomass waste into sustainable organic fertilizers. *Sustainability, 11*(8), 2266.

Chishti, Z., Hussain, S., Arshad, K. R., Khalid, K., & Arshad, A. (2013). Microbial degradation of chlorpyrifos in liquid media and soil. *Journal of Environmental Management, 114*, 372–380.

Cycoń, M., Piotrowska-Seget, Z., Kaczyńska, A., & Kozdrój, J. (2006). Microbiological characteristics of a loamy sand soil exposed to tebuconazole and l-cyhalothrin under laboratory conditions. *Ecotoxicology, 15*, 639–646.

Cycoń, M., Piotrowska-Seget, Z., & Kozdroj, J. (2010). Responses of indigenous microorganisms to a fungicidal mixture of mancozeb and dimethomorph added to sandy soils. *International Biodeterioration and Biodegradation, 64*, 316–323.

Das, P., Pal, R., & Chowdhury, A. (2007). Effect of novaluron on microbial biomass, respiration, and fluorescein diacetate-hydrolyzing activity in tropical soils. *Biology and Fertility of Soils, 44*, 387–391.

Defo, M. A., Njine, T., Nola, M., & Beboua, F. S. (2011). Microcosm study of the long term effect of endosulfan on enzyme and microbial activities on two agricultural soils of Yaounde-Cameroon. *African Journal of Agricultural Research, 6*, 2039–2050.

Desneux, N., Decourtye, A., & Delpuech, J. M. (2007). The sublethal effects of pesticides on beneficial arthropods. *Annual Review of Entomology, 52*, 81–106.

Di, H., Cameron, K. C., Shen, J.-P., Winefield, C. S., O'Callaghan, M., & Bowatte, S. (2010). Ammonia-oxidizing bacteria and archaea grow under contrasting soil nitrogen conditions. *FEMS Microbiology Ecology, 72*, 386–394.

Dick, W. A., Cheng, L., & Wang, P. (2000). Soil acid and alkaline phosphatase activity as pH adjustment indicators. *Soil Biology and Biochemistry, 32*, 1915–1919.

Dominati, E., Patterson, M., & Mackay, A. (2010). A framework for classifying and quantifying the natural capital and ecosystem services of soils. *Ecological Economics, 69*(9), 1858–1868.

Eilers, K. G., Lauber, C. L., Knight, R., & Fierer, N. (2010). Shifts in bacterial community structure associated with inputs of low molecular weight carbon compounds to soil. *Soil Biology & Biochemistry, 42*, 896–903.

Eilers, K. G., Debenport, S., Anderson, S., & Fierer, N. (2012). Digging deeper to find unique microbial communities: The strong effect of depth on the structure of bacterial and archaeal communities in soil. *Soil Biology & Biochemistry, 50*, 58–65.

Eldridge, B. F. (2008). *Pesticides application and safety training for applicators of public health pesticides*. Vector Borne Diseases Section.

Feketeová, Z., Hrabovský, A., & Šimkovic, I. (2021). Microbial features indicating the recovery of soil ecosystem strongly affected by mining and ore processing. *International Journal of Environmental Research and Public Health, 18*, 3240. https://doi.org/10.3390/ijerph18063240

Fierer, N., Bradford, M. A., & Jackson, R. B. (2007). Towards an ecological classification of soil bacteria. *Ecology, 88*, 1354–1364.

Filimon, M. N., Roman, D. L., Caraba, I. V., & Isvoran, A. (2021). Assessment of the effect of application of the herbicide s-metolachlor on the activity of some enzymes found in soil. *Agriculture, 11*(6), 469.

Franklin, O., Hall, E. K., Kaiser, C., Battin, T. J., & Richter, A. (2011). Optimization of biomass composition explains microbial growth-stoichiometry relationships. *American Naturalist, 177*, 29–42.

Gangan, N., Vaidya, V., & Sheehan, J. (2015). Impact of cardiovascular complications among patients with Type 2 diabetes mellitus: A systematic review. *Expert Review of Pharmacoeconomics & Outcomes Research, 15*(3), 487–497.

Gianfreda, L., & Rao, M. (2008). Interactions between xenobiotics and microbial and enzymatic soil activity. *Environmental Science & Technology, 38*, 269–310.

Gostkowska, K., Furczak, J., Domzal, H., & Bielinska, E. J. (1998). Suitability of some biochemical and microbiological tests for the evaluation of the degradation degree of podzolic soil on the background of it differentiated usage. *Polish Journal of Soil Science (Poland), 31*, 69.

Griffiths, B. S., & Philippot, L. (2013). Insights into the resistance and resilience of the soil microbial community. *FEMS Microbiology Reviews, 37*(2), 112–129.

Gundi, V., Viswanath, B., Chandra, M. S., Kumar, V. N., & Reddy, B. R. (2007). Activities of cellulase and amylase in soils as influenced by insecticide interactions. *Ecotoxicology and Environmental Safety, 68*, 278–285.

Hamilton, G. C. (2001). *2001 control recommendations for household and structural insect pests*. Rutgers Cooperative Extension.

Hans, R. K., & Faroq, M. (2000). Dissipation and accumulation kinetics of lindane in soils and earthworms. *Pollution Research, 19*, 394–396.

Hart, M. M., Powell, J. R., Gulden, R. H., Dunefield, K. E., Pauls, K. P., Swanton, C. J., Klironomos, J. N., Antunes, P. M., Koch, A. M., & Trevors, J. T. (2009). Separating the effect of crop from herbicide on soil microbial communities in glyphosate resistant corn. *Pedobiologia, 52*, 253–262.

Hibbett, D. S., Binder, M., Bischoff, J. F., Blackwell, M., Cannon, P. F., & Eriksson, O. E. (2007). A higher-level phylogenetic classification of the Fungi. *Mycology Research, 111*, 509–547.

Hobbie, E. A. (2006). Carbon allocation to ectomycorrhizal fungi correlates with total below ground allocation in culture studies. *Ecology, 87*, 563–569.

Jastrzebska, E. (2011). The effect of chlorpyrifos and teflubenzuron on the enzymatic activity of soil. *Polish Journal of Environmental Studies, 20*, 903–910.

Jeschke, P. (2016). Propesticides and their use as agrochemicals. *Pest Management Science, 72*(2), 210–225.

Joseph, R., Reed, S., Jayachandran, K., Clark-Cuadrado, C., & Dunn, C. (2010). Endosulfan has no adverse effect on soil respiration. *Agriculture, Ecosystems and Environment, 138*, 181–188.

Kalam, A., Tah, J., & Mukherjee, A. K. (2004). Pesticide effects on microbial population and soil enzyme activities during vermicomposting of agricultural waste. *Journal of Environmental Biology, 25*, 201–208.

Kalyani, S. S., Sharma, J., Dureja, P., Singh, S., & Lata. (2010). Influence of endosulfan on microbial biomass and soil enzymatic activities of a tropical Alfisol. *Bulletin of Environmental Contamination and Toxicology, 84*, 351–356.

Klose, S., Acosta-Martinez, V., & Ajwa, H. A. (2006). Microbial community composition and enzyme activities in a sandy loam soil after fumigation with methyl bromide or alternative biocides. *Soil Biology and Biochemistry, 38*, 1243–1254.

Lamberth, C., Jeanmart, S., Luksch, T., & Plant, A. (2013). Current challenges and trends in the discovery of agrochemicals. *Science, 341*(6147), 742–746.

Lü, S., Chen, J., Zhang, Z., Zhao, X., Li, X., Ning, P., & Liu, M. (2018). Environmentally friendly fertilizers: A review of materials used and their effects on the environment. *Science of the Total Environment, 613*, 829–839.

Madhavi, S., Sennu, P., Arun, N., Aravindan, V., & Lee, Y. S. (2019). All carbon based high energy lithium-ion capacitors from biomass: The role of crystallinity. *Journal of Power Sources, 414*, 96–102.

Madhuri, R. J., & Rangaswamy, V. (2002). Influence of selected insecticides on phosphatase activity in groundnut (Arachishypogeae L.) soils. *Journal of Environmental Biology, 23*, 393–397.

Małgorzata, T., Anna, L., Izabela, S., & Dorota, D. (2021). The effect of biologically active compounds in the mucus of slugs Limaxmaximus and Arionrufus on human skin cells. *Scientific Reports, 11*(1), 1–9.

Mariane, I., Suryadi, S., Firdausijah, T., & Ilhami, R. (2020). Pemberdayaan Masyarakat Melalui Public Private Partnership Bagi Peningkatan Pariwisata Desa Warnasari Kecamatan Pangalengan Kabupaten Bandung. *Kaibon Abhinaya: Jurnal Pengabdian Masyarakat, 2*(2), 71–76.

Martinez-Toledo, M. V., Salmeron, V., & Gonzalez-Lopez, J. (1992). Effect of the insecticides methylpyrimifos and chlorpyrifos on soil microflora in an agricultural loam. *Plant and Soil, 147*(1), 1573–5036.

Mergel, A., Timchenko, A., & Kudeyarov, V. (1998). Role of plant root exudates in soil carbon and nitrogen transformation. In J. E. Box (Ed.), *Root demographics and their efficiencies in sustainable agriculture, grasslands and forest ecosystems: Developments in plant and soil sciences* (Vol. 82, pp. 43–54). Springer.

Min, H., Ye, Y. F., Chen, Z. Y., Wu, W. X., & Du, Y. F. (2002). Effects of butachlor on microbial enzyme activities in paddy soil. *Journal of Environmental Sciences (China), 14*, 413–417.

Moharram, A. M., Abdel-Hafez, S. I., El-Said, A. H., & Saleem, A. (2004). Effect of two systemic fungicides on cellulose decomposing fungi of tomato plants and on some enzymatic activities. *Acta Microbiologica et Immunologica Hungarica, 51*(4), 403–430.

Monkiedje, A., & Spiteller, M. (2002). Effects of the phenylamide fungicides, mefenoxam and metalaxyl, on the microbiological properties of a sandy loam and a sandy clay soil. *Biology and Fertility of Soils, 35*, 393–398.

Monkiedje, A., Ilori, M. O., & Spiteller, M. (2002). Soil quality changes resulting from the application of the fungicides mefenoxam and metalaxyl to a sandy loam soil. *Soil Biology and Biochemistry, 34*, 1939–1948.

Muñoz-Leoz, B., Garbisu, C., Charcosset, J. Y., Sánchez-Pérez, J. M., Antigüedad, I., & Ruiz-Romera, E. (2013). Non-target effects of three formulated pesticides on microbially-mediated processes in a clay-loam soil. *Science of the Total Environment, 449*, 345–354.

Nannipieri, P., Ascher, J., Ceccherini, M. T., Landi, L., Pietramellara, G., Renella, G., et al. (2008). Effects of root exudates in microbial diversity and activity in rhizosphere soils. In C. S. Nautiyal & P. Dion (Eds.), *Molecular mechanisms of plant and microbe coexistence* (pp. 339–365). Springer.

Nannipieri, P., Giagnoni, L., Renella, G., Puglisi, E., Ceccanti, B., Masciandaro, G., et al. (2012). Soil enzymology: Classical and molecular approaches. *Biology and Fertility of Soils, 48*, 743–762.

Nataraj, S. K., Kulkarni, P., Balakrishna, R. G., Nagaraju, D. H., & Reddy, M. V. (2017). Nanostructured binary and ternary metal sulfides: Synthesis methods and their application in energy conversion and storage devices. *Journal of Materials Chemistry A, 5*(42), 22040–22094.

Niemi, R. M., Heiskanen, I., Ahtiainen, J. H., Rahkonen, A., Mäntykoski, K., Welling, L., Laitinen, P., & Ruuttunen, P. (2009). Microbial toxicity and impacts on soil enzyme activities of pesticides used in potato cultivation. *Applied Soil Ecology, 41*, 293–304.

Niewiadomska, A. (2004). Effect of carbendazim, imazetapir and thiram on nitrogenase activity, the number of microorganisms in soil and yield of red clover (Trifolium pretense L). *Polish Journal of Environmental Studies, 13*, 403–410.

Niewiadomska, A., & Klama, J. (2005). Pesticide side effect on the symbiotic efficiency and nitrogenase activity of Rhizobiaceae bacteria family. *Polish Journal of Microbiology, 54*, 43–48.

Ogbodo, E. N., & Onwa, N. C. (2013). Impact of long-term application of agrochemicals on the agro-ecology of the lower Anambra river basin southeast Nigeria. *Journal of Environment and Earth Science, 3*(5), 32–37.

Oliveira, C. M., Auad, A. M., Mendes, S. M., & Frizzas, M. R. (2014). Crop losses and the economic impact of insect pests on Brazilian agriculture. *Crop Protection, 56*, 50–54.

Omar, S. A., & Abdel-Sater, M. A. (2001). Microbial populations and enzyme activities in soil treated with pesticides. *Water, Air, and Soil Pollution, 127*, 49–63.

Onder, M., Cevhan, E., & Kahraman, A. (2011). Effects of agricultural practices on environment. *Biology, Environment and Chemistry, 24*, 28–32.

Perucci, P., Dumontet, S., Bufo, S. A., Mazzatura, A., & Casucci, C. (2000). Effects of organic amendment and herbicide treatment on soil microbial biomass. *Biology and Fertility of Soils, 32*, 17–23.

Philippot, L., Hallin, S., & Schloter, M. (2007). Ecology of denitrifying bacteria in agricultural soil. *Advances in Agronomy, 96*, 249–305.

Princy, S., & Prabagaran, S. R. (2020). Reduction of Cr (VI) by Bacillus species isolated from tannery effluent contaminated sites of Tamil Nadu, India. *Materials Today: Proceedings*.

Puglisi, E., Nicelli, M., Capri, E., Trevisan, M., & Del Re, A. A. M. (2005). A soil alteration index based on phospholipid fatty acids. *Chemosphere, 61*, 1548–1557.

Puglisi, E., Vasileiadis, S., Demiris, K., Bassi, D., Karpouzas, D. G., Capri, E., Cocconcelli, P. S., & Trevisan, M. (2012). Impact of fungicides on the diversity and function of non-target ammonia-oxidizing microorganisms residing in a litter soil cover. *Microbial Ecology, 64*(3), 692–701.

Quiquampoix, H., & Burns, R. G. (2007). Interactions between proteins and soil mineral surfaces: Environmental and health consequences. *Elements, 3*, 401–406.

Radivojevic, L. J., Gasic, S., Santric, L. J., & Stankovic-Kalezic, R. (2008). The impact of atrazine on several biochemical properties of chernozem soil. *Journal of the Serbian Chemical Society, 73*(10), 951–959.

Rao, C. S., Grover, M., Kundu, S., & Desai, S. (2017). Soil enzymes. In *Encyclopedia of soil science* (3rd ed., pp. 2100–2107). Taylor & Francis. https://doi.org/10.1081/E-ESS3-120052906

Rasool, N., & Reshi, Z. A. (2010). Effect of the fungicide Mancozeb at different application rates on enzyme activities in a silt loam soil of the Kashmir Himalaya, India. *Tropical Ecology, 51*, 199–205.

Ratcliff, A. W., Busse, M. D., & Shestak, C. J. (2006). Changes in microbial community structure following herbicide (glyphosate) additions to forest soils. *Applied Soil Ecology, 34*, 114–124.

Relyea, R. A. (2005). The impact of insecticides and herbicides on the biodiversity and productivity of aquatic communities. *Ecological Applications, 15*(2), 618–627.

Riah, W., Laval, K., Laroche-Ajzenberg, E., Mougin, C., Latour, X., & Trinsoutrot-Gattin, I. (2014). Effects of pesticides on soil enzymes: A review. *Environmental Chemistry Letters, 12*(2), 257–273.

Romero, E., Fernandez-Bayo, J., Diaz, J. M. C., & Nogales, R. (2010). Enzyme activities and diuron persistence in soil amended with vermicompost derived from spent grape marc and treated with urea. *Applied Soil Ecology, 44*, 198–204.

Ross, M. A., & Childs, D. J. (1996). *Herbicide mode of action summary*. Purdue University.

Sawicka, B., Pszczółkowski, P., Krochmal-Marczak, B., Barbaś, P., & Özdemir, F. A. (2020). The effects of variable nitrogen fertilization on amino acid content in sweet potato tubers (Ipomoea

batatas L.[Lam.]) cultivated in central and eastern Europe. *Journal of the Science of Food and Agriculture, 100*(11), 4132–4138.
Scherer, H. W., Konrad, M., Gunter, K., & Karl, S. (2009). Fertilizers. In *Ullmann's encyclopedia of industrial chemistry*. Wiley.
Schneider, K., Turrion, M. B., Grierson, P. F., & Gallardo, J. F. (2001). Phosphatase activity, microbial phosphorus, and fine root growth in forest soils in the Sierra de Gata, western central Spain. *Biology and Fertility of Soils, 34*, 151–155.
Sebiomo, A., Ogundero, V. W., & Bankole, S. A. (2012). The impact of four herbicides on soil minerals. *Research Journal of Environmental and Earth Sciences, 4*(6), 617–624.
Sekaran, U., Santhi, R., Dey, P., Meena, S., & Maragatham, S. (2019a). Validation of soil test and yield target based fertilizer prescription model developed for pearl millet on Inceptisol. *Research on Crops, 20*(2), 266–274.
Sekaran, U., McCoy, C., Kumar, S., & Subramanian, S. (2019b). Soil microbial community structure and enzymatic activity responses to nitrogen management and landscape positions in switchgrass (Panicum virgatum L.). *GCB Bioenergy, 11*, 836–851. https://doi.org/10.1111/gcbb.12591
Shan, M., Fang, H., Wang, X., Feng, B., Chu, X. Q., & Yu, Y. L. (2006). Effect of chlorpyrifos on soil microbial populations and enzyme activities. *Journal of Environmental Sciences, 18*(1), 4–5.
Sharma, I. D., Chandel, R. S., Kaushal, R., Patyal, S. K., & Sharma, P. L. (2010). Effect of pesticides on biochemical properties of tomato (Solanumlycopersicon) cropped soil. *Indian Journal of Agricultural Sciences, 80*, 640–644.
Sharma, A., Kumar, V., Shahzad, B., Tanveer, M., Sidhu, G. P. S., Handa, N., & Thukral, A. K. (2019). Worldwide pesticide usage and its impacts on ecosystem. *SN Applied Sciences, 1*(11), 1–16.
Sofo, A., Scopa, A., Dumontet, S., Mazzatura, A., & Pasquale, V. (2012). Toxic effects of four sulphonylureas herbicides on soil microbial biomass. *Journal of Environmental Science and Health. Part. B, 47*, 653–659.
Steinauer, K., Chatzinotas, A., & Eisenhauer, N. (2016). Root exudate cocktails: The link between plant diversity and soil microorganisms? *Ecology and Evolution, 6*, 7387–7396.
Suciu, N., Vasileiadis, S., Puglisi, E., Pertile, G., Tourna, M., Karas, P. A., et al. (2019). Azadirachtin and trifloxystrobin had no inhibitory effects on key soil microbial functions even at high dose rates. *Applied Soil Ecology, 137*, 29–38.
Sukul, P. (2006). Enzymatic activities and microbial biomass in soil as influenced by metalaxyl residues. *Soil Biology and Biochemistry, 38*, 320–326.
Tabatabai, M. A. (1994). Soil enzymes. In W. A. Dick (Ed.), *Methods of soil analysis, 2. Microbiological and biochemical properties* (pp. 775–833). Soil Science Society of America.
Tan, M., Zhao, H., Li, G., & Qu, J. (2020). Assessment of potentially toxic pollutants and urban livability in a typical resource-based city, China. *Environmental Science and Pollution Research, 27*(15), 18640–18649.
Tejada, M. (2009). Evolution of soil biological properties after addition of glyphosate, diflufenican and glyphosate plus diflufenican herbicides. *Chemosphere, 76*, 365–373.
Tejada, M., Gomez, I., Garcia-Martinez, A. M., Osta, P., & Parrado, J. (2011). Effects of Prochloraz fungicide on soil enzymatic activities and bacterial communities. *Ecotoxicology and Environmental Safety, 74*, 1708–1714.
Van Horn, D., Sinsabaugh, R., Takacs-Vesbach, C., Mitchell, K., & Dahm, C. (2011). Response of heterotrophic stream biofilm communities to a gradient of resources. *Aquatic Microbial Ecology, 64*, 149–161.
Vavoulidou, E., Avramides, E., Wood, M., & Lolos, P. (2009). Response of soil quality indicators to the pesticide cadusaphos. *Communications in Soil Science and Plant Analysis, 40*, 419.
Vischetti, C., Perucci, P., & Scarponi, L. (2000). Relationship between rimsulfuron degradation and microbial biomass content in a clay loam soil. *Biology and Fertility of Soils, 31*, 310–314.

Vischetti, C., Marucchini, C., Leita, L., Cantone, P., Danuso, F., & Giovanardi, R. (2002). Behaviour of two sunflower herbicides (metobromuron, aclonifen) in soil. *European Journal of Agronomy, 16*(3), 231–238.

Vischetti, C., Coppola, L., Monaci, E., Cardinali, A., & Castillo, M. D. (2007). Microbial impact of the pesticide chlorpyrifos on Swedish and Italian biobeds. *Agronomy for Sustainable Development, 27*(3), 267–272.

de Vries, F. T., Williams, A., Stringer, F., Willcocks, R., McEwing, R., Langridge, H., et al. (2019). Changes in root-exudate-induced respiration reveal a novel mechanism through which drought affects ecosystem carbon cycling. *The New Phytologist, 224*, 132–145.

Wang, C., Lü, Z., Li, J., Cao, Z., Wei, B., Li, H., Shang, M., & Su, C. (2020). Efficient use of waste carton for power generation, tar and fertilizer through direct carbon solid oxide fuel cell. *Renewable Energy, 158*, 410–420.

Wassila, R., Laval, K., Laroche-Ajzenberg, E., Mougin, C., Latour, X., & Trinsoutrot-Gattin, I. (2014). Effects of pesticides on soil enzymes: A review. *Environmental Chemistry Letters, 12*(2), 257–273.

Weaver, M. A., Krutz, L. J., Zablotowicz, R. M., & Reddy, K. N. (2007). Effects of glyphosate on soil microbial communities and its mineralization in a Mississippi soil. *Pest Management Science, 63*, 388–393.

Wood, T. J., & Goulson, D. (2017). The environmental risks of neonicotinoid pesticides: A review of the evidence post 2013. *Environmental Science and Pollution Research, 24*, 17285–17325.

Xia, X. M., Zhao, M., Wang, H. Y., & Ma, H. (2011). Influence of butachlor on soil enzymes and Microbial growth. *Journal of Food, Agriculture and Environment, 9*, 753–756.

Xie, H., Goa, F., Tan, W., & Wang, S. G. (2011). A short term study on the interaction of bacteria, fungi and endosulfan in soil microcosm. *Science of the Total Environment, 412–413*, 375–379.

Yadav, I. C., Devi, N. L., Syed, J. H., Cheng, Z., Li, J., Zhang, G., & Jones, K. C. (2015). Current status of persistent organic pesticides residues in air, water and soil and their possible effect on neighboring countries. *The Science of the Total Environment, 511*, 123–137.

Yan, H., Wang, D. D., Dong, B., Tang, F. F., Wang, B. C., Fang, H., & Yu, Y. L. (2011). Dissipation of carbendazim and chloramphenicol alone and in combination and their effects on soil fungal: Bacterial ratios and soil enzyme activities. *Chemosphere, 84*, 634–641.

Yao, X. H., Min, H., Lu, Z. H., & Yuan, H. P. (2006). Influence of acetamiprid on soil enzymatic activities and respiration. *European Journal of Soil Biology, 42*, 120–126.

Ye, S., & Peng, B. (2019). Effects of foliar application of zinc fertilizer on yield and fruit quality of pear-jujube (Ziziphus jujube Mill.). *Bangladesh Journal of Botany, 48*(3 Special), 673–679.

Zabaloy, M. C., Garland, J. L., & Gomez, M. A. (2008). An integrated approach, to evaluate the impacts of the herbicides glyphosate, 2,4-D and metsulfuron-methyl on soil microbial communities in the Pampas region, Argentina. *Applied Soil Ecology, 40*, 1–12.

Zabaloy, M. C., Gomez, E., Garland, J. L., & Gomez, M. A. (2012). Assessment of microbial community function and structure in soil microcosms exposed to glyphosate. *Applied Soil Ecology, 61*, 333–339.

Zayed, S., Farghaly, M., Mahdy, F., & Soliman, S. M. (2008). Impact of repeated pesticide applications on the binding and release of methyl (14)C-monocrotophos and U-ring labelled (14)C-carbaryl to soil matrices under field conditions. *Journal of Environmental Science and Health. Part. B, 43*, 595–604.

Zeng, J., Liu, X., Song, L., Lin, X., Zhang, H., Shen, C., & Chu, H. (2016). Nitrogen fertilization directly affects soil bacterial diversity and indirectly affects bacterial community composition. *Soil Biology and Biochemistry, 92*, 41–49.

Zhang, T.'a., Chen, H. Y. H., & Ruan, H. (2018). Global negative effects of nitrogen deposition on soil microbes. *The ISME Journal, 12*(7), 1817–1825.

Effects of Aquatic (Freshwater and Marine) Pollution on Microbial Enzyme Activities

Gabriel Gbenga Babaniyi, Babatunde Oyemade, and Damilola Orija

Abstract Water pollution has a dual impact on nature, as it is harmful to both humans and the environment. Contamination from distributed sources is difficult to control, and despite significant advances in the construction of modern sewage-treatment plants, dispersed sources continue to be a significant source of water pollution. Bacterial activity is the most important process in the hydrolysis of organic contaminants. The study recognized the location, type, and size of the water body to which it habituates, either natural water body like lake, sea, or artificial water body like the wetlands, so as to give a holistic enumeration effectiveness of the preferred solution. Physical remediation and bioremediation techniques are the methods that can be used for the mitigation and improvement of water quality. Study revealed that microbial agents or photosynthetic bacteria and microalgae-bacteria medium degrades organic matter in water significantly and also reduces the level of chemical oxygen demand (COD), biochemical oxygen demand (BOD), and nutrients aeration, precipitation and ion-exchange or addition of nutrients and activators to the water is an eco-friendly solution to improve the water quality aids during the activities of microbial enzymes. The effective remediation is the best practice to weighing both pros and cons toward the effect of pollution on microbial enzymes activities.

G. G. Babaniyi (✉)
Department of Agricultural Development and Management, Agricultural and Rural Management Training Institute (ARMTI), Ilorin, Nigeria

B. Oyemade
Bioresources Development Centre, Ogbomoso, Nigeria

D. Orija
Federal University of Technology Akure, Akure, Nigeria

N. R. Maddela et al. (eds.), *Ecological Interplays in Microbial Enzymology*, Environmental and Microbial Biotechnology,
https://doi.org/10.1007/978-981-19-0155-3_17

1 Introduction

Pathogenic microorganisms, putrescible organic waste, plant nutrients, poisonous compounds, sediments, heat, petroleum (oil), and radioactive substances are all examples of pollutants that can pollute water bodies. Nonetheless, home sewage is the primary source of pathogens (disease-causing bacteria) and putrescible organic compounds. Putrescible organic matter poses a variety of threats to the water quality of a body of water. The dissolved oxygen concentration of the water is diminished as organics fester organically in the sewage by bacteria and other microbes. This jeopardizes the quality of lakes and streams, where fish and other aquatic species require high quantities of oxygen to survive, according to Kraus (2019). Furthermore, microbial populations play critical roles in sediment ecology, with microorganisms responsible for the amount of energy flow and mineral cycling, according to Kraus (2019). Nonetheless, movements of nutrients, resources, and predators often link the fates of food webs in nearby ecosystems. In freshwater ecosystems, aquatic insects that metamorphosis from aquatic larvae to terrestrial adults perform crucial roles as prey subsidies that transport nutrients and energy from aquatic to terrestrial food webs, according to Allen and Wesner (2016).

Trace metal contamination and neonicotinoid insecticides, on the other hand, significantly diminish adult aquatic insect emergence and are thought to be contributing to the global decline in insect biomass (Hallmann et al., 2014; Kraus et al., 2014; Morrissey et al., 2015; Cavallaro et al., 2017). As a result, many terrestrial insectivores, such as spiders, young waterfowl, aerial insectivorous birds, and bats, have lost an essential prey supply (Hallmann et al., 2014; Morrissey et al., 2015). Other contaminants, like as persistent organic pollutants and organometals, on the other hand, do not always diminish emergence production at a location, despite the fact that they bioaccumulate in insect larvae and stay in their tissues during metamorphosis (Walters et al., 2008; Tweedy et al., 2013; Moy et al., 2016; Kraus et al., 2017; Richmond et al., 2018). When highly polluted adult aquatic insects fall prey to terrestrial insectivores, they become contaminant vectors, reducing insectivore reproduction, health, and juvenile success.

Meanwhile, chemical monitoring of water, according to Reddy and Rawat (2013), aids in determining the degree of infectivity. Biomarker response, on the other hand, can be used to monitor and evaluate the effects of pollution on numerous organisms, including fish. Most biomarkers are restricted in their expression, whereas histopathology evaluates a wide range of conditions. Histopathological changes in animal tissues are dependable and direct markers of environmental stresses, as well as the most straightforward means of evaluating both acute and long-term harmful consequences.

1.1 Sources of Water Pollution

Water contaminants arise from either point sources or distributed sources, according to Richmond et al. (2018). A pipe or channel, such as those used for discharge from an industrial facility or a city sewerage system, is referred to as a point source. A scattered source, also known as a nonpoint source, is a large, unconfined area from which a range of contaminants, such as runoff from an agricultural area, enters a water body. Because the polluted water has been collected and delivered to a single location where it may be cleaned, point sources of water pollution are easier to control than distributed sources. Contamination from distributed sources is difficult to control, and despite significant advances in the construction of modern sewage-treatment plants, dispersed sources continue to be a significant source of water pollution. According to Owa (2014), water pollution in Nigeria is caused by a variety of activities, including agricultural pollution, radioactive substances, river dumping, marine dumping, sewage leakages, high population density, oil spillage, pollution of ground water through drilling activities, and flooding during the rainy season, which transports waste deposits into our waters. Building lavatories and visionaries over running water or even the sea, as some riverine areas do, radioisotopes, heavy metal, combustion, toxic waste disposal at sea, mineral processing plants (e.g., coal production), eroded sediments, deforestation, mining, littering, pesticides, herbicides, and fertilizers, failing septic systems, household chemicals, and animal wastes are all issues that need to be addressed.

Water contamination is mainly caused by human actions carried out for the betterment of oneself. These could be viewed as part of the different actions that lead to pollution that man engages in. Highly contaminated rivers, according to Owa (2013), have an unpleasant odor and contain little or no flora or fauna. The outflow of hot water from cooling engines in industry is another source of water hazardous waste. This raises the demand for oxygen by raising the water temperature and lowering the metabolic rate of organisms. Meanwhile, pollution poses a serious threat to human life, particularly when water is utilized for drinking, cooking, and other domestic functions. As a result, contaminated streams can spread diseases like cholera, typhoid, and tuberculosis to humans. Oil spilled in significant quantities from tankers or damaged oil pipes from oil industry has been a major water contaminant, resulting in the deaths of sea weeds, mollusks, marine birds, crabs, fishes, and other aquatic animals that provide food for humans. Thus, the effects of water pollution have been so severe in some locations that they have irreparably altered aquatic ecosystems, posing a threat to plants and animals, including humans.

Sediment (e.g., silt) Surface runoff can carry this into water bodies as a result of soil erosion. Suspended sediment obstructs sunlight penetration and disrupts a body of water's natural balance. As a result, it can disrupt fish and other forms of life's reproductive cycles, and when it settles out of suspension, it can suffocate bottom-dwelling species.

Toxic waste is defined as waste that is poisonous, radioactive, explosive, carcinogenic (causes cancer), mutagenic (causes chromosome damage), teratogenic (causes birth defects), or bioaccumulative (causes cancer) (i.e., increasing in concentration at the higher ends of food chains). Toxic chemicals come from a variety of places, including inadequately discarded wastewater from industrial plants and chemical processing facilities (lead, mercury, chromium), as well as surface runoff including pesticides used on agricultural fields and residential lawns (chlordane, dieldrin, heptachlor).

2 Aquatic Microbial Enzymes

Marine microorganisms have been recognized as potential sources of novel enzymes because they are relatively more stable than the corresponding enzymes derived from plants and animals. Enzymes from marine microorganisms also differ from homologous enzymes in terrestrial microorganisms based on salinity, pressure, temperature, and lighting conditions. Marine microbial enzymes can be used in diverse industrial applications (Fig. 1) (Nguyen & Nguyen, 2017).

2.1 *Protease*

Proteases are a type of enzyme that hydrolyzes peptide bonds in water and then synthesizes them in a nonwatery environment. On the basis of peptide chain catalysis, proteases are classified as endopeptidases or exopeptidases. Serine endopeptidase, cysteine endopeptidase, aspartic endopeptidases, and metallopeptidases

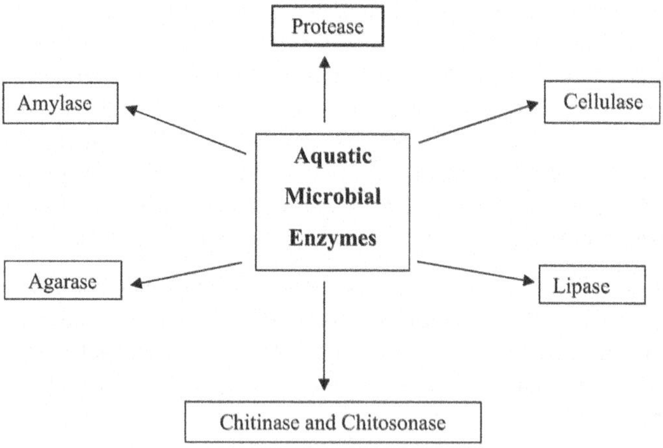

Fig. 1 Aquatic microbial enzymes with few examples

are the different types of endopeptidases depending on the active site position. Exopeptidases only work near the chain's terminal amino or carboxylic position; aminopeptidase reacts on free amino acids, while carboxypeptidase reacts on carboxyl terminals (Chandrakant & Shwetha, 2011). Microorganisms manufacture extracellular proteases to digest big polypeptides in the media into peptides and amino acids before cellular absorption. Proteases can be found in all types of organisms, including prokaryotes, eukaryotes, viruses, and microorganisms. Proteases either break particular peptide bonds (limited proteolysis) or totally degrade a peptide into amino acids (complete proteolysis) (unlimited proteolysis). Its action can either activate a function or a signaling pathway, or it can cause a destructive change (such as removing a protein's function or digesting it down to its basic components).

Mechanism of protease catalysis is by either of these two: activating water molecule which in turn does a nucleophilic attack on the peptide bond (hydrolysis) or covalent linking of protease to substrate protein by nucleophilic residue. The covalent acyl-enzyme intermediate is then hydrolyzed by activated water to complete catalysis. Proteases could be highly specific (Renicke & Taxi, 2016) or has wide range of protein substrates hydrolyzed (Rodriguez & Redman, 2008). The activities of proteases could be inhibited by protease inhibitors (Puente et al., 2004). Bacterial and fungal proteases are particularly important to the global carbon and nitrogen cycles in the recycling of proteins, and such activity tends to be regulated by nutritional signals in these organisms.

2.2 *Chitinase and Chitosanase*

Chitin is an insoluble amino polysaccharide composed of β-1,4-*N*-acetylglucosamine (Thakur et al., 2019). They are present in a variety of microorganisms which include bacteria and fungi (Rathore & Gupta, 2015). Aquatic environments are always rich in chitinous materials hence the abundance of chitinolytic enzymes in that habitat; however, researches on chitinolytic enzymes are scarce (Cottrell et al., 1999). Chitinases are chemically hydrolytic enzymes liable to the degradation of chitin. Chitinases are divided into endochitinase and exochitinase. Endochitinase cleaves chitin at internal sites to general multimers of GlcNAc; exochitinase, on the other hand, facilitates the swift hydrolysis of chitin to produce GlcNAc, chitobiose, or chitotriose (Saima et al., 2013). Bacteria like *Serratia marcescens*, *Aeromonas* spp., *Pseudomonas aeroginosa*, *Bacillus*, *Chromobacterium*, *Erwinia*, *Vibrio*, *Streptomyces* Enterobacter can synthesize different chitinases. Fungal strains *Trichoderma harzianum* and *Aspergillus niger* are also prospective chitinase-producing strains (Dahiya et al., 2005; Fadhil et al., 2014; Yuriy et al., 2015).

2.3 *Amylases*

Amylases catalyze the hydrolysis of starch. They are broadly classified into α, β, and γ. α and β are the most widely studied. α-amylase acts faster than β-amylase. Amylases are called glycoside hydrolases because they act on α-1,4 glycosidic bonds. The first amylase was isolated by Anselme Payen in 1833 (Guzmàn-Maldonado et al., 1995; Gupta et al., 2003). Amylases make up approximately 25% of the enzyme market (Mojsov, 2012). Amylases can be divided into endoamylases and exoamylases. Endoamylases catalyze hydrolysis in a random manner within the starch molecule which forms linear and branched oligosaccharides of various chain lengths. Exoamylases hydrolyze the substrate from the nonreducing end; this results in shorted end-products (Gupta et al., 2003).

Further, bacteria, yeast, and fungi produce microbial amylases; however, amylases production from bacteria is cheaper and faster than from other microbial sources. Also, highly amenable production of recombinant enzymes through genetic engineering is easier in bacteria than other microbial sources. *Bacillus* spp, *Lactobacillus fermentum*, *Lactobacillus manihotivorans*, and *Pseudomonas stutzeri* all secret amylases (Kumarevel et al., 2008a, b, c). Amylases from fungal sources have the advantage of being secreted externally. *Aspergillus oryzae*, *Aspergillus niger*, *Aspergillus fumigatus*, *Aspergillus flavus*, *Aspergillus kawa chii*, *Penicillum* spp, Streptomyces, Mucor spp. are all amylase-producing fungi (Hussain et al., 2013; Gupta et al., 2003; Sundarram & Murthy, 2014; de Souza 2010).

2.4 *Agarase*

Agarases are enzymes that hydrolyze agar. Agar is composed of agarose and agaropectin. Mode of action of agarases on agarose makes agarases to be classified into two groups namely: α-agarase and β-agarase. α-agarase hydrolyzes α-1,3 linkages, while β-agarase hydrolyzes β-1,4 linkages in agarose. They are majorly obtained from bacteria from aquatic environment. Organisms in aquatic environment that secrete agarose are majorly red algae (Schroeder et al., 2003), agarolytic bacteria like *Vibrio* (Zhang & Sun, 2007), *Pseudomonas* (Ryu et al., 2001), *Alteromonas* (Wang et al., 2006), *Microbulbifer* (Ohta et al., 2004), and *Pseudoalteromonas* (Ma et al., 2007).

2.5 *Lipase*

Lipases degrade lipids derived from a large variety of microorganisms, animals, and plants (Riffaldi et al., 2006). Lipases can catalyze various reactions like hydrolysis, interesterification, esterification, alcoholysis, and aminolysis (Prasad & Manjunath,

2011). First microbial lipase was found in Peniccilum oxalicum and Aspergillus flavus by Kirsh in 1935 (David, 1935). Lipases can be classified into two types on the basis of criteria such as (a) enhancement in enzyme activity as soon as the triglycerides form an emulsion and (b) lipases with a loop of protein (lid) covering on the active site. Lipases are ubiquitous, and they catalyze hydrolysis of triacylglycerols to glycerols and free fatty acids. Lipolytic reactions occur at lipid–water interface, where lipolytic substrates usually form equilibrium between monomers, micelles, and emulsions (Sharma et al., 2011).

Achromobacter sp., *Alcaligenes* sp., *Arthrobacter* sp., *Pseudomonas* sp., *Staphylococcus* sp., and *Chromobacterium* sp. have been exploited for the manufacturing of lipases between the bacteria; fungi employed in lipase production include but not limited to *Rhizopus*, *Aspergillus*, *Penicillium*, *Mucor*, *Ashbya*, *Geotrichum*, *Beauveria*, *Humicola*, *Rhizomucor*, *Fusarium*, *Acremonium*, *Alternaria*, *Eurotrium*, and *Ophiostoma*. Other species such as *Candida rugosa*, *Candida antarctica*, *T. lanuginosus*, *Rhizomucor miehei*, *Pseudomonas*, *Mucor*, and *Geotrichum* (Chandra et al., 2020).

2.6 Cellulases

Cellulases are inducible enzymes secreted by various microorganisms mainly bacteria and fungi, when they grow on cellulosic materials (Kubicek, 1993; Lee & Koo, 2001). The microorganisms can either be aerobes, anaerobes, mesophiles, or thermophiles. The mostly studied producers of cellulase are *Clostridium*, *Cellulomonas*, *Thermomonospora*, *Trichoderma*, and *Aspergillus*. Cellulases are composed of independently, folding, structurally, and functionally discrete units called domains or modules, making up cellulases modules (Carvalho et al., 2003). Based on mechanism of action, cellulase is a family of at least three groups of enzymes (Percival Zhang et al., 2006), endo-(1,4)-β-D-glucanase (EC 3.2.1.4) exo-(1,4)-β-D-glucanase (EC 3.2.1.91), and β-glucosidases (EC 3.2.1.21).

However, the structure of cellulose systems and cellulosomes of fungi are simpler than that of bacteria (Percival Zhang et al., 2006). Short polylinker region joins the domains of fungal cellulases to the catalytic domain at the N-terminal. The domains of the fungal cellulases are catalytic domain (CD) and cellulose binding module (CBM). CBM comprise about 35 amino acids with the linker region rich in serine and threonine. The cellulosomes-cohesin containing scaffolding and dockering containing enzymes is the major difference between cellulosomes and free cellulase enzymes. The free cellulase contains cellulose binding domains (CBMs) that are replaced by a dockerin in cellulosomal complex, and a single scaffolding-born CBM directs the entire cellulosomes complex to cellulosic biomass (Carvalho et al., 2003; Bayer et al., 2004). Mechanism of cellulose degradation by anaerobe bacteria involves mediating adherence of anaerobic cellulolytic bacteria to the substrate, followed by supramolecular reorganization, then finally the redistribution of cellulosomal subunits to interact with the different target substrates (Bayer et al., 2004).

3 Importance of Aquatic Microbial Enzymes

3.1 *Protease*

Protease enzymes are one of the most important industrial enzymes; they are widely used in pharmaceutical, brewing, protein hydrolysis, detergent, photographic, baking, meat, and leather production. Protease enzymes are used in producing dipeptide aspartame (a noncalorific artificial sweetener) used in food industries. Clostridial collagenase or Subtilisin, when combined with broad-spectrum antibiotics is used in burns and wounds treatment. Alkaline proteases are used in leather industries to remove hairs and parts that are present on the animal skin (Chandrakant & Shwetha, 2011; Rao et al., 1998).

3.2 *Chitinase*

Chitinases play a major role in production of single-cell protein, growth factors, mosquito control, ethanol and fertilizer production, biocontrol agent of fungal pathogen, isolation of fungal of fungal protoplast, and antifungal drugs (Vega & Kalkum, 2011; Das et al., 2015; Khatri et al., 2017; Veliz et al., 2017). More so, chito-oligosaccharides' water solubility and being biologically more active (compared to polymer chitosan) makes it a more preferable option in food and pharmaceutical products (Ekowati et al., 2009). However, reduction of environmental hazard and complementary production of industrial important value-added products have made a need to increase production of microbial chitinase (Karunya et al., 2011). Thus, most natural chitin degradation by various microorganisms and it occurs in aquatic habitat and other places like soil, animal guts, etc. Microbial chitin degradation occurs by either chitinoclastic or deacetylation mechanisms. In chitinoclastic, chitin degradation occurs solely by glycosidic bonds hydrolysis; chitinolytic system (chitinases and β-*N*-acetylglucosaminidases) act on the substrate. Deacetylation involves conversion of chitin into chitosan; this happens mostly in fresh water or soil sediment (Anupama, 2020).

3.3 *Amylase*

Amylases are used in production of syrups (corn, maltose, and glucose), juices, alcohol fermentation, and baking (Mojsov, 2012). Amylases are used as food additives. Amylases play a major role in beer and liquor brewing from sugar. In brewing, the alcohol content, flavor, and mouth-feel of the end product can be varied by altering the temperature and conditions for α- and β-amylase activities (Gopinath et al., 2017).

3.4 *Agarase*

Agarases are majorly used in the production of oligosaccharides from agar. Oligosaccharides from agar have the following properties: hepatoprotective property (Chen et al., 2006), antioxidative property (Wang et al., 2004). Agarases have potential applications in food and cosmetic and medical industries; they are generally regarded as safe as food additives (Kobayashi et al., 1997). Agarases can be used in the preparation of protoplasts by degrading the cell walls of marine algae (Dipakkore et al., 2005). Agarases could also be used to reclaim DNA from agarose gel.

3.5 *Lipase*

Lipases are used in detergents, paper production, cosmetics, food flavoring, organic synthesis, and other industrial applications (Kobayashi et al., 2008; Chi et al., 2009). Lipases hydrolyze triglyceride to diacylglycerol, monoacylglycerol, glycerol, and fatty acids. Monoacylglycerol is used as emulsifying agent in food, cosmetic, and pharmaceutical products. Glycerols and fatty acids are also widely used as raw materials.

3.6 *Cellulase*

Cellulases are used in various industries in different ways:

Paper and Pulp Industry

The use of cellulases in pulping (biomechanical pulping) saves substantial energy by about 20–40% during refining. This causes improved hand-sheet strength properties (Singh et al., 2007; Pere et al., 2001; Bhat, 2000). Cellulase and hemicellulose improves drainage and beatability in paper mills before or after beating the pulp (Dienes et al., 2004). Cellulase gives a better result in bleachability of softwood kraft pulp compared to xylanase treatment (Singh et al., 2007; Ruohonen et al., 1993). Cellulases alone, or combined with xylanases, are used in deinking of various paper wastes (Chander Kuhad et al., 2010). Cellulases are also used in the production of easily biodegradable cardboards (Buchert et al., 1998), production of soft papers and serviette papers (Salonen, 1990; Hsu & Lakhani, 2002) and removal of adhered paper.

Textile Industry

Cellulases are used in biostoning of jeans and biopolishing of cotton and other cellulosic fabrics (Sukumaran et al., 2005; Singh et al., 2007; Galante et al., 1998). The activities of cellulase remove short fibers, surface fuzziness; create smooth and glossy appearance; and improve brightness, hydrophilicity, and moisture absorbance (Bhat, 2000).

Bioethanol Industry

In biofuel production, cellulases are involved in enzymatic saccharification of lignocellulosic material; this leads to bioconversion of lignocellulosic materials into useful and higher value products.

Wine and Brewing Industry

Main purpose of using enzymes in making wine are better maseration, improved color extraction, easy clarification, easy filtration, improved wine quality, and improved stability (Galante et al., 1998).

Food Processing

Macerating enzymes (cellulase and pectinase) have wide range of applications in food processing especially in extraction and clarification of fruit and vegetable juice (de Carvalho et al., 2008).

Animal Feed

The use of cellulase and hemicellulose in animal feed improves feed value and performance of animals (Dhiman et al., 2002). The nutritional value of agricultural silage and grain can be improved by pretreating them with cellulases or xylanases. Cellulases also remove antinutritional factors in feed grains, they degrade certain feed constituents to improve nutritional value and provide supplementary digestive enzymes like proteases, amylases, and glucanases (Ali et al., 1995).

Agriculture

Combination of cellulases, with other enzymes like hemicellulases and pectinases, can enhance growth of crops and control plant diseases (Bhat, 2000). Cellulases can aid the improvement of soil quality and reduce dependence on mineral fertilizers (Ortiz Escobar & Hue, 2008; Tejada et al., 2008).

Waste Management

Celluloses (unutilized and underutilized) are present in wastes generated from sources like agro-industries, forest, agricultural activities, and industries which cause environmental pollution (Milala et al., 2005; Abu et al., 2000). But these days, the wastes are now being converted to more valuable products like enzymes, sugars, chemicals, biofertilizer, biofuels, cheap energy sources for fermentation, improved animal feeds, and human nutrients (Kuhad et al., 2010; Karmakar & Ray, 2011; Gupta et al., 2011a, b, 2009).

Hence, other applications of cellulases range from uses in olive oil extraction, carotenoid extraction. When combined with proteases and lipases, they are used in detergent production (Singh et al., 2007).

4 Effects of Pollution on Aquatic Microbial Enzymes

Water pollution has a dual impact on nature, as it is harmful to both humans and the environment. Pollution has a wide range of consequences for humans and aquatic communities. According to Owa (2013), water pollution causes roughly 14,000 deaths per day, the majority of which are caused by untreated sewage contaminating drinking water in underdeveloped countries. Biomass and community diversity are expected when huge amounts of harmful compounds are dumped into streams, lakes, and coastal waters in the ocean. Sewage, in which organic waste predominates, is responsible for a large portion of aquatic pollution. Industrial chemicals and petroleum hydrocarbons pollute soil and water, which is a major setback for the modern world. They are discovered as environmental pollutants in a variety of aquatic and terrestrial environments as a result of their widespread use. Furthermore, Amoatey and Baawain (2019) propose that using bioremediation technology to remove these toxins is a safe and cost-effective alternative to traditional physical–chemical treatment. Bacterial activity is the most important process in the hydrolysis of organic contaminants. If polluting discharges to a river are only made on a regular basis, the river can often return to a clean and unpolluted state, as the pollutants are washed out and taken down to the sea. River water, on the other hand, has some capacity for self-purification due to the organisms there, unless too many of these species are killed off too soon.

Furthermore, Rao et al. (2010) claimed that trash can boost secondary productivity while changing the aquatic community's character. The majority of fishes, particularly those sought for as food by humans, are among the most vulnerable species, disappearing with the least concentrated pollution. As a result, water pollution causes harm to human health, affecting drinking water and posing a health risk. Direct harm to plant and animal nutrition, on the other hand, has an impact on human health. Plant nutrients such as nitrogen, phosphorus, and other compounds that stimulate aquatic plant growth may be in excess, resulting in algal blooms and weed growth. This imparts fragrance, flavor, and, in certain cases, color to the water. As a result, Karigar and Rao (2011) concluded that a body of water's ecological balance has been disrupted. Sulfur dioxide and nitrogen oxides generate acid rain, which reduces soil PH values, while carbon dioxide emissions cause ocean acidification, which is the continual decline in the PH of the earth's oceans as CO_2 dissolves.

In the meantime, some industrial effluents that are emitted in high numbers can be dangerous. Food sector effluents, for example, are not very harmful, but due to their organic content and enormous volume, they can place a significant oxygen demand on the environment in the discharge area. Oxidative coupling is used by a variety of bacteria, fungi, and higher plants to detoxify harmful chemical molecules. Oxidoreductases are involved in this process. Microbes obtain energy by cleaving chemical bonds and assisting the flow of electrons from a reduced organic substrate (donor) to another chemical molecule, which is mediated by these enzymes (acceptor). According to Karigar and Rao (2011), pollutants are eventually converted to innocuous molecules during such oxidation–reduction events.

Oxidase is an oxidoreductase enzyme that participates in the oxidation of reduced substrates by transferring oxygen from molecular oxygen (O_2) with the help of FAD/NADH/NADPH as a cosubstrate. On the basis of the number of oxygen atoms utilized for oxygenation, oxygenases are divided into two categories: monooxygenases and dioxygenases. They play an important role in the metabolism of organic molecules by increasing their reactivity, water solubility, or causing aromatic ring cleavage. Oxygenases are active against a wide range of chemicals, including chlorinated aliphatics, and have a large substrate range. Generally, when oxygenase introduces O_2 atoms into an organic molecule, the aromatic rings are cleaved. Historically, bacterial mono- or dioxygenases have been the most investigated enzymes in bioremediation. Enzymes, on the other hand, have a number of advantageous properties; they are the primary effectors of all biota processes. They are catalysts with either narrow (chemo-, region-, and stereo-selectivity) or wide (chemo-, region-, and stereo-selectivity) specificity. As a result, they can be used with a wide variety of various substances in mixtures. They have the potential to change the structural and toxicological features of pollutants, as well as convert them entirely into harmless inorganic end products. They could carry out processes for which no effective chemical transformations have been found, for example (Karigar & Rao, 2011). Furthermore, enzymes can work intracellularly, extracellularly, free (i.e., soluble in solution) or immobilized (i.e., connected to a solid matrix by various connections) and catalysis can be homogeneous or heterogeneous (Gianfreda & Rao, 2004).

Furthermore, esterases, amidases, and proteases may break down esteric, amidic, and peptidic linkages, resulting in compounds with low or no toxicity. Bacterial hydrolases such as carbamate or parathion hydrolases from Achromobacter, Pseudomonas, Flavobacterium, Nocardia, and Bacillus cereus, for example, have proven successful in the transformation of pollutants like carbofuran and carbaryl or parathion, diazinon, and coumaphos (Gianfreda & Rao, 2004). Carbohydrases, depolymerases, proteases, and phosphatases, which are produced by a variety of bacteria and fungi, can also be used to convert intractable materials like carbohydrates, plastics, and proteins (van Wyk, 1999; Nakamura et al., 2001; Singh, 2003).

A class of enzymes engaged in the transformation of nitrile compounds with a –CN functional group plays an intriguing role (Banerjee et al., 2002; Singh et al., 2006). In chemical nomenclature, the prefix cyano is used to denote the presence of a nitrile group in a molecule. The –CN group is also known as a cyanide group or cyano group, and substances containing it are sometimes referred to as cyanides. Many cyanide-containing compounds are extremely toxic and lethal poisons; however, some nitriles may be rather safe. Plants, fungi, bacteria, algae, insects, and sponges all produce nitrile compounds.

5 Types of Remediation

5.1 Engineered/Physical Remediation Techniques

Aeration

Aeration is one of the best physical methods that proffer solution to the decrease of microbial activity in water. Aeration is used to significantly remove vaporous substances and gases in water. Aeration when applied through aerators does not need chemical for remediation, and it improves the quality of water in an eco-friendly manner (Lokhande & Dixit, 2017). They are easy to employ, operate, and sustain; but they are costly to implement. The effectiveness of aeration on microbes is affected by the type of aeration used (mobile aeration and fixed point aeration) and the rate at which they are applied. Therefore, it should be applied properly, several previous studies have demonstrated that the application of aeration techniques effectively improved the water quality of some rivers, such as the Oeiras River in Portugal, the Emsche River in Germany, the Thames River in the United Kingdom, and the Homewood Canal in the United States (Rogers, 2000) and removed the black color, odor, chemical oxygen demand (COD i.e the amount of oxygen consumed when the water sample is chemically oxidized), and biochemical oxygen demand (BOD i.e. the amount of dissolve oxygen consumed by biological organisms when they decompose orgaic matter in water) of river water in Busan (South Korea), Qing River, Guancheng downstream and Shanghai Suzhou River (Wang et al., 1999).

Ecological Floating Beds

Floating beds is relatively new technologies that do not employ soil for plant growth medium. It is a type of phytoremediation that uses a synthetic buyout mat that acts as a substrate for the growth of the plant and the roots the plant extends into the water body (Cui et al., 2018). Ecological floating beds is a form of engineering method but the main mechanisms by which it removes pollutant is through phytoremediation (i.e., the usage of plant to remediate either metals or other pollutants through uptake by absorption), microbial biodegradation of the organic chemicals then the removal of N and P by absorption and sedimentation processes. Ecological floating beds are getting more popular for the remediating of river water treatment, due to their relatively cheap cost and the degree of their effectiveness in removal of pollutant in water (Zheng & Wang, 2017). Their further ecological advantages are that they can be moved anywhere in the river for usage once constructed and these beds serves has a niche for birds and fishes. In addition, they disrupt water wave and erosion in the river banks there by ingresses vegetation in the areas and in turn leads to the increase in activity of microbes and their enzymes (Zhao et al., 2012). To say the least, this method of remediation in rivers promotes the aquatic life in general and also helps remediate pollutant.

Addition of Nutrients Activators

With increase in pollution and reduction of water quality, microbial enzymes are majorly impacted by the limited availability of nutrients. Therefore, adding nutrients salts in polluted water can improve the metabolic activities of microbes. Levy et al., worked on a neurotransmitter of reduction agent for pollutants and used it to control both organic and inorganic pollution in water. Apart from microbial agents, there are also nutritional agents that can remediate water polluted by diesel, oil, and gasoline (Gu et al., 2015). As reported by Stanish and Monbouquette (2001) also use biostimulant produced by EIT in the United States to experiment on radiation and found out that it promote degrading of organic matters in water and removes pollutant like ammonia, nitrogen, and phosphorous (Cheng, 2014).

Precipitation and Ion Exchange

pH has a direct impact on the activity of microbial enzymes, and with increase in pollutant like heavy metals in water, the level of pH increases exponentially and affects the quality of water. To mitigate the effect of increase in pH on the enzymes activities, precipitation, and ion-exchange method is a great tool to this end. Precipitation is the use of a nondirected physio–chemical complexation reaction between dissolved pollutant and charged cellular components, while ion-exchange removes the ions from the liquid phase by exchanging the anion or cation between the pollutants. Although pH and some oxidant can affect the exchange, a research by Hossain and Rezaul (2020) shows that anthraquinone compounds and humic acids react as an electron acceptor in the process of degrading organic matters.

5.2 *Bioremediation Technique*

Aquatic Animal Remediation

Aquatic animals can be used for the mitigation and improvement of water quality. For example, eutrophication of water due to algae bloom negatively affects the water quality by decreasing both the level of DO and BOD present in water for microbial activities, stocking zooplankton, filter-feeding fish, clams, snails, silver carp; common carp can reduce the negative effect of algae in water (Cheng, 2014). Also Xiao et al. (2010) reported that *Anodonta* has the capability to remediate Pb^{2+}, Cu^{2+}, Cr^{2+}, but this method is known to be time consuming and cost ineffective. Although enhanced animal remediation method is chiefly used as an indicator of pollution of heavy metal in water than as a remediating tool, for example, fresh water gastropod as reported by Li et al. (2018) was successfully used as bioindicator to estimate the level of contamination of some heavy metals.

Phytoremediation Process

Aquatic plants is by no means a new technology for remediation of pollutants in water, its eco-friendly, lower cost and range of pollutant to which it can remediate and made it an important choice as a remediating tool in modern era (Zhao et al., 2018). These plants when applied either in river or streams can purify water through several methods depending on the type of plant used; it can either be through absorption, adsorption, accumulation, and degradation of different pollutant. An aquatic plant has also shown to have the capacity to remove nutrients like N and P from the water with the plants root serving as a rhizosphere system that promote growth of bacteria and microbes in water. Phytoremediation can be applied along the river banks or at the discharge point source. The most recognized aquatic plants as stated by Hossain and Rezaul (2020) for remediation purpose in water are: water hyacinth (*Eichhornia crassipes*), alligator weed (*Alternanthera philoxeroides*) (Sato et al., 2008), water lettuce (*Pistia stratiotes*) (Zimmels et al., 2008), Whorl-leaf watermilfoil (*Myriophyllum verticillatum*), pondweed (*Potamogeton* spp.), cattail (*Typha latifolia*), duckweed (*Lemna gibba*), and canna (*Canna indica*) (Allam et al., 2016). It is important to note that the success of this method is dependent on the type of plant used, and its variable tolerance to pollutant loading rates, also the coverage of plant growth area plays a vital role in its effectives for remediation.

Application of Microbial Agents

Due to the target organisms involved, selected remediation methods are more favored to use than others. But the usage of microbial agents as a remediating tool serves the purpose. From literatures, it is well proven that microbial agents or photosynthetic bacteria and microalgae–bacteria medium degrades organic matter in water significantly and also reduces the level of chemical oxygen demand (COD), biochemical oxygen demand (BOD), and nutrients (Liu et al., 2017). Also noted by Zhang et al. (2010) that the mixing of microbes with water directly increases the level of dessolved oxygen (DO), and reduces mildly the NH_3-N and chemical oxygen demand (COD). The use of microbial agents as a solution to pollution water is a simple and viable long-term operation, and their processes do not necessary need additional construction that may affect habitat of the targeted organisms. Improved and more efficient results can be achieved when microbial agents are deployed together with other engineering/physical techniques, although the success and cost of such application is dependent on the types of microbes used for the process. The usage of carriers (AquaMats and semiflexible supports) with the aid of aeration process has also been reported as a vital role in remediating polluted rivers (Hossain & Rezaul, 2020). Shan et al., in their work stated that microbial agents containing nitrobacteria, mixed bacteria, and humic acid can efficiently remove TN, TP, NH_4-N, and COD in water.

Some other less popular method to aid aquatic microbial enzymes are:

1. **Intergrowth:** degradation of many pollutants is as a result of cometabolism of various microbes. One microbe most often cannot directly use the energy produced by its own supersession. Microbes of intergrowth help each other substrate environment to growth. And this mechanism is used for degradation for lots of pollutant.
2. **Alternation:** also similar to intergrowth, alternation can be used to replenish the population microbial enzymes in water, since the metabolized outcomes of one microbe create an ecological position to another microbe. That is, in a controlled environment, new microbe can be introduced to help satisfy the needs of nutrition of the other microbes present in the polluted water.

6 Possible Solutions to the Effect Pollution on Microbial Enzymes Activities

The general condition of the environment is intrinsically linked to the quality of life on earth. Unfortunately, as science, technology, and industry improve, a vast amount of trash ranging from raw sewage to radioactive waste is released or thrown into the ecosystem, posing a serious threat to mankind's survival on earth. New garbage disposal technologies, such as high-temperature incineration and chemical breakdown, have emerged. They can be quite successful at reducing a wide range of pollutants, but they also have a number of disadvantages. These procedures are complicated, inefficient, and unpopular (Karigar & Rao, 2011). However, with advancement in science and the increase in population, there has been rise in the anthropogenic activity of humans which has led to increase in xeno-materials to the water bodies, and has thus affect the quality of water both for human and aquatic life—aquatic microbes inclusive. Improving the quality of water, therefore, directly impacts positively the life and activities of microbial enzymes. But the method to remediate polluted water and not affect or be detrimental to life and culture of aquatic life in general must be taken to account.

Furthermore, according to Karigar and Rao (2011), bioremediation is the transformation or breakdown of pollutants by microorganisms into nonhazardous or less-hazardous compounds. It has been stated that diverse species such as bacteria, fungi, algae, and plants can be used to efficiently bioremediate contaminants (Vidali, 2001; Leung, 2004). Phytoremediation refers to the use of plants in the bioremediation of contaminants. Phytoremediation is a new green technology that allows hazardous compounds to be removed or degraded from soils, sediments, groundwater, surface water, and the air (RTDF). Plants that have been genetically modified are also in use. Arsenic, for example, is phytoremediated by Arabidopsis thaliana, a genetically engineered plant that produces two bacterial genes. One of these genes permits the plant to convert arsenate to arsenite, while the other binds and stores the transformed arsenite in the vacuoles (Leung, 2004).

More so, microbial communities exposed to plastic-contaminated areas can adapt, build dense biofilms on the plastic surface, and create active catalytic enzymes, according to Anjana et al. (2020). Synthetic polymers can be degraded by these enzymes. Microbial enzymes can be used to break down man-made polymers because of their high catalytic activity. Nonetheless, laboratory studies have found diverse impacts of microbes on several types of polymers, mainly through enzymatic hydrolysis or oxidation, according to Krueger et al. (2015). Most popular plastics, on the other hand, have proven to be quite resistant to microbial deterioration, even under settings known to favor microbial destruction. Environmental degradation knowledge is even scarcer. Similarly, da Costa Waite et al. (2016) argue that biomarkers such as esterase and dehydrogenase enzymatic activity should be used to measure resistance and tolerance in populations that have been exposed to stressors earlier.

According to a recent study, Jambeck et al. (2015) opined that an annual intake of 4.8–12.7 million metric tons of plastic garbage into the ocean contributes 60–80% of marine macro- and mega debris (Gregory & Ryan, 1997). On the other hand, the lack of direct toxicity of polymeric substances may explain the carelessness with which synthetic polymers have been discarded, and the complacency with which plastic pollution, such as that of the marine environment, has been tolerated for decades. Plastic materials, more than any other class of pollutants, make it clear that our planet is out of balance when it comes to their environmental intake and disposal. It has even been proposed that all plastic, with the exception of that which has been destroyed, may still exist (Thompson et al., 2005). The current state of global plastic pollution, which is already considered unacceptable, along with unabatedly expanding production rates necessitates greater control of synthetic polymer use and handling.

None withstanding, to proffer solutions to the effect of pollution on aquatic microbial enzymes, one must first identify the location, type, and size of the water body to which it habituates, either natural water body like lake, sea, or artificial water body like the wetlands, so as to give an holistic enumeration, both in cost and effectiveness of the preferred solution. Possible solutions can be either engineered/physical method (aeration, precipitation and ion-exchange or addition of nutrients and activators to the water) or it can be done through bioremediation method. The bioremediation method, peradventure is the most preferred and less cost effective, though it takes time for it to be effective, but when employed properly, it is an eco-friendly solution to improve the water quality aids the activities of microbial enzymes.

Whereas this method of remediation can also be done either in-situ water treatment or it can be done from point source. As slatted earlier, there are several methods that can be applied for remediation of polluted water, but usage of a single method is not quite effective for complete remediation. Hybrid techniques, i.e., applying one or more engineered/physical methods with bioremediation processes, have proven to be effective, fast, less costly, and environmentally sustainable. Therefore, several bioremediation and engineered/physical methods will be discussed and accessed, weighing their pros and cons toward effectively remediating the effect of pollution on microbial enzymes activities.

7 Conclusion

Presently, there is much attention focused on effects of aquatic (freshwater and marine) pollution on microbial enzyme activities. However, pathogenic microorganisms such as putrescible organic waste, plant nutrients, poisonous compounds, sediments, heat, petroleum (oil), and radioactive substances are all good examples of contaminants that are capable/responsible for water body's pollution. So, marine microorganisms being relatively stable than the equivalent enzymes derived from plants and animals have been recognized as potential sources of novel enzymes. Though, it has been established that water pollution has a dual impact on nature, as it is harmful to both humans and the environment. This pollution has a wide range of effects on humans and aquatic communities. Therefore, one of the best physical methods consider to give solution to the decrease of microbial activity in water is aeration among others. Consequently, for mitigation and improvement of water quality, the use of aquatic animals can be considered necessary as aquatic plants is by no means a new technology for remediation of water contaminants. This method is eco-friendly, has a lower cost and a range of impurity to which it can remediate, made it an important choice as a remediating tool in modern age. Hence, as a way of proffering solutions to the effect of pollution on aquatic microbial enzymes, it is imperative to identify the location, type, and size of the water body to which it habituates first, so as to give a holistic account, both in cost and effectiveness of the preferred solution.

References

Abu, E. A., Onyenekwe, P. C., Ameh, D. A., Agbaji, A. S., & Ado, S. A. (2000). Cellulase production from sorghum bran by *Aspergillusniger* SL:1: An assessment of pretreatment methods. In *Proceedings of the International Conference on Biotechnology: Commercialization and Food Security (ICBCFS '00),* Abuja, Nigeria (pp. 153–159).

Ali, S., Hall, J., & Soole, K. L. (1995). Targeted expression of microbial cellulases in transgenic animals. *Progress in Biotechnology, 10*, 279–293.

Allam, A., Tawfik, A., El-Saadi, A., & Negm, A. M. (2016). Potentials of using duckweed (Lemna gibba) for treatment of drainage water for reuse in irrigation purposes. *Desalination and Water Treatment, 57*, 459–467.

Allen, D. C., & Wesner, J. S. (2016). Synthesis: Comparing effects of resource and consumer fluxes into recipient food webs using meta-analysis. *Ecology, 97*, 594–604.

Amoatey, P., & Baawain, M. S. (2019). Effects of pollution on freshwater aquatic organisms. *Water Environment Research, 91*(10), 1272–1287. https://doi.org/10.1002/wer.1221

Anjana, K., Hinduja, M., Sujitha, K., & Dharani, G. (2020). Review on plastic wastes in marine environment–Biodegradation and biotechnological solutions. *Marine Pollution Bulletin, 150*, 110733.

Anupama, S. (2020). *Microbial degradation of Chitin (enzymes, steps, mechanisms).* Microbenotes.com.

Banerjee, A., Sharma, R., & Banerjee, U. (2002). RETRACTION: The nitrile-degrading enzymes: Current status and future prospects. *Applied Microbiology and Biotechnology, 100*, 7359.

Bayer, E. A., Belaich, J. P., Shoham, Y., & Lamed, R. (2004). Thecellulosomes: Multienzyme machines for degradation of plant cell wall polysaccharides. *Annual Review of Microbiology, 58*, 521–554.

Bhat, M. K. (2000). Cellulases and related enzymes in biotechnology. *Biotechnology Advances, 18*(5), 355–383.

Buchert, J., Oksanen, T., Pere, J., Siika-aho, M., Suurnakki, A., & Viikari, L. (1998). Applications of *Trichoderma reesei* enzymes in the pulp and paper industry. In G. F. Harman & C. P. Kubicek (Eds.), *Biological control and commercial applications: Trichoderma & Gliocladium—Enzymes* (Vol. 2, pp. 343–363).

Carvalho, L. M. J., Deliza, R. C., Silva, C. A. B., Miranda, R. M., & Maia, M. C. A. (2003). Identifying the adequate process conditions by consumers for pineapple juice using membrane technology. *Journal of Food Technology, 1*, 150–156.

de Carvalho, L. M. J., de Castro, I. M., & da Silva, C. A. B. (2008). A study of retention of sugars in the process of clarification of pineapple juice (*Ananascomosus*, L. Merril) by micro- and ultra-filtration. *Journal of Food Engineering, 87*(4), 447–454.

Cavallaro, M. C., Morrissey, C. A., Headley, J. V., Peru, K. M., & Liber, L. (2017). Comparative chronic toxicity of imidacloprid, clothianidin, and thiamethoxam to Chironomus dilutus and estimation of toxic equivalency factors. *Environmental Toxicology and Chemistry, 36*, 372–382.

Chander Kuhad, R., Mehta, G., Gupta, R., & Sharma, K. K. (2010). Fed batch enzymatic saccharification of newspaper cellulosics improves the sugar content in the hydrolysates and eventually the ethanol fermentation by *Saccharomyces cerevisiae*. *Biomass and Bioenergy, 34*(8), 1189–1194.

Chandra, P., Enespa, Singh, R., & Arora, P. K. (2020). Microbial lipases and their industrial applications: A comprehensive review. *Microbial Cell Factories, 19*, 169. https://doi.org/10.1186/s12934-020-01428-8

Chandrakant, S. K., & Shwetha, S. R. (2011). Role of microbial enzymes in the bioremediation of pollutants: A review. *Enzyme Research, 2011*, 805187. https://doi.org/10.4061/2011/805187. 11 pages.

Chen, H. M., Yan, X. J., Zhu, P., & Lin, J. (2006). Antioxidant activity and hepatoprotective potential of agaro-oligosaccharides in vitro and in vivo. *Nutrition Journal, 5*, 31.

Cheng, J. (2014). Bioremediation of contaminated water-based on various technologies. *Open Access Library Journal, 1*, e056. https://doi.org/10.4236/oalib.preprints.1200056

Chi, Z., Chi, Z., Zhang, T., Liu, G., Li, J., & Wang, X. (2009). Production, characterization and gene cloning of the extracellular enzymes from the marine-derived yeasts and their potential applications. *Biotechnology Advances, 27*, 236–255.

da Costa Waite, C. C., da Silva, G. O. A., Bitencourt, J. A. P., Sabadini-Santos, E., & Crapez, M. A. C. (2016). Copper and lead removal from aqueous solutions by bacterial consortia acting as biosorbents. *Marine Pollution Bulletin, 109*(1), 386–392.

Cottrell, M. T., Moore, J. A., & Kirchman, D. L. (1999). Chitinases from uncultured marine microorganisms. *Applied and Environmental Microbiology, 65*(6), 2553–2557.

Cui, N. X., Chen, G. F., Liu, Y. Q., Zhou, L., Cai, M., Song, X. F., & Zou, G. Y. (2018). Comparison of two different ecological floating bio-reactors for pollution control in hyper-eutrophic freshwater. *Scientific Reports, 8*, 14306.

Dahiya, N., Tewari, R., Tiwari, R. P., & Hoondal, G. S. (2005). Chitinase from *Enterobacter* sp. NRG4: Itspurification, characterization and reaction pattern. *Electronic Journal of Biotechnology, 8*, 134–145. https://doi.org/10.2225/vol8-issue2-fulltext-6

Das, M. P., Devi, P. V., & Yasmine, Y. (2015). A study on antagonistic potential of bacteria against phytopathogenic fungi. *International Journal of Pharmaceutical Sciences Review and Research, 34*, 191–193. Retrieved from https://pdfs.semanticscholar.org/44c6/f66e45b2d38fef4f80f3b3b847bbc9e818cc.pdf

David, K. (1935). Lipase production by *Penicillium oxalicum* and *Aspergillus flavus*. *Botanical Gazette, 97*, 321.

Dhiman, T. R., Zaman, M. S., Gimenez, R. R., Walters, J. L., & Treacher, R. (2002). Performance of dairy cows fed forage treated with fibrolytic enzymes prior to feeding. *Animal Feed Science and Technology, 101*(1–4), 115–125.

Dienes, D., Egyházi, A., & Réczey, K. (2004). Treatment of recycled fiber with *Trichoderma* cellulases. *Industrial Crops and Products, 20*(1), 11–21.

Dipakkore, S., Reddy, C. R. K., & Jha, B. (2005). Production and seeding of protoplasts of *Porphyraokhaensis* (Bangiales, Rhodophyta) in laboratory culture. *Journal of Applied Phycology, 17*, 331–337.

Ekowati, C., Dewi, S. Z., & Agustinus, R. U. (2009). Screening and characterization of bacterial chitosanase from marine environment. *The Journal of Coastal Development, 12*(2), 64–72.

Escobar, M. O., & Hue, N. V. (2008). Temporal changes of selected chemical properties in three manure–amended soils of Hawaii. *Bioresource Technology, 99*(18), 8649–8654.

Fadhil, L., Kadim, A., & Mahdi, A. (2014). Production of chitinase by *Serratiamarcescens* from soil and its antifungal activity. *Journal of Natural Science Research, 4*, 80–86. https://doi.org/10.7176/JNSR/2014.8.12635.4

Galante, Y. M., De Conti, A., & Monteverdi, R. (1998). Application of *Trichoderma* enzymes in food and feed industries. In *Biological control and commercial applications: Trichoderma and Gliocladium—Enzymes* (Vol. 2, pp. 311–326). Taylor & Francis.

Gianfreda, L., & Rao, M. A. (2004). Potential of extra cellular enzymes in remediation of polluted soils: A review. *Enzyme and Microbial Technology, 35*, 339–354.

Gopinath, S. C. B., Periasamy, A., Md Arshad, M. K., Thangavel, L., Chun, H. V., Uda, H., & Suresh, V. C. (2017). Biotechnological processes in microbial amylase production. *BioMed Research International, 2017*, 1272193. https://doi.org/10.1155/2017/1272193. 9 pages.

Gregory, M. R., & Ryan, P. G. (1997). Pelagic plastics and other seaborne persistent synthetic debris: A review of southern hemisphere perspectives. *Marine Debris*, 49–66.

Gu, D. G., Xu, H., He, Y., Zhao, F., & Huang, M. S. (2015). Remediation of urban river water by Pontederia Cordata combined with artificial aeration: Organic matter and nutrients removal and root-adhered bacterial communities. *International Journal of Phytoremediation, 17*, 1105–1114.

Gupta, R., Gigras, P., Mohapatra, H., Goswami, V. K., & Chauhan, B. (2003). Microbial α-amylases: A biotechnological perspective. *Process Biochemistry, 38*(11), 1599–1616.

Gupta, R., Sharma, K. K., & Kuhad, R. C. (2009). Separate hydrolysis and fermentation (SHF) of Prosopisjuliflora, a woody substrate, for the production of cellulosic ethanol by *Saccharomyces cerevisiae* and *Pichiastipitis*-NCIM 3498. *Bioresource Technology, 100*(3), 1214–1220.

Gupta, R., Khasa, Y. P., & Kuhad, R. C. (2011a). Evaluation of pretreatment methods in improving the enzymatic saccharification of cellulosic materials. *Carbohydrate Polymers, 84*, 1103–1109.

Gupta, R., Mehta, G., Khasa, Y. P., & Kuhad, R. C. (2011b). Fungal delignification of lignocellulosic biomass improves the saccharification of cellulosics. *Biodegradation, 22*(4), 797–804.

Guzmàn-Maldonado, H., Paredes-Lòpez, O., & Biliaderis, C. G. (1995). Amylolytic enzymes and products derived from starch: A review. *Critical Reviews in Food Science and Nutrition, 35*(5), 373–403.

Hallmann, C. A., Foppen, R. P., van Turnhout, C. A., de Kroon, H., & Jongejans, E. (2014). Declines in insectivorous birds are associated with high neonicotinoid concentrations. *Nature, 511*, 341.

Hossain, M. A., & Rezaul, C. (2020). Remediation of polluted water by biological, chemical and engineering processes. *Sustainability, 12*, 7017. https://doi.org/10.3390/su12177017

Hsu, J. C., & Lakhani, N. N. (2002). *Method of making absorbed tissue from recycled waste paper*. US patent 6413363.

Hussain, S. F., Mahmood, M. S., & Ahmed, S. I. (2013). A review of the microbiological aspect of α-amylase production. *International Journal of Agriculture and Biology, 15*(5), 1029–1034.

Jambeck, J. R., Geyer, R., Wilcox, C., Siegler, T. R., Perryman, M., Andrady, A., . . . Law, K. L. (2015). Plastic waste inputs from land into the ocean. *Science, 347*(6223), 768–771.

Karigar, C. S., & Rao, S. S. (2011). Role of microbial enzymes in the bioremediation of pollutants: A review. *Enzyme Research, 2011*, 1–11. https://doi.org/10.4061/2011/805187

Karmakar, M., & Ray, R. R. (2011). Current trends in research and application of microbial cellulases. *Research Journal of Microbiology, 6*(1), 41–53.

Karunya, S. K., Reetha, D., Saranraj, P., & Milton, D. J. (2011). Optimization and purification of chitinase produced by *Bacillus subtilis* and its antifungal activity against plant pathogens. *International Journal of Pharmaceutical & Biological Archive, 2*, 1680–1685.

Khatri, D. K., Tiwari, D. N., & Bariya, H. S. (2017). Chitinolytic efficacy and secretion of cell wall-degrading enzymes from Trichoderma spp. in response to phytopathological fungi. *Journal of Applied Biology and Biotechnology, 5*, 1–8. https://doi.org/10.7324/JABB.2017.50601

Kobayashi, R., Takisada, M., Suzuki, T., Kirimura, K., & Usami, S. (1997). Neoagarobiose as a novel moisturizer with whitening effect. *Bioscience, Biotechnology, and Biochemistry, 61*, 62–63.

Kobayashi, T., Koide, O., Mori, K., Shimamura, S., Matsuura, T., Miura, T., Takaki, Y., Morono, Y., Nunoura, T., Imachi, H., Inagaki, F., Takai, K., & Horikoshi, K. (2008). Phylogenetic and enzymatic diversity of deep subseafloor aerobic microorganisms in organics- and methane-rich sediments off Shimokita Peninsula. *Extremophiles, 12*, 519–527.

Kraus, J. M. (2019). Contaminants in linked aquatic–terrestrial ecosystems: Predicting effects of aquatic pollution on adult aquatic insects and terrestrial insectivores. *Freshwater Science, 38*(4), 919–927.

Kraus, J. M., Schmidt, T. S., Walters, D. M., Wanty, R. B., Zuellig, R. E., & Wolf, R. E. (2014). Cross-ecosystem impacts of stream pollution reduce resource and contaminant flux to riparian food webs. *Ecological Applications, 24*, 235–243.

Kraus, J. M., Gibson, P. P., Walters, D. M., & Mills, M. A. (2017). Riparian spiders as sentinels of polychlorinated biphenyl contamination across heterogeneous aquatic ecosystems. *Environmental Toxicology and Chemistry, 36*, 1278–1286.

Krueger, M. C., Harms, H., & Schlosser, D. (2015). Prospects for microbiological solutions to environmental pollution with plastics. *Applied Microbiology and Biotechnology, 99*(21), 8857–8874.

Kubicek, C. P. (1993). From cellulose to cellulase inducers: Facts and fiction. In P. Suominen & T. Reinikainen (Eds.), *Proceedings of the 2nd Symposium Trichoderma reesei Cellulases and*

Other Hydrolases (TRICEL '93) (Vol. 8, pp. 181–188). Foundation for Biotechnical and Industrial Fermentation Research.

Kuhad, R. C., Gupta, R., & Khasa, Y. P. (2010). *Bioethanol production from lignocellulosic biomass: An overview. Wealth from waste.*

Kumarevel, T., Nakano, N., & Ponnuraj, K. (2008a). Crystal structure of glutamine receptor protein from *Sulfolobustokodaii* strain 7 in complex with its effector l-glutamine: Implications of effector binding in molecular association and DNA binding. *Nucleic Acids Research, 36*(14), 4808–4820.

Kumarevel, T., Sakamoto, K., Gopinath, S. C. B., Shinkai, A., Kumar, P. K. R., & Yokoyama, S. (2008b). Crystal structure of an archaeal specific DNA-binding protein (Ape10b2) from *Aeropyrumpernix* K1. *Proteins, 71*(3), 1156–1162.

Kumarevel, T., Tanaka, T., & Nishio, M. (2008c). Crystal structure of the MarR family regulatory protein, ST1710, from *Sulfolobus tokodaii* strain 7. *Journal of Structural Biology, 161*(1), 9–17.

Lee, S. M., & Koo, Y. M. (2001). Pilot-scale production of cellulase using Trichoderma Reesei rut C-30 fed-batch mode. *Journal of Microbiology and Biotechnology, 11*(2), 229–233.

Leung, M. (2004). Bioremediation: Techniques for cleaning up a mess. *BioTeach Journal, 2*, 18–22.

Li, D. L., Pi, J., Zhang, T., Tan, X., & Fraser, D. J. (2018). Evaluating a 5-year metal contamination remediation and the biomonitoring potential of a freshwater gastropod along the Xiangjiang River, China. *Environmental Science and Pollution Research, 25*, 21127–21137.

Liu, J. Z., Wu, Y. H., Wu, C. X., Muylaert, K., Vyverman, W., Yu, H. Q., Muñoz, R., & Rittmann, B. (2017). Advanced nutrient removal from surface water by a consortium of attached microalgae and bacteria: A review. *Bioresource Technology, 241*, 1127–1137.

Lokhande, S., & Dixit, S. (2017). Natural and effective ways of purifying lake water. *Annals of Civil and Environmental Engineering, 1*, 49–54. https://doi.org/10.29328/journal.acee.1001006

Ma, C. P., Lu, X. Z., Shi, C., Li, J. B., Gu, Y. C., Ma, Y. M., Chu, Y., & Han, F. (2007). Molecular cloning and characterization of a novel β-Agarase, AgaB, from marine *Pseudoalteromonas* sp. CY24. *The Journal of Biological Chemistry, 282*, 3747–3754.

Milala, M. A., Shugaba, A., Gidado, A., Ene, A. C., & Wafar, J. A. (2005). Studies on the use of agricultural wastes for cellulase enzyme production by *Aspergillusniger*. *Research Journal of Agriculture and Biological Sciences, 1*, 325–328.

Mojsov, K. (2012). Microbial α-amylases and their industrial applications: A review. *International Journal of Management, IT and Engineering, 2*, 583–609.

Morrissey, C. A., Mineau, P., Devries, J. H., Sanchez-Bayo, F., Liess, M., Cavallaro, M. C., & Liber, K. (2015). Neonicotinoid contamination of global surface waters and associated risk to aquatic invertebrates: A review. *Environment International, 74*, 291–303.

Moy, N. J., Dodson, J., Tassone, S. J., Bukaveckas, P. A., & Bulluck, L. P. (2016). Biotransport of algal toxins to riparian food webs. *Environmental Science and Technology, 50*, 10007–10014.

Nakamura, K., Tomita, T., Abe, N., & Kamio, Y. (2001). Purification and characterization of an extra cellular poly(L-lactic acid) depolymerase from a soil isolate, Amycoatopsis sp. Strain K104-1. *Applied and Environmental Microbiology, 67*, 345–353.

Nguyen, T. H., & Nguyen, V. D. (2017). Characterization and applications of marine microbial enzymes in biotechnology and probiotics for animal health. In *Advances in Food and Nutrition Research* (Vol. 80, pp. 37–74). Academic Press.

Ohta, Y., Hatada, Y., Nogi, Y., Miyazaki, M., Li, Z., Akita, M., Hidaka, Y., & Goda, S. (2004). Enzymatic properties and nucleotide and amino acid sequences of a thermostable β-agarase from a novel species of deep-sea Microbulbifer. *Applied Microbiology and Biotechnology, 64*, 505–514.

Owa, F. D. (2013). Water pollution: Sources, effects, control and management. *Mediterranean Journal of Social Sciences, 4*(8), 65–65.

Owa, F. W. (2014). Water pollution: Sources, effects, control and management. *International Letters of Natural Sciences, 3*.

Percival Zhang, Y. H., Himmel, M. E., & Mielenz, J. R. (2006). Outlook for cellulase improvement: Screening and selection strategies. *Biotechnology Advances, 24*(5), 452–481.

Pere, J., Puolakka, A., Nousiainen, P., & Buchert, J. (2001). Action of purified *Trichodermareesei* cellulases on cotton fibers and yarn. *Journal of Biotechnology, 89*(2–3), 247–255.

Prasad, M. P., & Manjunath, K. (2011). Comparative study on biodegradation of lipid-rich wastewater using lipase producing bacterial species. *Indian Journal of Biotechnology, 10*(1), 121–124.

Puente, M. E., Bashan, Y., Li, C. Y., & Lebsky, V. K. (2004). Microbial populations and activities in the rhizoplane of rock-weathering desert plants. I. Root colonization and weathering of igneous rocks. *Plant Biology, 6*(05), 629–642.

Rao, M. B., Tanksale, A. M., Ghatge, M. S., & Deshpande, V. V. (1998). Molecular and biotechnological aspects of microbial proteases. *Microbiology and Molecular Biology Reviews, 62*(3), 597–635.

Rao, M. A., Scelza, R., Scotti, R., & Gianfreda, L. (2010). Role of enzymes in the remediation of polluted environments. *Journal of Soil Science and Plant Nutrition, 10*(3), 333–353.

Rathore, A. S., & Gupta, R. D. (2015). Chitinases from bacteria to human: Properties, applications, and future perspectives. *Enzyme Research, 2015*, 91907. https://doi.org/10.1155/2015/791907

Reddy, P. B., & Rawat, S. S. (2013). Assessment of aquatic pollution using histopathology in fish as a protocol. *International Research Journal of Environment Sciences, 2*(8), 79–82.

Renicke, C., & Taxis, C. (2016). Biophotography: Concepts, applications and perspectives. *Applied Microbiology and Biotechnology, 100*(8), 3415–3420.

Richmond, E. K., Rosi-Marshall, E. J., Walters, D. M., Fick, J., Brodin, T., Sundelin, A., & Grace, M. R. (2018). A diverse suite of pharmaceuticals contaminates stream and riparian food webs. *Nature Communications, 9*, 1–9.

Riffaldi, R., Levi-Minzi, R., Cardelli, R., Palumbo, S., & Saviozzi, A. (2006). Soil biological activities in monitoring the bioremediation of diesel oil-contaminated soil. *Water, Air, and Soil Pollution, 170*(1–4), 3–15.

Rodriguez, R., & Redman, R. (2008). More than 400 million years of evolution and some plants still can't make it on their own: Plant stress tolerance via fungal symbiosis. *Journal of Experimental Botany, 59*(5), 1109–1114.

Rogers, G. R. (2000). Water quality management at Santa Cruz Harbor. *Aire-O2 News, 7*, 4–5.

Ruohonen, L., Koivula, A., Reinikainen, T., Valkeajärvi, A., Teleman, A., Claeyssens, M., & Teeri, T. (1993). Active site of T. reesei cellobiohydrolase II. In *2nd Tricel symposium on Trichoderma reesei cellulases and other hydrolases (TRICEL93)* (pp. 87–96).

Ryu, S. K., Cho, S. J., Park, S. R., Lim, W. J., Kim, M. K., Hong, S. Y., Bae, D. W., & Park, Y. W. (2001). Cloning of the cel9A gene and characterization of its gene product from marine bacterium. *Pseudomonas* sp. SK38. *Applied Microbiology and Biotechnology, 57*, 138–145.

Saima, K., Roohi, M., & Ahmad, I. Z. (2013). Isolation of novel chitinolytic bacteria and production optimization of extracellular chitinase. *Journal, Genetic Engineering & Biotechnology, 11*, 39–46. https://doi.org/10.1016/j.jgeb.2013.03.001

Salonen, S. M. (1990). *Method for manufacturing paper or cardboard and product containing cellulase*. US patent 4980023.

Sato, K., Sakui, H., Sakai, Y., & Tanaka, S. (2008). Long-term experimental study of the aquatic plant system for polluted river water. *Water Science and Technology, 46*, 217–224.

Schroeder, D. C., Jaffer, M. A., & Coyne, V. E. (2003). Investigation of the role of a beta(1–4) agarase produced by *Pseudoalteromonasgracilis* B9 in eliciting disease symptoms in the red alga *Gracilariagracilis*. *Microbiology, 149*, 2919–2929.

Sharma, D., Sharma, B., & Shukla, A. K. (2011). Biotechnological approach of microbial lipase: A review. *Biotechnology, 10*(1), 23–40.

Singh, C. J. (2003). Optimization of an extracellular protease of Chrysosporium keratinophilum and its potential in bioremediation of keratinic wastes. *Mycopathologia, 156*(3), 151–156.

Singh, H. P., Batish, D. R., Kaur, S., Arora, K., & Kohli, R. K. (2006). α-Pinene inhibits growth and induces oxidative stress in roots. *Annals of Botany, 98*(6), 1261–1269.

Singh, R. J., Chung, G. H., & Nelson, R. L. (2007). Landmark research in legumes. *Genome, 50*(6), 525–537.

de Souza, M. P. (2010). de Oliveira e and Magalhães. Application of microbial-amylase in industry-a review. *Brazilian Journal of Microbiology, 41*, 850–861.

Stanish, I., & Monbouquette, H. G. (2001). Engineering A23187/EDTA-containing metal-sorbing vesicles for selective extraction of divalent heavy metal ions. *Journal of Membrane Science, 192*(1–2), 99–113.

Sukumaran, R. K., Singhania, R. R., & Pandey, A. (2005). Microbial cellulases—Production, applications and challenges. *Journal of Scientific and Industrial Research, 64*(11), 832–844.

Sundarram, A., & Murthy, T. P. K. (2014). α-amylase production and applications: A review. *Journal of Applied & Environmental Microbiology, 2*(4), 166–175.

Tejada, M., Gonzalez, J. L., García-Martínez, A. M., & Parrado, J. (2008). Application of a green manure and green manure composted with beet vinasse on soil restoration: Effects on soil properties. *Bioresource Technology, 99*(11), 4949–4957.

Thakur, N., Nath, K., Chauhan, A., Parmar, S. C., & Pandey, H. (2019). Chitinases from microbial sources, their role as biocontrol agents and other potential applications. *Journal of Entomology and Zoology Studies, 7*, 837–843.

Thompson, A. R., Doelling, J. H., Suttangkakul, A., & Vierstra, R. D. (2005). Autophagic nutrient recycling in Arabidopsis directed by the ATG8 and ATG12 conjugation pathways. *Plant physiology, 138*(4), 2097–2110.

Tweedy, B. N., Drenner, R. W., Chumchal, M. M., & Kennedy, J. H. (2013). Effects of fish on emergent insect-mediated flux of methyl mercury across a gradient of contamination. *Environmental Science & Technology, 47*, 1614–1619.

Van Wyk, J. P. H. (1999). Saccharification of paper products by cellulase from Penicillium funiculosum and Trichoderma reesei. *Biomass and Bioenergy, 16*(3), 239–242.

Vega, K., & Kalkum, M. (2011). Chitin, Chitinase responses, and invasive fungal infections. *International Journal of Microbiology, 2011*, 1–10. https://doi.org/10.1155/2012/920459

Veliz, E. A., Hidalgo, P. M., & Hirsch, A. M. (2017). Chitinase-producing bacteria and their role in biocontrol. *AIMS Microbiology, 3*, 689–705. https://doi.org/10.3934/microbiol.2017.3.689

Vidali, M. (2001). Bioremediation. an overview. *Pure and Applied Chemistry, 73*(7), 1163–1172.

Walters, D. M., Fritz, K. M., & Otter, R. R. (2008). The dark side of subsidies: Adult stream insects export organic contaminants to riparian predators. *Ecological Applications, 18*, 1835–1841.

Wang, C. X., Ling, H., & Shi, K. H. (1999). Oxygen restoration of polluted water with pure oxygen aeration. *Shanghai Academy of Environmental Sciences, 18*, 411–413. (In Chinese). Available online: https://en.cnki.com.cn/Article_en/CJFDTotal-SHHJ199909012.htm. Accessed on 26 Aug 2020.

Wang, J. X., Jiang, X. L., Mou, H. J., & Guan, H. S. (2004). Anti-oxidation of agar oligosaccharides produced by agarase from a marine bacterium. *Journal of Applied Phycology, 16*, 333–340.

Wang, J. X., Mou, H. J., Jiang, X. L., & Guan, H. S. (2006). Characterization of a novel β-agarase from marine *Alteromonas* sp. SY37–12 and its degrading products. *Applied Microbiology and Biotechnology, 71*, 833–839.

Xiao, L. J., Ouyang, H., Li, H. M., Chen, M. R., Lin, Q. Q., & Han, B. P. (2010). Enclosure study on phytoplankton response to stocking of silver carp (Hypophthalmichthys molitrix) in a eutrophic tropical reservoir in South China. *International Review of Hydrobiology, 95*, 428–439.

Yuriy, M. S., Atanas, I. P., & Albert, I. K. (2015). Chitinase biotechnology: Production, purification, and application. *Engineering in Life Sciences, 15*, 30–38.

Zhang, W. W., & Sun, L. (2007). Cloning, characterization, and molecular application of a beta-agarase gene from Vibrio sp. strain V134. *Applied and Environmental Microbiology, 73*(9), 2825–2831.

Zhang, Y., Gao, B., Lu, L., Yue, Q., Wang, Q., & Jia, Y. (2010). Treatment of produced water from polymer flooding in oil production by the combined method of hydrolysis acidification-dynamic membrane bioreactor–coagulation process. *Journal of Petroleum Science and Engineering, 74*(1–2), 14–19.

Zhao, G., Huang, X., Tang, Z., Huang, Q., Niu, F., & Wang, X. (2018). Polymer-based nanocomposites for heavy metal ions removal from aqueous solution: A review. *Polymer Chemistry, 9*(26), 3562–3582.

Zhao, Y., Yang, Z. F., Xia, X. H., & Wang, F. A. (2012). shallow lake remediation regime with Phragmites australis: Incorporating nutrient removal and water evapotranspiration. *Water Research, 46*, 5635–5644.

Zheng, L. G., & Wang, H. P. (2017). Effect of combined ecological floating bed for eutrophic lake remediation. *AIP Conference Proceedings, 1839*, 020058. Retrieved May 8, 2017, from https://aip.scitation.org/doi/pdf/10.1063/1.4982423

Zimmels, Y., Kirzhner, F., & Malkovskaja, A. (2008). Application and features of cascade aquatic plants system for sewage treatment. *Ecological Engineering, 34*, 147–161.

In Silico Analysis of Biochemical Pathways in Bacterial Enzyme Synthesis

Blaise Manga Enuh, Belma Nural Yaman, Pınar Aytar Çelik, and Ahmet Çabuk

Abstract Comprehension of bacteriology and its physiology and subsequently its application to better the lives of humans have ever been at the core of live science. The development of bacterial strains with optimized desirable functions has been the interest of microbiologists for many years. This has fuelled research interests to understand bacterial physiology up to the molecular level in order to obtain an in-depth knowledge of various biochemical processes within the cells. Over the years, a large amount of data on bacterial genomics, proteomics, and metabolomics have been generated. This, alongside the development of high through-put next-generation sequencing technologies and an increase in computational power have paved the way for the development of robust in silico analysis techniques employing sophisticated mathematical model-based algorithms, which perform multi-omics level modelling of complex bacterial systems. Many of these models can be one of

B. M. Enuh
Department of Biotechnology and Biosafety, Graduate School of Natural and Applied Sciences, Eskişehir Osmangazi University, Eskişehir, Turkey

B. Nural Yaman
Department of Biotechnology and Biosafety, Graduate School of Natural and Applied Sciences, Eskişehir Osmangazi University, Eskişehir, Turkey

Department of Biomedical Engineering, Faculty of Engineering and Architecture, Eskişehir Osmangazi University, Eskişehir, Turkey

P. Aytar Çelik (✉)
Department of Biotechnology and Biosafety, Graduate School of Natural and Applied Sciences, Eskişehir Osmangazi University, Eskişehir, Turkey

Environmental Protection and Control Program, Eskişehir Osmangazi University, Eskişehir, Turkey
e-mail: paytar@ogu.edu.tr

A. Çabuk
Department of Biotechnology and Biosafety, Graduate School of Natural and Applied Sciences, Eskişehir Osmangazi University, Eskişehir, Turkey

Department of Biology, Faculty of Science and Letter, Eskişehir Osmangazi University, Eskişehir, Turkey

N. R. Maddela et al. (eds.), *Ecological Interplays in Microbial Enzymology*, Environmental and Microbial Biotechnology,
https://doi.org/10.1007/978-981-19-0155-3_18

chemical kinetics-based, constraint-based, or network-based models. They offer the possibility to reconstruct biochemical pathways, predict gene editing targets or the design of completely new biochemical pathways. Furthermore, over the years, the accuracy and efficiency of bacterial engineering have improved significantly as well as the reduction in cost and average time for developing beneficial microbial strains and metabolites.

This chapter will explore the developments and applications of various in silico analysis techniques that are being used for biochemical pathway analysis as applicable to soil bacterial enzyme synthesis, important applications, the challenges that are currently faced and also future prospects of these highly promising methods in synthetic biology.

1 Introduction

The desire to engineer microbial genes has been long around for about three decades beginning with random gene mutation methodologies improving over the years to the use of sophisticated computers to predict mutation targets.

Metabolic engineering is a discipline that deals with the modification of the metabolism machinery to achieve the expression of a desired phenotype such that it accomplishes the synthesis of a product of human interest. It's been about three decades since metabolic engineering has been around (Woolston et al., 2013). The early beginnings rooted within genetic engineering and the idea that genes of interest could be expressed in bacteria led to increase research to produce at industrial scale products of interest to humans (Cohen et al., 1973). The engineering of biological systems involves the design, construction and optimization of the natural routes of synthesis of a product in the natural producers or completely introduce new routes to achieve the same aim. Despite the successes of introducing single genes to engineer products like insulin in *E. coli*, other products such as ethanol required engineering more than one gene and hence better comprehension of entire pathways (Woolston et al., 2013). Over the years, therefore, metabolic engineers have sought not just to engineer single genes but to engineer entire metabolic pathways to produce more quantities of desired product.

Developments have quite advanced, but the cost effectiveness of the procedures is still research concerns and important milestones to achieve. Despite the short time required to develop metabolic pathways for engineering, which is usually within months, it requires much more time and effort to make significant improvements that can endorse commercial application. Far beyond genetic engineering alone, there was a need for an interdisciplinary approach integrating developments from other seemingly unrelated disciplines (Stephanopoulos, 2012). One such discipline was mathematics, which can develop formalisms for understanding how biological systems behave. These formalisms can provide a framework through which we can integrate both available knowledge and experimental data (Saa & Nielsen, 2017). In metabolism, enzymes control the metabolic flux of biochemical pathways.

The metabolic flux is the rate at which molecules are converted in a hypothetical metabolic pathway. Looking at a reaction, it is possible to derive a rate law that can enable the representation of the reaction mathematically. Kinetic models employ this methodology to model reaction dynamics. Fluxes are traditionally measured with the aid of stable isotope tracers such as ^{13}C for tracing hydrocarbon metabolism which can be time consuming and resource demanding (Chan et al., 2013). The determination of these fluxes is a key aspect in metabolic engineering and there are special conditions to be met to generate reliable flux estimates which are to measure tracer enrichments accurately as well as assessment of acceptable confidence intervals. In order to know which enzymes, the major controllers of rates of reactions in a pathway are, it is important to view these fluxes as differences from the base states (Stephanopoulos, 2012). There was need to develop easier methods to determine fluxes that support maximum growth (Chan et al., 2013).

Tools used to analyse metabolic fluxes including mathematical models have been improved over the years likewise interests towards understanding and redirecting metabolic fluxes to produce more products of industrial and medical value (Yang et al., 1998). Larger models layering omics data and flux data have been developed and are being improved every day to enhance their effectiveness. Genome-scale models which generate various maximum flux possibilities directed towards maximum growth have attracted interests. They analyse fluxes and produce possible profiles that optimize product yield. Today, metabolic engineering has evolved significantly going beyond just pathway design. Though there are still limitations such as the inadequacy of kinetic data for enzymes and regulatory models to allow for a more globalized pathway design (Stephanopoulos, 2012). The colossal amount of data generated over the years on biological systems has enabled progress in the development of these models now requiring more computational power for analysis (Kim et al., 2015). There is no longer a great need for intuitive methods as in the past to understand the metabolism of microbes. We now have plenty of data which can enable us to visualize and better understand these systems. Exploiting this data on computers to produce rational strategies for optimal phenotype expression has led to the development of in silico metabolic engineering, which can therefore be defined as the modelling, optimization, and simulation of biological systems to obtain computationally the required information in order to design the most effective intervention strategies on our research problems (Chan et al., 2013).

2 Mathematical Models in Metabolic Engineering

In the past, genetic engineering by gene mutation had been used with the aim of improving strain quality. These methods produced some important industrial successes such as the achievement of 2000-fold increase in penicillin production. Despite these successes, limitations such as the introduction of random mutations in the organisms were undesirable. Mathematical expressions and algorithms that rapidly analyse information in combination with genetic engineering tools have been developed to obtain better accuracy and effective predictions. Computational techniques employing these algorithms are now an important component of the Design-Build-Test-Learn cycle in pathway engineering (St. John & Bomble, 2019). More in silico methods have been developed which describe behaviours of microorganisms at dynamic and steady states (Badri et al., 2017) with each method producing very useful data and predictions. Within the field of metabolic engineering, a broad classification can be made of these methods in to; kinetics-based models describing dynamic states, constraint-based analysis models assuming steady states, and methods based on network analysis. Methods to integrate all these in a single model are also currently being researched fervently. We will examine the formulation of some kinetic models in metabolic engineering and their applications in studying various soil enzyme biochemical reactions.

2.1 Kinetic Models

Brief History

Biological systems are dynamic systems and the science of modelling their behaviour with mathematical formulae dates back in the early 1900s (Badri et al., 2017). The bedrock of kinetic modelling is the fields of molecular biology and enzymology and has advanced speedily from enzyme studies to studies involving large metabolic pathways. This has allowed for the development of concepts used today such as the biochemical system theory and the metabolic control analysis which are being applied for the development of kinetic models (Saa & Nielsen, 2017).

The Structure Kinetic Models in Metabolic Engineering

Kinetic models describing metabolism are made from non-linear rate laws with mathematical expressions having one or more parameters (Kerkhoven et al., 2015; Saa & Nielsen, 2017). They employ differential equations to represent processes such as cell growth, enzyme kinetics, substrate consumption and the synthesis of metabolites. Mathematical formalisms like the canonical laws (e.g. mass action laws) mechanistic rate laws (e.g. elementary reaction mass law) and approximate kinetics such as the Michaelis–Menten have been used for modelling of enzyme-catalysed reactions (Saa & Nielsen, 2017). They require the definition of parameters, collection of experimental data, testing and appraisal of the mathematical expression then according to estimates of the parameters choose that which is the closest to the actual live biological system. It is important to know factors that affect or control these processes in order to predict various reaction output in different conditions (Badri et al., 2017). When enzymes are saturated during conditions of very high substrate concentration, the rate of reaction is maximum; it can be said that the velocity of catalysis is directly proportional to enzyme concentration. It is assumed that this velocity can represent the amount of actively present enzymes. In an experimental situation, a substrate concentration corresponding to 100-fold of the K_m value is required to achieve about a 99% saturation of enzyme-binding sites. Assuming that the system is linear, it is possible to apply linear regression to know the enzyme activity or the amount of active enzymes in the reaction at the early in the progress curve. In stopped assays, however, there is more need to check if these assumptions hold (Boeckx et al., 2017).

In order to analyse kinetic models, we can modify metabolite concentration or change the rate of reaction then simulate the effect on the system through a response analysis approach, or in systems involving a large number of parameters, to check the robustness of the model output with large variations in input through sensitivity analysis and in pathways where we need to know the effect of each step on the pathway using metabolic control analysis. By analysing the coefficients representing component properties of the system while considering that enzymes in that system are in a steady state and single connected unit, we can evaluate the correlation between the system's properties and its components. Flux control coefficients and concentrations control coefficients are commonly used for this purpose. Thus, an enzyme with the highest flux control coefficients can be said to be the one that has the highest control over the pathway flux. Post-MCA methods such as universal methods and metabolic design analysis and lin-log approach can also be used to strengthen the output of kinetic models (Badri et al., 2017). It is also possible to apply kinetic models in the optimization of soil enzyme assays through a systematic analysis of metabolic reaction parameters (Boeckx et al., 2017). A detailed review of kinetic model, their formulation, construction and analysis can be found elsewhere (Saa & Nielsen, 2017).

The soil is a very dynamic environment involving a host of microbial reactions affecting human lives and the planet as a whole in many different ways. Kinetic models have helped us to have a deep understanding of these processes and how human activities on the soil also affect the environment. Kinetic modelling has been used to assess the effects of Azadirachtin, a plant-derived pest control chemical on the activity of soil urease; it was shown that soil urease exhibited typical Michaelis–Menten kinetic behaviour, i.e. high doses caused increased Km and maximum reaction rates (Vm) were reduced. It showed that pesticide doses were effective in influencing the behaviour of urease enzymes. Urease enzyme catalyses the hydrolysis of urea to ammonia by releasing nitrogen from soil colloids. Using this, they were able to conclude that soil health and quality can be monitored using the generated kinetic parameters (Kizilkaya et al., 2015). Using the same model but integrating different enzymes such as amylase, invertase, protease, urease and dehydrogenase to assess the metabolic response of soil, the kinetic parameters of these enzymes were determined and it was shown that the parameters could be used as early sensitive indicators of changes in the properties of soil as a result of human activities (Kujur & Kumar Patel, 2014). In another study, adsorbed laccase and free laccase were used to compare their biodegradation abilities, using the lin-log modelling and analysing thermal deactivation kinetics, it was inferred that immobilized laccase improved soil ecology and can be used for remediation of contaminated soil

(Wang et al., 2019). Given the limitations of the Michaelis–Menten equation in determining reaction parameters through the progress curve assay, which works mostly only when there is a huge excess in substrates over enzymes, and guarantee to identify parameters, Bayesian approach with total quasi-steady state approximations was proposed with which an experiment with better parameter estimation could be designed without any initial information. Using a minimal amount of time course data, enzymes with different catalytic efficiencies such as chymotrypsin, fumarase and urease can be estimated in a short time (Choi et al., 2017) works employing this approach are quite limited.

Extracellular Enzymes

The structures of the earth's environments are so diverse, with heterogeneous soil structures which support growth of microorganisms in different ways. Soil structures affect distribution of organic matter therefore microbial population dynamics will vary with respect to these environments. Small functional niches within community members can aid in the survival of microbial communities by providing carbon sources effectively and maintaining soil health. Microorganisms use extracellular enzymes to acquire nutrients from complex chemicals in their immediate environment also playing an important role in nutrient cycling.

The activity of extracellular enzymes can be modelled using partial differential equations (Traving et al., 2015). A kinetic model representing how microbes grow on soil carbon sources was developed representing substrate and enzyme dynamics. Cell physiology and microbial diversity was simulated considering optimal allocation of resources towards the synthesis of enzymes. A difference in adaptation dynamics was observed in different bacteria having membrane bound hydrolases and those that release hydrolases. But it was observed that their co-existence was mutually beneficial because the relative differences in enzyme build up times fulfilled needs for cellulose degradation as carbon source during basal and exponential growth (Resat et al., 2012). On the other hand, environments with land pollution of synthetic organic compounds could find kinetic models useful in bioremediation. Understanding how these compounds are mineralized could help devise better strategies for interventions. Kinetic models describing co-metabolisms were developed using *Pseudomonas acidovorans* and *Salmonella typhimurium* on phenol, glucose and arabinose as substrates. Compounds that don't support growth, usually human introduced can be found at low concentrations in the soil. They can be metabolized by bacteria using them as nutrients or by those that have an excess of their required nutrients. Using 12 kinetic models, the effect of second substrates on the kinetics of mineralization organic compounds in low concentration was examined. The dynamics of mineralization was best represented by first-order kinetic models or models describing the growth kinetics of bacteria growing on other readily available substrates (Schmidt et al., 1985). This model was thus recommended for the bioremediation by bacteria of compounds with very low and even high concentrations in the soil.

On a grand scale, kinetic models of soil enzyme metabolisms have seen applications helpful to the Earth System Model (ESM) which is still being developed since its start about 100 years ago. The ESM seeks to numerically simulate the earth's atmosphere, oceans, soil in order to produce predictions of the future nature of the earth's systems (Randall et al., 2018). Within this context, biogeochemistry models play an important role in land modelling. Nutrient cycling and biotransformation carried out mostly by microorganisms is at the core of biogeochemistry. Soil enzymes can help us infer the degree at which land has been used. In a study, the Michaelis–Menten model kinetic parameters for amylase, invertase, protease, urease and dehydrogenase were used to evaluate microbial responses in forest soils and fresh mine soils to gradual accumulations of organic matter. Forest soil enzymes were found to have smaller K_m values as compared to fresh mine soil enzymes. This implies that enzymes in the forest soils had a higher substrate affinity and less used compared to those in the fresh mine. This can help with further information about the land management system of an environment and provide early indication of the changes in the properties and health of soil (Kujur & Kumar Patel, 2014). Nitrogen cycling is also an important process in the biogeochemical cycle as well as in global warming implying that their incorporation into earth models are also important. N_2O oxides accumulated and emitted from soils by pre- and de novo synthesized enzymes were modelled based on a two-substrate usage and Monod kinetics. The denitrifications metabolic steps were incorporated with enzyme synthesis and activation to account for regulation. The model parameters were successful in accurately predicting the dynamics of N_2O. The enzyme activity function provided a value to differentiate the activities of both enzymes used (Zheng & Doskey, 2015). Warming being an issue of future concern, it is as well important to understand how future temperatures could affect soil enzymes conditions. The effects of different warming rates on the enzyme kinetics of carbon, phosphorus and nitrogen in subtropical soils have been examined. It was shown within the laboratory that enzyme kinetics could be affected by contrasting warming rates. An application of this in actual field conditions can provide actual enzyme kinetic information helpful to biogeochemical models (Sihi et al., 2019). Important to soil as well are affinity parameters to measure how substrate bind to enzymes in soil and its application to soil respiration have also been evaluated (Tang & Riley, 2019). Other hydrolytic enzymes important in the global carbon cycle such as cellobiohydrolase, β-glucosidase, β-xylosidase, α-glucosidase and *N*-acetyl-β-D-glucosaminidase degrading soil organic matter have shown temperature sensitivity. Looking at temperature sensitivity of enzymes from higher and lower latitudes, a Michaelis–Menten model showed that V_{max} and K_m were temperature sensitive at Q10 1.53–2.27 and 0.9–1.57, respectively. On an overall scale, taking into account the temperature sensitivities in soil organic matter decomposition, extracellular hydrolytic enzymes were shown to adapt locally to temperatures (German et al., 2012). To obtain better comprehension of the influence on soil carbon mineralization by organic matter in wetlands at freshwater tide, field manipulation of organic matter and the corresponding changes were observed in the production of carbon dioxide and methane. The community structure of microbes as well as the activity of C, N and P acquisition enzymes were considered. Carbon dioxide production rates did not correlate with microbial community structure. To explain this, it was suggested that microorganisms that produce carbon dioxide were functionally redundant. On the other hand, methane production was correlated with community structure and the kinetic parameters (V_{max} and K_m) for C (3-1,4-glucosidase, 1,4-P-cellobiosidase and β-D-xylosidase) and N acquisition (leucyl aminopeptidase). Given that methanogenic characteristics are generally attributed to the archaea, it could be concluded that community structure was linked to methane production (Morrissey et al., 2014). Another issue to consider within climatic zones is the activation energy of enzymes. This can also help in quantification of enzyme kinetics within climatic zones. Areas such as the artic, tropics or even at the subsurface of the soil are very diverse given the kind of reactions going on, activation energies could be very determinant factor for the kind of reactions that take place within these areas. One of lignocellulose degrading enzymes, peroxidase showed low activation energies when in the artic and subarctic zones (Steinweg et al., 2013). Kinetic models of extracellular enzymes are providing vital information that could help us improve the environment and preserve biodiversity.

Immobilized Enzymes

Soil enzymes are often bound to microbial cell walls or different soil matrixes from which they carryout numerous biochemical functions that affect the ecosystem in different ways.

The use of immobilized enzymes has increased in various technological areas requiring in-depth comprehension of enzyme kinetic properties for the effective design of systems where each enzyme can be efficiently exploited. Immobilized enzyme systems can also be applied in the modelling of enzymes that carry out their functions while bound in living organisms. This therefore enables a broad application of the system (Engasser & Horvath, 1976). Immobilized enzymes are used in many industrial processes in order to allow for reuse and better stability during processes. Though the activity of the enzyme is not the same as in situations when they are free in solution, kinetic models can still enable the measurement of the effect immobilization has on the activity of the enzymes. The effects of iron, aluminium, magnesium oxides and clay minerals on the adsorption and dynamic characteristics of an acid phosphatase were evaluated. The experiment showed that as the ability of the inorganic compounds to adsorb the enzyme increases, the V_{max}/K_m ratio decreases. It is suggested that the catalytic efficiency was determined by the ability of the enzyme to adsorb inorganic components (Shindo et al., 2002). Using these methodologies, we can understand the kinetics of extracellular enzymes especially those bound on the cell walls.

Limitations of Kinetic Approaches in the Modelling of Enzyme Activity

Kinetic methods are certainly very powerful in predicting the outcomes of biological systems and are even considered the best by some to achieve such scientific aim (Saa & Nielsen, 2017). Most especially for soil environments, there are many hurdles during modelling bacteria dynamics such as the absence of a 3D structure that mimic soil distribution of soluble, particulate and gaseous substrates, interactions of molecules and microbes with surfaces of particles, nutrient transport and cellular activities, and microbial interspecies interactions (Resat et al., 2012).

Kinetic models provide valuable information on the maximum enzyme activity only along the area of the curve that is linear. This area corresponds to periods of enzyme reactions when the substrate saturates the enzymes active site. There is a nonlinear final phase of the reaction because enzymes are not always saturated as products are being produced from substrates (Boeckx et al., 2017).

The requirement for accurate detail and precision in the development of kinetic models makes it quite tedious. There is a need for high-quality experimental data when developing kinetic models. During the fitting of the model using conventional methods, there is also a difficulty to capture uncertainty especially when the model is a large-scale model (Saa & Nielsen, 2017).

In cases of highly accurate parameter values, it can still be observed that kinetic models cover only a little part of entire reaction pathways. This implies that other important components of the cell that can affect pathways such as cofactor availability are not considered (Chen et al., 2014). There are efforts to create genome-scale kinetic models. But the need for large amounts of reaction parameters make their generation very complicated (Kerkhoven et al., 2015).

2.2 *Constraint-Based Analysis Methods*

The difficulties encountered in the design of effective kinetic models and the growing amount of data on genome annotations and other omics promoted the development of stoichiometry-based models as an alternative approach for in silico modelling of biological systems. Using this method, it is possible to make complicated and fast computations on complex metabolic networks with the assumption that they are in a steady state (Martins Conde et al., 2016). In theory, an organism, in order to assure its survival in an environment governed by certain constraints, adjusts its metabolism to better adapt to new conditions imposed as a result of those constraints. This method though doesn't depend on the reaction dynamics within the metabolism; it still employs reaction stoichiometry, physicochemical and biological constraints to generate accurate predictions. They often are a representation of all known cellular biochemical reactions built from genome annotation and detailed biochemical reactions in a bottom-up fashion (building the model layers beginning from simple metabolic reactions up to the annotated genome). The name

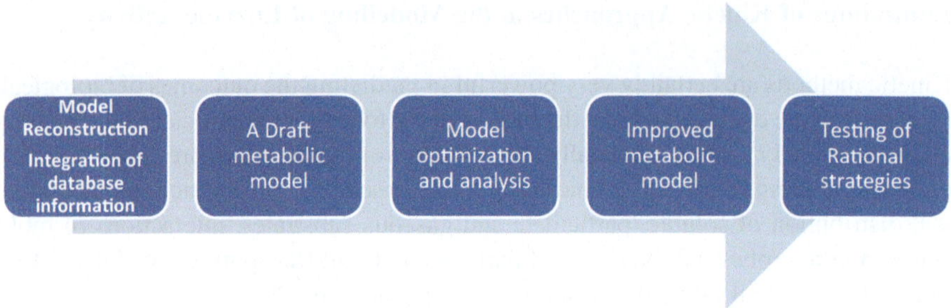

Fig. 1 Steps for building constraint-based models

genome-scale metabolic network reconstruction is used to refer to such in silico systems. Various algorithms designed to analyse these types of networks can produce a range of possible pathway outcomes that can bring about the phenotype we desire to express (Badri et al., 2017). Genome-scale metabolic model reconstructions begin with compiling all related stoichiometric reactions and annotated genomes, after setting linear mass balance equations, the gaps within the network is filled with information from the literature and KEGG database. Strain improvements simulations are carried out and validated against experimental data. These models are now very important in strain development providing virtual microbes to facilitate this process. Genome-scale models keep increasing with the increase in information generated from high-throughput technologies. Given its requirement for highly technical knowledge, several automated and semi-automated tools such as AutoKeGGRec, AuReMe, CarveMe, MetaDraft, Raven, ModelSEED (Mendoza et al., 2019) have been developed to increase their user friendliness. Algorithms used to analyse these models are flux balance analysis, the minimization of metabolic adjustments, regulatory on and off minimizations, flux scanning based on enforced object and bi-level optimization methods (Badri et al., 2017). Constraint-based models integrating kinetic data called dynamic flux balance models are in development with the aim of increasing their predictive ability. Figure 1 summarizes steps for building constraint-based models.

In order to measure carbon use efficiency in soil, Saifuddin and co-workers analysed 13 manually curated genome-scale constraint-based metabolic models using flux balance analysis to generate theoretical predictions for carbon use efficiency for over 200 microbial taxa, they found that it is possible to have over >300% variations in carbon use efficiency within taxa as a result of intrinsic physiological differences. According to their research, their findings provide a framework for the analysis of the impact of shifts in bacteria communities on the cycling of carbon; in addition, they used a recent model representing heterotrophic respiration in the soil, and it showed that variations in microbial physiology within taxa have enormous effects on the estimates of carbon emissions in the soil. With large genome-scale models, when identifying engineering targets that will produce the optimal production, there is a need to generate many pathway possibilities as possible. To help with this issue, bi-level approaches can be used to more efficiently identify the mutations

Table 1 Summary of constraint-based methodologies

Method	Objectives
FBA	Predictions of the distribution of fluxes considering steady states.
MOMA	Predictions of the distribution of metabolic fluxes in primitive mutants
ROOM	Predictions of how metabolic networks behave after gene knockouts at steady states
Dynamic FBA	Extension of FBA, allows the model to update and change over time, more dynamic
Dynamic ROOM	Extension of ROOM but more flexible
OptGene	Prediction of knockout targets using FBA
OptKnock	Prediction of knockout targets with bi-level optimization framework
BiMOMA	Improvement of MOMA using a bi-level optimization framework
SimOptStrain	Upgraded OptStrain.
OptStrain	Prediction based on deletions and addition of foreign reactions.

that will generate optimal production. Bi-level approaches are used by OptKnock which establishes possible deletions alongside biomass and production (Chan et al., 2013). Other approaches OptStrain, SimOptStrain, OptStrain and MOMA consist of many steps that establish foreign reactions that can be used for overproduction. A summary of these methods and their objectives is shown on the Table 1 below, which was adapted from Chan et al. (2013).

Omics Databases for Soil Bacteria

The importance of previously generated data has already been emphasized in the context of designing in silico approaches to tackle metabolic engineering questions. Databases are available in enormous amounts on the internet. They could cover broad or specific fields or even particular organisms as well as they could be designed just for specific purposes. In this section, there will be an examination of the main databases that are being employed when carrying out in silico analysis. They are structured into genomics, proteomics, metabolomics and repositories. An increase in high-throughput methods, development of meta-omics methods and decrease in costs is expected to grow the data even more in the coming years.

Genomics Databases

Genomics datasets enable us to structure and store information from genomes of organisms providing a platform to access and use this information when need arises. Some are specific having information from specific class of organisms, while others are non-specific and collect information on a wide class of organisms. The multitude of genomics databases present in world have their data in large part from the health and industrial biotechnology sectors, as a result, the soil microbial communities are

Table 2 DNA sequence and genome annotation databases

Database		Objectives	
DDBJ.	http://www.ddbj.nig.ac.jp/	Existing since 1987, contains annotated nucleotide sequences from researchers	Mashima et al. (2017)
EMBL	http://www.ebi.ac.uk/embl/	Annotated nucleotide sequences and protein	Baker et al. (2000)
GenBank	http://www.ncbi.nlm.nih.gov/Genbank/	DNA sequences from public sources	Benson et al. (1998)
Integr8	http://www.ebi.ac.uk/integr8/	Genome and proteome information	Kersey et al. (2005)
CMR	http://cmr.jcvi.org/	DNA sequence for bacteria and archaea	Davidsen et al. (2010)
IMG	http://img.jgi.doe.gov/	DNA sequence information including plasmids and viruses	Markowitz et al. (2012)
SEED	http://seed-viewer.theseed.org/	Integrated and annotated genomic data	Overbeek et al. (2014)

quite poorly represented. Due to this, problems of genomic annotation can arise because the information is not compatible with microbial life in soil. There are a limited number of databases therefore to enable the effective study of soil microbiology. In order to solve this, projects such as the Earth microbiome have been developed. The aim of the project was to provide data comprising about 200,000 samples based on microbial diversity and functional potential (Choi et al., 2016; Gilbert et al., 2014). More recently, a curated RefSoil database for soil bacteria representative cultures has been developed as a subsidiary of the NCBI database. This is expected to provide a framework on which soil investigations will be able to obtain accurate annotations, better understand sequencing information and fill a significant gap in soil microbiology. RefSoil is said to contain 922 genomes of 888 and 34 different bacteria and archaea, respectively (Choi et al., 2016). An extension of RefSoil known as RefSoil+ has also been reported as a reference database for keeping information on genes as well as soil plasmids (Dunivin et al., 2019). Some resources (Table 2) to get genomic data for soil bacteria in silico experiments are given below.

Proteomics Databases

A lot of databases exist for retrieval of information for proteomics studies. The importance of proteins in the functions of organisms and as important molecules that reveal key information on metabolism and structure of microorganisms cannot be over emphasized. As a result, developing databases to store this information is as well important. Improvements in next-generation sequencing technology have permitted the sequencing of a myriad of genes allowing simultaneous generation of proteomic data. This has enabled scientists to make numerous discoveries and uncovered many cellular functions previously unknown (Chen et al., 2017). A table of prominent proteomics databases is also provided below (Table 3).

Table 3 Protein and enzyme databases modified from Durot et al. (2009)

Database		Objectives	
TransportDB	http://www.membranetransport.org/	Database for membrane transport proteins and membrane channels from sequenced genomes	Ren et al. (2007)
BRENDA	http://www.brenda-enzymes.info/	Information on reactions, structure, classification and nomenclature of enzymes.	Schomburg et al. (2004)
ENZYME	http://www.expasy.ch/enzyme/	Nomenclature of enzymes	Bairoch (2000)
UniProt	http://www.ebi.ac.uk/uniprot/	Information on protein sequence and functional annotation.	The Universal Protein Resource (UniProt) (2008)
PSORTdb	http://db.psort.org/	Bacterial subcellular localization and computational analyses of proteins	Rey et al. (2005)
Prolinks.	http://prolinks.mbi.ucla.edu/	Provides inference for protein function and linkages.	Bowers et al. (2004)
STRING	http://string.embl.de/	Contains known and predicted protein-protein associations	Szklarczyk et al. (2017)

Metabolomics Databases

Metabolomics databases comprise information regarding metabolic reactions, parameters for enzymes stoichiometries, metabolic pathways etc. Most of the information within these databases is obtained from researchers directly. Some metabolic reaction databases are given in Table 4.

Data Repositories

Experimental Data Repositories

Because *E. coli* industrially established and used for production of enzymes from other soil bacteria, we also include some resources (Table 5) that could be beneficial for this purpose.

Genome-Scale Model Repositories

Genome-scale models for many organisms have been developed and are being updated continuously with the availability of more data. Developed genome-scale models are uploaded and stored in repositories from which templates can be obtained when needed. A table of sources of genome-scale metabolic models is shown below (Table 6).

Table 4 Metabolic reaction databases

Database		Objectives	
CheBI	http://www.ebi.ac.uk/chebi/	Dictionary for small chemical compounds of biological interest.	Degtyarenko et al. (2008)
Pubchem	http://pubchem.ncbi.nlm.nih.gov/	Chemical molecules and their biological activities	Kim et al. (2016b)
LipidMaps	http://www.lipidmaps.org/	Structures and annotated biological lipids.	Sud et al. (2007)
KEGG	http://www.genome.jp/kegg/	Database for genes and their functions	Kanehisa and Goto (2000)
UM-BBD	http://umbbd.msi.umn.edu/	Catabolism related enzymes, pathways important in chemical manufacture and biodegradation.	Burgard and Maranas (2002)
UniPathway	http://www.grenoble.prabi.fr/	Database for metabolic pathways and enzyme catalysed and spontaneous reactions.	Morgat et al. (2012)
	obiwarehouse/unipathway/		
BioCyc	http://www.grenoble.prabi.fr/	Enzymes and metabolic pathways for small molecules obtained experimentally	Caspi et al. (2016)
	obiwarehouse/unipathway/		

Table 5 Sources of experimental data

Database		Objectives	
IntAct	http://www.ebi.ac.uk/intact/	Data for protein–protein interactions from curation and direct deposition.	Kerrien et al. (2012)
DIP	http://dip.doe-mbi.ucla.edu/	The protein–protein interaction networks in living organisms.	Xenarios et al. (2000)
Array Express	http://www.ebi.ac.uk/aerep/	Database of microarray data and annotated gene expression profiles	Parkinson et al. (2007)
GEO	http://www.ncbi.nlm.nih.gov/geo/	Contains both functional genomics and high-throughput transcriptomics data	Clough and Barrett (2016)
ASAP	http://asap.ahabs.wisc.edu/	Functionally characterized genome sequence information	Glasner et al. (2003)
Systomonas	http://www.systomonas.de/	Omics data and regulatory networks of Pseudomonads.	Choi et al. (2007)

Table 6 Sources of genome-scale metabolic models

Database		Objectives	
BiGG	http://bigg.ucsd.edu/	Database for genome-scale metabolic models.	Schellenberger et al. (2010)
BioModels	http://www.ebi.ac.uk/biomodels/	Database for annotated quantitative kinetic models.	Li et al. (2010)

Information Management and Curation

Extracting information from databases is an important task, information within databases is not always harmonized and therefore the algorithms used must be sensitive enough to pick up this required information and probe expansive databases. Platforms such as MATLAB are used for analysis and simulations within which other tools can be integrated to perform additional analysis.

Visualization, Analysis and Simulations Systems

Systems Biology Markup Language is generally used to represent metabolic models. It is used for storage and exchange of information on models used for biological systems. Software analysing metabolic networks can import or export data using this file format (Jing et al., 2014). Some software for analysis of constraint-based models is given in Table 7.

Elementary Flux Modes

For metabolic engineering to be efficient, there is a need to develop very powerful theoretical methods in which an in-depth information on metabolic network topologies are considered. The metabolism of organisms during steady states consist of elementary flux modes (EFM), which is the least amount of enzymes required to obtain valid steady states (Schuster, 1999). However, in the presence to environmental stimuli, flux distributions can vary in response to stimuli. This is observed during microbial growth on single substrates. There is a repression of genes that code for enzymes that function in other alternative pathways. When more than one carbon substrate is provided, the substrate that is most efficiently utilized is used up initially because of catabolite repression. EFM can be considered as the minimum reactions that take place in a steady state considering that all the irreversible reactions are going on towards the correct direction. It is very useful when there is a metabolic competition to synthesize the metabolite of interest (Badri et al., 2017). An EFM considers pseudo steady state, feasibility and nondecomposability. It is mainly limited by the complexity to identify computationally EFMs in a metabolic network. Simply, an EFM is the simplest metabolic pathway at steady state that can bring about the conversion of a substrate to a product of interest. Using EFMs, the structure and function of a metabolic network can be inferred making it suitable for a wide range of areas in biotechnology (Beurton-Aimar et al., 2011).

Table 7 Various software for analysis of constraint-based models

Name	FBA and related	Integrated data analysis	FBA-based strain design	Analysis of EMs	Analysis of © MCSs	License	Dependencies
CellNetAnalyzer	Yes	Yes	No	Yes	Yes	Free academic	Matlab
COBRA	Yes	Yes	Yes	No	No	GNU GPLv3	Matlab
EFMtool	No	No	No	Yes	No	BSD	none
FASIMU	Yes	Yes	Yes	No	No	GNU GPL	none
Metatool	No	No	No	Yes	No	Free academic	none
MicrobesFlux	Yes	No	No	No	No	Free	none
OptFlux	Yes	Yes	Yes	Yes	No	GNU GPLv3	none
YANA	No	Yes	No	Yes	No	Free academic	none

Carbon 13 Metabolic Flux Analysis

Flux analysis has been used for several years mainly to understand the metabolism of living organisms. By labelling substrates with ^{13}C and following their metabolism at steady states, it provided a very useful and direct method for flux quantification. It made it possible to have a direct insight into microbial metabolism (He et al., 2016). The labelled substrate is redistributed in metabolic intermediates within a metabolic network and measured with either NMR (determination of carbon enrichment and positional isotopomers) or mass spectrometry (for the determination of mass isotopomers). Over the years, it has shown to be very effective for modelling of metabolic pathways of microbial cells (Beurton-Aimar et al., 2011). Better flux methodologies have been developed and tested in *Pseudomonas putida* and *P. aeruginosa*. Discovery of new glucose metabolism pathways was reported due to the good resolution of these methods (Kohlstedt & Wittmann, 2019). To improve computational user friendliness simpler interfaces are being developed to enable easy use (He et al., 2016). These methods are generally used to investigate the carbon cycles within the soil. Carbon flux patterns in microbial communities have been measured using labelled pyruvate (Dijkstra et al., 2011). Modelling of intra-species exchanges has also been described with labelled amino acids and peptides providing information on diverse metabolic pathways within microbial communities (Ghosh et al., 2014). Difficulties of obtaining high-quality isotopomers (isotopic isomers differing by the presence of an isotope in one of the molecules) and maintaining cell cultures in steady states within culture media (He et al., 2016).

Applications of Constraint-Based Methods

Carbon Use Efficiency and Cycling

Metabolic processes especially respiration by soil microbes is amongst the largest fluxes of carbon from the soil. It is amongst the largest carbon exchange vehicles from the soil to the atmosphere which is about 98 ± 12 Pg C/year in the form of carbon dioxide. Respiration processes in the soil are affected by microbial physiology and metabolism. Microbial metabolism is the means through which microorganisms allocate various fractions of absorbed carbon to respiration, cell growth, enzyme synthesis etc. (Sinsabaugh et al., 2016). Understanding carbon metabolism is critical especially in the context of climate change. Furthermore, it could be of interest to know if carbon use efficiency can be predicted at the genomic level. Substrates within the soil are made up of different types of with variation in their supply rates. Measuring carbon use efficiency across different microbial species could help elucidate how various species impact carbon use within an environment. Species of organisms with large genomes can potentially metabolize a wider variety of carbon sources (Saifuddin et al., 2019).

Constraint-based methods can quantitatively describe the functional organizational structure of microbial communities in soil environment which can help to improve the description of carbon use efficiency in metabolism and in ecosystem in silico models (Sinsabaugh et al., 2016).

Nitrogen Use and Cycling

Nitrogen is an important component of fertilizers which is delivered through ammonium salts. Nitrogen oxides are potent global warming gases that are released from the soil through nitrifications and aerobic ammonia oxidations. Genome-scale models modelling the dynamics of nitrification by co-culturing of *Nitrosomonas europaea* and *Nitrobacter winogradskyi* integrating biotic and abiotic components showed that during microaerobic conditions, biotic and abiotic factors all play a major role as sources and sinks of nitrogen oxides (Mellbye et al., 2018). Given that some nitrite oxidizing bacteria are metabolically versatile, it is quite important to understand the various mechanisms involved that can enable the improvement of nitrogen fertilizer formulations (Koch et al., 2015).

Remediation and Detoxification of Soils

Information on microbial communities within contaminated soils can provide information on possible bio-stimulation strategies. Atrazine is a potent herbicide which disrupts microbial community structures in soil. The metabolism of *Arthrobacter* sp. and four other degraders were modelled and demonstrated that atrazine degradation was enhanced by a consortium of direct degraders and non-degraders (Xu et al., 2019). With this, an effective bio-stimulation consortium could be developed as a natural solution to soil quality degradation as a result of atrazine.

Identifying Optimum Production Genes and Pathways

Soil microorganisms such as *Streptomyces* have developed efficient metabolic strategies to survive within the soil given the scarcity of resources. Scarcity of resources is an important constraint factor that causes organisms to develop various strategies to adapt within their ecosystems. In economic terms, investigations of these metabolisms can lead to discovery of potential human interest metabolite candidates (Fondi et al., 2017). Their huge ability to diversify secondary metabolites has made them very desirable in the industrial production. Genome-scale models are continuously being developed to integrate various data to improve the process of industrial strain development. *Streptomyces coelicolor* A3(2) is reported to be the best studied strain in this genus with over 20 secondary metabolite encoding genes mainly antibiotics used in humans and animals (Bentley et al., 2002). Others including *Streptomyces tsukubaensis* (Huang et al., 2013), *Streptomyces hygroscopicus* var. *ascomyceticus* for ascomycin production by engineering ethylmalony-CoA pathway were studied (Wang et al., 2017). Models now increasing the ease for integration of multi-omics data have been developed enabling better flux predictions (Amara et al., 2018; Kim et al., 2016a; Toro et al., 2018). *Pseudomonas putida* has been modelled to produce chemicals (Stephan et al., 2006), biomaterials (Liu & Chen, 2007). The synthesis of vitamin C by *Bacillus megaterium* WSH002 was studied (Zou et al., 2013). In silico research in soil, bacteria are very promising for the industrial production of substances relevant to other areas than the soil.

Community Analysis

Constraint-based methods have also gone multicellular, modelling from two cells to ecosystems is certainly complex but it is providing a lot of understanding and functioning of microbial interactions such as symbiosis, mutualism etc. Given the scarcity of resources with the soil, such associations are very common to ensure competition and survival within ecosystems. In silico, ecosystems have been developed simulating processes of coproduction. *Ketogulonicigenium vulgare* and *Bacillus megaterium* interaction was modelled and showed potential to identify potential features that can enable improvement of vitamin C production (Ye et al., 2014). Soil nitrification has also been modelled for *Nitrosomonas europaea* and *Nitrobacter winogradskyi* giving a comprehensive framework for development of ammonium salt fertilizer formulations (Mellbye et al., 2018). Communities of degraders and non-degraders of atrazine have been modelled and shown promising results for the improvement of bio-stimulation approaches (Xu et al., 2019).

Community modelling is new and quite a time-consuming process. Ecosystems are made of many different species, and various data are lacking for some species resorting sometimes to the need for culturing and experimentation. Automatic model generation has been proposed to solve this shortcoming (Bosi et al., 2017). Combined models usually come from different sources and software which requires harmonization, definition of community functions mathematically, accounting for growth requirements supplied by other members of the community (Gottstein et al., 2016). Approaches to help fill the gaps such as to predict species composition and changes in abundance within ecosystems were proposed (Chan et al., 2017). Ensuring that communities stay stable could be particularly desirable in an industrial fermentation if we intend to maintain stable conditions for the expression of particular phenotypes.

Limitations of Constraint-Based Approaches

Although GEMs are applied in various metabolic engineering workflows, there are still various limitations that are associated with the use of this tool. Firstly, these models assume steady states of the biological system. This is often not true for example in industrial processes where conditions such as high temperatures, high concentrations of acids that benefit the process can interfere with these states. Though at the level of the metabolites homeostasis can be achieved at a fast rate, changes at the level of transcription will generally take longer periods and therefore causing an error in the representation of the system.

Constraint-based models are only limited to the information available to build the model. This implies model results will always be as accurate as there is enough information to build it (Badri et al., 2017).

When the assumptions used for building constraint-based models don't hold, there are higher chances for predictions to be false. In cases where metabolites happen to accumulate in the organism, it disrupts the mass balance assumption (Durot et al., 2009).

The applications of the predictions from GEMs are not always straight forward given that regulatory networks are often not taken into account. Metabolic networks are highly regulated at various points in order to improve efficiency of the system. To help with this, it is recommended to combine modelling data with experimental data. During experiments, it is possible to accurately determine areas that are being regulated therefore open avenues for further investigations (Kerkhoven et al., 2015).

2.3 Combined Kinetic and Constraint-Based Modelling Methods

In order to overcome the limitations of both kinetic and constraint-based models, more robust models integrating components from both models are being developed. The parameters from kinetic models can enable us to produce constraints which can reduce the solution space lower than those produced by constraint-based models alone thereby increasing the accuracy. ODEs and stoichiometric matrices representing various metabolic reactions are formulated to represent the system. This is particularly useful in designing metabolic engineering interventions in natural environments where conditions are more dynamic compared to the situation in batch cultures. These conditions affect the growth of microorganisms differently hence the need for different model types to represent growth in these environments (Ang et al., 2018; Kim et al., 2018). First mentions of dynamic FBA were for the representation of diauxic growth in *E. coli* (Mahadevan et al., 2002). Dynamic FBA has been applied very often for analysis of soil microbial communities. In a study, dynamic genome-scale model of the competitive interaction between *Rhodoferax* and *Geobacter* under different environmental conditions in anoxic subsurface environment was developed. The predicted metabolic states showed impact on the efficiency of uranium bioremediation. The results showed promising application in the design of effective bioremediation interventions (Zhuang et al., 2011).

2.4 Methods Based on Graphical Network Analysis

These methods are developed to image graphically the soil environment and characterize microbial communities. Given the complexity of soil environments, the heterogeneity and physical environments of microbial communities, it is important for soil microbiologists to acquire this knowledge to understand these as a system. These methods take advantage of fluorescence in situ hybridization (FISH), FISH-micro autoradiography (MAR), Nano-secondary ion mass spectrometry (Nano-SIMS) and X-ray tomography techniques which enable physical and functional identification of microorganisms in their natural environment. In order to quantify these methods, algorithms modelling microbial heterogeneity and dynamics have been developed. Individual-based and general individual-based approaches are used but reported to be limited in soil environments. A variation of individual-based model called the lattice Boltzmann can model characteristic soil processes. They can be used to model colonies of cells in biological systems (O'Donnell et al., 2007). Information on microbial community diversity, mechanisms of dynamics and inter-species interactions can be obtained using these models. It can be helpful to determine the effects on microbial communities of abiotic factors over time (Kim & Or, 2016), identification of communities of operational taxonomic units within the soil and their relationship with land use (Lupatini et al., 2014). The scalability of these methods is worth taking note of, nationwide studies have been possible and possibility to know and predict bacteria species richness on vast areas of land (Terrat et al., 2017).

3 Comparative Advantages and Disadvantages of Applied Modelling Approaches

In silico metabolic engineering is certainly a state-of-the-art versatile and useful approach. But there are numerous challenges that have limited its use in certain instances. There is a low availability of data on different species of microorganisms, absence of harmonization of computational platforms, few improvements of community models, the need to develop more sophisticated computational systems that can enable full understanding of biological systems, effectively fill knowledge gaps and improve models for more effective applications in synthetic biology (Badri et al., 2017). In Table 8 below, we summarize the advantages and disadvantages of the methods we have discussed so far.

Table 8 Advantages and disadvantages of modelling methods

Modelling method	Advantages	Disadvantages
Kinetic modelling	Suitable for quantitative outcome prediction	Many assumptions cannot be accounted for in natural environment.
	Relatively simple with lower sophistication	They are generally small representing a few reactions
		Need for a lot of experimental data, therefore methods are limited by the amount of experimental data available
Constraint-based modelling	Larger models	Steady-state assumptions are not applicable in nature
	Can produce quantifiable predictions	Does not take into account dynamics
	No need for experimental data	Needs a lot of technical knowledge
		Usually limited to a few organisms due to information gaps, and lack of uniformity
		Does not predict the effects of gene regulation
Hybrid kinetic and constraint-based	More accurate than kinetic and constraint-based methods alone	Highly complex
Graphical network methods	They are large models	Very complex
	No need for experimental data	
	Can measure dynamics over time in the environment	

4 Future Trends and Perspectives

Kinetic models will realize the long-awaited vision of predicting how the genetic material and environmental conditions dynamically form metabolic phenotypes. As such, we envisage a future where kinetic models are not merely symbolic cell metabolism abstractions, but instrumental in understanding the dynamic regulation of metabolic networks. Those mathematical representations will also be the basis for driving genetic engineering discoveries. Furthermore, an increase in the complexity and degree of integration of additional metabolic regulatory layers enabled by recent developments in frameworks modelling is expected. The development of large-scale kinetic models will require some critical aspects to be considered: (1) larger scale model development, (2) capturing and representation of uncertainty and (3) effective model communication (Saa & Nielsen, 2017).

Combining GEMs and kinetic models could help in the elucidation of some metabolic bottlenecks and more understanding of some particular enzymatic steps (Kerkhoven et al., 2015). Continuous increase in information, simplification of modelling frameworks and increase in technical knowledge is expected to make modelling more exploitable in the future (Durot et al., 2009).

Models for more bacteria need to be developed to enable large-scale studies involving complex microbial communities. The progress in that direction will be useful to understand the interaction of different members in a community leading to better functional interpretation of metagenomic data. A huge data repository will be formed through which novel genetic constructions and hypothesis will be generated that will be directed towards the selection of required strains for particular engineering objectives (Durot et al., 2009).

5 Concluding Remarks

In silico metabolic engineering has a lot of merits; many biochemical processes in the soil have been modelled and have increased our understanding of many processes within the soil. Kinetic models are important for analysing dynamic reactions in the soil and providing a framework for studying changes in soil with various environmental or human inputs. Constraint-based GEMs can be considered as 'systems-level' layers of algorithms for analysis of knowledge obtained so far on the biochemistry of organisms encoded in their genome and subsequent comparison of the knowledge with the already known physiology of a species or with experimental data. They form solid bedrock for metabolic engineering and systems biology by bringing closer phenotypes and genotypes opening the way for a variety of possible in silico analysis. In silico metabolic engineering is providing a fast and reliable method to research systems and make predictions, with more data provided in future many more discoveries and breakthroughs are expected.

References

Amara, A., Takano, E., & Breitling, R. (2018). Development and validation of an updated computational model of Streptomyces coelicolor primary and secondary metabolism. *BMC Genomics, 19*, 519.

Ang, K. S., Lakshmanan, M., Lee, N.-R., & Lee, D.-Y. (2018). Metabolic modeling of microbial community interactions for health, environmental and biotechnological applications. *Current Genomics, 19*, 712–722. https://doi.org/10.2174/1389202919666180911144055

Badri, A., Srinivasan, A., & Raman, K. (2017). In silico approaches to metabolic engineering. In *Current developments in biotechnology and bioengineering* (pp. 161–200). Elsevier. https://doi.org/10.1016/B978-0-444-63667-6.00008-0

Bairoch, A. (2000). The ENZYME database in 2000. *Nucleic Acids Research, 28*, 304–305. https://doi.org/10.1093/nar/28.1.304

Baker, W., van den Broek, A., Camon, E., Hingamp, P., Sterk, P., Stoesser, G., & Tuli, M. A. (2000). The EMBL nucleotide sequence database. *Nucleic Acids Research, 28*, 19–23.

Benson, D. A., Boguski, M. S., Lipman, D. J., Ostell, J., & Ouellette, B. F. (1998). GenBank. *Nucleic Acids Research, 26*, 1–7.

Bentley, S. D., Chater, K. F., Cerdeño-Tárraga, A.-M., Challis, G. L., Thomson, N., James, K. D., Harris, D. E., Quail, M. A., Kieser, H., & Harper, D. (2002). Complete genome sequence of the model actinomycete Streptomyces coelicolor A3 (2). *Nature, 417*, 141.

Beurton-Aimar, M., Beauvoit, B., Monier, A., Vallée, F., Dieuaide-Noubhani, M., & Colombié, S. (2011). Comparison between elementary flux modes analysis and 13C-metabolic fluxes measured in bacterial and plant cells. *BMC Systems Biology, 5*, 95. https://doi.org/10.1186/1752-0509-5-95

Boeckx, J., Hertog, M., Geeraerd, A., & Nicolai, B. (2017). Kinetic modelling: An integrated approach to analyze enzyme activity assays. *Plant Methods, 13*, 69. https://doi.org/10.1186/s13007-017-0218-y

Bosi, E., Bacci, G., Mengoni, A., & Fondi, M. (2017). Perspectives and challenges in microbial communities metabolic modeling. *Frontiers in Genetics, 8*, 88. https://doi.org/10.3389/fgene.2017.00088

Bowers, P. M., Pellegrini, M., Thompson, M. J., Fierro, J., Yeates, T. O., & Eisenberg, D. (2004). Prolinks: A database of protein functional linkages derived from coevolution. *Genome Biology, 5*, R35. https://doi.org/10.1186/gb-2004-5-5-r35

Burgard, A. P., & Maranas, C. D. (2002). Review of the biocatalysis/biodegradation database (UM-BBD). *Metabolic Engineering, 4*, 111–113. https://doi.org/10.1006/mben.2001.0222

Caspi, R., Billington, R., Ferrer, L., Foerster, H., Fulcher, C. A., Keseler, I. M., Kothari, A., Krummenacker, M., Latendresse, M., Mueller, L. A., Ong, Q., Paley, S., Subhraveti, P., Weaver, D. S., & Karp, P. D. (2016). The MetaCyc database of metabolic pathways and enzymes and the BioCyc collection of pathway/genome databases. *Nucleic Acids Research, 44*, D471–D480. https://doi.org/10.1093/nar/gkv1164

Chan, W. H., Mohamad, M. S., Deris, S., & Illias, R. M. (2013). A review of computational approaches for in silico metabolic engineering for microbial fuel production. *Current Bioinformatics, 8*, 253–258. https://doi.org/10.2174/1574893611308020013

Chan, S. H. J., Simons, M. N., & Maranas, C. D. (2017). SteadyCom: Predicting microbial abundances while ensuring community stability. *PLoS Computational Biology, 13*, e1005539. https://doi.org/10.1371/journal.pcbi.1005539

Chen, Y., Bao, J., Kim, I. K., Siewers, V., & Nielsen, J. (2014). Coupled incremental precursor and co-factor supply improves 3-hydroxypropionic acid production in Saccharomyces cerevisiae. *Metabolic Engineering, 22*, 104–109. https://doi.org/10.1016/j.ymben.2014.01.005

Chen, C., Huang, H., & Wu, C. H. (2017). Protein bioinformatics databases and resources. *Methods in Molecular Biology, 1558*, 3–39. https://doi.org/10.1007/978-1-4939-6783-4_1

Choi, C., Münch, R., Leupold, S., Klein, J., Siegel, I., Thielen, B., Benkert, B., Kucklick, M., Schobert, M., Barthelmes, J., Ebeling, C., Haddad, I., Scheer, M., Grote, A., Hiller, K., Bunk, B., Schreiber, K., Retter, I., Schomburg, D., & Jahn, D. (2007). SYSTOMONAS — An integrated database for systems biology analysis of Pseudomonas. *Nucleic Acids Research, 35*, D533–D537. https://doi.org/10.1093/nar/gkl823

Choi, J., Yang, F., Stepanauskas, R., Cardenas, E., Garoutte, A., Williams, R., Flater, J., Tiedje, J. M., Hofmockel, K. S., Gelder, B., & Howe, A. (2016). RefSoil: A reference database of soil microbial genomes. *Ecology*. https://doi.org/10.1101/053397

Choi, B., Rempala, G. A., & Kim, J. K. (2017). Beyond the Michaelis-Menten equation: Accurate and efficient estimation of enzyme kinetic parameters. *Scientific Reports, 7*, 1–11. https://doi.org/10.1038/s41598-017-17072-z

Clough, E., & Barrett, T. (2016). The gene expression Omnibus database. *Methods in Molecular Biology, 1418*, 93–110. https://doi.org/10.1007/978-1-4939-3578-9_5

Cohen, S. N., Chang, A. C., Boyer, H. W., & Helling, R. B. (1973). Construction of biologically functional bacterial plasmids in vitro. *Proceedings. National Academy of Sciences. United States of America, 70*, 3240–3244. https://doi.org/10.1073/pnas.70.11.3240

Davidsen, T., Beck, E., Ganapathy, A., Montgomery, R., Zafar, N., Yang, Q., Madupu, R., Goetz, P., Galinsky, K., White, O., & Sutton, G. (2010). The comprehensive microbial resource. *Nucleic Acids Research, 38*, D340–D345. https://doi.org/10.1093/nar/gkp912

Degtyarenko, K., de Matos, P., Ennis, M., Hastings, J., Zbinden, M., McNaught, A., Alcántara, R., Darsow, M., Guedj, M., & Ashburner, M. (2008). ChEBI: A database and ontology for chemical entities of biological interest. *Nucleic Acids Research, 36*, D344–D350. https://doi.org/10.1093/nar/gkm791

Dijkstra, P., Blankinship, J., Selmants, P., Hart, S., Koch, G., Schwartz, E., & Hungate, B. (2011). Probing carbon flux patterns through soil microbial metabolic networks using parallel position-specific tracer labeling. *Soil Biology and Biochemistry, 43*, 126–132. https://doi.org/10.1016/j.soilbio.2010.09.022

Dunivin, T. K., Choi, J., Howe, A., & Shade, A. (2019). RefSoil+: A reference database for genes and traits of soil plasmids. *mSystems, 4*, e00349. https://doi.org/10.1128/mSystems.00349-18

Durot, M., Bourguignon, P.-Y., & Schachter, V. (2009). Genome-scale models of bacterial metabolism: Reconstruction and applications. *FEMS Microbiology Reviews, 33*, 164–190. https://doi.org/10.1111/j.1574-6976.2008.00146.x

Engasser, J.-M., & Horvath, C. (1976). Diffusion and kinetics with immobilized enzymes. In L. B. Wingard, E. Katchalski-Katzir, & L. Goldstein (Eds.), *Applied biochemistry and bioengineering, immobilized enzyme principles* (pp. 127–220). Elsevier. https://doi.org/10.1016/B978-0-12-041101-6.50009-1

Fondi, M., Pinatel, E., Talà, A., Damiano, F., Consolandi, C., Mattorre, B., Fico, D., Testini, M., De Benedetto, G. E., Siculella, L., De Bellis, G., Alifano, P., & Peano, C. (2017). Time-resolved transcriptomics and constraint-based modeling identify system-level metabolic features and overexpression targets to increase spiramycin production in Streptomyces ambofaciens. *Frontiers in Microbiology, 8*, 835. https://doi.org/10.3389/fmicb.2017.00835

German, D. P., Marcelo, K. R. B., Stone, M. M., & Allison, S. D. (2012). The Michaelis–Menten kinetics of soil extracellular enzymes in response to temperature: A cross-latitudinal study. *Global Change Biology, 18*, 1468–1479. https://doi.org/10.1111/j.1365-2486.2011.02615.x

Ghosh, A., Nilmeier, J., Weaver, D., Adams, P. D., Keasling, J. D., Mukhopadhyay, A., Petzold, C. J., & Martín, H. G. (2014). A peptide-based method for 13C metabolic flux analysis in microbial communities. *PLoS Computational Biology, 10*, e1003827. https://doi.org/10.1371/journal.pcbi.1003827

Gilbert, J. A., Jansson, J. K., & Knight, R. (2014). The Earth Microbiome project: Successes and aspirations. *BMC Biology, 12*, 1–4. https://doi.org/10.1186/s12915-014-0069-1

Glasner, J. D., Liss, P., Plunkett, G., Darling, A., Prasad, T., Rusch, M., Byrnes, A., Gilson, M., Biehl, B., Blattner, F. R., & Perna, N. T. (2003). ASAP, a systematic annotation package for community analysis of genomes. *Nucleic Acids Research, 31*, 147–151. https://doi.org/10.1093/nar/gkg125

Gottstein, W., Olivier, B. G., Bruggeman, F. J., & Teusink, B. (2016). Constraint-based stoichiometric modelling from single organisms to microbial communities. *Journal of the Royal Society Interface, 13*, 20160627. https://doi.org/10.1098/rsif.2016.0627

He, L., Wu, S. G., Zhang, M., Chen, Y., & Tang, Y. J. (2016). WUFlux: An open-source platform for 13C metabolic flux analysis of bacterial metabolism. *BMC Bioinformatics, 17*, 444. https://doi.org/10.1186/s12859-016-1314-0

Huang, D., Li, S., Xia, M., Wen, J., & Jia, X. (2013). Genome-scale metabolic network guided engineering of Streptomyces tsukubaensis for FK506 production improvement. *Microbial Cell Factories, 12*, 52. https://doi.org/10.1186/1475-2859-12-52

Jing, L. S., Shah, F. F. M., Mohamad, M. S., Hamran, N. L., Salleh, A. H. M., Deris, S., & Alashwal, H. (2014). Database and tools for metabolic network analysis. *Biotechnology and Bioprocess Engineering, 19*, 568–585. https://doi.org/10.1007/s12257-014-0172-8

Kanehisa, M., & Goto, S. (2000). KEGG: Kyoto encyclopedia of genes and genomes. *Nucleic Acids Research, 28*, 27–30.

Kerkhoven, E. J., Lahtvee, P.-J., & Nielsen, J. (2015). Applications of computational modeling in metabolic engineering of yeast. *FEMS Yeast Research, 15*, 1–13. https://doi.org/10.1111/1567-1364.12199

Kerrien, S., Aranda, B., Breuza, L., Bridge, A., Broackes-Carter, F., Chen, C., Duesbury, M., Dumousseau, M., Feuermann, M., Hinz, U., Jandrasits, C., Jimenez, R. C., Khadake, J., Mahadevan, U., Masson, P., Pedruzzi, I., Pfeiffenberger, E., Porras, P., Raghunath, A., …

Hermjakob, H. (2012). The IntAct molecular interaction database in 2012. *Nucleic Acids Research, 40*, D841–D846. https://doi.org/10.1093/nar/gkr1088

Kersey, P., Bower, L., Morris, L., Horne, A., Petryszak, R., Kanz, C., Kanapin, A., Das, U., Michoud, K., Phan, I., Gattiker, A., Kulikova, T., Faruque, N., Duggan, K., Mclaren, P., Reimholz, B., Duret, L., Penel, S., Reuter, I., & Apweiler, R. (2005). Integr8 and genome reviews: Integrated views of complete genomes and proteomes. *Nucleic Acids Research, 33*, D297–D302. https://doi.org/10.1093/nar/gki039

Kim, M., & Or, D. (2016). Individual-based model of microbial life on hydrated rough soil surfaces. *PLoS One, 11*, e0147394. https://doi.org/10.1371/journal.pone.0147394

Kim, B., Kim, W. J., Kim, D. I., & Lee, S. Y. (2015). Applications of genome-scale metabolic network model in metabolic engineering. *Journal of Industrial Microbiology & Biotechnology, 42*, 339–348. https://doi.org/10.1007/s10295-014-1554-9

Kim, M., Yi, J. S., Lakshmanan, M., Lee, D.-Y., & Kim, B.-G. (2016a). Transcriptomics-based strain optimization tool for designing secondary metabolite overproducing strains of Streptomyces coelicolor. *Biotechnology and Bioengineering, 113*, 651–660. https://doi.org/10.1002/bit.25830

Kim, S., Thiessen, P. A., Bolton, E. E., Chen, J., Fu, G., Gindulyte, A., Han, L., He, J., He, S., Shoemaker, B. A., Wang, J., Yu, B., Zhang, J., & Bryant, S. H. (2016b). PubChem substance and compound databases. *Nucleic Acids Research, 44*, D1202–D1213. https://doi.org/10.1093/nar/gkv951

Kim, O. D., Rocha, M., & Maia, P. (2018). A review of dynamic modeling approaches and their application in computational strain optimization for metabolic engineering. *Frontiers in Microbiology, 9*, 1690. https://doi.org/10.3389/fmicb.2018.01690

Kizilkaya, R., Samofalova, I., Mudrykh, N., Mikailsoy, F., Akça, İ., Sushkova, S., & Minkina, T. (2015). Assessing the impact of azadirachtin application to soil on urease activity and its kinetic parameters. *Turkish Journal of Agriculture and Forestry, 39*, 976.

Koch, H., Lücker, S., Albertsen, M., Kitzinger, K., Herbold, C., Spieck, E., Nielsen, P. H., Wagner, M., & Daims, H. (2015). Expanded metabolic versatility of ubiquitous nitrite-oxidizing bacteria from the genus Nitrospira. *PNAS, 36*, 11371–11375. https://doi.org/10.1073/pnas.1506533112

Kohlstedt, M., & Wittmann, C. (2019). GC-MS-based 13C metabolic flux analysis resolves the parallel and cyclic glucose metabolism of Pseudomonas putida KT2440 and Pseudomonas aeruginosa PAO1. *Metabolic Engineering, 54*, 35–53. https://doi.org/10.1016/j.ymben.2019.01.008

Kujur, M., & Kumar Patel, A. (2014). Kinetics of soil enzyme activities under different ecosystems: An index of soil quality. *Chilean Journal of Agricultural Research, 74*, 96–104. https://doi.org/10.4067/S0718-58392014000100015

Li, C., Donizelli, M., Rodriguez, N., Dharuri, H., Endler, L., Chelliah, V., Li, L., He, E., Henry, A., Stefan, M. I., Snoep, J. L., Hucka, M., Le Novère, N., & Laibe, C. (2010). BioModels database: An enhanced, curated and annotated resource for published quantitative kinetic models. *BMC Systems Biology, 4*, 92. https://doi.org/10.1186/1752-0509-4-92

Liu, W., & Chen, G.-Q. (2007). Production and characterization of medium-chain-length polyhydroxyalkanoate with high 3-hydroxytetradecanoate monomer content by fadB and fadA knockout mutant of Pseudomonas putida KT2442. *Applied Microbiology and Biotechnology, 76*, 1153–1159. https://doi.org/10.1007/s00253-007-1092-8

Lupatini, M., Suleiman, A. K. A., Jacques, R. J. S., Antoniolli, Z. I., de Siqueira Ferreira, A., Kuramae, E. E., & Roesch, L. F. W. (2014). Network topology reveals high connectance levels and few key microbial genera within soils. *Frontiers in Environmental Science, 2*, 10. https://doi.org/10.3389/fenvs.2014.00010

Mahadevan, R., Edwards, J. S., & Doyle, F. J. (2002). Dynamic flux balance analysis of diauxic growth in Escherichia coli. *Biophysical Journal, 83*, 1331–1340. https://doi.org/10.1016/S0006-3495(02)73903-9

Markowitz, V. M., Chen, I.-M. A., Palaniappan, K., Chu, K., Szeto, E., Grechkin, Y., Ratner, A., Jacob, B., Huang, J., Williams, P., Huntemann, M., Anderson, I., Mavromatis, K., Ivanova,

N. N., & Kyrpides, N. C. (2012). IMG: The integrated microbial genomes database and comparative analysis system. *Nucleic Acids Research, 40*, D115–D122. https://doi.org/10.1093/nar/gkr1044

Martins Conde, P. R., Sauter, T., & Pfau, T. (2016). Constraint based modeling going multicellular. *Frontiers in Molecular Biosciences, 3*, 3. https://doi.org/10.3389/fmolb.2016.00003

Mashima, J., Kodama, Y., Fujisawa, T., Katayama, T., Okuda, Y., Kaminuma, E., Ogasawara, O., Okubo, K., Nakamura, Y., & Takagi, T. (2017). DNA data bank of Japan. *Nucleic Acids Research, 45*, D25–D31. https://doi.org/10.1093/nar/gkw1001

Mellbye, B. L., Giguere, A. T., Murthy, G. S., Bottomley, P. J., Sayavedra-Soto, L. A., & Chaplen, F. W. R. (2018). Genome-scale, constraint-based modeling of nitrogen oxide fluxes during coculture of Nitrosomonas europaea and Nitrobacter winogradskyi. *mSystems, 3*, e00170. https://doi.org/10.1128/mSystems.00170-17

Mendoza, S. N., Olivier, B. G., Molenaar, D., & Teusink, B. (2019). A systematic assessment of current genome-scale metabolic reconstruction tools. *Genome Biology, 20*, 158. https://doi.org/10.1186/s13059-019-1769-1

Morgat, A., Coissac, E., Coudert, E., Axelsen, K. B., Keller, G., Bairoch, A., Bridge, A., Bougueleret, L., Xenarios, I., & Viari, A. (2012). UniPathway: A resource for the exploration and annotation of metabolic pathways. *Nucleic Acids Research, 40*, D761–D769. https://doi.org/10.1093/nar/gkr1023

Morrissey, E. M., Berrier, D. J., Neubauer, S. C., & Franklin, R. B. (2014). Using microbial communities and extracellular enzymes to link soil organic matter characterisitics to greenhouse gas production in a tidal freshwater wetland. *Biogeochemistry, 117*, 473–490.

O'Donnell, A. G., Young, I. M., Rushton, S. P., Shirley, M. D., & Crawford, J. W. (2007). Visualization, modelling and prediction in soil microbiology. *Nature Reviews. Microbiology, 5*, 689–699. https://doi.org/10.1038/nrmicro1714

Overbeek, R., Olson, R., Pusch, G. D., Olsen, G. J., Davis, J. J., Disz, T., Edwards, R. A., Gerdes, S., Parrello, B., Shukla, M., Vonstein, V., Wattam, A. R., Xia, F., & Stevens, R. (2014). The SEED and the Rapid Annotation of microbial genomes using Subsystems Technology (RAST). *Nucleic Acids Research, 42*, D206–D214. https://doi.org/10.1093/nar/gkt1226

Parkinson, H., Kapushesky, M., Shojatalab, M., Abeygunawardena, N., Coulson, R., Farne, A., Holloway, E., Kolesnykov, N., Lilja, P., Lukk, M., Mani, R., Rayner, T., Sharma, A., William, E., Sarkans, U., & Brazma, A. (2007). ArrayExpress—A public database of microarray experiments and gene expression profiles. *Nucleic Acids Research, 35*, D747–D750. https://doi.org/10.1093/nar/gkl995

Randall, D. A., Bitz, C. M., Danabasoglu, G., Denning, A. S., Gent, P. R., Gettelman, A., Griffies, S. M., Lynch, P., Morrison, H., Pincus, R., & Thuburn, J. (2018). 100 Years of earth system model development. *Meteorological Monographs, 59*, 12.1–12.66. https://doi.org/10.1175/AMSMONOGRAPHS-D-18-0018.1

Ren, Q., Chen, K., & Paulsen, I. T. (2007). TransportDB: A comprehensive database resource for cytoplasmic membrane transport systems and outer membrane channels. *Nucleic Acids Research, 35*, D274–D279. https://doi.org/10.1093/nar/gkl925

Resat, H., Bailey, V., McCue, L. A., & Konopka, A. (2012). Modeling microbial dynamics in heterogeneous environments: Growth on soil carbon sources. *Microbial Ecology, 63*, 883–897. https://doi.org/10.1007/s00248-011-9965-x

Rey, S., Acab, M., Gardy, J. L., Laird, M. R., de Fays, K., Lambert, C., & Brinkman, F. S. L. (2005). PSORTdb: A protein subcellular localization database for bacteria. *Nucleic Acids Research, 33*, D164. https://doi.org/10.1093/nar/gki027

Saa, P. A., & Nielsen, L. K. (2017). Formulation, construction and analysis of kinetic models of metabolism: A review of modelling frameworks. *Biotechnology Advances, 35*, 981–1003. https://doi.org/10.1016/j.biotechadv.2017.09.005

Saifuddin, M., Bhatnagar, J. M., Segrè, D., & Finzi, A. C. (2019). Microbial carbon use efficiency predicted from genome-scale metabolic models. *Nature Communications, 10*, 1–10. https://doi.org/10.1038/s41467-019-11488-z

Schellenberger, J., Park, J. O., Conrad, T. M., & Palsson, B. Ø. (2010). BiGG: A biochemical genetic and genomic knowledgebase of large scale metabolic reconstructions. *BMC Bioinformatics, 11*, 213. https://doi.org/10.1186/1471-2105-11-213

Schmidt, S. K., Simkins, S., & Alexander, M. (1985). Models for the kinetics of biodegradation of organic compounds not supporting growth. *Applied and Environmental Microbiology, 50*, 323–331.

Schomburg, I., Chang, A., Ebeling, C., Gremse, M., Heldt, C., Huhn, G., & Schomburg, D. (2004). BRENDA, the enzyme database: Updates and major new developments. *Nucleic Acids Research, 32*, D431–D433.

Schuster, S. (1999). Detection of elementary flux modes in biochemical networks: A promising tool for pathway analysis and metabolic engineering. *Trends in Biotechnology, 17*, 53–60. https://doi.org/10.1016/S0167-7799(98)01290-6

Shindo, H., Watanabe, D., Onaga, T., Urakawa, M., Nakahara, O., & Huang, Q. (2002). Adsorption, activity, and kinetics of acid phosphatase as influenced by selected oxides and clay minerals. *Soil Science and Plant Nutrition, 48*, 763–767.

Sihi, D., Inglett, P. W., & Inglett, K. S. (2019). Warming rate drives microbial nutrient demand and enzyme expression during peat decomposition. *Geoderma, 336*, 12–21.

Sinsabaugh, R., Turner, B., Talbot, J., Waring, B., Powers, J., Kuske, C., Moorhead, D., & Follstad Shah, J. (2016). Stoichiometry of microbial carbon use efficiency in soils. *Ecological Monographs, 86*, 172–189. https://doi.org/10.1890/15-2110.1

St. John, P. C., & Bomble, Y. J. (2019). Approaches to computational strain design in the multiomics era. *Frontiers in Microbiology, 10*, 597. https://doi.org/10.3389/fmicb.2019.00597

Steinweg, J. M., Jagadamma, S., Frerichs, J., & Mayes, M. A. (2013). Activation energy of extracellular enzymes in soils from different biomes. *PLoS One, 8*, e59943. https://doi.org/10.1371/journal.pone.0059943

Stephan, S., Heinzle, E., Wenzel, S. C., Krug, D., Müller, R., & Wittmann, C. (2006). Metabolic physiology of Pseudomonas putida for heterologous production of myxochromide. *Process Biochemistry, from Biochemical Engineering to Systems Biology, 41*, 2146–2152. https://doi.org/10.1016/j.procbio.2006.06.022

Stephanopoulos, G. (2012). Synthetic biology and metabolic engineering. *ACS Synthetic Biology, 1*, 514–525. https://doi.org/10.1021/sb300094q

Sud, M., Fahy, E., Cotter, D., Brown, A., Dennis, E. A., Glass, C. K., Merrill, A. H., Murphy, R. C., Raetz, C. R. H., Russell, D. W., & Subramaniam, S. (2007). LMSD: LIPID MAPS structure database. *Nucleic Acids Research, 35*, D527–D532. https://doi.org/10.1093/nar/gkl838

Szklarczyk, D., Morris, J. H., Cook, H., Kuhn, M., Wyder, S., Simonovic, M., Santos, A., Doncheva, N. T., Roth, A., Bork, P., Jensen, L. J., & von Mering, C. (2017). The STRING database in 2017: Quality-controlled protein–protein association networks, made broadly accessible. *Nucleic Acids Research, 45*, D362–D368. https://doi.org/10.1093/nar/gkw937

Tang, J., & Riley, W. J. (2019). A theory of effective microbial substrate affinity parameters in variably saturated soils and an example application to aerobic soil heterotrophic respiration. *Journal of Geophysical Research – Biogeosciences, 124*, 918–940. https://doi.org/10.1029/2018JG004779

Terrat, S., Horrigue, W., Dequietd, S., Saby, N. P. A., Lelièvre, M., Nowak, V., Tripied, J., Régnier, T., Jolivet, C., Arrouays, D., Wincker, P., Cruaud, C., Karimi, B., Bispo, A., Maron, P. A., Prévost-Bouré, N. C., & Ranjard, L. (2017). Mapping and predictive variations of soil bacterial richness across France. *PLoS One, 12*, e0186766. https://doi.org/10.1371/journal.pone.0186766

The Universal Protein Resource (UniProt). (2008). *Nucleic Acids Research, 36*, D190–D195. https://doi.org/10.1093/nar/gkm895

Toro, L., Pinilla, L., Avignone-Rossa, C., & Ríos-Estepa, R. (2018). An enhanced genome-scale metabolic reconstruction of Streptomyces clavuligerus identifies novel strain improvement strategies. *Bioprocess and Biosystems Engineering, 41*, 657–669. https://doi.org/10.1007/s00449-018-1900-9

Traving, S. J., Thygesen, U. H., Riemann, L., & Stedmon, C. A. (2015). A model of extracellular enzymes in free-living microbes: Which strategy pays off? *Applied and Environmental Microbiology, 81*, 7385–7393. https://doi.org/10.1128/AEM.02070-15

Wang, J., Wang, C., Song, K., & Wen, J. (2017). Metabolic network model guided engineering ethylmalonyl-CoA pathway to improve ascomycin production in Streptomyces hygroscopicus var. ascomyceticus. *Microbial Cell Factories, 16*, 169. https://doi.org/10.1186/s12934-017-0787-5

Wang, Z., Ren, D., Zhao, Y., Huang, C., Zhang, S., Zhang, X., Kang, C., Deng, Z., & Guo, H. (2019). Remediation and improvement of 2,4-dichlorophenol contaminated soil by biochar-immobilized laccase. *Environmental Technology, 42*, 1679–1692. https://doi.org/10.1080/09593330.2019.1677782

Woolston, B. M., Edgar, S., & Stephanopoulos, G. (2013). Metabolic engineering: Past and future. *Annual Review of Chemical and Biomolecular Engineering, 4*, 259–288. https://doi.org/10.1146/annurev-chembioeng-061312-103312

Xenarios, I., Rice, D., Salwinski, L., Joubert, M., Marcotte, E., & Eisenberg, D. (2000). DIP: The database of interacting proteins. *Nucleic Acids Research, 28*, 289–291. https://doi.org/10.1093/nar/28.1.289

Xu, X., Zarecki, R., Medina, S., Ofaim, S., Liu, X., Chen, C., Hu, S., Brom, D., Gat, D., Porob, S., Eizenberg, H., Ronen, Z., Jiang, J., & Freilich, S. (2019). Modeling microbial communities from atrazine contaminated soils promotes the development of biostimulation solutions. *The ISME Journal, 13*, 494–508. https://doi.org/10.1038/s41396-018-0288-5

Yang, Y.-T., Bennett, G. N., & San, K.-Y. (1998). Genetic and metabolic engineering. *Electronic Journal of Biotechnology, 1*, 134–141. https://doi.org/10.2225/vol1-issue3-fulltext-3

Ye, C., Zou, W., Xu, N., & Liu, L. (2014). Metabolic model reconstruction and analysis of an artificial microbial ecosystem for vitamin C production. *Journal of Biotechnology, 182–183*, 61–67. https://doi.org/10.1016/j.jbiotec.2014.04.027

Zheng, J., & Doskey, P. V. (2015). Modeling nitrous oxide production and reduction in soil through explicit representation of denitrification enzyme kinetics. *Environmental Science & Technology, 49*, 2132–2139. https://doi.org/10.1021/es504513v

Zhuang, K., Izallalen, M., Mouser, P., Richter, H., Risso, C., Mahadevan, R., & Lovley, D. R. (2011). Genome-scale dynamic modeling of the competition between Rhodoferax and Geobacter in anoxic subsurface environments. *The ISME Journal, 5*, 305–316. https://doi.org/10.1038/ismej.2010.117

Zou, W., Zhou, M., Liu, L., & Chen, J. (2013). Reconstruction and analysis of the industrial strain Bacillus megaterium WSH002 genome-scale in silico metabolic model. *Journal of Biotechnology, 164*, 503–509. https://doi.org/10.1016/j.jbiotec.2013.01.019

Microbial Enzymes for Sustainable Development: Future Guidelines

Sesan Abiodun Aransiola, Munachimso Odenakachi Victor-Ekwebelem, and Naga Raju Maddela

Abstract Pollution of the environment is a significant threat to the health of humans and other living things. Traditional pollutant removal methods are ineffective at reducing pollution levels to acceptable levels. For pollutant remediation, biological methods are preferred due to their greater efficiency and biocompatibility. Bioremediation is the term for these low-cost, environmentally friendly methods of reducing pollution. Enzymes play the most important role in bioremediation methods. PAHs, azo dyes, polymers, organocyanides, lead, chromium, and mercury are among the organic and inorganic pollutants that enzymes can help to eliminate. Various enzymes from various species have been isolated. Recently, various enzymes isolated from various species have been used for pollutant bioremediation. Cytochrome P450s, laccases, hydrolases, dehalogenases, dehydrogenases, proteases, and lipases are some of the most common enzymes involved in bioremediation, and they have shown promise in the degradation of polymers, aromatic hydrocarbons, halogenated compounds, dyes, detergents, agrochemical compounds, and others. Mechanisms like oxidation, reduction, elimination, and ring-opening have aided recent advancements in the use of microbial enzymes for bioremediation.

S. A. Aransiola
Bioresources Development Centre, National Biotechnology Development Agency (NABDA), Ogbomoso, Nigeria

M. O. Victor-Ekwebelem
Department of Microbiology, Alex Ekwueme Federal University Ndufu-Alike (AE-FUNAI), Abakaliki, Nigeria

N. R. Maddela (✉)
Departamento de Ciencias Biológicas, Facultad de Ciencias de la Salud, Universidad Técnica de Manabí, Portoviejo, Ecuador

Instituto de Investigación, Universidad Técnica de Manabí, Portoviejo, Ecuador
e-mail: raju.maddela@utm.edu.ec

N. R. Maddela et al. (eds.), *Ecological Interplays in Microbial Enzymology*, Environmental and Microbial Biotechnology,
https://doi.org/10.1007/978-981-19-0155-3_19

1 Introduction

Bioremediation of pollutants is one of the most recent advances in the environmental application of microbial enzymes. Genetic engineering has recently improved the environmental application of microbial enzymes in bioremediation; the enzymes are being engineered to improve their stability and efficiency for specific conditions or substrates (Festa et al., 2008; Theerachat et al., 2012; Seyyed et al., 2021; Bhandari et al., 2021).

Various physical and chemical methods for cleaning up pollutants have been used over the years, including the use of oxidizing agents, electrochemical treatments, adsorption of pollutants, ion exchange, and membrane filtration (Ufarte et al., 2015). However, because of some drawbacks such as high cost, nonspecificity, the potential for secondary contamination, and the inability to reduce contamination levels to regulatory limits (Malik, 2004; Seyyed et al., 2021), eco-friendly, cost-effective, and biological methods, known as bioremediation, have gained popularity in recent years. This method reduces pollution at the source by accumulating pollutants intracellularly or transforming them enzymatically, posing less of a threat to the environment and human health. Enzymes are the most powerful bioremediation tools because they accelerate all chemical changes in pollutants. This chapter, therefore, reviews the recent advances in the environmental applications, sustainability, and future direction of microbial enzymes.

2 Recent Advances in the Environmental Applications of Microbial Enzymes

The specificity of enzymes is usually broad enough to allow them to act on a variety of molecules with similar structures. Enzymes can be used in bioremediation in two ways: as isolated enzymes that are added to the contaminated area, or as whole cells, such as bacteria, fungi, or algae (Rayu et al., 2012; Eibes et al., 2015). Individual enzymes have higher specificity, are easier to handle and store, have standardizable activity, are more mobile due to their smaller size, are active in the presence of high concentrations of toxic compounds, and are biodegradable, which reduces persistence and recalcitrance (Eibes et al., 2015; Seyyed et al., 2021). Extracellular enzymes and cofactor-independent enzymes benefit greatly from this method (Seyyed et al., 2021). In situ or ex situ enzymatic bioremediation is possible. The free or immobilized enzyme (adsorbed enzymes on mineral supports that minimize the loss of enzymatic activity) is added to the soil in in situ methods with the least environmental disruption. Because no excavation or soil transportation is required, the method is less expensive.

Mono- and dioxygenases, halogenases, peroxidases, phosphotriesterases, hydrolases, transferases, and oxidoreductases from various species of bacteria, fungi, algae, and plants have been used for the bioremediation of pollutants (Seyyed et al., 2021) (Table 1). Their general mechanisms of action were summarized in Fig. 1.

2.1 Some Microbial Enzymes Used in Bioremediation

Peroxidases

Peroxidases that break down lignin (Ligninolytic enzymes) are a type of enzyme that has a wide range of applications in the environment, including bioremediation. It degrades a wide range of recalcitrant compounds due to its high nonspecificity and nonstereoselectivity (Kaur et al., 2016). They use a free-radical–based chain reaction with H_2O_2 and molecular oxygen to degrade chemicals in a pseudo-first–order kinetic (Kaur et al., 2016). Laccase (LAC), lignin peroxidase (LiP), manganese peroxidase (MnP), and versatile peroxidase are the four main enzymes involved in ligninolysis (VP).

Laccase

Laccase is an oxygen oxidoreductase that is used in the environment to oxidize phenolic compounds, PAHs, dyes, and pesticides such as benzenediol. Their substrates go through the following pathways as an oxidase: (1) aromatic ring cleavage, (2) polymerization, and (3) covalent bond degradation between monomers. The reaction is primarily composed of four copper atoms, with oxygen serving as the final electron recipient (Ufarte et al., 2015; Chauhan et al., 2017). Laccase comes in two colors: white and blue. For the degradation of nonphenolic substrates, the blue laccase requires a "mediator." Laccase oxidizes the intermediator and produces oxidized radicals that react with substrates with a high redox potential. Effective mediators include ABTS (2,2′-azino-bis (3-ethylbenzothiazoline-6-sulfonic acid) and N-heterocycles with N-OH, such as violuric acid, N-hydroxybenzotriazole, and N-hydroxy-N-phenylacetamide (Chauhan et al., 2017). Every year, approximately 7×10^4 to 1×10^7 tons of dyes enter the environment (Chauhan et al., 2017). Laccase is a dye-remediation enzyme. In the presence of acetosyringone as a mediator, a *Bacillus licheniformis* LS40-derived laccase can decolorize azo, indigo, and anthraquinone dyes by 80% in 1 h (Seyyed et al., 2021). PAHs are a major environmental concern due to their toxicity, persistence, mutagenicity, and carcinogenicity in nature, and are produced as a result of incomplete combustion of fossil fuels and various industrial wastes (Bhandari et al., 2021). Because of their low solubility and slow degradation rate, PAHs are classified as xenobiotic pollutants. Laccase converts PAHs to quinine, a less toxic form, and CO_2 (Table 1). According to a study, both the purified and crude forms of the recombinant CotA laccase from *E. coli* can decolorize seven structurally different dyes as well as simulated textile effluents STE (Wang & Zhao, 2017). Laccase produced from a recombinant strain of

Table 1 Microbial enzymes involved in bioremediation and their function

Enzymes	Mechanism	Function
Cytochrome P450	Performs electron transfer reactions and catalysis by reduction or oxidation of heme iron. Utilizes pyridine nucleotides as electron donors producing carbon substrates and oxidized products.	Synthesis and metabolism of various molecules and chemicals within cells oxidize steroids, fatty acids, and xenobiotics
	$NAD(P)H + O_2 + R \longrightarrow NAD(P) + RO + H_2O$	
Laccase	Reduction of the O_2 molecule, including the oxidation of one electron with a wide range of aromatic compounds.	Ring cleavage in aromatic compounds and reduce one molecule of oxygen in the water and produce free radicals
Dehalogenase	Mainly occurred through three mechanisms:	Cleaves the carbon-halogen bond and eliminates the halogens
	(1) Hydrolytic mechanism: water molecule serves as a cofactor; halogen substituent is replaced in SN reaction by the hydroxyl group	
	(2) Oxygenlytic mechanism: catalyzed by mono/dioxygenase incorporating one/two atoms of molecular oxygen into the substrate	
	(3) Reductive mechanism: it is related to the carbamide family; in this course, halogen is substituted by hydrogen under aerobic conditions, where organohalides are used as the terminal electron acceptors	
Dehydrogenase	Catalyze the reactions with coenzymes such as $NAD^+/NADP^+$ or flavin such as FAD and FMN as an electron acceptor. It transfers two hydrogen atoms from organic compounds to electron acceptors.	Oxidizing organic compounds and generating energy
Hydrolase	In triglyceride hydrolysis, 1-mol triglyceride (T) reacts with 3 mol of water (W) to produce 1-mol glycerol (G), and 3-molfatty acids (P) peptide bond of protein is broken down by hydrolyzing.	Degradation of fats and proteins
Protease	Catalyze the breakdown of peptide bonds of proteins	Degradation of proteins like keratin, casein, etc., leather dehairing, and wastewater treatment

(continued)

Table 1 (continued)

Enzymes	Mechanism	Function
Lipase	The transfer of a proton between the aspartate, the histidine, and the serine residues of the lipase followed by hydroxyl residue of the serine attacks the carbonyl group of the substrate. In the deacylation step, nucleophile attacks the enzyme regenerating the enzyme and releasing the product.	Catalyzes the hydrolysis of mono-, di-, and triglycerides into fatty acids and glycerol. Also, catalyze the esterification and transesterification reactions.

Source: Bhandari et al. (2021)

Bacillus licheniformis has also been reported to degrade carmine and reactive black completely in 1 h. At pH 9.0, the purified recombinant laccase decolorizes more than 93% of the dyes tested in just 4 h (Lu et al., 2013; Bhandari et al., 2021). The recombinant laccase from *B. vallismortis* strain fmb103 has been used in aquaculture wastewater bioremediation (Sun et al., 2017).

Lignin Peroxidase

Lignin peroxidases (LiPs) are a class of monomeric enzymes with high redox potential that catalyze the degradation of phenolic and nonphenolic compounds by breaking alpha and beta carbon bonds, demethylation, and opening the aromatic ring of dyes (Falade et al., 2017). When H_2O_2 is used as an electron acceptor, the activity of LiP increases. In the presence of high H_2O_2 concentrations, however, LiPs may be damaged (Wang et al., 2018).

Manganese Peroxidase

Manganese peroxidases (MnPs) are glycol proteins that contain heme. MnP, like other ligninolytic peroxidases, oxidizes Mn^{2+} to Mn^{3+} using H_2O_2. Mn^{2+} oxidation rate can be induced by aliphatic organic acids like lactate and oxalate, and Mn^{3+}-acid chelates have a higher redox potential. In the presence of glutathione and unsaturated fatty acids like tween 80, MnP activity rises. Making calcium alginate beads and carbon nanotubes, for example, has been used to immobilize and improve the efficacy of MnP bioremediation (Bilal & Asgher, 2015; Bilal et al., 2016). PAHs and nitroaromatic compounds (Qin et al., 2014) azo dyes and endocrine-disrupting chemicals like bisphenol A and alkylph can all be remedied with MnP. MnP is also capable of oxidizing nonphenolic structures with the help of mediators like lipid and thiyl radicals. *Phanerochaete chrysosporium*, *Trametes versicolor*, Irpex lacteus, *Dichomitus squalens*, and *Ganoderma lucidum* are just a few of the fungi that can produce MnP (Xu et al., 2017; Seyyed et al., 2021).

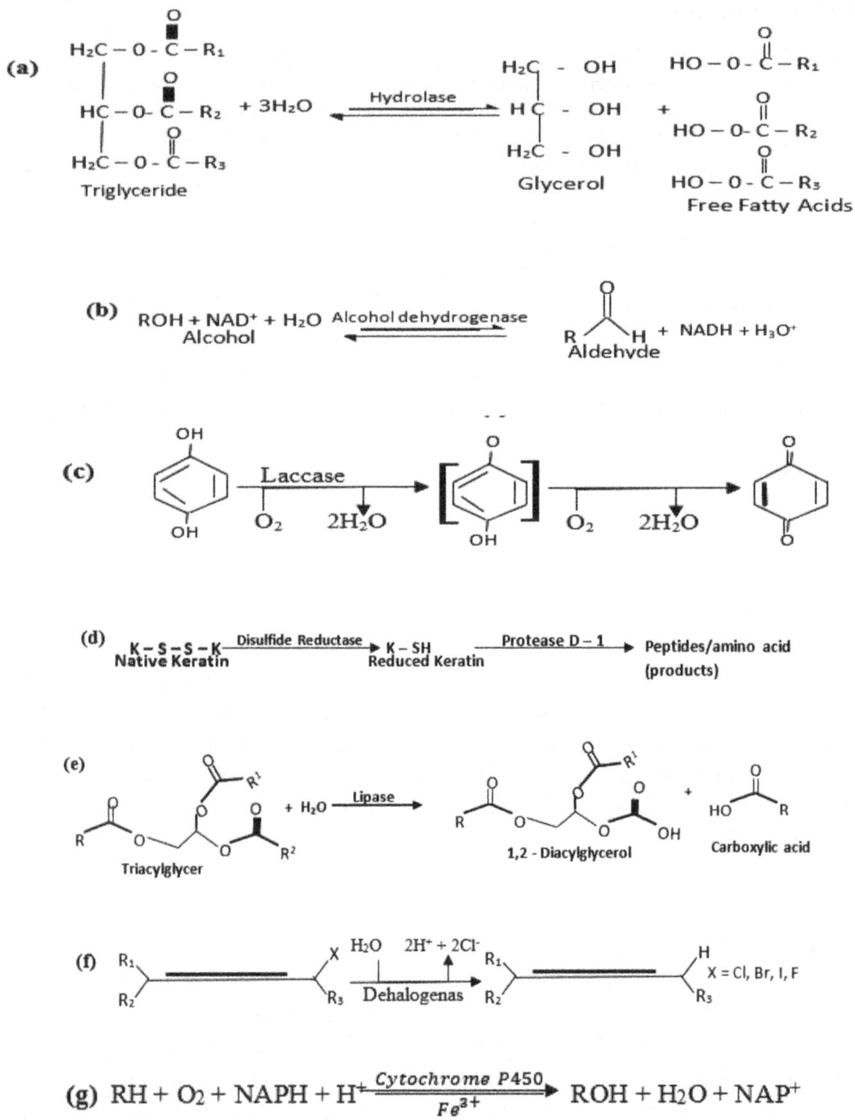

Fig. 1 Some microbial enzymes used in bioremediation; general enzymatic reactions catalyzed by (**a**) hydrolase, (**b**) dehydrogenase, (**c**) laccase, (**d**) protease, (**e**) lipase, (**f**) dehalogenase and (**g**) cytochrome P450 (Regrouped extracts from Bhandari et al. 2021)

Versatile Peroxidase (VP)

Versatile peroxidase is a heme-containing ligninolytic enzyme that functions similarly to LiP and MnP. Because VP has two active sites, it can oxidize both Mn^{2+} and veratryl alcohol via a mechanism similar to MnP and LiP (Wang et al., 2018). VP can oxidize phenolic and nonphenolic compounds, as well as low and high redox potential compounds, polycyclic aromatic hydrocarbons, azo dyes, high molecular weight aromatics, and both phenolic and nonphenolic compounds and pollutants (Wang et al., 2018; Knop et al., 2016). *Pleurotus* spp. and *Bjerkandera* spp. (Wang et al., 2018) are examples of species that contain it.

Cytochrome P450

Cytochrome P450 (EC 1.14.14.1) is a ubiquitous heme enzyme that performs a variety of functions, including the biotransformation of toxic chemicals in our environment (Li et al., 2020). Chemical transformations such as aliphatic hydroxylations and epoxidations, dealkylations, dehalogenation, and various mechanism-based inactivations are central to bioremediation chemistry, and they have the ability to degrade xenobiotics. PAHs are metabolized by CYP101, CYP102, CYP1A1, CYP1A2, and CYP1B1, with CYP1A1 showing the most activity toward dibenzo-*p*-dioxin (DD) and mono-, di-, and trichloro-DDs. P450 from *Bacillus megaterium*, CYP102A1 (P450BM3), demonstrated its ability to oxidize PAHs such as phenanthrene, fluoranthene, and pyrene to a mixture of phenols and quinones, while other microbial P450s showed potential for bioremediation of organic pollutants and hydrocarbons (Bhandari et al., 2021). Kumar et al. (2012) studied engineered CYP102A1, has showed enhanced activity toward PAHs, polychlorinated biphenyls (PCBs), and linear alkanes often used in the bioremediation of toxic compounds, detoxification of gaseous alkanes, and terpenes (Kumar et al., 2012).

P450s can degrade recalcitrant halogenated pollutants, which are resistant to dioxygenases that the mutants F87W, F98W, Y96F, and V247L of heme monooxygenase CYP101A1 (P450cam) from *Pseudomonas putida* showed activity with insoluble pentachlorobenzene, without the need of surfactants or organic solvents. So, the rational re-engineering of wild-type CYP101A1 provides active site mutants with a vastly improved ability to oxidize polychlorinated benzenes into chlorophenol products. Hence, the CYP101A1 mutants could form the basis of novel bioremediation systems for polychlorinated benzenes (Bhandari et al., 2021). Similarly, Chakraborty and Das (2016) have reported several microorganisms such as *Rhodococcus*, *Gordonia*, *Mycobacterium*, and *Pseudomonas* harbor catabolic genes, plasmids, and genomes expressing P450s for the degradation and removal of POPs from the environment.

Awad and Mohamed (2019) found that P450 BM3 (CYP102A1) from *Bacillus megaterium*, engineered from *E. coli* BL21, is useful in the degradation of various organic gases pollutants by immobilizing hollow nanosphere particles of Pt/TiO_2-Cu under solar radiation where the degradation of isopropanol was found high (95%) at a pH of 7.0, ambient temperature, and concentration of 20 mg/L with the continuous supply of electrons via nanoparticles. Bisphenol A (BPA), plasticizer, was degraded by strain YC-JY1 isolated from *Sphingobium* sp, is strain could utilize 4-hydroxybenzaldehyde and 4′-hydroxyacetophenone as a sole carbon source. Strain YC-JY1Δ*bisdB* was constructed in *E. coli* to explore the role of the P450 (Jia et al.,

2020). Also, Kan et al. (2020) isolated *Rhodococcus* sp. P14 from crude oil-contaminated sediments where the regulatory expression of CYP108J1 resulted in PAHs' degradation, which can be used as the sole source of carbon and energy. Further mutational analysis showed that NarL (nitrate-dependent two-component regulatory factor) acts as a novel repressor for the expression of CYP108J1 during PAHs degradation (Kan et al., 2020). The most essential enzymes for the bioremediation of pollutants in the environment were summarized in Table 1.

3 Microbial Enzymes for Environmental Sustainability: Challenges and Future Directions

Microbial enzymes have several advantages toward environmental sustainability. For example, enzyme-mediated biotransformation doesn't produce any toxic by-products, which quite commonly occurred either in the chemical or microbial process of detoxification (Gianfreda et al., 2016). If needed, addition of cosolvents or surfactants is easier in enzyme-mediated transformation compared to reaction systems using whole cells. However, future research is greatly warranted in some areas of enzyme applications. For example, depending on the objective of the enzyme application, there may be a need for a high amount of enzymes with high activity and high stability, such requirements greatly limit the applications of enzymes in different environmental settings. Recombinant-DNA technology may be a suitable technology to overcome these problems. Site-directed mutagenesis, truncation, and terminal fusions are recombinant DNA strategies, which can overcome the problems related to catalytic efficiency, activity, and stability (Yang et al., 2017). Recombinant enzymes have a huge global market; it has been estimated that 50–60% of the world enzyme market is recombinant enzymes (Sanchez & Demain, 2017). Future research should also emphasize on how to use enzyme preparations for multistep processes. In general, mineralization of pollutants is a multistep process, which is not possible when a single enzyme is used. Furthermore, certain enzymes require cofactors for their activity. Therefore, more research is needed to design effective enzyme preparations that can mediate multistep reactions, as well as preparations that carry cofactors. Another biggest problem with the enzyme usage is rapid degradation of cell-free enzyme proteases released by microorganisms present in the soil/water environment. Enzyme deactivation can be prevented by incorporating the enzymes into humic-like substances, adsorption on clay mineral particles, enzyme immobilization in/on synthetic materials. Investigation in these lines not only protects the enzymes from proteolytic inactivation, but also prolongs the life of the enzymes (Nannipieri & Bollag, 1991; Ó'Fágáin, 2003; Mateo et al., 2007). Immobilization of laccase on montmorillonite (Ahn et al., 2002), Mn-peroxidase on nonoclays (Acevedo et al., 2010) has prolonged the transformation activity (of PAHs) and thermal stability of two enzymes. Surprisingly, encapsulated laccase on polyacrylamide hydrogel and pectin has shown 90% higher BPA transformation than free enzyme (Gassara et al., 2013). Additionally, enzyme stability and protection of enzymes from inactivation by noncompetitive inhibition have been observed by encapsulation of enzymes.

A new research area in enzyme application is "nanobiocatalysis" where there is an integration of nanotechniques in enzyme immobilization. Several engineered nanomaterials (ENMs) are available in the form of nanostructures, such as nanoporous media, nanofibers, carbon nanotubes, nanoparticles, etc. These types of ENMs offer additional advantages over traditional immobilizing materials (Kim et al., 2006), for example, high stability against many deactivating factors, enzyme activity against larger range of pH and temperatures; as a result, application of enzymes is made possible in a wider area, including detoxification and monitoring of contaminated sites (Hu et al., 2007; Garcia et al., 2011). However, one of the main problems in the nanobiocatalysis is interactions between enzymes and ENMs; this needs future investigations (Kim et al., 2006, 2008).

The field of microbial enzymes is breaking new grounds in different environmental settings with the emergence of new techniques. New technical tools allow us to use the enzyme preparations in the form of "crystalline catalysts" which have an ability to recycle the cofactors (Sanchez & Demain, 2017). On the other side, genetically modified enzymes, which have a capacity to function in different solvents with multiple activities greatly help in the better understanding of structure–function relationships of enzymes. In addition, there is a need to search for enzymes from exotic environments to find out enzymes with better catalytic properties.

Importantly, there are different driving factors that control the enzyme activities in different environments like terrestrial, freshwater, and marine systems (Arnosti et al., 2014). For example, sorption and/or occlusion of enzymes in terrestrial environments (i.e., soil and sediment) are important phenomenon either in the stabilization of enzymes or inhibition of enzyme activities; therefore, it is necessary to understand the nature and properties of surfaces in the soil and sediment environments for knowing the behavior of enzymes in that environment. In freshwater systems, variations in pH and temperature greatly affects the enzyme lifetimes. In general, enzymes are produced by microorganisms under one condition, but they do not function if the condition is changed. There are many unanswered questions related to the environmental factors that govern the lifetime of enzymes in different environments. For example, how long enzymes are catalytically effective? What are the most important parameters that determine the enzyme lifetimes? Are lifetime determining parameters same in different environments (i.e., terrestrial, freshwater, marine system)? What is the fate of extracellular enzymes?

There should also be an emphasy on the basic understandings of enzyme induction in microorganisms. We now know more about enzyme induction in model organism such as *E. coli*, but we do not know much about enzyme induction in microorganisms that grow in carbon-poor environment (e.g., oligotrophic environment like ocean water) (Sanchez & Demain, 2017). Also, in-depth understanding on the capabilities of microbial communities is needed. Because, in certain situations, though certain microorganisms are few in number in a community but produce biogeochemically important enzymes (Beier & Bertilsson, 2011); conditions which do favor production of enzymes by a small number of microorganisms should be explored. Communications (i.e., quorum sensing; Maddela et al., 2019) among the members of a community help in the emergence of different functions of a community. Research results of above new directions can be expected to provide efficient enzyme applications in reducing the harm to the environment. Nevertheless, compared to chemical reactions, enzyme-catalyzed processes will support more specific and cleaner technologies, which will help in achieving sustainability in agriculture, industries (food, dairy, detergent, and chemical), and treatment facilities (water and wastewater). Thus, in this section, an attempt was made to highlight the principal limitations and knowledge gaps in the sustainable applications of microbial enzymes; subsequently, possible future directions of research were also suggested.

4 Conclusion

Environmental applications especially the role of some enzymes in the bioremediation of pollutants were discussed. While many physical and chemical methods of treating contaminated soil and water are not efficient enough, bioremediation opens a new way to clean up toxic pollutants. Enzymes as practical tools of living organisms are an eco-friendly and bio-based strategy for bioremediation. Microorganisms exposed to contaminated sites and specific pollutants are fascinating sources for the isolation of active enzymes against those pollutants. Overall, using enzymes for pollutant bioremediation seems to be a cost-effective, efficient, practical approach, and sustainable. Further research on enzyme activity, mechanism of action and isolation of new enzymes would be a promising way for future direction, to reduce pollutants and make a healthier environment for humans and all other species.

References

Acevedo, F., Pizzul, L., Castillo, M. D., González, M. E., Cea, M., Stenström, J., Gianfreda, L., & Diez, M. C. (2010). Degradation of polycyclic aromatic hydrocarbons by free and nanoclay-immobilized manganese peroxidase from Anthracophyllum discolor. *Chemosphere, 80*, 271–278.

Ahn, M. Y., Dec, J., Kim, E. Y., & Bollag, J.-M. (2002). Treatment of 2,4-dichlorophenol polluted soil with free and immobilized laccase. *Journal of Environmental Quality, 31*, 1509–1515.

Arnosti, C., Bell, C., Moorhead, D. L., Sinsabaugh, R. L., Steen, A. D., Stromberger, M., Wallestein, M., & Weintraub, M. N. (2014). Extracellular enzymes in terrestrial, freshwater, and marine environments: Perspectives on system variability and common research needs. *Biogeochemistry, 117*(1), 5–21.

Awad, G., & Mohamed, E. F. (2019). Immobilization of P450 BM3 monooxygenase on hollow nanosphere composite: Application for degradation of organic gases pollutants under solar radiation lamp. *Applied Catalysis B: Environmental, 253*, 88–95.

Beier, S., & Bertilsson, S. (2011). Uncopling of chitosan activity and uptake of hydrolysis products in freshwater bacterioplankton. *Limnology and Oceanography, 54*(4), 1179–1188. https://doi.org/10.4319/lo.2011.56.4.1179

Bhandari, S., Darbin, K. P., Rishab, M., Sonika, D., Karan, K., Sitaram, P., Shreesti, S., Santosh, G., Kusum, B., Uddhav, K., & Niranjan, P. (2021). Microbial enzymes used in bioremediation. *Journal of Chemistry, 2021*, 8849512. https://doi.org/10.1155/2021/8849512

Bilal, M., & Asgher, M. (2015). Dye decolorization and detoxification potential of Ca-alginate beads immobilized manganese peroxidase. *BMC Biotechnology, 15*(1), 111–118.

Bilal, M., Asgher, M., Iqbal, M., Hu, H., & Zhang, X. (2016). Chitosan beads immobilized manganese peroxidese catalytic potential for detoxification and decolorization of textile effluent. *International Journal of Biological Macromolecules, 89*, 181–189.

Chakraborty, J., & Das, S. (2016). Molecular perspectives and recent advances in microbial remediation of persistent organic pollutants. *Environmental Science and Pollution Research, 23*(17), 16883–16903.

Chauhan, P. S., Goradia, B., & Saxena, A. (2017). Bacterial laccase: Recent update on production, properties and industrial applications. *3 Biotech, 7*(5), 323.

Eibes, G., Arca-Ramos, A., Feijoo, G., Lema, J. M., & Moreira, M. T. (2015). Enzymatic technologies for remediation of hydrophobic organic pollutants in soil. *Applied Microbiology and Biotechnology, 99*(21), 8815–8829.

Falade, A. O., Nwodo, U. U., Iweriebor, B. C., Green, E., Mabinya, L. V., & Okoh, A. I. (2017). Lignin peroxidase functionalities and prospective applications. *Microbiology Open, 6*, 394–400.

Festa, G., Autore, F., Fraternali, F., Giardina, P., & Sannia, G. (2008). Development of new laccases by directed evolution: Functional and computational analyses. *Proteins: Structure, Function, and Bioinformatics, 72*(1), 25–34.

Garcia, J., Zhang, Y., Taylor, H., Cespedes, O., Webb, M. E., & Zhou, D. (2011). Multilayer enzyme-coupled magnetic nanoparticles as efficient, reusable biocatalysts and biosensors. *Nanoscale, 3*, 3721–3730.

Gassara, F., Brar, S. K., Verma, M., & Tyagi, R. D. (2013). Bisphenol A degradation in water by ligninolytic enzymes. *Chemosphere, 92*, 1356–1360.

Gianfreda, L., Rao, M. A., Scelza, R., & De la Luz Mora, M. (2016). Role of enzymes in environment cleanup/remediation. In *Agro-industrial wastes as feedstock for enzyme production* (pp. 133–155). Academic Press.

Hu, X., Zhao, X., & Hwang, H. (2007). Comparative study of immobilized Trametes versicolor laccase on nanoparticles and kaolinite. *Chemosphere, 66*, 1618–1626.

Jia, Y., Eltoukhy, A., Wang, J., Li, X., Hlaing, T. S., Aung, M. M., Nwe, M. T., Lamraoui, I., & Yan, Y. (2020). Biodegradation of bisphenol A by Sphingobium sp. YC-JY1 and the essential role of cytochrome P450 monooxygenase. *International Journal of Molecular Sciences, 21*(10), 3588.

Kan, J., Peng, T., Huang, T., Xiong, G., & Hu, Z. (2020). NarL, a novel repressor for CYP108j1 expression during PAHsdegradation in Rhodococcus sp. P14. *International Journal of Molecular Sciences, 21*(3), 983–989.

Kaur, H., Kapoor, S., & Kaur, G. (2016). Application of ligninolytic potentials of a white-rot fungus *Ganoderma lucidum* for degradation of lindane. *Environmental Monitoring and Assessment, 88*(10), 88.

Kim, J., Grate, J., & Wang, P. (2006). Nanostructures for enzyme stabilization. *Chemical Engineering Science, 61*, 1017–1026.

Kim, J., Grate, J., & Wang, P. (2008). Nanobiocatalysis and its potential applications. *Trends in Biotechnology, 26*, 639–646.

Knop, D., Levinson, D., Makovitzki, A., et al. (2016). Limits of versatility of versatile peroxidase. *Applied and Environmental Microbiology, 82*(14), 4070–4080.

Kumar, S., Jin, M., & James, L. (2012). Cytochrome P450-mediated phytoremediation using transgenic plants: A need for engineered cytochrome P450 enzymes. *Journal of Petroleum & Environmental Biotechnology, 3*(5), 10–21.

Li, Z., Jiang, Y., Guengerich, F. P., Ma, L., Li, S., & Zhang, W. (2020). Engineering cytochrome P450 enzyme systems for biomedical and biotechnological applications. *The Journal of Biological Chemistry, 295*(3), 833–849.

Lu, L., Wang, T. N., Xu, T. F., Wang, J. Y., Wang, C. L., & Zhao, M. (2013). Cloning and expression of thermo-alkali-stable laccase of Bacillus licheniformis in Pichia pastoris and its characterization. *Bioresource Technology, 134*, 81–86.

Maddela, N. R., Sheng, B., Yuan, S., Zhou, Z., Villamar-Torres, R., & Meng, F. (2019). Roles of quorum sensing in biological wastewater treatment: A critical review. *Chemosphere, 221*, 616–629.

Malik, A. (2004). Metal bioremediation through growing cells. *Environment International, 30*(2), 261–278.

Mateo, C., Palomo, J. M., Fernandez-Lorente, G., Guisan, J. M., & Fernandez-Lafuente, R. (2007). Improvement of enzyme activity, stability and selectivity via immobilization techniques. *Enzyme and Microbial Technology, 40*, 1451–1463.

Nannipieri, P., & Bollag, J.-M. (1991). Use of enzymes to detoxify pesticide-contaminated soils and waters. *Journal of Environmental Quality, 20*, 510–517.

Ó'Fágáin, C. (2003). Enzyme stabilization – Recent experimental progress. *Enzyme and Microbial Technology, 33*, 137–149.

Qin, X., Zhang, J., Zhang, X., & Yang, Y. (2014). Induction, purification and characterization of a novel manganese peroxidase from *Irpex lacteus* CD2 and its application in the decolorization of different types of dye. *PLoS One, 9*(11), 1–7.

Rayu, S., Karpouzas, D. G., & Singh, B. K. (2012). Emerging technologies in bioremediation: Constraints and opportunities. *Biodegradation, 23*(6), 917–926.

Sanchez, S., & Demain, A. L. (2017). Useful microbial enzymes—An introduction. In *Biotechnology of microbial enzymes* (pp. 1–11). Academic Press.

Seyyed, M. M., Seyyed, A. H., Seyed, M. I., Navid, R., Ahmad, G., Chin, W. L., Wei-Hung, C., Navid, O., Khadije, Y., & Gity, B. (2021). Recent advances in enzymes for the bioremediation of pollutants. *Biochemistry Research International, 2021*, 1–12. https://doi.org/10.1155/2021/5599204

Sun, J., Zheng, M., Lu, Z., Lu, F., & Zhang, C. (2017). Heterologous production of a temperature and PH-stable laccase from Bacillus vallismortis fmb-103 in Escherichia coli and its application. *Process Biochemistry, 55*, 77–84.

Theerachat, M., Emond, S., & Cambon, E. (2012). Engineering and production of laccase from *Trametes versicolor* in the yeast *Yarrowia lipolytica*. *Bioresource Technology, 125*, 267–274.

Ufarte, L., Laville, E., Duquesne, S., & Potocki-Veronese, G. (2015). Metagenomics for the discovery of pollutant degrading enzymes. *Biotechnology Advances, 33*(8), 1845–1854.

Wang, T. N., & Zhao, M. (2017). A simple strategy for extracellular production of CotA laccase in *Escherichia coli* and decolorization of simulated textile effluent by recombinant laccase. *Applied Microbiology and Biotechnology, 101*(2), 685–696.

Wang, X., Yao, B., & Su, X. (2018). Linking enzymatic oxidative degradation of lignin to organics detoxification. *International Journal of Molecular Sciences, 19*(11), 3373.

Xu, H., Guo, M. Y., Gao, Y. H., Bai, X. H., & Zhou, X. W. (2017). Expression and characteristics of manganese peroxidase from *Ganoderma lucidum* in *Pichia pastoris* and its applicationin the degradation of four dyes and phenol. *BMC Biotechnology, 17*(1), 19–12.

Yang, H., Li, J., Du, G., & Liu, L. (2017). Microbial production and molecular engineering of industrial enzymes: Challenges and strategies. In *Biotechnology of microbial enzymes* (pp. 151–165). Academic Press.

The manufacturer's authorised representative in the EU is Springer Nature Customer Service Centre GmbH, Europaplatz 3, 69115 Heidelberg, Germany. If you have any concerns regarding our products, please contact ProductSafety@springernature.com

Printed and bound by CPI Group (UK) Ltd, Croydon, CR0 4YY
30/09/2026
02225239-0019